Mathematical Modelling

Mathematical Modelling

Simon Serovajsky

al-Farabi Kazakh
National University, Kazakhstan

CRC Press
Taylor & Francis Group
Boca Raton London New York

CRC Press is an imprint of the
Taylor & Francis Group, an **informa** business

A CHAPMAN & HALL BOOK

First edition published 2022
by CRC Press
6000 Broken Sound Parkway NW, Suite 300, Boca Raton, FL 33487-2742

and by CRC Press
2 Park Square, Milton Park, Abingdon, Oxon, OX14 4RN

© 2022 Taylor & Francis Group, LLC

CRC Press is an imprint of Taylor & Francis Group, LLC

Library of Congress Cataloging-in-Publication Data

ISBN: 9780367474300 (hbk)
ISBN: 9781032147871 (pbk)
ISBN: 9781003035602 (ebk)

DOI: 10.1201/9781003035602

Publisher's note: This book has been prepared from camera-ready copy provided by the authors.

To my comrades in the Problem Laboratory of Mathematical Modelling of the Kazakh University, from where my scientific activity began.

Contents

Preface   xv

List of Figures   xix

List of Tables   xxiii

1   Foundations of mathematical modeling   **1**
Lecture 1
   1   Cognition and modeling 1
   2   Natural Sciences and Mathematics 2
   3   Content or form? 3
   4   Copernicus or Ptolemy? 6
   5   Mathematical model of a body falling 7
   6   Principles for determining mathematical models 11
   7   Classification of mathematical models 14
Appendix 16
   1   Probe movement 16
   2   Missile flight 18
   3   Glider flight 19
Notes 20

I   Systems with lumped parameters   **23**

2   Approximate solving of differential equations   **27**
Lecture 27
   1   Conception of approximate solution 27
   2   Euler method 28
   3   Probe movement 30
   4   Missile flight 31
   5   Glider flight 31
Appendix 32
   1   Runge–Kutta method 33
   2   Two-body problem 33
   3   Predator–pray model 35
Notes 36

3   Mechanical oscillations   **39**
Lecture 39
   1   Determination of the pendulum oscillation equation 39
   2   Solving of the pendulum oscillation equation 42
   3   Pendulum oscillation energy 43
   4   Oscillation of a pendulum with friction 45

    5   Equilibrium position of the pendulum 47

    6   Forced oscillations of the pendulum 49

   Appendix 50

    1   Spring oscillation 50

    2   Large pendulum oscillations 51

    3   Problems of nonlinear oscillation theory 52

   Notes 53

4 Electrical oscillations   57

   Lecture 57

    1   Electrical circuit 57

    2   Energy of circuit 60

    3   Circuit with resistance 61

    4   Forced circuit oscillations 62

   Appendix 66

    1   Forced oscillations of spring 66

    2   Circuit with nonlinear capacity 67

    3   Van der Pol circuit 67

   Notes 69

5 Elements of dynamical system theory   71

   Lecture 71

    1   Evolutionary processes and differential equations 71

    2   General notions of dynamic systems theory 72

    3   Change in species number with excess food 74

    4   Oscillations of pendulum 75

    5   Stability of the equilibrium position 77

    6   Limit cycle 80

   Appendix 82

    1   Exponential growth systems 82

    2   Brussellator 83

    3   System with two limit cycle 84

   Notes 84

6 Mathematical models in chemistry   89

   Lecture 89

    1   Chemical kinetics equations 89

    2   Monomolecular reaction 93

    3   Bimolecular reaction 95

    4   Lotka reaction system 96

   Appendix 99

    1   Brusselator 99

    2   Oregonator 100

    3   Chemical niche 100

    4   Laser healing model 101

   Notes 102

7 Mathematical model in biology **105**
Lecture 105
  1   One species evolution 105
  2   Biological competition model 109
  3   Predator–prey model 112
  4   Symbiosis model 116
Appendix 119
  1   Models of chemical and physical competition 119
  2   Fluctuations in yield and fertility 120
  3   Ecological niche model 120
  4   SIR model for spread of disease 121
  5   Antibiotic resistance model 122
Notes 123

8 Mathematical model of economics **129**
Lecture 129
  1   One company evolution 129
  2   Economic competition model 131
  3   Economic niche model 134
  4   Free market model 138
  5   Monopolized market model 140
Appendix 141
  1   Ecological niche model 141
  2   Inflation model 142
  3   Model of economic cooperation 143
  4   Racketeer—entrepreneur model 145
  5   Solow model of economic growth 146
Notes 146

9 Mathematical models in social sciences **149**
Lecture 149
  1   Political competition 149
  2   Political niche 150
  3   Bipartisan system 151
  4   Trade union activity 156
  5   Allied relations 157
Appendix 158
  1   Competition models 158
  2   Niche models 159
  3   Predator–prey models 160
Notes 161

II Systems with distributed parameters **165**

10 Mathematical models of transfer processes. 1 **169**
Lecture 169
  1   Heat equation 169
  2   First boundary value problem for the homogeneous heat equation 174
  3   Non-homogeneous heat equation 178
Appendix 181
  1   Generalizations of the heat equation 181

2   Second boundary value problem for the heat equation 183
3   Diffusion equation 184
Notes 187

11 Mathematical models of transfer processes. 2   191
Lecture 192
1   Heat equation and similarity theory 192
2   Goods transfer equation 193
3   Finite difference method for the heat equation 195
4   Diffusion of chemical reactants 197
5   Stefan problem for the heat equation 199
Appendix 200
1   Overview of transfer processes 201
2   Finite difference method: Implicit scheme 204
3   Competitive species migration 206
4   Hormone treatment of the tumor with hormone resistance 207
Notes 209

12 Wave processes   213
Lecture 213
1   Vibration of string 213
2   Vibrations of string with fixed ends 218
3   Infinitely long string 220
4   Electrical vibrations in wires 225
Appendix 229
1   Energy of vibrating string 229
2   Mathematical models of wave processes 232
3   Beam vibrations 232
4   Maxwell equations 234
5   Finite difference method for the vibrating string equation 235
Notes 236

13 Mathematical models of stationary systems   239
Lecture 239
1   Stationary heat transfer 239
2   Spherical and cylindrical coordinates 242
3   Vector fields 244
4   Electrostatic field 247
5   Gravity field 250
Appendix 253
1   Stationary fluid flow 253
2   Steady oscillations 253
3   Bending a thin elastic plate 254
4   Variable separation method for the Laplace equation in a circle 255
5   Establishment method 257
Notes 258

14 Mathematical models of fluid and gas mechanics — 261

Lecture 261
 1   Material balance in a moving fluid 261
 2   Ideal fluid movement 263
 3   Ideal fluid under the gravity field 265
 4   Viscous fluid movement 268
Appendix 270
 1   Burgers equation 271
 2   Surface wave movement 272
 3   Boundary layer model 273
 4   Acoustic problem 273
 5   Thermal convection 275
 6   Problems of magnetohydrodynamics 276
Notes 276

15 Mathematical models of quantum mechanical systems — 279

Lecture 279
 1   Quantum mechanics problems 279
 2   Wave function 280
 3   Schrödinger equation 281
 4   Particle movement under an external field 283
 5   Potential barrier 284
Appendix 288
 1   Wave function normalization 288
 2   Particle movement in a well with infinitely high walls 289
Notes 290

III   Other mathematical models — 293

16 Variational principles — 297

Lecture 297
 1   Brachistochrone problem 297
 2   Lagrange problem 299
 3   Shortest curve 301
 4   Body falling problem and the concept of action 302
 5   Principle of least action 303
 6   Vibrations of string 304
Appendix 306
 1   Law of conservation of energy 306
 2   Fermat's principle and light refraction 307
 3   River crossing problem 309
 4   Pendulum oscillations 312
 5   Approximate solution of minimization problems 313
Notes 315

17 Discrete models — 319

Lecture 319
 1   Discrete population dynamics models 319
 2   Discrete heat transfer model 322
 3   Transportation problem 323
 4   Traveling salesman problem 324

    5   Prisoner's dilemma   326
  Appendix   329
    1   Discrete model of epidemic propagation   329
    2   Potential method for solving a transportation problem   333
    3   Production planning   335
    4   Concepts of game theory   335
  Notes   336

18 Stochastic models   **339**
  Lecture   339
    1   Stochastic model of pure birth   339
    2   Monte Carlo method   344
    3   Stochastic model of population death   346
    4   Stochastic Malthus model   349
  Appendix   352
    1   Malthus model with random population growth   352
    2   Models with random parameters   354
    3   Discrete model of selling goods   355
    4   Passage of a neutron through a plate   356
  Notes   357

IV   Additions   **361**

19 Mathematical problems of mathematical models   **365**
  Lecture   365
    1   Cauchy problem properties for differential equations   365
    2   Properties of boundary value problems   367
    3   Boundary value problems for the heat equation   368
    4   Hadamard's example and well-posedness of problems   369
    5   Classical and generalized solution of problems   370
  Appendix   372
    1   Nonlinear boundary value problems   372
    2   Euler's elastic problem   372
    3   Bénard problem   374
    4   Generalized model of stationary heat transfer   375
    5   Sequential model of stationary heat transfer   376
  Notes   378

20 Optimal control problems   **381**
  Lecture   382
    1   Maximizing the shell flight range   382
    2   Maximizing the missile flight range   383
    3   General optimal control problem   385
    4   Solving of the maximization problem of the missile flight range   386
    5   Time-optimal control problem   389
  Appendix   391
    1   Maximizing the probe's ascent height   391
    2   Approximate methods for solving optimality conditions   392
    3   Gradient methods   394
  Notes   395

21 Identification of mathematical models **399**
   Lecture 400
     1   Problem of determining the system parameters 400
     2   Inverse problems and their solving 401
     3   Heat equation with data at the final time 403
     4   Differentiation of functionals and gradient methods 405
     5   Solving of the heat equation with reversed time 406
   Appendix 408
     1   Boundary inverse problem for the heat equation 408
     2   Inverse problem for the falling of body 409
     3   Inverse gravimetry problem 411
     4   Well-posedness of optimal control problems 411
   Notes 412

Epilogue **415**

Bibliography **417**

Index **435**

Preface

Mathematics has a special place in human culture. It is radically different from physics, biology, history, psychology, and other sciences. Having abstract concepts as a subject of research such as numbers and functions, equations and sets that are absent in the real world and are a strange product of the human brain, it nevertheless turns out to be surprisingly adapted to comprehend any phenomena of nature and society. The amazing applicability of Mathematics for the analysis of all kinds of events in the surrounding world is due to the presence of an incredible relation between objective reality and the dry abstraction of mathematical constructions. There is a magic bridge that allows us to mysteriously transfer from the usual real world to the fantastic land of Mathematics and come back after a fascinating journey through its endless expanses with the accumulated baggage of knowledge. This mysterious bridge is mathematical modelling, which is a specific form of cognition of the surrounding world, capable of translating the laws of nature studied by individual sciences into mathematical language, and then seeing real life events behind dispassionate mathematical formulas.

The subject of this book is mathematical modelling. The author did not set himself the goal to either teach the reader some individual laws of nature and society, or acquaint them with mathematical methods to solve applied problems. Our aim is to learn how to build bridges between Mathematics and the outside world. We want to show how completely different natural phenomena can be described mathematically using the same techniques. Here both strict mechanics, thoroughly saturated with mathematical ideas, and shaky psychology, seemingly not at all amenable to formalization, find themselves in an equal position.

The choice of research objects remains entirely on the author's conscience. The breadth of the covered of the subject determined the relatively shallow depth of its study from the standpoint of both mathematics and special sciences. There are multivolume respected books devoted to the application of mathematical methods for the analysis of processes exclusively related to the field of a particular science activity, or even its small section. If any of the serious readers is interested in mathematical economics or mathematical ecology, not to mention mathematical physics, then they are advised not to waste their invaluable time studying this modest book, but refer to the relevant literature.

The book consists of an introduction and four parts. The introduction, which is the subject of the first chapter, discusses the general principles of mathematical modelling. The first part, consisting of eight chapters, is devoted to lumped-parameter systems. It deals with dynamical systems described by ordinary differential equations and related to physics, chemistry, biology, economics and social sciences. For a better perception of the material and the performance of the tasks given here, basic information is also given on the methods of qualitative and quantitative analysis of such systems. The second part, consisting of six chapters, examines distributed-parameter systems described by partial differential equations with various practical interpretations. The third part is devoted to other types of models. In particular, variational principles, discrete systems and stochastic systems related to various subject areas are studied here. The final part contains additional material, particularly, mathematical problems related to the considered models, optimal control problems and identification of mathematical models.

It should be noted that although each chapter has a self-sufficient subject, the materials of different chapters are rather closely intertwined. On the one hand, the same mathematical models can have qualitatively different interpretations. For example, the same equations describe the coexistence of different biological species on the same territory, and the competition of firms in the free market, and the struggle of political parties for votes. The model describing the heat distribution also characterizes the processes of diffusion and electrical conductivity, as well as the a biological species migration, the epidemic spread and commercial activity in a certain area. On the other hand, a specific process under certain conditions can be described qualitatively in different ways. For instance, a continuous model based on ordinary differential equations or a discrete model based on difference equations can be used to describe the evolution of a biological species. If the species migrates over a territory, then partial differential equations should be used to describe this process. If the process is influenced by some random factors, then stochastic models should be used. For describing the body movement, one can use both the equations of motion and variational principles.

Each chapter of the book includes several levels. First of all, it is a lecture which contains the main material of this chapter. Those who want to get only a general idea of the subject under discussion can limit themselves to this, boldly ignoring all the rest of the material. This is followed by an appendix, which provides additional information to allow the reader to broaden their understanding of the subject of the study. In addition, there are notes that perform several functions at once. It provides a literature review on the discussed issue. It reveals the serious connections of the considered subject with the results of other chapters, which increases the integrity of the book subject. Some remarks are made there, allowing a deeper understanding of the material under consideration. Some mathematical transformations from the lecture and the appendix are also taken there. This facilitates the perception of the material for the reader who does not want to go deeply into technical details, but provides such an opportunity for the interested reader. Thus, to some extent, it is possible to maintain a balance between material availability and mathematical rigor. Finally, in the course of the lecture and the appendix, tasks are given with the necessary methodological instructions. They can be used as assignments for independent work of students, as well as for individual readers engaged in self-education.

To understand the presented material, the reader does not require particularly serious knowledge in the field of mathematics and special sciences. Among the readers, one would like to see students, graduate students and young specialists of various specialties, and, possibly, schoolchildren interested in the use of mathematical methods and computing technology to solve problems of a wide profile, as well as everyone who has a taste for research work and experiencing love and respect for Mathematics, but do not want to focus on to mathematical abstractions. The book is written on the basis of a course of lectures given by the author for many years at the Faculty of Mechanics and Mathematics of the al-Farabi Kazakh National University, Almaty. Some of the above models given are associated with the implementation of various scientific research developments in which the author was involved. I would like to express my deep gratitude to Sh. Smagulov, reviewer of the first version of the book, and also to Yu. Gasimov, V. Romanovsky and M. Ruzhansky, who supported the publication of its new, substantially revised version. In the process of working on individual chapters of the book, many specialists rendered invaluable assistance to me. In particular, various physics problems were discussed with I. Zherebyat'ev, A. Karimov, V. Kashkarov, N. Kosov, Yu. Kulakov, L. Kurlapov and S. Kharin, chemistry problems — with P. Itskova, A. Lukyanov and Z. Mansurov, biology problems — with A. Sarmurzina and V. Shcherbak, economics problems — with A. Ashimov, medical problems — with A. Ilyin and R. Islamov, optimal control problems — with S. Aisagaliev, A. Antipin, V. Boltyanskiy, A. Butkovsky, F. Vasiliev, M. Dzhenaliev, A. Egorov, V. Litvinov, K. Lurie, V. Neronov, T. Sirazetdinov, stochastic models — with K. Shakenov, inverse and ill-posed problems — with S. Kabanikhin, approximate calculation methods — with Sh. Smagulov and N. Danaev. I am grateful to D. Nurseitov, with whom we worked together on solving a number of applied problems of mathematical modelling, as well as A. Berlizev, A. Gavrilov, N. Lyskovskaya, N. Myasnikov, N. Popova and E. Popov, with whom we developed training software packages for mathematical modelling. I express my deep gratitude to many employees and students of al-Farabi Kazakh National University, who to one degree or another helped me in my work. I am also extremely grateful to the staff of CRC Press/ Taylor & Francis Group C. Frazer, M. Kabra and S. Kumar, who actively supported me throughout the work on the book. I am also grateful to the artists V. Shcherbak, I. Saitov and B. Tasov for the book cover. I express special gratitude to A.A. Teplov, without whom this book would hardly have been written. Finally, I am deeply grateful to my wife Larissa Ananyeva for her understanding and support.

The book is dedicated to my comrades in the problem laboratory of mathematical modelling of the Kazakh University, in which my scientific activity began.

List of Figures

1.1   Simplified process of cognition.   2
1.2   Characteristics of the body falling.   11
1.3   Forces acting on the missile.   18
1.4   Forces acting on the glider.   19

2.1   Derivative approximation.   29
2.2   Forces acting on the missile.   34

3.1   Mathematical pendulum.   40
3.2   Calculation of the angular velocity.   41
3.3   Solution of the problem (3.3) and (3.4) is periodic function.   43
3.4   Solution of the equation of body movement under a friction force.   46
3.5   Damped oscillations of pendulum.   47
3.6   Oscillation of the spring.   50
3.7   Ball oscillation in pit.   54

4.1   Electrical circuit.   58
4.2   Electrical oscillations in circuits.   59
4.3   Circuit with resistor.   61
4.4   Charge in the case of the forces circuit oscillations.   63
4.5   Charge in the case of the forces circuit oscillations with resistor.   64
4.6   Dependence of the oscillation amplitude from the frequency of forced oscillations.   64
4.7   Phase curve for the forced oscillations of spring.   67
4.8   Nonlinear dependence of voltage on the capacitor from the charge.   67
4.9   Phase curve for the Van der Pol equation.   68

5.1   Evolution of a process.   73
5.2   Phase velocity.   74
5.3   Phase velocities for the equation (5.1).   74
5.4   Family of integral curves for the equation (5.1).   75
5.5   Phase curves and phase velocities for the pendulum oscillation equality with identity frequency.   76
5.6   Stability and non-stability of equilibrium positions.   77
5.7   Equilibrium positions on the plane.   78
5.8   Integral curves in the neighborhood of the zero equilibrium position.   79
5.9   Phase curves for the system (5.6).   81
5.10   Phase curves for the system with two limit cycles.   85

6.1   Concentration change for a monomolecular reaction.   94
6.2   Influence of parameters for the monomolecular reaction.   94
6.3   Concentration change for a bimolecular reaction.   96
6.4   Solutions of the Volterra–Lotka equations.   98

7.1   Variants of evolution of the species number. 106

7.2   Species number tends to x^* if the quantity of food is bounded. 108

7.3   Variants of the system evolution for the Verhulst model. 109

7.4   Change of species number of biological competition model. 111

7.5   Variants of system evolution for the competition model. 112

7.6   Directions of system evolution for the predator–pray model. 114

7.7   Solutions of system (7.9) for $u_0 = 0.3$, $v_0 = 0.8$, $m = 2$. 115

7.8   State functions of the predator–pray model are periodic. 115

7.9   Directions of the system evolution in the symbiosis model. 117

7.10 Phase curves for the symbiosis model. 118

8.1   Phase curves for the competition model. 132

8.2   Possible outcomes for the competition model. 133

8.3   Directions of the system evolution for the economic niche model. 136

8.4   Change in income and prices in the free market with parameter values $x_{10} = 0.3$, $x_{20} = 0.8$, $\varepsilon_1 = 1$, $\varepsilon_2 = 1$, $\gamma_1 = 1$, $\gamma_2 = 1$. 138

9.1   Oscillations of trade union influence and social positions. 156

10.1 Directions of the heat flux. 171

10.2 The body temperature for the first boundary value problem. 177

10.3 The body cools the faster, the greater the coefficient a. 177

10.4 The temperature of the body under the heat source influence. 180

10.5 The set Ω with boundary S and directions of external normal n. 183

10.6 The body temperature for the second boundary value problem. 184

10.7 Geometric sense of the mean theorem. 188

11.1 Heat flux occurrence. 202

11.2 Diffusion flux occurrence. 203

12.1 Transverse vibrations of string. 214

12.2 The action of tension forces on a segment. 215

12.3 Position of the string and its velocity at different points in time. 221

12.4 Initial form of the string. 223

12.5 String profiles at different points in time. 224

12.6 Initial form of the string for Task 12.2. 225

12.7 The analogy of a spring, string, circuit, and wire. 225

12.8 Part of wire with inductance and resistance. 226

13.1 Internal and external problems. 242

13.2 Spherical coordinates. 243

13.3 Cylindrical coordinates. 244

13.4 Dot and cross products of vectors. 246

13.5 Fluid movement through the surface. 247

13.6 Potential of a point charge field. 249

13.7 Potential of the field of an infinite wire. 250

13.8 Sphere surrounding the particle. 251

14.1 Volume of fluid flowing through the cross section. 262

14.2 Action of the pressure force. 264

14.3 Transformation of the equality (14.10). 266

14.4 Steady flow of viscous fluid between the plates. 270

15.1 Classical particle with small energy does not pass the potential barrier. 285
15.2 Potential barrier. 286
15.3 Squared modulus of the solution of equation (15.10). 287
15.4 Square of the modulus of the equation solution for infinite barrier or for $\hbar \to 0$. . 288

16.1 Movement of the body for the brachistochrone problem. 298
16.2 Calculating of arc length. 298
16.3 Light refraction. 308
16.4 Movement of the boat. 309

17.1 Sequence $\{y_n\}$. 321
17.2 Graph of a symmetric traveling salesman problem. 325
17.3 Graph with cycles not forming a route. 326
17.4 Relationship between population compartments of the epidemic model. 331

18.1 Probability distribution in the pure birth model. 342
18.2 Normal distribution probability density. 345
18.3 Distribution of probabilities in the death model. 348
18.4 Expected value and variance in the model of population death. 349
18.5 Probability density of a lognormal distribution. 353
18.6 Estimation of the probability of population extinction. 353
18.7 Movement of a neutron between two interactions. 358

19.1 Phase velocity and change in the function are inverse directions. 366
19.2 Set of solutions of Cauchy problem is infinite. 367
19.3 Relationship between classical and generalized solutions. 371
19.4 Solutions of the problem (19.15). 373
19.5 Eulers problem has non-unique solution. 374
19.6 Formation of convective fluid flows. 375
19.7 Bénard cells. 375

20.1 Flight of shell. 382
20.2 Flight of missile. 384
20.3 Graph of the function φ. 388
20.4 Phase curves for the system (20.16). 390
20.5 Solutions of time-optimal control problem. 391
20.6 Direction of the algorithm is determined by the sign of the derivative. 395

List of Tables

1.1	Stages of model constructing.	10
1.2	Stages of model constructing.	14
1.3	Mathematical model classes.	16
3.1	Elements of the pendulum oscillations model.	42
4.1	Comparison of mechanical and electrical oscillations.	65
5.1	Systems with exponential growth of state.	83
6.1	Analogy between processes.	101
7.1	Interpretation of the Verhulst equation.	107
7.2	Analogy between models of the predator–prey type.	116
7.3	Characterization of predator–prey models.	120
8.1	Interpretation of the Verhulst equation.	130
8.2	Evolution the outcome (a) for the competition model. First company goes bankrupt.	133
8.3	Evolution the outcome (b) for the competition model. Second company goes bankrupt.	133
8.4	Evolution the outcome (c) for the competition model. Both firms persist.	134
8.5	Analogy between models of the symbiosis type	144
8.6	Analogy between predator–prey type models.	145
8.7	Characteristic of "predator–prey" models.	146
9.1	Characterization of competition models.	159
9.2	Predator–prey models.	161
9.3	Characteristics of two-subject models.	162
10.1	Analogy between thermal conductivity and diffusion.	186
11.1	Comparative characteristic of transfer processes.	202
12.1	Stages of deriving of state equations.	217
12.2	Comparison of lumped and distributed models.	228
12.3	Wave processes.	232
13.1	Analogy between electrostatic and gravitational fields.	252
14.1	Analogy between transfer processes.	268
17.1	Discrete and continuous Malthus models.	320

17.2   Interpretations of the prisoner's dilemma. 329

17.3   Baseline plan. 334

17.4   Baseline plan. First iteration 334

18.1   Deterministic and stochastic pure birth models. 343

18.2   Stochastic models of pure birth. 343

18.3   Deterministic and stochastic Malthus models. 351

Chapter 1

Foundations of mathematical modeling

In Chapter 1 of our course, we try to imagine a mathematical modeling as a specific form of cognition of the world. Any idea of the researcher about the studied object is its model. We are only interested in models formulated in mathematical language. They are unique links between abstract Mathematics and the real World. We try to identify some features of mathematical models and understand the general principles of their constructing. As an example, one considers the classical problem of the fall of a body. The most important types of mathematical models are described. Appendix contains simple, but non-trivial mathematical models of the movement of bodies in a field of gravitational force.

Lecture

1 Cognition and modeling

The basis of this course is the concept of the mathematical model. This is an amazing form of representing all kinds of phenomena occurring in nature. In itself, modeling presupposes the active interaction of Man with the World around him and inevitably leads us to the general problem of cognition. The process of cognition is certainly determined by the presence of an object of cognition, i.e., recognizable thing and a researcher that is a knowing subject.

The researcher studies the object of interest, observes the course of events, and, possibly, actively intervenes in them, asking certain questions and receiving corresponding answers. Having established information about the subject of research, he forms an idea of this object, his vision of the phenomenon being studied. This idea, based on the available objective information about the subject of knowledge and reflecting the subjective point of view of the researcher, is the *model* of the studied object; on its basis, conclusions can be drawn about the properties of the considered object, see Figure 1.1. Thus, the process of cognition, which actually amounts to collecting, storing, and processing all kinds of *information* about the considered object, is at the same time a modeling process. Apparently, for us there is no particular need to distinguish between these two, it would seem completely different terms.

There are various types of cognition of the World, various ways of perceiving reality by us. This includes science, philosophy, religion, literature, art. It would seem that there could be something in common between them? The scientist is persistently studying a structure of the atom and the emergence of serfdom. The philosopher is stubbornly trying to comprehend the meaning of Being and the place of Man in this wonderful World. The religious thinker is intensely seeking the manifestation of the Eternal in Man and is meditatingly reflecting on the revelation of God. The writer composes a detective novel or high tragedy. The artist depicts a sunset by the sea and draws a portrait of a lover. The composer writes a heroic symphony, a frivolous hit or a funeral march.

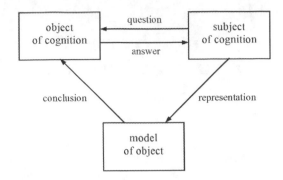

FIGURE 1.1: Simplified process of cognition.

Each of them perceives reality in his own way and tries to express his own vision of the world with available means. All of them voluntarily or involuntarily engage in modeling, and their professional activity invariably comes down to the processing of some information. The existing differences are determined only by the choice of the subject of research and the form of presentation of information. The problems directly facing us in this course relate to only one of the forms of world perception, one of the many types of modeling.

The model of an object is an idea of it,
reflecting the subjective point of view of a researcher
based on the available objective information.

2   Natural Sciences and Mathematics

Of course, we interest in exclusively scientific cognition of various objects on the base of the mathematical apparatus. In this regard, we want, at first, to try to some extent to understand the features of Mathematics and its relationship with other sciences[1].

Physics and chemistry, biology and psychology, sociology and archeology have specific more or less clearly defined areas of application. Laser radiation and the synthesis of acetylene, the cell structure and the principle of private ownership, soil erosion and the sunset of the Roman Empire are associated with real and quite tangible objects. They can and should be studied, because behind them is the World around us, perceived by our senses. Mathematicians, on the contrary, deal exclusively with abstract concepts, for example, number, function, set, operation, etc., which in themselves do not exist in nature, but are only peculiar products of the human brain. If the subject of ordinary sciences is the objectively existing material reality, then Mathematics operates exclusively with an ideal world of human ideas[2].

Each of the usual sciences has relatively clear boundaries that separate it from everything else. The physicist, remaining a pure physicist, is not able to study the relationships between different forms of ownership. The historian does not have the opportunity to investigate chemical reactions. The botanist is not able to apply his professional skills to analyze Roman law. At the same time, the serious mathematician does not stop in front of any obstacles and does not tolerate boundaries artificially separating various scientific disciplines. Remaining in its essence a pure mathematician, he is able to boldly invade any form of human activity and achieve tangible success. So it turns out that Mathematics in some incomprehensible way manages to study almost any material objects, while operating exclusively with their own abstract concepts. Do we have the right after all this to consider that the term "science" applies to Mathematics to the same extent as to physics, sociology or geography? Is it not more logical to assume that we are dealing with some other form of worldview?

A natural question arises, how can one explain the grandiose successes of abstract mathematical methods in solving specific applied problems? Why the numbers and functions, equations and operators, which are absent in the nature, allow us to carefully reveal the secrets of planetary motion, the interaction of chemical elements, the transfer of genetic information, and the mechanism of pricing in the free market? How undoubtedly abstract Nothing turns into completely tangible concrete Something? There must certainly exist some kind of connecting link between Mathematics and the reality surrounding us. It is a specific type of model, on the one hand, capable of containing rich information about a particular subject of study, and on the other hand, formulated using abstract mathematical constructions and, therefore suitable for applying a powerful mathematical apparatus. This is the **mathematical model** of the studying phenomenon, which serves as a kind of translation of the laws identified by a particular science into a rigorous mathematical language. Mathematical modeling turns out to be a grandiose bridge uniting two different worlds that are the objective reality perceived by our senses and studied by means of individual sciences, and the abstract world of human ideas, where mathematical laws dominate[3].

Our problem is precisely to assess, to some extent, the features of mathematical modeling of various processes, get acquainted with the principles of their construction and methods of their research, and therefore, learn to build bridges between worlds.

Mathematical objects relate to the ideal world of human ideas.
Mathematics is able to study almost all objects of material world.
Mathematical model is a bridge
between the surrounding reality and the ideal world of Mathematics.

3   Content or form?

When considering a particular work of literature or art, we evaluate both its content and form. When one talks about the content, one is directly interested in the plot of this work, those specific goals that the creator pursued, those thoughts that he consciously put into his work. Form, however, rather characterizes the means by which the author brought to life his innermost thoughts. If the content carries, first of all, objective information about the subject or phenomenon under consideration, then the form largely corresponds to the personal tastes of the author, and therefore it turns out to be deeply subjective.

It seems quite natural that the goal, idea, thought precede the form, i.e., the subjective view of the researcher should not obscure the true nature of the studying phenomenon. However, for some reason, genuine art and literature can be fully justified only when the form of the work is at least adequate to its content, when the means used are worthy of the chosen goal.

Any person is able to tell more or less distinctly about a particular episode of his life, characterize one of his close acquaintances, and share his own impressions of some famous event. However, will we get a real work of art as a result? Such a story, with all its possible informativeness, is likely to be of interest only to a very narrow group of people who have a direct personal relationship with this quite ordinary subject. However, the poet gives his narrative a unique poetic size, supplies it with unexpected rhymes, rich metaphors that would seem to carry no clear semantic load, not only not clarifying, but sometimes, and obscuring the immediate plot of the work. Any person can get a completely adequate image of the subject, using a camera or smartphone. But with rare exceptions, the result is unlikely to cause particular interest among the masses. However, a real artist takes up the work, and as a result, a picture is obtained that continues to excite the viewer through dozens, if not hundreds of years. However, the composer conveys his impressions through a harmony of sounds that are not very clear from an information point of view. As a result, we meet with Art that will outlive its creator and will excite many generations of completely different people that are infinitely far from the problems that once worried the author. For every cultured person, these divine creations evoke its individual associations.

So, is it really necessary to achieve extremely high accuracy of information reproduction when constructing of a model of the considered phenomenon? Is, in reality, a serious writer who talks about an event that interests him reproduce exactly what really happened? Is a real artist who paints a picture, mainly cares about its maximum proximity to the existing original? According to Auguste Rodin, an artist who is content with an accurate depiction of reality and slavishly reproduces even the most insignificant details will never become a true master. However, the portrait of the brush of a talented painter for some reason speaks much more about the inner world of man than the clearest photograph. Analogically, the story of a talented writer makes an incomparably deeper impression than the most accurate report of a strict forensic expert of the highest qualification. For some strange reason, these ghostly foggy forms invariably turn out to be surprisingly informative.

Now we ask ourselves about the relationship between content and form in scientific research and, in particular, in mathematical modeling. Since the goal of any modeling is to reproduce information about the object under consideration, we, apparently, should pay main attention to the content of the model, considering its form to be of secondary importance. From the mathematical model, we have the right to expect maximum objectivity and adequacy. Considering it permissible, and sometimes even necessary, a certain deviation from the accuracy of reproduction of information in literature, art, religion, we, as if, should demand maximum accuracy from Science, and even more so, from Mathematics. After all, it is here that we are entitled to expect extremely clear and convincing answers to all correctly posed questions.

But is the reliability of Science so high? First of all, we note that this is only one of many forms of cognition of the World. If it can give complete, comprehensive answers to all the questions that concern us, then hardly anyone had the idea of turning to religion or art. Unfortunately (or fortunately?), the possibilities of Science are very limited. Indeed, what is the root cause of all that exists? What is life and why did it arise? What is the mind and why are we endowed with it? Alas, these most important problems invariably confronting humanity are, in fact, outside the scope of Science.

However, we will not ask Eternal Questions. Turn to simple everyday events, which, without a doubt, can and should be studied by scientific methods, and preferably with the active involvement of the mathematical apparatus. Consider the usual process of the movement of a body in space. The problem of physics (more precisely, mechanics) is to clearly indicate where a given body is at a concrete time and how fast it moves.

However, sadly, in modern physics the mediocre question of what actually happens in each particular case can be devoid of a clear meaning. Which of the two events actually happened before? Where exactly is an electron located at a given time and what momentum does it have? According to the special relativity theory, the first question sometimes turns out to be incorrect without specifying the frame of reference, and the choice of a concrete frame of reference is a researchers personal affair. Quantum mechanics imposes a fundamental ban on the possibility of a complete and final answer to the second question.

Is it possible to precisely establish the trajectory of a specific single molecule in a fluid flow? Besides, is there an urgent need for this? Is it worth achieving an absolute correspondence between the results of a full-scale experiment and the proposed model if the most reliable equipment inevitably makes measurements with some error? The results of the experiment can be affected (sometimes very significantly) by both high solar activity, and the nearby air conditioner, and a heavy dump truck driving along the street at this moment, not enough cleaned laboratory test tube, and family troubles of the experimenter. So should serious research sound every random noise? Any process to one degree or another is invariably influenced by various kinds of random factors, which are often not only impossible, but generally not necessary to be taken into account. The researchers problem is to understand the essence of the studying phenomenon and not to display all kinds of interference. In the same way, a writer or artist freely departs from the original, boldly discard the secondary, in their opinion, details and reproduces the object under study as they see fit. So is the difference between strict Science and free Art so great?

As for the supposedly sinless Mathematics with its proud claim to absolute reliability and objectivity, is its foundation really so reliable? Do the laws of mathematics really serve as a standard of absolute accuracy and reliability? It would seem that the properties of commutativity and associativity of multiplication, which are behind the experience of millennia, are natural and trivial. However, replacing from numbers to matrices, for some reason we lose commutativity, and for distributions a violation of the associativity of multiplication is possible[4]. It would seem that the axiomatics of Euclidean geometry is absolutely flawless and verified by many generations of mathematicians. But we ventured to replace one axiom with another, and the entire magnificent building, contrary to expectations, did not collapse at all, but took on a different, completely unexpected, but completely austere look. The very specific question of how many lines parallel to a given line passes through a point taken outside the line cannot have a clear, unambiguous answer without specifying the chosen system of axioms. At the same time, every researcher has the right to choose the axiomatics that he likes best, so long as it turns out to be consistent and gives a clear answer to a specific question. There is an obvious analogy to the choice of a reference frame in relativistic physics. Is it worth it to be surprised? The Mathematics is what the real World it describes is. What kind of objectivity can we talk about here, if the principal result substantially depends on the will of the researcher?

Moreover, what about the famous paradoxes of set theory, if almost every self-respecting mathematician once proposed his own, the only true, in his opinion, way out of that deep quagmire, in which, as if, the almighty Mathematics was firmly stuck. Then the entire mathematical world split into many scientific schools that were deadly warring among themselves, as if it was not about respectable ministers of the most rigorous science, but about stubborn philosophers of ancient Greece or China. So is the Cantor continuum hypothesis true or not? Should the choice axiom be included in the fundamental principles of set theory? Is the principle of exclusion of the third is an undeniable absolute truth? Finally, remember the legendary Gödel theorem, which finally puts an end to attempts to create a fairly meaningful complete consistent mathematical theory. It is not a sin to recall here Tarski's theorem, according to which the concept of truth in a particular formalized theory cannot be characterized by the means of this theory itself. And there is also a curious Levenheim–Skolem theorem, according to which, roughly speaking, any theory, designed to describe a predetermined set of concepts, will describe objects lying outside the specified set[5].

However, we will not climb into the hopeless wilds of the foggy foundations of Mathematics. In this case, we are interested in a very specific question, is adequacy always the most important requirement for a mathematical model? Is content really the decisive factor in the process of mathematical modeling, and is the form something auxiliary and secondary?

Mathematics and science as a whole do not answer all questions.
Mathematics is like the world that it reflects.
When modeling, it is not worth voicing every random noise.

4   Copernicus or Ptolemy?

Consider a concrete example. Let us try to understand what is the fundamental difference between the cosmological theories of Copernicus and Ptolemy? Why did Copernicus prevail in this long tense debate?

It would seem, well, what is there, actually, to think for a long time. A natural answer immediately arises. The clever Copernicus correctly described the structure of the Solar system, and poor Ptolemy was seriously mistaken. And after some temporary misunderstandings, in the end, justice triumphed.

Unfortunately, we cannot be satisfied with such an obvious answer. Indeed, both theories considered cannot pretend to be absolute truth and are actually just models of the Solar system. In reality, the model is not objective reality itself, but only a certain idea of a particular researcher about the studying object being. Thus, the Copernican system cannot be recognized as absolutely correct. It just more or less adequately describes the motion of the planets around the Sun. Do we have the right to attribute to her absolute infallibility? Let us recall at least that, according to Copernicus, the Sun is the center of the Universe, and the planets move around it in circular orbits, which seems to be not very true.

On the other hand, was this Ptolemy so bad with its "bad" geocentric system? Are we aware that Claudius Ptolemy is rightfully one of the leading mathematicians, astronomers and geographers of antiquity? He was one of the creators of trigonometry. One and a half thousand years before Fermat, Descartes, and Euler, he talked about three dimensions of space. He was one of the first to doubt the evidence of the fifth postulate of Euclid. He was the inventor of astrolabe. And, perhaps, his place in the history of world culture is no less honorable than the place of Nikolai Copernicus. We also note that among the predecessors of Ptolemy in the development of the geocentric system were such brilliant thinkers as Eudoxus and Hipparchus. Moreover, if the Ptolemy system was so hopeless, then would really the best astronomers of the whole world use it for one and a half thousand years?

Try to clarify our answer. Apparently, both considered theories describe the studying phenomena. For a long time, astronomers were content with an acceptable, but relatively crude model of Ptolemy, until it was replaced by a more advanced theory. The Copernican system, of course, turned out to be significantly more accurate, which explains its undeniable successes. However, this seemingly perfectly flawless answer, unfortunately, does not fully correspond to reality. In fact, Ptolemy's theory is practically not inferior to the Copernican system in the accuracy of describing the motion of the planets or the eclipse of the Sun and Moon. By the time the heliocentric system appeared, astronomy actually did not know the phenomena that it would describe with a greater degree of accuracy than the Ptolemy geocentric system. Despite this, the full and unconditional victory of Copernicus was inevitable. Why?

As we have already noted, modeling invariably boils down to the collection, storage, and processing of information. Therefore, of the two models of the same object, the one that performs the conversion of information most simply and efficiently will be preferable[6]. We can offer some abstract mathematical formula for describing a specific object and achieve a sufficiently high accuracy of prediction of the studied single phenomenon with the help of numerous sophisticated corrections. However, do we get the right to recognize the mathematical model established because of such dubious tricks quite satisfactory? The Ptolemy's model relatively accurately describes the movement of the planets of the Solar system. However, such high exactness is achieved solely through the use of extremely bulky artificial structures. The Copernican system (especially after its improvement by Kepler), achieving the same high accuracy, turns out to be significantly simpler, more convenient and beautiful. It is curious that the sentence to the Ptolemy system was actually made by its author himself, who owns the brilliant idea that astronomy should aim for a simpler mathematical model.

Thus, the cosmological theories of Copernicus and Ptolemy at a certain stage in the development of science were practically not inferior to each other in the sense of the accuracy of the description of the phenomena studied, i.e., turned out to be more or less equivalent in the sense of the content contained in them. The inevitable victory of Copernicus was due precisely to the choice of a more perfect form. His model is much more elegant and from an aesthetic point of view is much more preferable. The geocentric system at the early stage of its development was also quite simple and beautiful. Indeed, in the center of the Universe is the Earth, around which the Sun, Moon, and planets revolve in circular orbits. But unfortunately it was not so accurate. In the process of its refinement, due to additional artificial structures, the model gradually lost its aesthetic appeal. Therefore, after the advent of a less accurate, but more elegant and efficient Copernicus system, as it turned out that the future belongs to the latter.

As a result, we again return to discussions about the relationship between form and content in scientific research. Of course, the content of scientific theory, the informativeness of the mathematical model is an extremely important factor. However, we will receive a genuine work of Science only in that indispensable case, when its content is expressed through a perfect form. Since the time of Pythagoras, harmony in science is considered no less important requirement than in painting and poetry, music and religion. The world is structured in such a way that for some reason a more perfect mathematical model from an aesthetic point of view is always more effective[7]. Note that the purpose of Mathematics lies precisely in study of forms, and the endowment of this form with specific content remains under the jurisdiction of special sciences. Actually, something similar was expressed by Plato at one time.

> *Form of the mathematical model cannot be neglected.*
> *More aesthetic mathematical model is more effective.*

5 Mathematical model of a body falling

After a lengthy general discussion, we turn to something much more concrete. Perhaps one of the first in the historical sense of mathematical models that fully meets modern scientific requirements, characterizes the process of falling bodies under the influence of its own weight and is associated with the name of Galileo[8]. It is well known that a body raised above the ground, for some reason, will certainly fall to the ground. We will try to describe this process mathematically[9].

First of all, we ask ourselves what exactly happens during the fall of the body? Obviously, its **height** above the ground changes over **time**. Thus, the studied process can be described by the **function** $y = y(t)$, which characterizes the height of the body at an arbitrary time instant t. If we can really determine this functional dependence and thereby predict with a sufficient degree of accuracy the law of change in the position of the body in space with time, then we have serious reasons to consider the practical problem before us to be solved.

Thus, the movement of the body is certainly accompanied by a change in its height. It is natural to ask how quickly this function changes. Therefore, to find the desired function y, we can try to estimate the rate of its change. From a mathematical point of view, the rate of change of the function y at the point t is its derivative $\dot{y}$ at this point. In mechanics, this is usually called a **velocity** and denoted by v.

If a velocity v of the body movement is given, then we can find the function $y = y(t)$ from the easy equality

$$\dot{y} = v. \tag{1.1}$$

The problem (1.1) is called the **differential equation**[10]. Thus, we complete the first step of our analysis. The resulting equation (1.1), as it seems to us, should describe the studied process, being its mathematical model.

The next stage of the analysis involves the extraction by mathematical means of the information contained in the resulting model in a hidden form. We know the relation between the unknown function y and the given value v described by means of equality (1.1). To find the desired functional dependence, we have to solve the obtained equation, i.e., transform the information contained in a mathematical model. If the body velocity does not change with time, then any function of the form

$$y(t) = vt + c, \tag{1.2}$$

satisfies equality (1.1), where c is an arbitrary constant[11].

The next step is an **identification** of the model. In particular, we try to interpret the result from the standpoint of this subject area. We have to find out whether the model we constructed really adequately describes the considered phenomenon. To do this, compare the results of the analysis of the mathematical model with the experiment. The non-uniqueness of the solution found immediately leads to the idea that the information we have at our disposal about the process is still not enough to fully describe it. Consequently, the mathematical model that we have proposed requires serious refinement. This means that we have to return to the initial stage and try to fill the gap in understanding the considered process.

It is obvious, that the position of the body, which moves with a given velocity, depends from its position at an initial time t_0. Two bodies with same velocity and different initial positions have different positions subsequently. Thus, mathematical models of a moving body should provide for the dependence of the law of movement not only on velocity, but also on the initial position of the body. Suppose the body has a height y_0 at a time $t = t_0$, i.e., we have the equality

$$y(t_0) = y_0, \tag{1.3}$$

that is called the **initial condition**. The refined mathematical model is characterized by the equalities[12] (1.1), (1.2), (1.3).

Now we can determine the solution of the equation (1.1), which satisfies the initial condition[13] (1.3). Determine $t = t_0$ in formula (1.2). As a result, we get the value $y(t_0) = vt_0 + c = y_0$. Then we find the constant $c = y_0 - vt_0$. Thus the solution of the equation (1.1) with initial condition (1.3) is

$$y(t) = y_0 + v(t - t_0), \tag{1.4}$$

Thus, we can really determine the position of the given body at an arbitrary at any time. Then we again turn to the interpretation of the results and are faced with a clear discrepancy between the calculated height of the falling body and its value, observed experimentally. This sad fact indicates a serious inaccuracy in our assumptions. Indeed, the simplest experiment shows that contrary to our hypothesis, in the process of falling, the velocity of the body changes significantly. Refining the model, we get the same problem, but with a variable v.

Integrating the equality (1.1) from t_0 to a value t with using initial condition (1.3), we obtain the following law of the body movement

$$y(t) = y_0 + \int_{t_0}^{t} v(\tau)d\tau, \tag{1.5}$$

If the velocity is constant here, then we get the known formula (1.4).

Now, it is necessary to compare again the found functional dependence with the results of observations. However, unfortunately, we do not really have such an opportunity, since we do not know the law of the change of body velocity. Thus, it turns out that the studied process is characterized not only by the coordinate, but also by the velocity of the falling body. Once again, we are convinced of the need to include some additional information in the mathematical model.

Let us try to evaluate how quickly the velocity of the falling body changes. The rate of change of the function v, i.e., its derivative $\dot{v}$, which is thereby the second derivative $\ddot{y}$ of the coordinate, is called the **acceleration** in mechanics and is denoted by a. As a result, to find the velocity, we obtain another differential equation

$$\dot{v} = a. \tag{1.6}$$

This is an analogue of the equation (1.1) and requires an initial condition. For simplicity, we can assume that at time $t = t_0$ the body is just starting to move, which means it has zero velocity (we lifted the body above the ground and let it go). Thus, the following equality holds

$$v(t_0) = 0. \tag{1.7}$$

Now we obtain the system of differential equations (1.1), (1.6) with initial conditions (1.3) and (1.7).

The problems (1.6) and (1.7) have a unique solution analogous to the function y defined by formulas (1.4) or (1.5) depending on whether the acceleration a is constant or variable. Thus, we can establish the law of change in the velocity of a falling body under the indispensable condition that its acceleration is already known. This circumstance leads to certain thoughts.

Finding the height of the body comes down to determining its derivative, i.e., velocity. To calculate the velocity, it is already necessary to find its derivative, i.e., acceleration. It would be very sad if it turned out that now we still have to look for the derivative of acceleration, then the derivative of this derivative, etc. Fortunately, a very reliable experiment shows that the acceleration of a falling body does not change with a sufficiently high degree of accuracy for some reason, and does not even depend on the specific features of the body in question, being a universal physical constant[14]. This amazing fact allows us, in the end, to complete the ongoing study and achieve the long-awaited result.

Note that in the process of movement, the height of the fallen body decreases, which, in accordance with relation (1.1), is possible only in the case of negative velocity v. A certain function v, equal to zero at the initial moment of time and changing with a constant velocity a, can become negative only if the inequality $a < 0$ is satisfied. The positive value $g = -a$ is called the **gravitational acceleration**. This is an absolute constant that does not depend (unless we go into special subtleties) on the conditions of process. Thus, the formula (1.6) can be written as follows

$$\dot{v} = -g. \tag{1.8}$$

Now we obtain the mathematical model of the body falling[15]. This is described by the equalities (1.1), (1.3), (1.7), (1.8). However, it is still necessary to establish applicability conditions for this model. The obtained equations of motion of the body seem to remain valid at any moment in time. However, in reality, the presented model describes the process under study only on the time interval from the origin to a certain point in time T, at which the body reaches the surface of the earth. If you do not introduce an appropriate restriction in the statement of the problem, it turns out that according to the existing equations, our body will fall through the earth safely. Thus, equations (1.1), (1.8) make sense only on a limited time interval $0 < t < T$, where the time instant T is characterized by the equality

$$y(T) = 0. \tag{1.9}$$

TABLE 1.1: Stages of model constructing.

Step	Hypothesis	Realization	Checking	Conclusion
1	body movement is determined by its velocity	equation (1.1)	model does not allow to uniquely determine the solution	additional information must be considered
2	movement is determined by a velocity and initial height	problem (1.1), (1.3)	velocity changes	it is necessary to consider a velocity change
3	movement is determined by a variable velocity and initial height	problem (1.1), (1.3) with variable velocity	velocity is unknown	it is necessary to know the law of velocity change
4	movement is determined by an acceleration and initial values of height and velocity	problem (1.1), (1.6) (1.3), (1.7)	acceleration is unknown	it is necessary to know acceleration
5	movement is determined by gravitational acceleration and initial values of height and velocity	problem (1.1), (1.8) (1.3), (1.7)	model is not always applicable	it is necessary to indicate applicability conditions of model
6	movement is determined by gravitational acceleration and initial values of height and velocity under assumptions	problem (1.1), (1.8) (1.3), (1.7) with conditions (1.9), (1.10)	modeling results are generally consistent with experiment	model is applicable at this stage of the study

Finally, the parameter y_0 of the equality (1.3), i.e., the initial height of the body is positive. Then it satisfies the inequality

$$y_0 > 0. \tag{1.10}$$

Thus, the mathematical model of the body's fall process includes the equations of state (1.1), (1.8) on the time interval $0 < t < T$ with the initial conditions (1.3), (1.7), equality (1.10) for finding the final time moment T and the condition (1.10), which the system parameter y_0 must satisfy. The considered stages of constructing the mathematical model are shown in Table 1.1.

The resulting mathematical model contains in a hidden form quite complete information about the movement of the given body. Solving the considered system, we can determine the direct law of movement. Here, without loss of generality, we can determine $t_0 = 0$, because we can choose any value for the initial moment of time. Then, solving equation (1.8) with the initial condition (1.7), we determine that the velocity of the body is

$$v(t) = -gt. \tag{1.11}$$

Put this value to the formula (1.5) with $t_0 = 0$; determine the function

$$y(t) = y_0 - gt^2/2. \tag{1.12}$$

Use the equality (1.9) for determining the time of movement. After choosing $t = T$ and equating the result to zero, we get $y_0 = gT^2/2$. Now we determine the dependence of the movement time from the initial body height

$$T = \sqrt{2y_0/g}. \tag{1.13}$$

Thus, the formulas (1.11) and (1.12) characterize the movement of the body up to the time T determined by the formula (1.13). Another characteristic of the process that is of practical importance is a velocity of the body at the final time. For a given initial body height, it is equal to $v(T) = \sqrt{2y_0 g}$. The obtained results are illustrated by Figure 1.2.

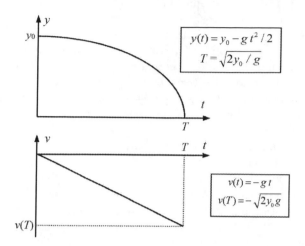

FIGURE 1.2: Characteristics of the body falling.

It should be noted that the obtained results correspond only to some specific level of research of the considered process. They can be refined if it is necessary. For example, we completely ignored the influence of the air resistance force, and equally the possible action of other forces, because of which, among other things, the movement of the body may no longer be straightforward. We were not interested in the shape, size and body weight, which, moreover, under certain conditions, themselves can change. Besides, the immutability of the acceleration of gravity should be recognized as true only with a relatively small difference in altitude.

Recent considerations do not at all cross out the analysis carried out above, but only indicate that any model is only some (most often a small) step on the long path to comprehending the Truth. Perhaps, based on the specifics of the problem being solved, to our great satisfaction, the obtained degree of accuracy will be completely satisfactory. However, it is possible that certain discrepancies between results of modeling and experimental data will be discovered again. Then we have yet to find any errors in the accepted hypotheses or factors not taken into account earlier and build a more accurate model. Some results in this direction is being made in the Appendix[16].

Process of body falling under its weight with a degree of accuracy is described by the movement equations with initial conditions.

6 Principles for determining mathematical models

Based on the above reasoning, one can try to identify general patterns in the construction of mathematical models. First of all, one should understand what is the direct **object** of study. As such, in the example considered above, a falling body acts. Moreover, the shape, size of the body and its internal structure do not bother us at all, i.e., we perceive it as a **material point**.

Then it is necessary to determine **state functions** of the system, i.e., values that, according to the researcher, most fully reflect the course of the process under consideration. Particularly, in the above example, we were only interested in the height of the body above the ground and the velocity of its movement. In solving the problems of thermophysics, we would like to know the body temperature[17]. In the problems of chemical kinetics, the concentrations of reacting substances and reaction products are especially important[18]. In the problem of distribution of manufactured products, of interest is the number of products sent from each point of production to any point of consumption[19]. In the study of a population of a biological species under the influence of random factors, the probability is determined that the population size takes one or another value[20]. Mathematical models, as a rule, are the problems of finding selected state functions. From a mathematical point of view, a state function can be a function of one or more variables, a vector function, a vector, a matrix, etc.

Studying the obtained mathematical model, solving the corresponding equations, we try to establish the dependence of state functions on certain parameters. First of all, these are **independent variables**, which are usually time and spatial coordinates. In the given example, we were interested not just in the position of the body, but in its height at one time or another. When studying the heat transfer process, one should not establish the body temperature in general, but its temperature at a concrete point. In the stochastic population model, it is not just the probability of the abundance of a species that is determined, but the fact that this probability takes on a specific value.

Having understood what are state functions and independent variables, we must also indicate the **coordinate system** within which the mathematical model is formulated. In particular, in the problem of the fall of a body beyond the origin (a point on the t, y plane), the moment of the beginning of motion and the surface of the earth were chosen. In this case, the time axis is directed to the future, and the height is from the surface of the earth up. When describing the speed for the reference point, we choose the zero speed corresponding to the state of rest. We chose the positive direction of the velocity such that it corresponds to an increase in the height of the body above the ground.

Now we can try to determine mathematical relations characterizing the law of change of the state system functions. For this, it is necessary to find out what specifically influenced the course of events, i.e., indicate the **reason for the system evolution**. In the considered example, the cause of the fall of the body was the presence of the body initially at a certain height and the force of gravity acting on it. Other possible effects, in particular, air resistance, were not taken into account in this model.

Having indicated the causes that led to the events of interest to us, we must identify a **causal relationship**, i.e., establish how each of these factors specifically influenced the course of the simulated process. It is this connection, which expresses the law of change of the functions of the state of the system under the influence of these reasons, and is the basis of the mathematical model. It is formulated on the basis of deep patterns established by one or another private science. In relation to the considered example, such a relationship is actually expressed by Newton's second law, according to the second derivative of the state function is proportional to the effective gravitational force, as well as the corresponding initial conditions.

Having written the mathematical relationships expressing the causal relationship for the considered phenomenon, we find that in addition to state functions and independent variables, the mathematical model includes characteristics of another nature. We are talking about **system parameters**, allowing among the entire immense class of phenomena of this nature, described by the relation established earlier, to choose exactly the specific case that is of immediate interest now to us. Particularly, the law of the fall of the body is common for an infinite class of studied objects. However, to find the position of a particular body in space, one should also indicate the specific values of its initial height and initial velocity, which turn out to be process parameters. In our case, we assumed that the body was initially at rest, and therefore limited to the only parameter of the system that is the height of the body at the initial time. When describing the heat transfer process, we must certainly take into account the thermophysical properties of the body.

Formally, the system parameters can vary arbitrarily. However, far from all options for specifying these quantities have a natural physical meaning. The **conditions of applicability** of the mathematical model should be indicated, i.e., the range of parameter changes (both independent variables and system parameters) for which the established model makes sense. In particular, the equation of movement of a falling body remains valid only until the moment of landing, and the initial height of the body must be positive.

Thus, a mathematical model usually includes three classes of characteristics. The process parameters in each case are considered fixed values and constitute **input information**[21]. Independent variables are not fixed and vary within certain limits, thereby forming the domain of definition of state functions. Thus, the mathematical model most often represents the problem of restoring the dependence of state functions on the corresponding variables for arbitrary admissible parameters of process.

After obtaining the necessary mathematical relations, we have to find out what information about the process is of practical interest to us in addition to state functions of the system, i.e., **output parameters** should be indicated. As such, in the problem of the body fall are the moment the body falls on the surface of the earth and the velocity of the body at this point in time.

Summing up, we conclude that the construction of a mathematical model of the process under study includes the following elements:

1. Object of study.
2. State functions.
3. Independent variables.
4. Coordinate system.
5. Reasons for the system evolution.
6. Causal relationships.
7. Input parameters.
8. Conditions of applicability of the mathematical model.
9. Output parameters of the system.

The characteristics of the basic elements of constructing a mathematical model of the process of body fall are given in Table 1.2.

All these items are solved by the means of a science (physics, chemistry, biology, etc.), to which the studied phenomenon belongs. After that, we can forget for some time the private sciences and turn to pure mathematics. Now we have at our disposal specific mathematical relationships that we have to analyze based on certain qualitative and quantitative methods. Having carried out the corresponding analysis exclusively by mathematical means (as a rule, using a computer), we again return to the phenomenon under consideration with the goal of physically interpreting the results obtained, clarifying the tasks before us and summing up the corresponding results.

> *Mathematical model determines a dependence of state functions*
> *on the corresponding variables and system parameters, reflecting,*
> *under certain conditions, the corresponding causal relationship.*

TABLE 1.2: Stages of model constructing.

Elements	Body falling process
object of study	body falling
state functions	height of the body, its velocity
independent variable	time
coordinate system	initial time moment of is zero, coordinate is directed vertically upward, origin is the surface of the earth, zero velocity corresponds to a state of rest, velocity increases in the direction of coordinate increasing
reasons for evolution	gravitational field and the initial location of the body above the surface of the earth
causal relationships	second derivative of the coordinate is proportional to the gravity force
input parameters	initial body height
applicability conditions	initial body height is positive model is applicable until the moment of landing
output parameters	time of movement, velocity at the time of landing

7 Classification of mathematical models

Depending on the properties of the characteristics that make up the mathematical models and the mathematical relationships that connect them, principles of their classification can be indicated. There are many ways to classify mathematical models[22]. At first, this is the property of the independent variables. In particular, **continuous systems** are considered in which independent variables continuously change, and **discrete systems** with independent variables that change with some step (constant or not), or this is a finite ensemble of values. In the first case, the state of the system is characterized by functions of one or many variables, and in the second, by vectors. Continuous models are often described by **differential** or **integral equations**[23]; it can be analyzed using **mathematical analysis**[24] and its applications. In some cases, the system is described by differential-difference equations[25]. The discrete models are characterized by the methods of **discrete mathematics**[26], in particular, **graph theory**[27], **combinatorics**[28], **Boolean algebra**[29], etc. The continuous systems describes many physical phenomena, for example, pendulum oscillation[30] and heat transfer. Discrete systems that are widely used in economics include, for example, the **travelling salesman problem**[31].

The following classification principle is determined by the properties of parameters that affect the system. Here, **stochastic models** are distinguished, in which any random factors can be influenced by the process, and **deterministic models**, where similar effects are not taken into account, and the state of the system is determined uniquely. **Probability theory** methods are widely used to analyze stochastic models[32]. In particular, electric vibrations are characterized by a deterministic model, and the movement of a molecule in a fluid flow is characterized by a stochastic model.

In the future, we will distinguish between **systems with lumped parameters**, in which the state function depends only on time, and **systems with distributed parameters**, where the dependence of the state function on spatial variables is also allowed. Systems with lumped parameters are **finite-dimensional** in the sense that their state at a particular moment in time is described by a finite set of numbers. Systems with distributed parameters are **infinite-dimensional** because an infinite set of numerical characteristics is required to describe their state[33]. In particular, the projectile flight under certain conditions is relatively well described by a system with lumped parameters, and gas diffusion in a certain volume is described by a system with distributed parameters. As systems with lumped parameters, ordinary differential equations most often act. Distributed parameter systems are usually characterized by **partial differential equations**[34].

Note also **dynamic systems** with functions of states that change with time, and **stationary systems** whose characteristics do not change with time. Thus, chemical reactions and the evolution of a biological population are described by dynamic systems, while electrostatic and gravitational fields are described by stationary systems.

There are also distinguished **linear systems** in which the response to the sum of the actions is equal to the sum of the responses to each effect, and **nonlinear systems** for which this property is violated. For example, string oscillation and gas diffusion are described by linear systems, and the movement of an incompressible viscous fluid and the coexistence of two competing biological species are described by nonlinear systems[35].

Various equations are often considered as mathematical models. However, in some cases, the system can be described by **variational inequalities**[36] or extremum problems. Thus, the movement of a material point in classical mechanics can be described not only using the movement equations, but also on the basis of the principle of least action[37], according to which the evolution of the system is carried out in such a way that the energy consumption is minimal. Problems of this type are solved using the **calculus of variations**[38].

In engineering and economics systems are often the situation, in which there is a certain freedom of choice of certain parameters. In this case, the problem arises of choosing such parameters that correspond to the achievement of a certain goal, for example, the least cost of producing a certain product or maximizing production. Similar problems are solved using the **optimal control theory**[39]. In economics and politics, game problems often arise in which there are two or more parties, each of which pursues its own goals. To solve them, **game theory**[40] methods are used.

Mathematical models can be described by **well-posed systems** with unique solution that continuously depends on the systems parameters. However, **ill-posed systems** are possible in which these properties are violated[41]. In particular, if according to the known body temperature at a given moment in time it is required to determine the temperature distribution at subsequent time instants, then the corresponding problem is well-posed. However, if we try to restore the background of the system, then an ill-posed problem is obtained. One considers also **direct problems** in which, using the known values of all system parameters, it is required to find the state functions of the system, as well as **inverse problems** for which some system parameters are not known and must be found from some information about state functions[42]. This is related to the problem of **system identification**[43].

Table 1.3 indicates the most important classes of mathematical models considered in this book.

There are some other principle to classify mathematical models. Besides, these various classification principles may be considered in conjunction. In particular, the heat transfer process is characterized by a correct continuous linear non-stationary deterministic system with distributed parameters. The mathematical model of the fall of the body considered above is described by a well-posed deterministic linear continuous dynamic system.

> *Mathematical models differ in features the quantities*
> *entering into them and the relations connecting them.*

Direction of further work. *The considered mathematical model of the fall of a body was characterized by a linear equation. However, we consider more difficult equations in the Appendix. Determination of the analytic solution is very difficult here. Therefore, the next Chapter is devoted to approximate methods for solving such problems. These questions open Part I of the course, devoted to the study of dynamical systems with lumped parameters. After that, we will begin to analyze mathematical models of specific processes related to different subject areas.*

TABLE 1.3: Mathematical model classes.

class	place	example
continuous systems	Parts I and II	mechanical oscillation
discrete systems	Chapter 17	travelling salesman problem
deterministic systems	Parts I and II	biologic competition
stochastic systems	Chapter 18	Brownian motion
lumped parameter systems	Part I	chemical kinetics
distributed parameter systems	Part II	diffusion
dynamic systems	Parts I and II	predator-prey model
stationary systems	Chapter 13	gravitational field
linear systems	Parts I and II	string vibrations
nonlinear systems	Chapters 6–9,14	incompressible viscous fluid
variational principles	Chapter 16	refraction of light
optimal control problems	Chapter 20	missile range maximization
game problems	Chapter 17	prisoner's dilemma
well-posed problems	Parts I and II	electrodynamics
ill-posed problems	Chapter 19	Euler elastic problem
direct problems	Parts I, II, III	antibiotic resistance
inverse problems	Chapter 21	inverse heat conduction problems

Appendix

Let us consider some phenomena related to classical mechanics. This is the movement of bodies in the gravitational field[44]. As in the considered above, one assumes that in the process of movement the shape and internal structure of the body do not change, so that all its points describe the same trajectories. Thus, the objects of study that interest us are still **material points**. An undoubted step forward in comparison with the previously considered problem of the fall of the body under the influence of its own weight here will be taking into account the effect on the body of not only gravity, but also of some other forces. To describe the process of movement, we again use the second Newton's law, according to which the rate of change of the momentum of a body is equal to the sum of the forces acting on it.

We consider below the problems of the movement of the probe, missile and glider, each of which has its own specifics. In particular, if for a missile (as well as for a falling body) we neglect the influence of the air resistance force, then for the probe and glider this effect will be taken into account. For a light glider, the linear nature of the dependence of resistance on speed is allowed, and for a massive probe, this dependence is assumed to be quadratic. Like a falling body, the probe moves rectilinearly, while the missile and glider fly in a vertical plane and are characterized by two spatial coordinates. Unlike the other considered objects, the mass changes during the movement of the probe. The missile and the probe fly under the influence of traction, while the movement of glider depends from the lifting force. The study of these problems will be continued in Chapter 3 after consideration approximate methods for solving differential equations.

1 Probe movement

Consider the movement of a probe launched in a vertical direction[45]. The state function $y = y(t)$ here characterizes the height of given object above the ground at the corresponding moment in time. We choose the coordinate system in the same way as in the problem of the fall of the body. The force F acting on the body is the sum of the thrust force R due to fuel combustion, the air drag force Q, and the weight P. The thrust force is always directed upward, i.e., in the direction of movement, weight is directed down, and the resistance force acts against the direction of movement. Thus, we obtain the equality $F = R - Q - P$.

By **Newton's second law**, the rate of change of momentum of a body, i.e., its derivative is equal to a force acting on this body[46]. Given that the **momentum** of the body is equal to the product of its mass and velocity, we arrive at the equation of movement

$$\frac{d}{dt}\left(m\frac{dy}{dt}\right) = R - Q - P.$$

Traction is caused by the combustion of fuel. We assume that the rate of combustion (and, therefore, reactive force) is constant over the entire time interval, while there is a fuel reserve. Then the value of R is constant and equal to R_0 until the time t_*, when the fuel burns out completely, where t_* is the constant thrust force, that is the task parameter. Thus, the traction force is R_0 if $t < t_*$ and 0 otherwise. The force of air resistance depends significantly on the body velocity. In particular, with increasing velocity, there is a rapid increase in resistance force. Under certain conditions, the dependence of the resistance force on speed can be considered quadratic. It should be borne in mind that the resistance force is always directed in the direction opposite to the direction of movement. As a result, we obtain the equality $Q = \mu v|v| = \mu \dot{y}|\dot{y}|$, where k is air resistance coefficient, which substantially depends on the shape of the body. When the probe rises, the velocity v is positive, which means that $Q > 0$. When the probe descends, we have $v < 0$, and therefore, $Q < 0$. In both cases, the resistance force slows down the movement. Body weight is $P = mg$. Substituting the corresponding values of the forces in the movement equation, we have

$$\frac{d}{dt}\left(m\frac{dy}{dt}\right) = R(t) - \mu\frac{dy}{dt}\left|\frac{dy}{dt}\right| - mg.$$

In the process of fuel combustion, the mass decreases. Therefore, the velocity of change of mass must be negative. Suppose

$$\dot{m} = -\beta,$$

where the β is β_0 if $t < t_*$ and 0 otherwise. Thus, as long as there is a fuel supply, the mass of the probe decreases at a constant velocity β_0. Once all the fuel has burned, the mass ceases to change.

The considered equations are completed by the initial conditions[47] $y(0) = 0$, $\dot{y}(0) = 0$, $m(0) = m_0$. In accordance with these relations, the probe rises under the action of traction. Over time, all the fuel burns out, after which the probe flies for some time by inertia, and then falls down due to gravity. The relations under consideration remain valid up to a certain point in time T, when the probe will be on the surface of the earth, which means that it will satisfy the condition $y(T) = 0$. Thus, the mathematical model of the process is described by the specified differential equations on the time interval from zero until the time T with initial conditions[48].

The system parameters are traction force R_0, fuel combustion time t_*, air resistance coefficient k, fuel combustion rate β_0, and the initial probe mass m_0. Among the output parameters of the process, i.e., additional characteristics of undoubted practical interest include the maximum height of the probe, the time of maximum height, the time of landing of the probe and its speed at the time of landing.

Note that not with all values of the process parameters the considered mathematical model makes sense. First of all, the probe must certainly take off. This means that the force acting on the body at the initial moment of time must be positive. Assuming $t = 0$ in the right-hand side of the equation of motion and taking into account the initial conditions, we have the inequality $R_0 > m_0 g$. Thus, the probe takes off only if the traction force exceeds its initial weight. Obviously, at the time of fuel combustion, the mass of the probe should be positive. Integrating the equation for mass from zero to t_*, we obtain $m(t_*) - m_0 = -\beta_0 g$. Now we get $m_0 > \beta_0 g$, guaranteeing the positive mass at $t = t_*$, i.e., useful mass of the probe. Thus, the mathematical model under consideration is applicable under the specified inequalities.

Task 1.1 *Probe flight*. Give a general description of the mathematical model of the probe flight by constructing a table similar to Table 1.3.

2 Missile flight

Consider the movement of a ballistic missile launched at a certain angle to the horizon[49]. In this case, the movement of the body is not straightforward, since it is influenced not only by the gravitational force directed strictly downward, but also by the traction force F acting at an angle, see Figure 1.3. The missile position is characterized by the horizontal coordinate x and vertical coordinate y. The missile launch point is selected as the origin. Then the function $x = x(t)$ is the distance from the projection of the point at which the missile is at a given moment of time on the horizontal plane (surface of the earth) to the origin, and the function $y = y(t)$ determines the height of the rocket above the ground. To derive the equations of motion, it is necessary to write the second Newton's law in vector form. Note that the horizontal component F_x of the traction force will act in the horizontal direction, and the vertical component of the traction force F_y and the weight P will act in the vertical direction, see Figure 1.3. In contrast to the previous case, we assume that the mass of fuel is much less than the mass of the missile, so that the change in mass during the movement of the missile can be neglected.

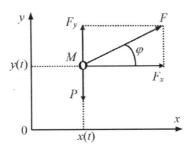

FIGURE 1.3: Forces acting on the missile.

By second Newton's law, the acceleration of the body in horizontal and vertical directions are proportional to the corresponding forces. Then we get the following system of equations

$$m\ddot{x} = F_x, \quad m\ddot{y} = F_y - P.$$

Determine the components of the traction force $F_x = F\cos\varphi$, $F_y = F\sin\varphi$, where $\varphi = \varphi(t)$ is an the angle between the direction of the traction force and the horizontal coordinate axe. Thus, we obtain the movement equations

$$\ddot{x} = a\cos\varphi, \quad \ddot{y} = a\sin\varphi - g,$$

where $a = F/m$ is the acceleration of the traction force.

Traction is acted until all fuel is burned. The fuel combustion time t_* is considered known. Then we have $a(t) = a_0$ if $t < t_*$ and 0 otherwise, where a_0 is the acceleration of the thrust force, which is a process parameter.

Suppose that at the initial moment of time the missile is at the origin and is at rest, which corresponds to zero values of the coordinates and velocities of the missile at the initial time. The movement of the missile continues up to a certain point in time T, at which it will land, and therefore, will have a zero vertical coordinate. Thus, we arrive at the condition $y(T) = 0$ for finding a landing time T.

The parameters of our mathematical model are the angle φ that can depend on the time, the value a_0, and the fuel combustion time t_*. The considered mathematical model makes sense only if its initial vertical acceleration is positive (the positivity of the horizontal component of the acceleration vector is obvious). The initial acceleration in the vertical direction is $a \sin \varphi(0) - g$. Thus, we have the following applicability condition of the model $a \sin \varphi(0) > g$.

Task 1.2 *Missile flight*. Give a general analysis of the mathematical model of the missile flight, see Table 1.3.

3   Glider flight

Consider the movement of a glider in the vertical plane. It occurs under the influence of gravitational force, air resistance, and also lifting force, see Figure 1.4. The resistance force F_r is considered proportional to the velocity of the glider and is directed in the opposite direction. The lifting force F_l is proportional to the square of the velocity vector and is directed perpendicular to the movement of the glider. As a result, we obtain the equations

$$\dot{x} = u, \quad \dot{u} = -\mu u - bv\sqrt{u^2 + v^2},$$

$$\dot{y} = v, \quad \dot{v} = -\mu v + bu\sqrt{u^2 + v^2} - g,$$

where u is the horizontal glider velocity, v its vertical velocity, μ is a resistance coefficient, and b is a coefficient of lifting force.

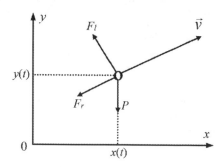

FIGURE 1.4: Forces acting on the glider.

At the initial moment of time, the glider rises from the ground with a velocity w at an angle φ to the horizon. The movement of the glider continues until the time moment T, when it lands, i.e., when equality is satisfied $y(T) = 0$. Note that glider movement does not occur for all initial states of the system.

Task 1.3 *Glider flight*. Based on the numerical solution of the considered model, task the following analysis:

1. Give a general analysis of the mathematical model of the glider flight.

2. Show the graphs of change of the vertical and horizontal glider coordinates and components of the velocity vector. Explain the results. The set of input parameters used for this should be considered as the main one.

3. Find values of system parameters in which the glider does not rise up.

4. Changing the system parameters, detect the "dead loop" mode.

5. Determine the influence of parameters on the maximal glider height.

Notes

[1] Various aspects of the interaction of Mathematics with the outside World are considered, for example, in [15], [42], [43], [64], [83], [105], [114], [124], [128], [140], [139], [180], [181], [238], [270], [295], [336], [337], [338], [357], [366], [371], [380].

[2] Other sciences also operate with ideal objects. Material point, absolutely black body, continuous medium, etc., also are abstractions that do not exist in the outside world. But mathematics operates exclusively with such objects, and this differs from other sciences.

[3] Problems of mathematical modeling, are considered, for example, in [8], [11], [25], [31], [82], [110], [116], [149], [230], [237], [240], [258], [276], [285], [300], [354], [374], [376].

[4] An example of the non-associativity of multiplication of distributions see [306].

[5] The philosophical meaning of these problems of the foundations of Mathematics is analyzed, for example, by M. Kline [180]. Problems of philosophy of mathematics see, for example, in [131], [187], [264].

[6] About the simplicity in the mathematical modeling of natural phenomena see [248].

[7] About the importance of aesthetics in scientific research see W. Heisenberg [139].

[8] It is clear that our arguments in deriving the mathematical model of the fall of a body have nothing to do with what Galileo was guided by.

[9] The process of a body falling under its own weight refers to classical mechanics. One can get acquainted with the problems of *classical mechanics* and the methods of their mathematical and computer modeling, for example, in [8], [47], [116], [115], [179], [183], [198], [197], [258], [285], [374], see also Chapter 3.

[10] This is the equation, because we have an equality with unknown value y. One calls it differential, because this contains a derivative of the unknown function. More exact, this is the ordinary differential equation unlike partial differential equations, see Part II. Some problems of the theory of differential equations will be considered in Chapters 2, 5, and 19, see also [14], [59], [132], [212], [275], [346], [383].

[11] The equality (1.2) gives the *general solution* of the equation (1.1). This means the function y is the solution of equation (1.1) whenever it satisfies the equality (1.2).

[12] We determined the *Cauchy problem* for the considered equation.

[13] We found a *particular solution* of the differential equation here.

[14] Actually, it was from such an experiment that Galileo repelled.

[15] In the future, the equation of the fall of the body will be derived using the variational principle of least action (see Chapter 16), which allows you to look at the problem under consideration from slightly different positions. These issues are discussed in more detail in [198], [197]. We also note that the fall of the body is due to the action of the gravitational field, see Chapter 13.

[16] We will return to the mathematical description of the body falling process in Chapter 16. There we will also consider a generalization of this problem, in which a curve is determined, moving along which, under the action of its own weight, the body will travel from one point to another in a minimum time. Chapter 21 will discuss the problem of falling of a body under its own weight, taking into account air resistance. At the same time, the problem of restoring the air resistance coefficient from the results of measuring the position of the falling body is analyzed.

[17] The simplest problems of thermophysics are considered in Chapters 10 and 11.

[18] Chapter 7 is devoted to the problems of chemical kinetics. Some mathematical models of chemical systems are considered in Chapters 11 and 18, see also [27], [56], [84], [90], [91], [115], [128], [171], [176], [194], [213], [245], [256], [277], [279], [282], [374].

[19] This is the transport problem discussed in Chapter 17. Models of economic systems are described in Chapters 9, 11, and 18, see also [2], [11], [18], [21], [50], [51], [60], [65], [75], [78], [82], [87], [92], [108], [111], [155], [160], [173], [184], [196], [238], [335], [363], [382].

[20] Chapter 7 is devoted to mathematical models of biological systems. Some mathematical models of biological systems are discussed in Chapters 11, 17, and 18, see also [7], [45], [50], [72], [73], [82], [85], [134],

[244], [245], [287], [289], [326], [342], [348], [355], [362], [372], [374]. Probability as a function of state is used in stochastic systems, see Chapter 18.

[21] The process parameters can be numbers, vectors, and functions that are considered known. For example, when describing the movement of a material point in space, the initial position is described by the coordinates of the body in space at the initial moment of time, see the mathematical models of rocket and glider movement in the Appendix to this chapter. When studying the heat transfer process, it is assumed to be known, which means that the initial temperature distribution is a parameter of the system. It changes from point to point of the considered body in general case, which means that this is characterized not by a number, or even by a vector, but by a function, see Chapter 10.

[22] Different principles for the classification of mathematical models are given, for example, in [115], [246], [265], [300], [332].

[23] The theory of integral equations can be found in [20], [189], [273].

[24] Mathematical analysis are described, for example, in [13], [61], [79], [148], [206], [344], [353].

[25] This applies in particular to the stochastic systems discussed in Chapter 18.

[26] Discrete mathematics is considered, for example, in [36], [167], [375].

[27] Graph theory see [41], [71], [174].

[28] Combinatorics problems are considered, for example, in [286], [359].

[29] One can get acquainted with Boolean algebra in [322], [369].

[30] Pendulum oscillation is the subject of Chapter 3.

[31] The travelling salesman problem is considered in Chapter 17.

[32] The basics of probability theory are described, for example, in [98], [120], [172].

[33] The apparatus of classical mathematical analysis is used to describe a finite-dimensional object, while infinite-dimensional systems are investigated using functional analysis, see, for example, [185], [283].

[34] The theory of partial differential equations is described, for example, in [96], [141], [147], [169], [192], [215], [235], [269], [349], [361].

[35] In particular, the mathematical model of the fall of the body considered above is a linear system, and the tasks on the flight of the probe and glider described in the Appendix are characterized by nonlinear systems.

[36] The variational inequalities, see [81], [113], [177].

[37] Variational principles in physics are described in [197], [198], and Chapter 16.

[38] The basics of calculus of variations can be found in [39], [89], [109], [169], [298], [378], see Chapter 16.

[39] The theory of optimal control is described, for example, in [102], [112], [156], [178], [213], [312], [313], [328], see also Chapter 20.

[40] Game theory, see [99], [158], [260], and Chapter 17.

[41] The general theory of ill-posed problems is considered in [4], [170], [207], [216], [350]. Examples of ill-posed problems see [53], [123], [141], and Chapter 19.

[42] Inverse problems of mathematical physics are investigated in [19], [133], [170], [207] see Chapter 21.

[43] The general theory of system identification is described in [93], [117].

[44] The problems of flight dynamics are considered in sufficient detail in [233].

[45] One optimization problem associated with the flight of the probe is considered in [210].

[46] We will return to the Newton's second law in Chapter 16.

[47] We have the second order differential equation and the first order equation. The *order of the differential equation* corresponds to the maximum order of the derivative of the unknown function here. To solve this equation, additional conditions are required in an amount equal to the order of the equation.

[48] In contrast to the model of the body fall, in this case we are dealing with a nonlinear system. We assumed that the dependence of the resistance force on velocity is quadratic. The procedure of squaring the unknown function is nonlinear. We will work with quadratic nonlinearities in subsequent chapters. In particular, this refers to the Verhulst equation, which describes the evolution of a biological species with limited food available, see Chapter 7. The determining of modulus is the nonlinear operation too.

[49] We will return to the mathematical model of missile movement in Chapter 21, where we will study the problem of determining the angle at which the missile will fly away as far as possible, see also [210]. Chapter 18 deals with the problem of missile flight in the presence of wind, the characteristics of which are random variables.

Part I

Systems with lumped parameters

Part I of this Course is devoted to systems with lumped parameters. For such systems, state functions depend on a single independent variable, as a rule, having the meaning of time. The mathematical models here are systems of linear or nonlinear ordinary differential equations with the corresponding initial conditions.

This Part consists of eight chapters. Two of them are auxiliary in nature and are devoted to the description of the simplest methods for the practical analysis of systems with lumped parameters. The remaining six chapters discuss concrete mathematical models of different systems.

In the Appendix to the previous Lecture, we came across quite difficult enough differential equations, the direct solution of which is not easy. In this regard, in Chapter 2 we have to familiarize ourselves with approximate methods for solving such problems. This will allow further analysis of the considered systems, in particular, to perform specific tasks given in subsequent chapters.

Chapter 3 in a certain sense continues the material presented in the Appendix to Chapter 1, where various problems on the movement of a material point in a gravitational field were considered, and Chapter 2, where the two-body problem was studied. We again turn to classical mechanics. In this case, the main object of study is the material point on which gravity acts. We consider the oscillations of a suspended body, i.e., the pendulum. The subject of the lecture is various problems related to mechanical oscillations. In the subsequent Chapter 4, mathematical models of electrical oscillations that are similar in form are considered.

These two lectures reveal some features of systems with lumped parameters. However, for a serious analysis of such problems, especially in the case of nonlinear equations, some information from the qualitative theory of differential equations is required. Chapter 5 is devoted to these questions. The results presented here, in addition to the materials of Chapter 2, serve as the foundation for the problems considered in subsequent lectures.

Chapter 6 discusses the simplest mathematical models of chemical processes. We consider the problems of chemical kinetics, in particular, we talk about the change of the concentrations of chemicals in the reaction process over time. Some models of biological systems are studied in Chapter 7. Here we consider the problems of population dynamics, in which populations of biological species with different forms of their coexistence are considered. Various dynamic models of economic systems are described in Chapter 8. Finally, the Chapter 9 is devoted to the simplest models of political science, sociology, and psychology.

Many of the described models can be extended to systems with distributed parameters, in which the dependence of state functions is allowed not only on time, but also on spatial variables. These issues are addressed in the next part of this book.

Chapter 2

Approximate solving of differential equations

In Chapter 1, we considered a mathematical model of the fall of a body under the influence of its own weight. It was based on a fairly simple ordinary differential equation, an analytic solution of which was found without much difficulty. However, the equations of the movement of probe, missile, and glider considered in the Appendix to that Chapter were too difficult so that finding their analytical solution is not possible. We are not be able to obtain explicit solutions to the vast majority of problems considered in the future. In this regard, before conducting a detailed review of mathematical models of physics, chemistry, biology, and other sciences, we turn to the description of approximate methods for solving of general differential equations[1].

The simplest and most natural method for solving the Cauchy problem for differential equations is the Euler method. Its idea is to split the interval on which the equations are considered into parts. On each of these parts, the derivative is replaced by the ratio of the difference between the values of the function at the ends of the interval to the length of this interval. The result is a formula that allows you to sequentially determine the values of this function at the boundaries of the selected sections, starting from the first, where you can use the specified initial condition[2].

The Euler method easily extends to the case of systems of differential equations, as well as to equations of higher orders. We use this method to study the previously described mathematical models of movement of the probe, missile, and glider.

The Appendix will describe a more complex, but also more accurate Runge–Kutta method. As an application, two famous models related to classical mechanics (the two-body problem) and biology (the predator–prey model) are considered.

Lecture

1 Conception of approximate solution

The subject of Part I is mathematical models characterized by systems with lumped parameters and usually described by ordinary differential equations. To begin with, we turn to the consideration of a general first-order equation[3] with an initial condition, i.e., to the **Cauchy problem**

$$\dot{x}(t) = f(t, x(t)),\ t > 0;\ \ x(0) = x_0, \tag{2.1}$$

where the function f and the number x_0 are known.

Sometimes, these equations can be solved analytically. Finding a solution to the problem as an explicit formula seems to be an ideal case. Using this formula, we can conduct a complete analysis of the considered problem, as was shown in the previous chapter for the equation of the fall of the body. However, as a rule, it is not possible to find the exact solution to the problem in explicit form. This applies, in particular, even to not so difficult models given in the Appendix to Chapter 1. Under these conditions, it remains for us to change the idea of what should be understood by solving the problem[4].

DOI: 10.1201/9781003035602-2

The approach described below is based on two deviations from the ideal at once. First, we do not seek to find the exact solution to the problem, satisfied with its approximate solution. Second, we do not try to find a solution to the problem at an arbitrary point in the domain of the state function, satisfied with its approximate value only in a finite set of points from its domain. How justified is this approach? Can we be satisfied with this situation?

It should be borne in mind that any model invariably relies on assumptions, which means that it only approximately describes the studied process[5]. In addition, any practically meaningful model includes parameters that must be determined in some way[6]. These parameters are known from a direct or indirect experiment[7], which in itself contains a certain error, often very significant. Besides, even explicit formulas of the analytical solution include integrals, infinite sums, special functions, etc.[8], which in practice can only be determined approximately. Thus, an analytical solution to the problem, even if it can be found, still cannot guarantee the receipt of accurate information about the behavior of a real system. As for finding a solution to the problem everywhere, in practice the state of the system will still be used only by a finite set of specific points.

Based on the foregoing, we will find not an exact, but an approximate solution to the problem if only the resulting error would be satisfactory based on the meaning of the specific problem being solved. We will find the solution not everywhere but only in a finite set of points if only there would be an acceptable number of points and they fit the needs of this particular case[9].

By significantly weakening the requirements for solving a problem, we significantly expand the class of problems that can be solved. These principles are fully implemented in the Euler method for the approximate solving of differential equations.

For difficult equations, it is impossible to find analytical solutions.
In the absence of analytical solution, we can find its approximate one.
Using of approximate methods is justified
if it is possible to find the problem solution
with satisfactory accuracy for a satisfactory number of points.

2   Euler method

Return to the direct analysis of the problem (2.1). The easiest method of approximate solving of this problem is based on the derivative definition. As well known, the **derivative** $\dot{x}$ of a function $x = x(t)$ at a point t is defined as the limit $[x(t+\tau) - x(t)]/\tau$ as $\tau \to 0$. If the value τ is small enough, then this derivative is approximately equal to the considered ratio. Geometrically, the derivative of the function at the point is equal to the slope tangent of the tangent line to the graph of the corresponding curve, while its approximation corresponds to the tangent of the corresponding secant, see Figure 2.1.

Let us analyze the system (2.1) on a segment $[0, T]$. Divide this segment into parts with a sufficiently small **step** τ that is the general parameter of the algorithm. Consider the given equation at the arbitrary point $t_i = i\tau$, $i = 1, ..., M$, where $M = T/\tau$ is the quantity of the used points. Replace the corresponding derivative here by that ratio, we get[10]

$$\frac{x(t_i + \tau) - x(t_i)}{\tau} = f(t_i, x(t_i)).$$

As a result, we determine the recurrence formula[11]

$$x(t_{i+1}) = x(t_i) + \tau f(t_i, x(t_i)), \ i = 0, ..., M - 1. \tag{2.2}$$

The number $x(t_0)$, i.e., the value of the function x at the point $t = 0$, is known from the

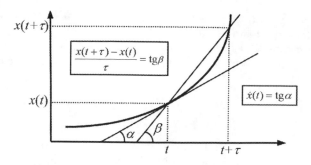

FIGURE 2.1: Derivative approximation.

initial condition. Then we find the approximate solution of the problem at the arbitrary point t_i, using the formula (2.2). This algorithm is called the ***Euler method***[12].

This result can be extended to the systems of differential equations. Let us consider the Cauchy problem

$$\dot{x}_j(t) = f_j(t, x_1(t), ..., x_n(t)), \ t > 0; \ x_j(0) = x_{0j}, \ j = 1, ..., n. \tag{2.3}$$

Using the previous idea, determine the analogue of the formula (2.2)

$$x_j(t_{i+1}) = x_j(t_i) + \tau f_j(t_i, x_1(t_i), ..., x_n(t_i)), j = 1, ..., n; \ i = 0, ..., M - 1. \tag{2.4}$$

Now we can find the values of the arbitrary function x_j at all considered points[13].

We describe the method of the solving the first-order differential equations and the systems of that equations. However, in Chapter 1, we considered second-order differential equations, as well as systems of such equations. Consider a differential equation of the arbitrary order r that is solvable with respect to its high derivative $x^{(r)}$

$$x^{(r)}(t) = f\big(t, x(t), \dot{x}(t), ..., x^{(r-1)}(t)\big), \ t > 0. \tag{2.5}$$

The Cauchy problem of this equation includes the value of the function x and its derivatives until $r - 1$ order at the left boundary of the given segment[14]

$$x(0) = x_0, \ \dot{x}(0) = x_1, \ ..., \ x^{(r-1)}(0) = x_{r-1}, \tag{2.6}$$

where numbers $x_0, x_1, ..., x_{r-1}$ are given. Particularly, the movement equations of Chapter 1 had second order. For their analysis, we used initial position and initial velocity, i.e., the unknown function and its first derivative at the initial time. Determine the functions $y_1 = x, y_2 = \dot{x}, ..., y_r = x^{(r-1)}$. Transformed the equation (2.5) to the system of first-order differential equations[15]

$$\dot{y}_1 = y_2, \ \dot{y}_2 = y_3, \ ..., \ \dot{y}_r = f(t, y_1, y_2, ..., y_r).$$

Then the initial conditions (2.6) are transformed to $y_j(0) = x_{j-1}, j = 1, ..., n$. Thus, the Cauchy problem (2.5), (2.6) is transformed to the partial case of the system (2.3). Therefore, we can use the previous technique for solving it.

The system of equations of arbitrary orders can be reduced in this way to the corresponding system of first order equations. Therefore, its approximate solution can also be found using the Euler method.

Thus, the Euler method, in principle, is applicable for solving the Cauchy problem for any ordinary differential equations and systems[16]. As an application of the algorithm under consideration, we consider the problem described in the Appendix to Chapter 1.

*Approximate solution of the Cauchy problem for a first order equation
is calculated using a recurrence formula by the Euler method.
System of first order equations is solved in a similar way.
Equations of arbitrary order are reduced to first order equations.*

3   Probe movement

Consider the movement of a probe launched in a vertical direction. This is carried out
under the influence of the traction force acting due to the combustion of fuel, air resistance,
and gravity. As noted in Chapter 1, the movement of the probe is characterized by its height
$y = y(t)$ above the surface of the earth and is described by the equation

$$\frac{d}{dt}\left(m\frac{dy}{dt}\right) = R(t) - \mu\frac{dy}{dt}\left|\frac{dy}{dt}\right| - mg,$$

where m is the mass of the probe, μ is the air resistance coefficient, g is the gravity acceleration, and the traction force $R(t) = R_0$ if $t < t_*$ and 0 otherwise. The mass of the probe
decreases by the fuel combustion by the formula

$$\dot{m} = -\beta(t),$$

where $\beta(t) = \beta_0$ if $t < t_*$ and 0 otherwise. The constant traction force R_0, the constant rate
of the fuel combustion β_0, and the fuel combustion time t_* are parameters of the system.

The probe is on the ground and is at rest at the initial time, and its initial mass is m_0.
Then we have the initial conditions

$$y(0) = 0, \; \dot{y}(0) = 0, \; m(0) = m_0.$$

Based on the physical meaning of the problem, the probe first rises under the action of
traction. However, after the combustion of all fuel, it eventually goes down, and at some
point in time T reaches the surface of the earth. Thus, the studied process is completed when
the condition $y(T) = 0$. This completes the mathematical model of the probe movement.

Task 2.1 *Probe movement.* Carry out the following analysis.

1. Determine the order of the system of differential equations for the probe movement.

2. Transform these equations to a system of first order differential equations.

3. Write the recurrence formulas of the Euler method for this system.

4. Prepare the corresponding computer program.

5. Choosing the system parameters in an arbitrary way, construct graphs of the change in time
of the height of the probe above the ground, its velocity and mass. Note the main stages of the probe
movement that are the active flight under the influence of traction, inertia ascent after turning off
the engines, descent with acceleration, and descent at an almost constant velocity. The set of input
parameters used here as the main one.

6. Show the graphs of changes in time with respect to traction and air resistance and explain
the results.

7. Changing the parameters of the system to detect two reasons for the violation of the conditions
of applicability of the model that are the insufficient traction and excessively high fuel consumption.

8. Show a graph of dependence of the maximum height of the probe from its initial mass and
fuel combustion velocity. To do this, fix the main set of input parameters. Then change one of
the input parameters, taking several values more and less than its main value, leaving the other
parameters unchanged. Explain the obtained results.

9. Show a graph of dependence of the time of the probe movement from the constant traction force and the air resistance coefficient, acting as in the previous step of the task. Explain the obtained results.

10. Show a graph of dependence of the probe velocity at the time of its landing from the time of fuel combustion and air resistance coefficient, acting as in the previous step of the task. Explain the obtained results.

4 Missile flight

Consider the movement of a ballistic missile launched at a certain angle to the horizon, see Chapter 1. The movement of the missile is carried out in a vertical plane under the action of a traction force acting until a certain moment of time t_* in the presence of gravity. The position of the missile at an arbitrary point in time is characterized by the horizontal coordinate x and vertical coordinate y, which are described by the equations

$$\ddot{x} = a \cos \varphi, \quad \ddot{y} = a \sin \varphi - g.$$

Here φ is the angle between the direction of the missile movement and the x axe that is the system parameter[17], and the traction force $a(t) = a_0$ if $t < t_*$ and $a(t) = 0$ otherwise, where a_0 is the acceleration of the traction force that is a process parameters.

Suppose that at the initial time the missile is at the origin and is at rest, which corresponds to the zero initial conditions. Obviously, under the influence of traction, the missile starts moving. After turning off the engines, the missile continues to move in a vertical plane by inertia and at some time T lands. The completion of the movement is characterized by equality $y(T) = 0$.

Task 2.2 *Missile flight*. Carry out the following analysis.

1. Write an algorithm for the approximate solving of the considered problem by the Euler method.

2. Prepare the computer program on the base of this algorithm.

3. Show the graphs of changes over time of the horizontal and vertical coordinates of the missile, as well as the components of the velocity vector for a certain set of system parameters, selected as the main.

4. Show the trajectory of the missile movement by determined the set of points $(x(t), y(t))$ for various times t for the main set of parameters.

5. Based on the output information, determine the three stages of missile flight that are active flight, inertial lift, and descent.

6. Determine the inapplicability of the model for small values of traction. Explain the results.

7. Show the graphs of the dependence of the maximal missile lift from the angle φ and acceleration a_0. Explain these results.

8. Show the graph of the dependence of the length of the missile flight, i.e., the value of its horizontal coordinate at the landing time from the angle φ and the fuel combustion time t_*. Explain these results.

9. Show the graph of the dependence of the module of the missile velocity vector at the landing time from the acceleration a_0 and the fuel combustion time t_*. Explain these results.

10. Show the graph of the dependence of the time of the missile movement from the angle φ and acceleration a_0. Explain these results.

5 Glider flight

Consider a glider movement in the vertical plane. This movement occurs under the influence of gravitational forces, air resistance, as well as lifting force. By Chapter 1, this process is described by a system of equations

$$\ddot{x} = -\mu\dot{x} - b\dot{y}\sqrt{\dot{x}^2 + \dot{y}^2}, \quad \ddot{y} = -\mu\dot{y} + b\dot{x}\sqrt{\dot{x}^2 + \dot{y}^2} - g$$

where x and y are horizontal and vertical coordinates of the glider. The air resistance coefficient μ and lifting force coefficient b are system parameters.

The glider rises from the ground at the initial moment of time with a velocity w at an angle φ to the horizon. The position of the glider at the initial time is chosen as the origin. Thus, we have the following initial conditions

$$x(0) = 0, \quad y(0) = 0, \quad \dot{x}(0) = w\cos\varphi, \quad \dot{y}(0) = w\sin\varphi.$$

The glider moves until the time moment T, when it lands, i.e., when the following equality holds $y(T) = 0$.

Task 2.3 *Glider flight.* Carry out the following analysis.

1. Prepare the computer program for analysis of the glider flight model based of the Euler method.

2. Show the graphs of changes over time of the horizontal and vertical coordinates of the glider, as well as the components of its velocity vector for a set of system parameters, selected as the main.

3. Show the trajectory of the missile movement by determined the set of points $(x(t), y(t))$ for various times t for the main set of parameters.

4. Choose a combination of system parameters so that the glider could not rise up. Explain the results.

5. By changing the system parameters, detect the dead loop mode.

6. Determine the influence of system parameters on the maximum height of the glider.

7. Determine the influence of system parameters on the length of flight.

Direction of further work. *Having learned methods for the approximate solution of differential equations, we can turn to the consideration of mathematical models of various processes described by these equations. In particular, in Chapter 3, the consideration of the problems of mechanics will be continued. The object of study here are mechanical oscillations.*

Appendix

The Euler method described in this chapter is quite simple and natural. However, it is not very accurate, because of which this method is extremely rarely used to solve real applied problems. In reality, algorithms that are more accurate are usually used in practice, among which the Runge–Kutta method considered below[18].

For practical development of the algorithm, we consider two extremely important problems related to classical mechanics and biology. We are talking, respectively, about the two-bodies problem and the predator–prey model. When considering them, we will be interested not so much in their physical meaning as in the possibility of practical solving to the corresponding differential equations. Of particular importance, here is the comparison of the Euler and Runge–Kutta methods for various values of the algorithm step. In subsequent chapters, we will no longer focus on algorithmic features, and in the tasks of the subsequent chapters of Part II, it is recommended to use the Runge–Kutta method.

1   Runge–Kutta method

The described Euler method is very natural and attractive in its simplicity. However, it is characterized by relatively low accuracy and is rarely used in practice. When solving applied problems, more efficient algorithms are usually used, among which the Runge–Kutta method is distinguished. Consider again the Cauchy problem

$$\dot{x} = f(t, x(t)),\ t > 0;\quad x(0) = x_0.$$

As in the case of the Euler method, the segment $[0, T]$ is divided by M equal parts with by step $\tau = T/M$ by the points $t_i = i\tau$, $i = 0, ..., M$. Suppose the value $x(t_i)$ is known. Using **Runge–Kutta method**, the subsequent value of the considered function is determined by the formula

$$x(t_{i+1}) = x(t_i) + \frac{\tau}{6}\left(k_1 + 2k_2 + 2k_3 + k_4\right),\ i = 0, ..., M - 1,$$

where

$$k_1 = f(t_i, x(t_i)),\quad k_2 = f\left(t_i + \tau/2, x(t_i) + \tau k_1/2\right),$$
$$k_3 = f\left(t_i + \tau/2, x(t_i) + \tau k_2/2\right),\quad k_4 = f\left(t_i + \tau, x(t_i) + k_3\right).$$

The Runge–Kutta method differs in a large volume of calculations in comparison with the Euler method, however it is significantly more accurate, which is due to its rather high popularity[19]. The Runge–Kutta method for a system of first order differential equations is characterized by the relations presented above, understood in a vector sense, similar to what it was for the Euler method. To use this algorithm in the case of high order equations, one should transform this equation to a system of first order equations and use the method indicated above.

Task 2.4 *Missile flight*. Carry out the following analysis.

1. Apply Runge–Kutta method for the missile flight mathematical model.

2. Prepare the computer program for the solving the missile flight equations on the base of this method.

3. Calculate the missile flight equations with some set of parameters.

4. Calculate the considered problem by the Euler method with the same parameters and algorithm step. Compare the results.

5. Increasing the step of partitioning the given segment, find a situation, where the Euler method leads to unsatisfactory results, while the Runge–Kutta method gives quite acceptable results.

2   Two-body problem

Return to the consideration of mathematical models of mechanics. The subject of our study will be the **two-body problem**[20]. It consists in finding the law of movement of two bodies interacting with each other[21]. Two bodies with mass m_1 and m_2 are considered. Bodies are considered material points, and their position in space is characterized by vectors r_1 and r_2 in a certain coordinate system. Using, as before in the problems of the movement of material points Newton's second law, we determine the following equations[22]

$$\begin{cases} m_1\ddot{r}_1 = F_{12}(r_1, r_2) \\ m_2\ddot{r}_2 = F_{21}(r_1, r_2) \end{cases}, \tag{2.7}$$

where F_{12} is a force acting on the first body from the side of the second one, F_{21} is a force acting on the second body from the side of the first one. Besides, the initial positions r_{10}

and r_{20} and the initial velocities v_1 and v_2 of the bodies are known. Therefore, we have the initial conditions

$$r_i(0) = r_{i0}, \ \dot{r}_i(0) = v_i. \tag{2.8}$$

As noted in the previous chapter, the model also includes limitations imposed on the system. In this case, this concerns the forces entering the right-hand sides of equations (2.7). Naturally, the dependence of the forces under consideration on the position of the bodies is determined by the nature of these forces, i.e., depends on the specific situation. However, there are some general considerations.

According to the **Newton's third law**, the forces of interaction between two bodies are equal in magnitude, directed in opposite directions and act along a straight line connecting these points. Thus, they turn out to be functions of the quantities r_1 and r_2 and satisfy the equality

$$F_{12}(r_1, r_2) = -F_{21}(r_1, r_2). \tag{2.9}$$

Moreover, these forces should not be determined by the absolute coordinates of the bodies under consideration, i.e., values of vectors r_1 and r_2, and their relative position, i.e., difference[23] (see Figure 2.2)

$$r = r_1 - r_2. \tag{2.10}$$

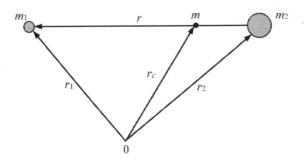

FIGURE 2.2: Forces acting on the missile.

Thus, we get

$$F_{12}(r_1, r_2) = F(r), \tag{2.11}$$

where the function F is determined by the concrete type of body interacting.

The equalities (2.7)–(2.11) are the mathematical model of the considered system. We show that due to conditions (2.9)–(2.11), the system (2.7) splits into two independent subsystems.

Adding the equalities (2.7) with using the condition (2.8), we get

$$m_1 \ddot{r}_1 + m_2 \ddot{r}_2 = F_{12}(r_1, r_2) + F_{21}(r_1, r_2) = 0.$$

It can be transformed to

$$\frac{d}{dt}(m_1 r_1 + m_2 r_2) = 0.$$

The value

$$r_c = \frac{m_1 r_1 + m_2 r_2}{m_1 + m_2}$$

characterizes the position of the **mass center** of the bodies, see Figure 2.2. Therefore, from the previous equality it follows the very easy equation

$$\ddot{r}_c = 0. \tag{2.12}$$

Using the equality (2.12), on determine that the velocity of the mass center movement is constant.

Determine the second equation. Subtract equalities (2.7) from one another, having previously divided them into m_1 and m_2, respectively. We have

$$\ddot{r}_1 - \ddot{r}_2 = \frac{F_{12}}{m_1} - \frac{F_{21}}{m_2}.$$

Using the conditions (2.9)–(2.11), transform this equality to

$$\ddot{r} = \left(\frac{1}{m_1} + \frac{1}{m_2}\right) F(r).$$

Denote by m^{-1} the value under the brackets, where m is called the **reduced mass**. We get

$$m\ddot{r} = F(r). \tag{2.13}$$

Thus, system (2.7) has been reduced to two independent problems with respect to the quantities r_c and r, which greatly simplifies the analysis of the system[24]. The initial conditions for these problems can be determined based on the initial conditions (2.8). If now solutions to these particular problems are found, then the law of change of the bodies in question can be determined by the formulas

$$r_1 = r_c + \frac{m_2}{m_1 + m_2} r, \quad r_2 = r_c - \frac{m_1}{m_1 + m_2} r. \tag{2.14}$$

Task 2.5 *Two-body problem*.

1. Write the algorithm of solving the two-body problem for the case of gravitational force and equality of masses on the basis of the Euler method.

2. Write the algorithm of solving the two-body problem for the case of gravitational force and equality of masses on the basis of the Runge–Kutta method.

3. Calculate the considered problem for some set of parameters the same for both methods. Compare the calculation results by two methods with each other with different steps of the algorithm. Compare the results for the different algorithm steps.

4. Calculate the law of movement of the mass center, using the equations (2.12) with the same values of parameters. Explain the results.

5. Calculate the law of change of the difference r, using the equations (2.13) with the same values of parameters. Explain the results.

6. Based on the obtained results, verify the validity of equalities (2.14).

3 Predator–pray model

Let us turn now to the well-known predator–prey model, which is related to biology. Two biological species existing on the same territory are considered. It is assumed that the first species is the food for the second one. The considered system is described by the following differential equations

$$\dot{x}_1 = \left(\varepsilon_1 - \gamma_1 x_2\right) x_1, \quad \dot{x}_2 = \left(\gamma_2 x_1 - \varepsilon_2\right) x_2,$$

where x_1 is the number of the first species (prey), x_2 is the number of the second species (predator), ε_1 is growth of prey in the absence of predators, ε_2 is the rate of predator extinction in the absence of food, i.e., the prey, and the coefficients γ_1 and γ_2 characterize the natural interaction of species with each other. The initial values of species numbers x_{10} are x_{20} are known.

We obtain the predator–prey model. Now we are not interested in biological interpreting the statement of the problem and its solution[25], but in the possibility of applying methods for its approximate solution.

Task 2.6 *Predator–pray model.* Carry out the following analysis.

1. Write the algorithm of solving the considered problem by the Euler method.

2. Write the algorithm of solving the considered problem by the Runge–Kutta method.

3. Calculate the problem for some sets of parameters, the same for both methods. Compare the calculation results by two methods with each other with different steps of the algorithm.

Notes

[1] Approximate methods for solving ordinary differential equations are described, for example, in [16], [125], [159], [195].

[2] For ordinary differential equations, **boundary value problems** also make sense, in which additional conditions are set simultaneously at both ends of the interval on which the problem is considered. Naturally, those discussed in this chapter are not applicable for such problems. For methods of solving boundary value problems, see [17] and Chapter 11.

[3] The general form of first-order ordinary differential equations is $f\big(t, x(t), \dot{x}(t)\big) = 0$. But we restrict ourselves to considering equations solvable with respect to the derivative.

[4] Note that the use of quantitative methods for finding an approximate solution to a problem in the absence of the possibility of obtaining its analytical solution is not the only possible approach. Under these conditions, in some cases, the possibility remains of a qualitative analysis of the system, i.e., obtaining certain information about the properties of the solution to the problem in the absence of an explicit formula for the analytical solution of the problem. These issues are addressed in Chapter 5.

[5] For example, analyzing the process of the body falling in Chapter 1, we consider the body as a material point, not taking into account its shape and size, neglected air resistance, etc.

[6] For example, mathematical models of the probe and missile movement included an air resistance coefficient, the direct measurement of which is practically impossible, see also Chapter 21.

[7] A direct experiment involves the direct finding of unknown parameter. An indirect experiment does not determine the desired parameter itself, but some characteristic related to the state of the system. Then the considered parameter is determined in such a way that the indicated characteristic obtained from the existing mathematical model and from the experiment coincides. This corresponds to the concept of the inverse problem, see Chapter 21.

[8] The analytic solutions from problems of Part II will include these elements.

[9] Naturally, if it is possible to obtain an analytical solution to the problem, then it needs to be obtained, since it contains complete information about the system, while an approximate solution gives only some part of the information.

[10] Of course, approximating the derivative we get some error. It is precisely because of this that we obtain an approximate rather than an exact solution to the problem. Obviously, the smaller the step of the partition τ, the more accurately the approximation of the derivative is carried out, which means that the error of the Euler method will be smaller. Thus, the higher the accuracy of solving the problem is required on a fixed interval $[0, T]$, the more it is required to use points from this interval. Thus, to obtain a higher degree of accuracy, it is necessary to increase the amount of computational work.

[11] A *recurrence formula* is an explicit formula for defining an arbitrary element a_n of sequence through a number of preceding elements of the same sequence and number n.

[12] From the standpoint of computational mathematics, the Euler method is explicit, one-step and has a first order. The method is *explicit* because it indicates the explicit dependence of the solution of the problem at a given point on its value in the previous point. It is *one-step*, because to move from one point to another, one action should be performed. Finally, this method has the *first order of accuracy*, because

the error of the method has the order of the step of dividing the studied interval into parts, i.e., to reduce the calculation error by n times, it is also necessary to reduce the step n times. Note that algorithms that are more accurate than the Euler method turn out to be implicit or multi-step, and often implicit and multi-step at the same time.

[13] Determine the vector functions $x = (x_1, ..., x_n)$, $f = (f_1, ..., f_n)$. Then the Cauchy problem (2.3) has the form (2.1), and the equalities (2.4) are transformed to (2.2) if the formulas (2.1) and (2.2) are the vector sense.

[14] The order of the equation coincides with the number of additional conditions that must be specified.

[15] The *order of system of differential equations* is equal to the sum of the orders of its constituent equations. In particular, in this case, an equation of order r reduces to a system of order r.

[16] In Chapter 11, we will consider the finite difference method, which is a generalization of the Euler method to the case of partial differential equations.

[17] In principle, the angle φ can be a variable. In particular, we consider in Chapter 21 the problem of finding such dependence of angle from time that maximize the length of the missile flight. However, in this case we suppose that the angle is constant.

[18] In fact, there is a whole class of Runge–Kutta methods, see [16], [125], [159], [195]. However, the most widespread of them was the algorithm considered in this chapter.

[19] If the Euler method is one-step, then in the Runge–Kutta method, four intermediate steps should be taken to move from the previous point to the next one. On the other hand, the Runge–Kutta method has a fourth order of accuracy compared to the first order of the Euler method. As a result, halving the step reduces the calculation error by a factor of 16 for the Runge–Kutta method, while for the Euler method in this case the error decreases only by half. Thus, if in the Runge–Kutta method an increase in accuracy by 16 times requires a double increase in the number of points under consideration, while the Euler method achieves the same degree of accuracy by increasing the number of points in 16 times. Algorithms exist with a higher order of accuracy, see, for example, [126], [159]. However, they are even more difficult. Note that both considered methods are explicit, since the transition from the previous point to the next one is carried out by explicit formulas. There are also *implicit methods* in which when passing from one step to another, it is required to solve a certain system of equations, see [126]. Naturally, this leads to a significant increase in the amount of computation. However, there are *stiff differential equation* for which the use of explicit methods leads either to a sharp increase in the volume of calculations at small step values, or to a sharp increase in the error at a step that is not small enough. Note that for the numerical solution of partial differential equations, simpler but less reliable explicit methods and more difficult but less reliable implicit methods are also used, see Chapter 11.

[20] The two-body problem is considered in any course of theoretical mechanics, see, for example, [47], [179], [183], [198].

[21] The two-body problem naturally arises in astronomy. This may be the movement of a satellite around a planet, a planet around the sun or a double star. Another example would be an electron moving around the nucleus of an atom, viewed from the perspective of classical mechanics.

[22] Since the position of the body in space is characterized by three coordinates, the system (2.7) is of the sixth order.

[23] The indicated property is called the *translational symmetry*.

[24] In fact, the two-body problem comes down to the analysis of two independent problems on the movement of one body. This circumstance explains the possibility of its full solution. Naturally, it is necessary to specify the force F for solving the problem. In particular, if the interaction between bodies is due to gravity, then in accordance with the *law of universal gravitation* we have the equality $F = -Gm_1 m_2 r|r|^{-3}$, where G is the gravitational constant, and $|r|$ is the module of the vector. The situation changes significantly when moving to three and even more so to a larger number of bodies, see [1]. The three-body problem, which describes the movements of three interacting objects (for example, the Sun, the Earth, and the Moon), with the exception of a number of special cases, can no longer be solved analytically in principle.

[25] The predator–prey model will be completely analyzed in Chapter 7.

Chapter 3

Mechanical oscillations

In Chapter 1, mathematical models of the movement of various bodies were considered as examples. These processes belong to mechanics. One of the most fascinating problems of classical mechanics, which has significant theoretical interest and serious practical applications, is associated with the process of oscillation of the pendulum. In accordance with the procedure described in Chapter 1, the equation of movement of the mathematical pendulum is derived. In the case of small oscillation, this equation is linear, and its solution is a periodic function. The energy of the pendulum is calculated. In order to clarify the results, the influence of the friction force is taken into account. The equilibrium position of the pendulum is determined. The forced oscillations of the pendulum are also investigated. The Appendix considers spring oscillations, as well as some specific problems of the theory of nonlinear oscillations.

Lecture

1 Determination of the pendulum oscillation equation

The main object of our study is a **pendulum**. It is a massive solid suspended on a long thin thread. The shape and size of the body, as well as the properties of the thread in further studies are not taken into account. Thus, we are dealing with a material point, denoted by M and located at rest at the lower end of the thread, see Figure 3.1. The movement of the pendulum is carried out under the action of gravity and occurs in the plane formed by the initial position of the thread and its vertical state that is the line OA. Neglecting the change in the length of the thread, we conclude that the point M is invariably at the same distance l from the point O of fastening the thread. The idealized object considered under these assumptions is called the **mathematical pendulum**[1].

Now we have to choose the function of the system state, i.e., some value characterizing, as it seems to us, the movement of the pendulum. There is a certain freedom of choice. As a state function, we could choose the deviation of the pendulum from the vertical axis, i.e., segment MB, the change in the height of the pendulum BM_0 above the ground compared with its position at point M_0, corresponding to the vertical position of the thread, as well as the angle θ between the segments OM and OA. All three of these options are equivalent, and the corresponding quantities are related by the following obvious relations

$$MB = L\sin\theta, \quad BM_0 = L(1 - \cos\theta).$$

In mechanics, it is usually accepted to choose the angle θ as a state function. This is called the **angular displacement**. Thus, we have to establish the law of its change over time. Relations, allowing to find the form of this functional dependence, is the mathematical model of the considered process.

The model construction, as in the previous chapter, is based on Newton's second law, according to which the acceleration of the pendulum a in the direction of its movement

DOI: 10.1201/9781003035602-3

39

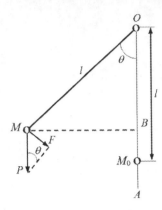

FIGURE 3.1: Mathematical pendulum.

is proportional to the acting force F. Then the equality $ma = F$ is true, where m is the mass of the pendulum. The movement of the pendulum at any time occurs in the direction perpendicular to the current position of the OM thread, see Figure 3.1. The force of interest to us is the projection of the weight P on the direction of the pendulum movement. As a result, we find the value

$$F = -P\sin\theta = -mg\sin\theta.$$

The minus sign here is explained by the fact that the pendulum moves in the direction opposite to its deviation from the vertical position, and the increase in the angle is usually counted in the counterclockwise direction.

Thus, the Newton's second law has the form

$$m\dot{v} = -mg\sin\theta,$$

where v is a velocity of the pendulum. Now we have the equation

$$\dot{v} = -g\sin\theta. \tag{3.1}$$

Therefore, the law of pendulum movement does not depend from its mass.

The equality (3.1) determines the relation between the velocity of the pendulum movement and the angle θ that is the state function of the system. However, the derivative of the function θ is its **angular velocity** ϑ, not the function v that is its linear velocity. Thus, in order to transform the resulting equality, it is necessary to establish a relationship between function v and the angular velocity of pendulum.

Suppose that at time t pendulum is located at point $M(t)$ at an angle $\theta(t)$ to the axis OA, and at time $t+\tau$ it is at point $M(t+\tau)$ at an angle $\theta(t+\tau)$, see Figure 3.2a. Denoting by $r(\tau)$ the distance between considered points, find the velocity of pendulum

$$v(\tau) = \lim_{\tau\to 0}\frac{r(\tau)}{\tau}.$$

The value $r(\tau)$ is determined by the equality (see Figure 3.2b)

$$\frac{r(\tau)}{2} = l\sin\frac{\delta(\tau)}{2},$$

where $\delta(\tau) = \theta(t+\tau) - \theta(t)$. Now we find

$$v(t) = \frac{dr}{d\tau}\Big|_{\tau=0} = \frac{d}{d\tau}\left[2l\sin\frac{\delta(\tau)}{2}\right]_{\tau=0} = \frac{l}{2}\cos\delta(0)\frac{d\delta(0)}{d\tau}.$$

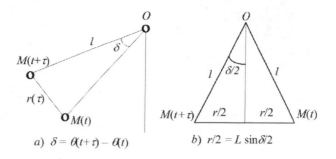

a) $\delta = \theta(t+\tau) - \theta(t)$
b) $r/2 = L\sin\delta/2$

FIGURE 3.2: Calculation of the angular velocity.

Using the obvious equalities

$$\delta(0) = 0, \quad \frac{d\delta(0)}{d\tau} = \dot{\theta}(t),$$

determine $v = l\dot{\theta}$. Putting it to the formula (3.1), we get the equation

$$l\ddot{\theta} = -g\sin\theta.$$

Determine the process parameter $\omega_0 = \sqrt{g/l}$, which is called **oscillation frequency** (more precisely, its **natural frequency**) of pendulum. We obtain the movement equation

$$\ddot{\theta} + \omega_0^2 \sin\theta = 0. \tag{3.2}$$

This is the second order differential equation[2] with respect to the state function θ. This equation is **nonlinear**, because the sum of its solutions and the product of a solution by number are not a solution of this[3]. The analysis of nonlinear systems is associated with serious difficulties[4]. In particular we cannot, as a rule, find its **analytic solution**, i.e., give a direct formula, which describes its dependence from the time and parameter ω_0.

The problem is greatly simplified for small oscillations of pendulum, where the sine value is arbitrarily close to the value of the angle itself, i.e., we have the condition $\sin\theta \approx \theta$. Then we get the equality

$$\ddot{\theta} + \omega_0^2 \theta = 0. \tag{3.3}$$

That is called the **equation of harmonic oscillator** or the **equation of pendulum oscillation** (more precisely, small free oscillations). This is the basis of our analysis. The equality (3.3) is a linear homogeneous ordinary second order differential equation[5]. It has different physical interpretations[6]. One of them is considered in Appendix. In the final subsection, we consider a non-homogeneous analogue of equation (3.3).

Differentiating the equality (3.3) and using the relation between angular displacement and angular velocity, we get

$$\ddot{v} + \omega_0^2 v = 0.$$

Thus, the angular velocity of the pendulum also satisfies the harmonic oscillator equation.[7]

To complete the mathematical description of the process, it remains to provide information on the initial state of the system. Suppose that at time $t = 0$ the pendulum is at an angle θ_0 to the vertical axis and has an angular velocity v_0. Then we have the initial conditions

$$\theta(0) = \theta_0, \quad \dot{\theta}(0) = v_0. \tag{3.4}$$

TABLE 3.1:   Elements of the pendulum oscillations model.

No	Elements	Characteristics
1	object of study	mathematical pendulum
2	state functions	angular displacement, angular velocity
3	independent variable	time
4	coordinate system	angle counts counterclockwise, origin is the equilibrium position
5	reasons for evolution	pendulum is suspended and has an initial position, gravity acts on it
6	causal relationships	Newton's second law
7	input parameters	pendulum length, initial angle, initial velocity
8	applicability conditions	pendulum length is positive, initial angle, and velocity are small enough
9	output parameters	amplitude, frequency, period, and phase of oscillation

The equalities (3.3), (3.4) that make up the Cauchy problem[8] are a mathematical model of the considered process. It includes the parameters ω_0 (oscillation frequency determined by the length of the pendulum), θ_0 (initial position) and v_0 (initial velocity). Table 3.1, similar to Table 1.2, presents the main elements of the mathematical model under consideration, and the specified output characteristics will be established in the process of solving the obtained problem below.

We now have to analyze the considered Cauchy problem and interpret the properties of its solution[9].

Movement of pendulum is described
by a second order nonlinear differential equation.
Small pendulum oscillation is described
by the linear equation of harmonic oscillator.

2   Solving of the pendulum oscillation equation

The second step of the research, as usual, involves a mathematical analysis of the resulting model in order to extract the information contained in it in a hidden form. In accordance with the classical theory of linear homogeneous ordinary differential equation with constant coefficients, determine the ***characteristic equation***[10]

$$\lambda^2 + \omega^2 = 0.$$

This is the algebraic (square) equation with respect to the parameter λ. Its solutions are $\lambda_1 = \omega_0 i$, $\lambda_2 = -\omega_0 i$, where i is the imaginary unit. Then the general solution of the equation (3.3) is

$$\theta(t) = c_1 \sin \omega_0 t + c_2 \cos \omega_0 t, \tag{3.5}$$

where c_1, c_2 are arbitrary constants.[11]

To interpret the results, we write the solution in the form

$$\theta(t) = a \sin(\omega_0 t + \varphi), \tag{3.6}$$

where the parameters a and φ are called the ***amplitude*** and the ***initial phase*** of the oscillation, respectively[12]. The concrete values a and φ are determined from conditions (3.4). Using the equalities

$$\theta(0) = a \sin \varphi = \theta_0, \ \ \dot{\theta}(0) = a\omega_0 \cos \varphi = v_0,$$

we find the amplitude and the initial phase of the oscillation

$$\varphi = \arctan\left(\frac{v_0}{\omega_0\theta_0}\right), \quad a = \frac{\theta_0}{\sin\varphi}.$$

The determined solution of the Cauchy problem (3.3), (3.4) is the periodic function[13] (see Figure 3.3) with the **period of oscillation** $T = 2\pi/\omega_0$. Thus, the pendulum oscillates around its equilibrium position[14]. The maximum deviation of the pendulum from its vertical position corresponds to its amplitude. The frequency, phase, amplitude, and period of oscillation are the most important characteristics of the oscillatory process. The information about them allows us to give an interpretation of the results that completes the third stage of the study of the problem.

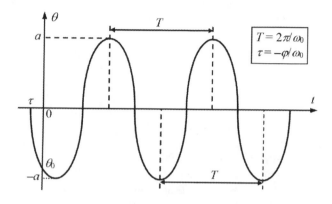

FIGURE 3.3: Solution of the problem (3.3) and (3.4) is periodic function.

Task 3.1 *Harmonic oscillation of pendulum*. Based on the numerical solving of the problem (3.3), (3.4), make the following analysis.

1. Make sure that the angle of deviation of the pendulum from the equilibrium position and its angular velocity change periodically with time.

2. Draw a graph of changes over time t points with coordinates $(\theta(t), \dot{\theta}(t))$ in the plane formed by the function of the state of the system, i.e., angle and angular velocity. This plane is called *phase*[15]. Verify that the resulting curve is closed. In particular, it is an ellipse.

3. By changing successively the length of the pendulum, its initial position and initial velocity, determine a change in the amplitude, frequency, and period of oscillation. Analyze the obtained results.

Solution of the pendulum oscillation equation is a periodic function.

3 Pendulum oscillation energy

In describing mechanical movement, the energy of a moving body is also of interest. To formulate the law of change of energy, we write the equation of pendulum oscillation (3.3) in the following form

$$L\ddot{\theta} + g\theta = 0.$$

Multiplying this equality by the derivative of function θ, we have

$$\frac{d}{dt}\left(\frac{l\dot{\theta}^2}{2} + \frac{g\theta^2}{2}\right) = 0.$$

Using the formula $l\dot\theta = v$, transform the previous equality to the form

$$\frac{d}{dt}\left(\frac{v^2}{2} + \frac{lg\theta^2}{2}\right) = 0.$$

Hence, after multiplying by the mass of pendulum m, we get the equality

$$\frac{mv^2}{2} + \frac{lLmg\theta^2}{2} = \text{const.} \qquad (3.7)$$

The first term on the left-hand side of equality (3.7) is the **kinetic energy** K. The potential energy of pendulum is $U = mgh$, where the value h corresponds to the length of the segment BM_0 in Figure 3.1, i.e., the difference between the length of pendulum l and the segment OM_0 that equal to $l\cos\theta$. Then we get $U = lmg(1 - \cos\theta)$. If the angle θ is small, we obtain $\cos\theta \approx (1 - \theta^2/2)$. Therefore, the second term in the left-hand side of equality (3.7) corresponds in the case of small pendulum oscillation to its **potential energy**. Denoting

$$K = \frac{mL^2\dot\theta^2}{2}, \quad U = \frac{lmg\theta^2}{2},$$

determine the equality

$$K(t) + U(t) = \text{const.} \qquad (3.8)$$

This is the **law of conservation of energy**. The concrete value of the constant on the right-hand side of this equality is determined by the initial values of the position and velocity of pendulum, which respectively set its potential and kinetic energy at the initial time.

From obtained results, it follows an interpretation of the pendulum movement from an energetic point of view. Suppose, for example, that at the initial moment of time, the pendulum has a deviation from the vertical position and zero initial velocity. Then its kinetic energy at the initial moment of time is equal to zero, and the potential one has a value, determined by its initial position. As the pendulum tends to a vertical position, its deviation from equilibrium decreases, and the potential energy too. Therefore, according to condition (3.8), the kinetic energy must increase, and hence velocity of movement. In a time equal to a quarter of the oscillation period, the pendulum will reach an equilibrium position characterized by zero potential energy. Thus, the entire initial potential energy has passed into kinetic, which corresponds to the maximum of movement velocity. Further, by inertia, the pendulum begin to deviate in the opposite direction, acquiring some potential energy. In accordance with the law of conservation of energy, its kinetic energy begins to decrease, and pendulum slows down movement. After half the period, the pendulum stops. Thus, its kinetic energy completely transforms into potential, exactly coinciding with the initial one, since under the assumptions made, energy is not lost in the process of movement. This means that the pendulum deviated at the same angle as at the initial time, but in the opposite direction. Then the process resumes, and every quarter of the period there is a transition from one type of energy to another.

Task 3.2 *Energy of pendulum oscillation*. Based on the numerical solving of the problem (3.3), (3.4), make the following analysis.

1. Ensure that the kinetic and potential energy of pendulum change periodically[16]. Pay attention to the relationship between the minimum and maximum values of these characteristics.

2. Based on the graph of the dependence of the total system energy on time, verify the validity of the energy conservation law.

3. Determine the influence of initial conditions and the mass of pendulum on the value of the total mechanical energy pendulum. Analyze the results.

Task 3.3 *Energy conservation law in the general case.* Establish the law of energy conservation for the general case, when the movement of the pendulum is described by equation (3.2).

Sum of the kinetic and potential energies of the pendulum is constant.

4 Oscillation of a pendulum with friction

By formula (3.6), the solution to the considered equation corresponds to harmonic oscillations of the pendulum. However, the simplest experiment shows that in reality another process is being realized. Instead of the expected periodic change in the deviation of the pendulum from the equilibrium position, the oscillations observed in practice over time for some reason persistently damp. This suggests that some additional force acts on the studied object, which significantly slows down its movement. We have already encountered a similar effect in the mathematical description of the problems of movement of both a probe and a glider, see Chapter 1. Thus, to obtain a more accurate mathematical model of the phenomenon under consideration, it is imperative to take into account the influence of the indicated force, which is called the **friction force**[17] and is denoted by F_f.

If only the friction force acts on the body, then the considered process in accordance with Newton's second law is described by the equation

$$m\ddot{x} = F_f.$$

The friction force acts in the direction opposite to the direction of movement, and depends on the velocity and body. In the simplest case, this dependence can be chosen linear (as before in the equation of glider flight). For small oscillation of the pendulum, this assumption is fairly accurate, although in some cases it is necessary to assume a more difficult dependence[18]. Thus, the friction force can be calculated by the formula $F_f = -\mu\dot{x}$, where the positive constant μ is a process parameter and is called the **friction coefficient**. Thus, we obtain the equation of body movement under the friction force

$$\ddot{x} + \tau^{-1}\dot{x} = 0, \tag{3.9}$$

where $\tau = m/\mu$ is the system parameter.

Let us try to clarify the physical sense of this coefficient. The state function x (length) is measured in meters. Its first derivative $\dot{x}$ (velocity) and second derivative $\ddot{x}$ (acceleration) are measured in meters per second and in meters divided by the second squared. In order for equality (3.9) to make sense, it is necessary that both terms on its left-hand side are measured in the same units[19]. As a result, we conclude that the parameter τ is measured in seconds, i.e., makes sense of time. It is called a **relaxation time**. This value shows how quickly the friction force inhibits the movement of the body.

Find the solution of equation (3.9). Determining the body velocity $v = \dot{x}$, we have the first order equation

$$\dot{v} + \tau^{-1}v = 0.$$

Its solution with initial condition $v(0) = v_0$ is (see Figure 3.4)

$$v(t) = v_0 \exp(-t/\tau). \tag{3.10}$$

Thus, the body velocity decreases with time under the action of the friction force, and the faster, the shorter the relaxation time or, equivalently, the greater the coefficient of friction (i.e., the greater the influence of the friction force). Solve the equation $\dot{x} = v$ with corresponding initial condition. Find the law of movement

$$x(t) = x_0 + \tau v_0\big[1 - \exp(-t/\tau)\big], \tag{3.11}$$

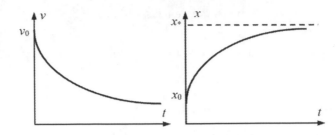

FIGURE 3.4: Solution of the equation of body movement under a friction force.

where x_0 is an initial body position. The function x increases, see Figure 3.4.

We define the properties of the problem solution with an unlimited increase in time. From the equalities (3.10), (3.11) it follows the values of the limits as t tends to infinity, i.e., $t \to \infty$

$$\lim_{t \to \infty} v(t) = 0, \quad \lim_{t \to \infty} x(t) = x_0 + \tau v_0.$$

Therefore, the velocity of the body decreases with time and tends to zero, and its position reaches a certain limit value $x_* = x_0 + \tau v_0$, see Figure 3.4. Thus, under the action of the friction force, the body eventually stops at a distance τv_0 from the point x_0, where it was at the initial moment of time.

If $\tau \to \infty$ or, which is the same, $\mu \to 0$, then the friction force tends to zero. Then it follows from formula (3.10) that the velocity of the body is $v(t) = v_0$, i.e., we are dealing with uniform movement. It is quite natural that, in the absence of any forces, the body velocity, of course, does not change.

Consider now the movement of pendulum in the presence of friction. In this case, in addition to gravity, the friction force F_f, directly proportional to the angular velocity of pendulum, acts on the considered object. As a result, we obtain the equality

$$ml\ddot{\theta} = -mgx - \mu\dot{\theta}.$$

Denoting $\tau = ml/\mu$, determine the movement equation

$$\ddot{\theta} + \tau^{-1}\dot{\theta} + \omega_0^2\theta = 0. \tag{3.12}$$

This transforms to formula (3.9) for $\omega_0 = 0$, and to the equation of harmonic oscillation (3.3) as $\tau \to \infty$. Note that the mass of the pendulum is the parameter of the system now, because it influences relaxation time.

Find a particular solution of equation (3.12). It can be assumed that it consists of a periodic function similar to the solutions (3.5) or (3.6) of equation (3.3) and an exponentially decreasing function similar to solution (3.11) of equation (3.9). Thus, you can try to find a solution to the problem in the form

$$\theta(t) = \exp(-\beta t)\sin(\omega t), \tag{3.13}$$

where parameters β and ω are chosen so that equality (3.12) is satisfied. It is not hard to make sure that the following equalities hold[20]

$$\beta = \tau/2, \quad \omega = \omega_0\sqrt{1 - 1/(2\omega_0\tau)^2}.$$

We determine the solution of the equation (3.12) with initial conditions $\theta(0) = 0$ and

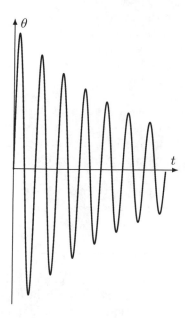

FIGURE 3.5: Damped oscillations of pendulum.

$\dot{\theta}(0) = \omega$. Then pendulum oscillations damp in the presence of friction, see Figure 3.5. In this case, the frequency of damped oscillations ω is less than the frequency of natural oscillations ω_0 and tends to it with an unlimited increase in the relaxation time.

Task 3.4 *Small pendulum oscillations with friction*. Based on the numerical solving of the problem (3.12), (3.4), make the following analysis.

1. Verify that pendulum oscillations are damped.

2. Make sure that on the phase plane (see Task 3.1) the spinning spiral corresponds to the considered movement.

3. Displaying the sum of the kinetic and potential energy, make sure that over time this value tends to zero.

4. To evaluate the effect of the mass of pendulum, its length and coefficient of friction on the velocity of vibration damping.

5. By setting sufficiently large values of the coefficient of friction, detect a monotonic damping of the oscillations.

Pendulum oscillations damp with time in the presence of friction.

5   Equilibrium position of the pendulum

Note a specific particular solution of the equation of a harmonic oscillator, which sharply differs in its properties from all others. If $\theta_0 = 0$, $\vartheta_0 = 0$, then from formula (3.6) for $t = 0$ it follows that

$$a \sin \varphi = 0, \quad a\omega_0 \cos \varphi = 0.$$

Then $a = 0$, and the solution of equation (3.3) is identically equal to zero. This curious result has a natural physical meaning. Pendulum, from the initial state of being upright and resting, invariably remains in this state. From a mathematical point of view, the state of a dynamic system that does not change with time is called the **equilibrium position**. In a qualitative study of a dynamic system, it is extremely important to know its possible equilibrium positions, since if the system nevertheless enters a state over time, then the latter will certainly be an equilibrium position.

In our case, we first distinguished the entire class of solutions to the problem, and then we found an equilibrium position among a great many particular solutions. It would be very sad if, to find the equilibrium position of the system, a priori knowledge of the solution of the problem would certainly be required. The fact is that the nonlinear differential equation, most of the parts that are found in applications, practically do not lend themselves to an analytical solution.

Fortunately, to find the equilibrium positions of a system, it is not necessary to know its analytical solution. Indeed, having now denoted by v the angular velocity of pendulum, we reduce the second order equation (3.3) to the equivalent system of two first order differential equations

$$\dot{\theta} = \vartheta, \quad \dot{\vartheta} = -\omega_0^2\theta.$$

Obviously, the position and velocity of pendulum will not change over time exclusively when their derivatives are equal to zero. It follows that the right-hand sides of the above equations vanish, which allows us to find the position of the pendulum equilibrium.

Equilibrium positions can very significantly in their properties[21]. In the case of an **asymptotically stable equilibrium position**, a small deviation of the system from this state leads to an indispensable return to this state. An example of such an equilibrium position is the zero state for the pendulum oscillation with friction[22]. Note that in the absence of friction, a pendulum having any small deviation from the lower vertical position or a nonzero initial velocity will oscillate around the equilibrium position, but will not return to it completely[23]. Thus, we are not dealing with an asymptotically stable equilibrium. At the same time, we have some confidence that the pendulum will forever remain in a certain neighborhood of the equilibrium position, and not move away from it. In this case, the equilibrium position is called **Lyapunov stable**. An example of an unstable equilibrium is given by the equation of nonlinear oscillations pendulum, see Appendix.

Task 3.5 *Equilibrium position of the equation of small pendulum oscillation*. Make the following analysis.

1. Solving numerically equation (3.3) with zero initial position and velocity of pendulum, make sure that it remains in equilibrium. Moreover, the state of the system does not change over time, and the corresponding phase curve degenerates to a point.

2. By setting small values of the initial position or initial velocity, make sure that the state of the system does not tend to the equilibrium position, but remains in some of its neighborhood, which corresponds to the Lyapunov stability of the equilibrium position.

3. Using the method described above, theoretically verify that for the zero values of the initial position and initial velocity of pendulum are the equilibrium positions of the system for the equation of small pendulum oscillations with friction (3.12).

4. Using the numerical analysis of equation (3.12), verify that the zero initial values of the position and velocity of the pendulum also correspond to the equilibrium position of the system.

5. Make sure that for non-zero initial values of the position and velocity and pendulum solution equation (3.12) tends to the equilibrium position, which corresponds to its asymptotic stability.

Equilibrium position is such a particular solution
of differential equation, which always remains unchanged.

6 Forced oscillations of the pendulum

In practice, forced oscillations of the pendulum are also of interest. Suppose a periodic external force $F_{ext}(t) = F_0 \sin \omega t$ acts on pendulum, where the amplitude of the force F_0 and the frequency of the forced oscillations ω are the parameters of the problem. The investigated process is described by equation

$$\ddot{\theta} + \omega_0^2 \theta = a_0 \sin \omega t, \tag{3.14}$$

where $a_0 = F_0/m$. This equation is **non-homogeneous** because of the presence non-zero term that does not depend on the unknown function. It is easy to verify that this equation has a particular solution

$$\theta(t) = \frac{a_0}{\omega_0^2 - \omega^2} \sin \omega t.$$

Thus, the oscillation of pendulum in the presence of a periodic external force is carried out with the frequency of the forced oscillations. In this case, the amplitude of the oscillation is greater, the closer the frequency of forced and natural oscillations. A sharp increase in the amplitude of oscillations as ω tends to ω_0 is called a **resonance**.

For $\omega = \omega_0$, the above formula does not make sense. However, the experiment shows that when the indicated frequencies coincide, we get oscillations with a finite amplitude. Thus, if the frequency ω is close to ω_0, the considered equation no longer describes the process with a sufficient degree of accuracy. To clarify the mathematical model, we consider the forced oscillations of the pendulum taking into account the friction described by the equation

$$\ddot{\theta} + \tau^{-1} \dot{\theta} + \omega_0^2 \theta = a_0 \sin \omega t. \tag{3.15}$$

It has the particular solution

$$\theta(t) = a \sin(\omega t + \varphi),$$

where

$$\varphi = \arctan \frac{\omega \tau^{-1}}{\omega_0^2 - \omega^2}, \quad a = \frac{a_0}{\sqrt{\left(\omega_0^2 - \omega^2\right)^2 - \left(\omega \tau^{-1}\right)^2}}.$$

For $\omega = \omega_0$, the oscillation amplitude is finite and is equal to $*= \tau/\omega_0$. Thus, with the coincidence of the frequencies of the forced and natural oscillations of the pendulum and the presence of friction, a resonance phenomenon is also observed. In this case, it is characterized by the fact that under these conditions the amplitude reaches its maximum value.

Task 3.6 *Forced oscillations of pendulum*. Make the following analysis.

1. Calculate equation (3.14) with initial condition (3.4), which corresponds to the mathematical model of forced oscillations of pendulum without friction for the case when the frequency of forced oscillations is three times greater and three times less than the natural frequency. Pay attention to the graph of the change in the position of the pendulum and its velocity over time, as well as the phase portrait of the system. Analyze the results.

2. Calculate equation (3.15) with the initial condition (3.4), which corresponds to the mathematical model of forced pendulum oscillations in the presence of friction. Pay attention to the behavior of the system over a sufficiently large time interval.

3. Calculate problem (3.15), (3.4) with various values of the frequency of forced oscillations. Having plotted the dependence of the oscillation amplitude on it, detect the resonance phenomenon.

> *Pendulum oscillation under a periodic external force*
> *is carried out with a frequency of forced oscillations.*
> *As the frequency of forced oscillations approaches*
> *to the frequency of natural oscillations, the amplitude increases,*
> *which corresponds to the phenomenon of resonance.*

Direction of further work. In a subsequent chapter, we consider electrical oscillations, which are described by an equation close to the equation of mechanical oscillations.

Appendix

The movement of the spring, which under natural assumptions is also described by the equation of the harmonic oscillator, is quite close to the pendulum oscillation[24]. Significantly, more difficult and sometimes unexpected solutions arise in nonlinear oscillation problems. Here we consider not only the general equation of oscillation of pendulum (3.2), but also some special problems of the theory of nonlinear oscillations[25].

1 Spring oscillation

Another classic example of mechanical oscillations is the movement of a spring. A spring is considered, one end of which is rigidly fixed, and a body of mass m is attached to the other. In its natural state, the spring is considered stationary. However, when it is compressed or stretched, a force arises that tends to return the spring to its original position, see Figure 3.6. The body movement on the spring is described by function $x = x(t)$, characterizing the deviation of the body from its equilibrium position at time t.

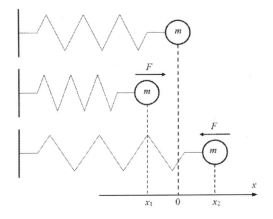

FIGURE 3.6: Oscillation of the spring.

For small deviations of the spring from the state of equilibrium, in accordance with *Hooke's law*, an *elasticity force* F_e will act on the body, directed toward the equilibrium and proportional to the deviation x (the more complex character of the dependence of the force on the position is used for the Duffing spring considered below). Thus, we obtain the equality $F_e = -kx$, where the constant k is called the *elasticity coefficient*, is a parameter of the problem. Substituting the value of force in the movement equation

$$m\ddot{x} = F_e,$$

we get

$$\ddot{x} + \omega_0^2 x = 0,$$

where $\omega_0 = \sqrt{k/m}$. This is the **equation of spring oscillations**[26]. It coincides with the equation of the harmonic oscillator up to the meaning of the quantities included in it[27]. Thus, the movement of the spring largely resembles the movement of a pendulum[28].

Task 3.7 *Spring oscillations.* Based on the numerical solving of the equation of oscillation, make the following analysis.

1. To establish the change with time of the position and velocity of the spring. Analyze the results. Pay attention to the analogy between spring oscillation and pendulum.

2. Determine analytically the law of conservation of energy for vibration of the spring. Establish a change over time with its kinetic, potential, and total energy.

3. Write down the equation for spring oscillations with friction and calculate it. Analyze the results.

2 Large pendulum oscillations

Now we return to the equation (3.2) of large pendulum oscillation

$$\ddot{\theta} + \omega_0^2 \sin\theta = 0.$$

It is very difficult to establish its analytical solution, because of its nonlinearity. However, this does not preclude its computer analysis based on well-known approximate methods of solving.

Task 3.8 *Nonlinear pendulum oscillation.* By approximate solving of problem (3.2), (3.4), make the following analysis.

1. Make sure that for small values of the initial position and initial velocity of the system, the solution of the equation of nonlinear pendulum oscillation is close enough to the solution of the equation of the harmonic oscillator.

2. By increasing the initial deviation or initial velocity of pendulum, to achieve large enough differences in the solutions of linear and nonlinear equations.

3. Determine the rotation of the pendulum around the suspension point for sufficiently large initial states of the system.

4. Set the law of conservation of energy in the case of nonlinear pendulum oscillation. To do this, we should carry out arguments similar to those given above, replacing equation (3.3) by (3.2) without approximating the cosine by a quadratic function.

5. Draw the graph of dependence of the kinetic, potential, and total energy from the time for the nonlinear oscillation of pendulum.

Determine the equilibrium position for equation (3.2). It can be transformed to the system

$$\dot{\theta} = \vartheta, \quad \dot{\vartheta} = -\omega_0^2 \sin\theta.$$

Equating the right-hand sides of these equations to zero, we find the values $\theta = n\pi$, $\vartheta = 0$. Thus, equilibrium position corresponds to any integer n. This striking result shows that in the analysis of nonlinear systems significant difficulties and amazing effects can arise. Nevertheless, it is the nonlinearity of the world around us that endows it with stunning, wonderful, and unique properties. Living in a linear world would be incredibly boring, not to mention the fact that life itself is associated exclusively with nonlinear systems.

Task 3.9 *Equilibrium positions for nonlinear pendulum oscillations.* Make the following analysis.

1. Solving numerically equation (3.2) with zero initial values of position and velocity of pendulum, make sure that it remains in equilibrium. Moreover, the state of the system does not change over time, and the phase curve degenerates to a point.

2. By choosing small values of the initial position or initial velocity, make sure that the state of the system does not tend to the equilibrium position, but remains in some of its neighborhood, which corresponds to the Lyapunov stability of the equilibrium position.

3. Determine the initial deviation of pendulum to π and zero initial velocity, which corresponds to finding pendulum at the extreme top point. Ensure that under these conditions pendulum remains in balance. One recommends observe the process at a short time interval.

4. Choosing a small deviation of the position of the pendulum from the above equilibrium position, make sure that the pendulum leaves the neighborhood of the equilibrium position, which corresponds to an unstable equilibrium position.

5. Determine again the initial deviation of the pendulum to π and zero initial velocity, but observe the system for a long period. Ensure that the pendulum moves out of equilibrium position over time. Explain the result.

The movement equation for the large pendulum oscillations with friction is

$$ml\ddot{\theta} + \mu\dot{\theta} + \omega_0^2 \sin\theta = 0, \tag{3.16}$$

that is the analogue of the equalities (3.2) and (3.12). This is approximated by the classic equation of harmonic oscillation for $\mu = 0$ and small enough initial data. Then pendulum oscillates around its equilibrium position $\theta = 0$. For sufficiently large initial data, a circular rotation of the pendulum is observed with increasing of the angle and periodic change of the velocity of rotation. The consideration of friction leads to a gradual deceleration of the rotation of the pendulum with the subsequent transition to the regime of damped oscillations.

Task 3.10 *Nonlinear pendulum oscillations with friction*. Make the following analysis on the base of approximate solving of equation (3.16) with initial condition (3.4).

1. Calculate the problem with small initial states of the system. Explain the results.

2. Calculate the problem with large enough initial states and small coefficient of friction.

3. For large enough initial states and a comparatively large value of the friction coefficient, detect a transition from rotation of the pendulum to its damped vibrations.

3 Problems of nonlinear oscillation theory

Consider others problems of nonlinear mechanical oscillations. The first example is the **Duffing spring**[29]. In contrast to the spring considered earlier, in which the elastic force is proportional to the deviation of the spring from the equilibrium position, in this case, this dependence is characterized by the following equality $F_e = -(ax + bx^3)$, where a and b are positive constants. Thus, the system is described by the equation

$$m\ddot{x} + ax + bx^3 = 0,$$

that is called the **Duffing equation**. It can be determined that, depending on the values of the coefficients included in it, the solution of this equation has very specific properties.

Another example of nonlinear oscillations is given by the **Froude pendulum**[30] with following dependence of the friction force on the angular velocity $F_e = a\dot{\theta}(b\dot{\theta}^2 - 1)$, where a and b are positive constants. We obtain the following equation

$$ml\ddot{\theta} + mg\sin\theta + a\dot{\theta}(b\dot{\theta}^2 - 1) = F$$

for large oscillations with exterior force F. The movement of Froude pendulum is significantly more difficult than the large oscillation of a usual pendulum.

Note also the pendulum with a force directed toward movement. Small oscillation of such pendulum with friction and external force are described by the equation

$$mL\ddot{\theta} + \mu\dot{\theta} + mgx = F\mathrm{sign}\dot{\theta}.$$

It turns out that for any initial data, the pendulum goes into oscillations with the same amplitude

$$A = \frac{1 + \exp(1 - hT)}{1 - \exp(1 - hT)} \frac{F}{mg},$$

where $h = \mu/2ml$, $T = \pi/\sqrt{g/l - h^2}$ is the half period of oscillation[31].

Notes

[1] Mathematical model of pendulum movement is considered in any course of classical mechanics, see, for example, [47], [179], [183], [198], [268] as well as in special literature on the oscillation theory [12], [66], [190], [223], [262].

[2] Differential equation has an **order** n if the maximal order of derivative of the unknown function here is n.

[3] More exact, the equation is **nonlinear** if this is not linear. The equation is called **linear** if it can be transformed to the equality $Ax = f$, where x is unknown value, A is a linear operator acting on it, and f is a known term. The operator, i.e., transformation, A is linear if for any values of x and y, on which it acts, and for any numbers α and β the following equality holds $A(\alpha x + \beta y) = \alpha Ax + \beta Ay$. In particular, we can denote $A\theta = \ddot{\theta} + \omega_0^2 \sin \theta$, $f = 0$ for the equation (3.2). The second derivative here is the linear operator. However, the sinus is the nonlinear transformation. The differential equations of the Appendixes of this and previous chapters are nonlinear too.

[4] Some peculiarities of nonlinear differential equations are given in Chapters 5 and 19. Their properties can be found in any differential equations course, see [14], [59], [132], [168], [212], [275], [346], [383].

[5] As noted earlier, an equation is linear if it can be represented in the form $Ax = f$, where A is a linear operator. Linear equation is **homogeneous** if $f = 0$. Linear homogeneous equations have the property that the sum of any two of its solutions and the product of any solution by a constant are also solutions of this equation. From an algebraic point of view, this means that the set of solutions of this equation forms a **vector** or **linear space**, see, for example, [153], [185], [283].

[6] Particularly, equation (3.3) describes also the electrical oscillations, see Chapter 4.

[7] The angular displacement and angular velocity of the pendulum are different physical characteristics. However, they satisfy the same equation. This is the first time we considered a situation where a concrete equation has different physical interpretations. In the Appendix, the process of spring oscillations will be considered, and in Chapter 4, electric oscillations, which are also characterized by the equation under consideration.

[8] The **Cauchy problem** for an ordinary differential equation of the n order consists of the considered equation with additional initial conditions. Initial conditions include the given values of the known function and all its derivatives up to the order $n - 1$ inclusive at some point in time (the value of independent variable), which is the origin. The Cauchy problem for the system of differential equations is defined in a similar way. The Cauchy problem for partial differential equations also makes sense, see Chapter 12.

[9] In Chapter 16, the equation of the pendulum oscillation will be determined from the principle of least action.

[10] The second order linear homogeneous differential equations with constant coefficients has the form

$$\ddot{x} + a\dot{x} + bx = 0,$$

where a and b are constants. Its characteristic equation is the square equation $\lambda^2 + a\lambda + b = 0$. If its roots λ_1 and λ_2 are real and different, then the general solution of this equation is $x(t) = c_1 e^{\lambda_1 t} + c_2 e^{\lambda_2 t}$, where c_1 and c_2 are arbitrary constants. If these roots coincide and are equal to a real number λ, then we get $x(t) = (c_1 + c_2 x)e^{\lambda t}$. Finally, if the characteristic equation has the complex roots $\lambda_1 = \alpha + \beta i$ and $\lambda_2 = \alpha - \beta i$, then the general solution is $x(t) = \left[c_1 \cos(\beta t) + c_2 \sin(\beta t)\right]e^{\alpha t}$. It is the latter case that is realized for equation (3.4) with $\alpha = 0$, $\beta = \omega_0$. As a result of this, the last formula takes the form (3.5). We will consider different forms of solutions of second order linear differential equation in Chapter 10 in the process of analyzing the heat equation.

[11] Some properties of the pendulum oscillation equation are considered in Chapter 5.

[12] It is obvious that formula (3.6) gives a solution for equation (3.3) for all a and φ. Thus, formulas (3.5) and (3.6) are equivalent.

[13] A function $x = x(t)$ is called **periodic** if there exists a number T called its period such that $x(t) = x(t+T)$ for all t.

[14] Differentiating equality (3.6), we find $v(t) = \dot{\theta}(t) = a\omega_0 \cos(\omega_0 t + \varphi)$. Thus, the pendulum velocity is the periodic function too with the same frequency and period, but, of course, with another amplitude. Actually, this is not surprising, since, as we know, the position of the pendulum and its velocity are described by the same equation.

[15] The phase plane is used for the analysis of general second order differential equations in Chapter 5.

[16] As can be seen from the above formulas, the potential energy is determined by the value of the angular displacement, and the kinetic energy is determined by the angular velocity. Since the angle and velocity are periodic functions, it is natural that kinetic and potential energies have the analogical properties.

[17] Different problems associated with the phenomenon of friction are described in [234].

[18] Nonlinear dependence of the friction force from the velocity of a moving body is considered, for example, in the probe movement problem described in Chapter 1, as well as for the Froude pendulum described in Appendix.

[19] When describing the motion of a body under the action of the friction force, it was noted that all terms included in the equation of state should have the same dimension. This natural requirement must certainly be satisfied for mathematical models of all phenomena.

[20] Determine the derivatives of the function θ

$$\dot{\theta} = -\beta \exp(-\beta t)\sin(\omega t) + \omega \exp(-\beta t)\cos(\omega t),$$

$$\ddot{\theta} = (\beta^2 - \omega^2)\exp(-\beta t)\sin(\omega t) - 2\beta \exp(-\beta t)\cos(\omega t).$$

Put these values to equality (3.12); we get

$$(\beta^2 - \omega^2 + \omega_0^2 - \beta/\tau)\exp(-\beta t)\sin(\omega t) + \omega(1/\tau - 2\beta)\exp(-\beta t)\cos(\omega t) = 0.$$

For the validity of this relation, it is necessary to nullify the coefficients before the trigonometric functions. The second term here is equal to zero if $\beta = \tau/2$. Determine the first term to zero. We get

$$\omega = \sqrt{\omega_0^2 - (1/2\tau)^2} = \omega_0\sqrt{1 - 1/(2\omega_0\tau)^2}.$$

[21] A more detailed study of the equilibrium positions of dynamical systems is considered in Chapter 5.

[22] The equilibrium position for free oscillations of pendulum in the presence of friction is classified as a stable focus, see Chapter 5.

[23] The equilibrium position for free oscillations of pendulum in the absence of friction is classified as a center, see Chapter 5. In subsequent chapter, we will meet with the concept of the center when considering the Volterra–Lotka equation, with diverse interpretations.

[24] Another example of mechanical oscillations, characterized in the simplest case by the harmonic oscillator equation, are the oscillations of a ball located in pit, see Figure 3.7 and [190].

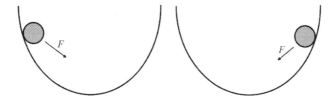

FIGURE 3.7: Ball oscillation in pit.

[25] Nonlinear oscillation problems are considered, for example, in [12], [116], [135], [164], [168], [175], [188],

[190], [239], [248], [250], [262]. We study nonlinear oscillations for chemical, biological, and other systems in subsequent chapters.

[26] We could consider the oscillation of the spring in the presence of friction, as well as its forced oscillations, see Chapter 4. A distributed (infinite-dimensional) analogue of spring oscillation is the process of oscillation of a string, see Chapter 13. In this case, the deviation of the string from the equilibrium position changes not only with time, but also from point to point, i.e., the state function of the system depends on two variables.

[27] The movement of the pendulum and the spring is due to the action of forces of a completely different nature (gravity and elastic force). However, these processes are described by exactly the same equation. Even more surprising is the fact that electrical oscillations are also characterized by the equation of a harmonic oscillator. Too much in common is revealed between these processes so that it can be attributed to chance coincidence. We will encounter such a striking situation more than once in the future.

[28] We considered forced spring oscillations in Chapter 4.

[29] About the Duffing spring and the properties of the equation of its movement see [188], [168], [239], and [262].

[30] About the Froude pendulum see [12], [175], [239], [248], and [262].

[31] This phenomenon is associated with the concepts of the limit cycle and self-oscillations, which will be considered in Chapter 5.

Chapter 4

Electrical oscillations

In the previous chapter, oscillations of mechanical systems were considered. Note that, although the pendulum moves under the action of gravitational force, and the spring moves by the elastic forces, both phenomena are described by the same harmonic oscillator equation. It is important that both the position of the pendulum and spring and its velocity are characterized by an identical equation. This circumstance suggests that, possibly, other types of oscillations lead to the same mathematical model.

Now we consider processes associated with electrical oscillations[1]. The immediate object of study here is the electrical circuit. It is a closed electrical circuit consisting of a capacitor and an inductor connected by wires. The studied phenomenon is characterized by equations for the current and charge on the capacitor, which exactly coincide with the equation of the harmonic oscillator. Here, a version of the energy conservation law is realized. However, if the sum of kinetic and potential energies is constant for mechanical oscillations, then in this case the sum of electric and magnetic energy does not change.

An analogue of friction for electrical oscillations is a resistance. Oscillations of the charge and current in the circuit in the presence of electrical resistance are damped. The presence in the circuit of an external voltage source that changes periodically leads to forced oscillations of the circuit, strikingly resembling the corresponding mechanical oscillations. In particular, a resonance phenomenon is also observed here.

Appendix discusses forced spring oscillations, as well as examples of nonlinear electrical oscillations.

Lecture

1 Electrical circuit

Consider a closed electrical network consisting of a capacitor and an inductor connected by wires, see Figure 4.1. In this case, the properties of the wires are not taken into account. This object is called the ***electrical circuit***. It can be verified that, over time, the ***charge*** q, ***current*** I, and ***voltage*** (potential difference) V in the circuit change. These characteristics can be chosen as state functions of the system. Circuit is understood as a lumped object, i.e., its properties change exclusively over time[2].

Consider arbitrary points A and B, located on opposite sides of inductor and capacitor, see Figure 4.1. The potential difference (voltage drop) on capacitor V_C is proportional to its charge q

$$V_C = \frac{1}{C}q, \tag{4.1}$$

where the parameter C is called ***electrical capacity*** and is a process parameter. The potential difference (voltage) across the inductor V_L is proportional to the velocity of change (i.e., the derivative) of the current and is directed so as to prevent a change in current. As a result, we obtain the equality

$$V_L = -L\dot{I}, \tag{4.2}$$

DOI: 10.1201/9781003035602-4

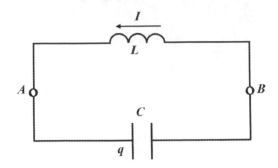

FIGURE 4.1: Electrical circuit.

where the coefficient of proportionality L is called the ***inductance*** or coefficient of self-induction. Then the current is the change velocity of charge, taken with the opposite sign, since the directions of current and the movement of electrical charge are considered to be opposite. Thus, we get the equality

$$I = -\dot{q}. \tag{4.3}$$

From formula (4.2), it follows that $V_L = L\ddot{q}$.

By the ***Kirchhoff law***, the voltage balance in a circuit is characterized by the equality

$$V_L + V_C = 0. \tag{4.4}$$

Using equality (4.1), we get

$$L\ddot{q} + \frac{1}{C}q = 0. \tag{4.5}$$

Denoting $\omega_0 = 1/\sqrt{LC}$, we obtain the second order linear homogeneous ordinary differential equation

$$\ddot{q} + \omega_0^2 q = 0. \tag{4.6}$$

Add the initial conditions

$$q(0) = q_0, \quad \dot{q}(0) = -I_0, \tag{4.7}$$

where the initial values of charge q_0 and current I_0 are known.

One cannot but be surprised by the fact that the obtained formula, up to the notation, coincides with the equations of the oscillations of pendulum and spring. In this regard, the parameter ω_0 retains the name of the oscillation ***frequency***, in this case electrical. A more complete comparison of mechanical and electrical oscillations is given in Table 4.1. By the way, differentiating equality (3.6) with formula (3.3) taken into account, we establish that current also satisfies a similar equation

$$\ddot{I} + \omega_0^2 I = 0. \tag{4.8}$$

As noted in the previous chapter, the harmonic oscillator equation has a periodic solution with frequency ω_0 and period $T = 2\pi/\omega_0$. In particular, we have the equality[3]

$$q(t) = a\sin(\omega_0 t + \varphi),$$

where the amplitude a and phase φ can be found using initial conditions (4.7). A striking feature of mathematical models is manifested here. Having established the coincidence of the equations of mechanical and electrical oscillations and knowing the solution of the equation of a harmonic oscillator, we are relieved of the need to repeat the procedure for finding a solution and we can immediately write down a ready-made formula.

Let us turn to the interpretation of the results. Suppose that at the initial moment of time, the capacitor has a charge, and the current in the circuit is absent, see Figure 4.2a. Then there is voltage on the capacitor, and therefore on the inductance too. As a result, the capacitor begins to discharge. Consequently, a current charge appears, i.e., current flows through the circuit, see Figure 4.2b. At time $t = T/4$, current reaches its maximum value and the capacitor is completely discharged, see Figure 4.2c. In accordance with relations (4.1), the capacitor reloads, along which the current through the inductor decreases, see Figure 4.2d. At $t = T/2$, the charge of the capacitor reaches its maximum value, up to the sign, coinciding with the original. At this time, current is zero (see Figure 4.2e). Then the capacitor is discharged, and current flows in the opposite direction to the previous one, see Figure 4.2f. At $t = 3T/4$, the charge of the capacitor is zero, and current has its maximum value, see Figure 4.2g. Then, the capacitor recharges with a decrease in the current, see Figure 4.2h, and at $t = T$ we return to the initial state, see Figure 4.2i. Next, the process resumes.

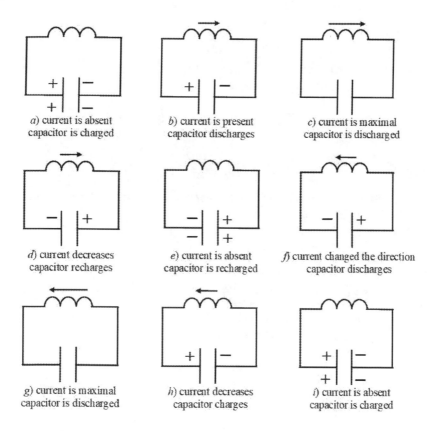

FIGURE 4.2: Electrical oscillations in circuits.

Task 4.1 *Electrical circuit.* Based on the numerical solving of problem (4.2)–(4.4), (4.7), make the following analysis.

1. Make sure that the current and charge in the circuit changes periodically with time.

2. Pay attention to the fact that in the phase plane such a movement corresponds to a closed curve (ellipse).

3. Assess the effect of capacity and inductance, as well as the initial states of the system on the process of electrical oscillations.

4. Pay attention to the analogy between oscillations of circuit, pendulum and spring.

> *Processes in the electrical circuit are described by the equation*
> *of harmonic oscillator with respect to current and charge.*
> *These characteristics change periodically with time.*
> *Charge and current in the circuit are electrical analogues*
> *of the position and velocity of pendulum or spring.*

2   Energy of circuit

As already noted, there is a serious analogy between electrical and mechanical oscillations. In a previous chapter, the law of conservation of energy was established for mechanical oscillations. We get a similar result for the electrical circuit. We return from equation (4.6) back to equality (4.5). Multiplying it by the derivative of the function q, we obtain the relation

$$\frac{L}{2}\frac{d}{dt}\dot{q}^2 + \frac{1}{2C}\frac{d}{dt}q^2 = 0.$$

Using formula (4.3), which expresses the relationship between current and charge, we transform the last equality to the following form

$$\frac{d}{dt}\left(\frac{L}{2}I^2 + \frac{1}{2C}q^2\right) = 0.$$

The values

$$M = \frac{L}{2}I^2, \quad E = \frac{1}{2C}q^2$$

are called, respectively, the **magnetic** and **electrical energies** of the circuit. Thus, the previous ratio can be written as

$$M(t) + E(t) = \text{const.} \tag{4.9}$$

Thus, the sum of electrical and magnetic energy of circuit does not change over time. Therefore, the last equality expresses the **law of conservation of energy** for the circuit.

Now turn to the interpretation of the processes depicted in Figure 4.2 from the standpoint of the law of energy conservation. The initial state of the system is characterized by a charged capacitor in the absence of current, see Figure 4.2a. This means that the magnetic energy of circuit is zero, while its electrical energy takes on a value determined by the initial charge circuit. As the capacitor discharges, current appears in the circuit, see Figure 4.2b. A decrease in charge on capacitor means a decrease in electrical energy. However, a variable electrical field generates a magnetic field on the inductor. Thus, the electrical energy begins to turn into magnetic, and the total energy remains unchanged. There comes a point in time when the capacitor is completely discharged, and the maximum current value is observed in the circuit, see Figure 4.2c. This means that electrical energy has completely transferred to magnetic one. The variable magnetic field on the inductor generates an electrical field in accordance with **Faraday law**. This starts the process of capacitor recharge, according to which the charge of the capacitor increases, and the current in the circuit decreases, see Figure 4.2d. There is a transition of magnetic energy to the electrical one. Over time, the charge of capacitor has a maximum value that differs from initial only in sign. In this case, the current is absent in the circuit, which means that all magnetic energy has passed into electrical one. The next half period corresponds to the next recharge of the capacitor.

The results are similar to the manifestation of the energy conservation law for undamped oscillations pendulum and spring.

Task 4.2 *Energy of circuit oscillation*. Based on the numerical solving of problem (4.6) and (4.7) with initial conditions make the following analysis.

1. Determine the change over time of electrical and magnetic energies.

2. Based on the graph of the dependence of the total energy system on time, verify the validity of the energy conservation law.

3. Pay attention to the analogy between oscillations of circuit, pendulum, and spring.

> *Sum of electrical and magnetic energies of circuit remains constant.*
> *Electrical and magnetic energies of circuit are electrical analogues*
> *of the kinetic and potential energy of pendulum and spring.*

3 Circuit with resistance

The considered properties of the electrical circuit are similar to free undamped oscillations of the pendulum or spring. It is logical to assume that there is a natural electrical analogue of mechanical oscillations in the presence of friction, see Table 4.1. Assume that the circuit also includes a resistor, see Figure 4.3. In this case, to determine the voltage balance between points A and B, it is also necessary to take into account the potential difference on resistor. In accordance with ***Ohm's law***, it is proportional to the current $V_R = RI$, where a positive coefficient R is a parameter of the problem called electrical ***resistance***. Then we get the following balance voltage $V_L + V_C = V_R$. Then the mathematical model of the circuit is characterized by the equation

$$\ddot{q} + \tau^{-1}\dot{q} + \omega_0^2 = 0, \qquad (4.10)$$

where the parameter $\tau = L/R$, as in the case of mechanical oscillations, is called the ***relaxation time***. The current satisfies the analogical equation. The resulting equality are strikingly consistent with equations oscillations pendulum or spring in the presence of friction. We are thus convinced that the resistor really plays the role of friction. Repeating the reasoning from the previous chapter, we conclude that with a sufficiently long relaxation time (i.e., with a relatively large inductance and low resistance), damped oscillation of the current and charge is observed in the circuit. For a small value of the relaxation time, there are no oscillations, and all the considered characteristics monotonically tend to zero[4].

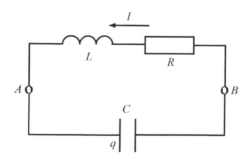

FIGURE 4.3: Circuit with resistor.

Task 4.3 *Circuit with resistor*. Based on the numerical solving of problem (4.10) with initial conditions (4.7), make the following analysis.

1. Verify that damped oscillations of charge and current are observed in the circuit. In this case, the phase curve is a twisting spiral.

2. At sufficiently large values of resistance, obtain a monotonic attenuation of electrical oscillations.

3. Pay attention to the analogy between damped oscillations of pendulum, spring, and circuit.

> *Electrical oscillations of circuit decrease over time in the presence of a resistor.*
> *Circuit oscillations with resistor are electrical analogue*
> *of the oscillations pendulum or spring in the presence of friction.*

4   Forced circuit oscillations

In the previous chapter, along with free mechanical oscillations characterized by a homogeneous equation of harmonic oscillator, forced oscillations due to the action of periodic external force were studied. We will establish an electrical analogue of this process. Consider the processes occurring in the electrical circuit in the presence of an external voltage. Assume that this voltage varies periodically by the law $V_{ext}(t) = V_0 \sin \omega t$, where the voltage amplitude V_0 and the frequency of forced oscillations ω are the parameters of the system. In this case, the voltage balance in the network instead of formula (4.4) is characterized by the equality $V_L + V_C = V_{ext}$. Using equalities (4.1)–(4.3), determine the equation

$$L\ddot{q} + \frac{1}{C}q = V_{ext}$$

that can be transformed to

$$\ddot{q} + \omega_0^2 q = a_0 \sin \omega t, \tag{4.11}$$

where $a_0 = V_0/L$.

The solution of equation (4.11) with initial conditions (4.7) with concrete values of parameters is given in Figure 4.4 for the values $L = 1$, $C = 1$, $q_0 = 1$, $I_0 = -1$, $a_0 = 1$, $\omega = 0.75$. As can be seen from the graph, the result is a superposition of two types of oscillations. These are natural oscillations, due to the internal properties of the circuit, and forced oscillations, determined by the actions of an external voltage source.

Consider a very important particular solution $q(t) = A \sin \omega t$ of equation (4.11). Putting it to equality (4.11), we get

$$-A\omega^2 \sin \omega t + A\omega_0^2 \sin \omega t = a_0 \sin \omega t.$$

Now we find the charge amplitude

$$A = \frac{a_0}{\omega_0^2 - \omega^2}.$$

Thus, the charge varies with frequency of forced oscillations by the formula

$$q(t) = \frac{a_0}{\omega_0^2 - \omega^2} \sin \omega t. \tag{4.12}$$

This is the solution of equation (4.5) with initial conditions

$$q(0) = 0, \quad \dot{q}(0) = \frac{a_0 \omega}{\omega_0^2 - \omega^2}.$$

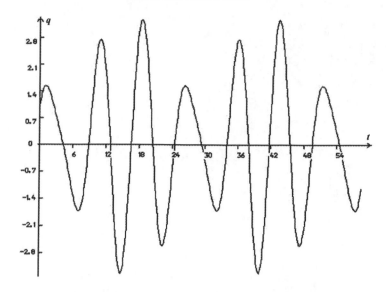

FIGURE 4.4: Charge in the case of the forces circuit oscillations.

As can be seen from formula (4.12), the amplitude oscillation increases sharply as frequency forced oscillations approach the frequency of natural oscillations. A similar phenomenon, called, as in the case of mechanical oscillations, resonance, is indeed observed experimentally. However, we note one very important circumstance. For $\omega = \omega_0$ formula (4.6) does not make sense. At the same time, the experiment shows that, assuming that the frequency of the forced oscillations is equal to the natural frequency ω_0, we get oscillations with a sufficiently large, but certainly finite amplitude. This suggests that if ω is close to ω_0, equation (4.5) can no longer accurately describe the studied process.

In order to refine the mathematical model, consider the forced electrical oscillations with resistor described by the equation

$$\ddot{q} + \tau^{-1}\dot{q} + \omega_0^2 q = a_0 \sin \omega t. \tag{4.13}$$

The solution of the corresponding Cauchy problem with parameters for $L = 2$, $C = 2$, $R = 0.2$, $q_0 = 1$, $I_0 = -0.5$, $a_0 = 0.05$, $\omega = 3$ is shown in Figure 4.5. Comparing the result with the previous graph, we pay attention to the influence of friction, which is the cause of the damping of oscillations. For sufficiently large values of the relaxation time, the influence of resistance is rather small, and equation (4.13) is close enough to equation (4.11).

Determine the particular solution of equation (4.13) by the formula

$$q(t) = A\sin(\omega t + \varphi).$$

Using equality (4.13), find the amplitude and the phase of oscillations[5]

$$\tan \varphi = \frac{\omega \tau^{-1}}{\omega_0^2 - \omega^2}, \quad A = \frac{a_0}{\sqrt{(\omega_0^2 - \omega^2)^2 + (\omega \tau^{-1})^2}}. \tag{4.14}$$

The dependence of the oscillation amplitude of the charge from the frequency of the forced oscillations for for $L = 2$, $C = 1$, $R = 0.5$, $a_0 = 0.75$ is given in Figure 4.6.

It is obvious, that for $\omega = \omega_0$ we determine the finite value of the amplitude

$$A_* = \frac{\tau a_0}{\omega_0}.$$

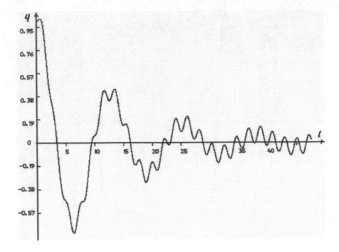

FIGURE 4.5: Charge in the case of the forces circuit oscillations with resistor.

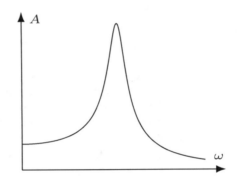

FIGURE 4.6: Dependence of the oscillation amplitude from the frequency of forced oscillations.

Note that this is not a maximum of the dependence of the amplitude A from the frequency ω of forced oscillations[6].

Task 4.4 *Circuit with external voltage without resistor*. Based on the numerical solving of problem (4.11) with initial conditions (4.7), make the following analysis.

1. Determine the change of charge and current in time, choosing the same values as the system parameters as in Figure 4.4.

2. Determine the phase curve for the specified set of parameters.

3. Carry out the above calculations with larger and smaller values of frequency of forced oscillations. Analyze the results.

4. Establish the influence of other system parameters on the process. To do this, change only one of the system parameters, leaving the rest unchanged. Analyze the results.

Task 4.5 *Circuit with external voltage and resistor*. Based on the numerical solving of problem (4.13) with initial conditions (4.7), make the following analysis.

1. Determine the change of charge and current in time, choosing the same values as the system parameters as in Figure 4.5.

TABLE 4.1: Comparison of mechanical and electrical oscillations.

Mechanical oscillations		Electrical oscillations	
deviation	x	charge	q
velocity	v	current	I
force	F	voltage	V
mass	m	inductance	L
elasticity coefficient	k	inverse capacity	$1/C$
velocity – deviation	$v = \dot{x}$	current – charge	$-I = \dot{q}$
Newton law	$F = m\dot{v}$	voltage on the inductor	$V_L = -L\dot{I}$
elasticity force	$F_e = -kx$	voltage on the capacitor	$V_C = q/C$
force balance	$m\dot{v} = F_e$	voltage balance	$V_L + V_C = 0$
natural frequency	$\omega_0 = \sqrt{k/m}$	natural frequency	$\omega_0 = \sqrt{1/LC}$
oscillation equation	$\ddot{x} + \omega_0^2 x = 0$	oscillation equation	$\ddot{q} + \omega_0^2 q = 0$
kinetic energy	$K = mv^2/2$	magnetic energy	$M = LI^2/2$
potential energy	$U = kx^2/2$	electrical energy	$E = q^2/2C$
energy conservation law	$U + K = const$	energy conservation law	$E + M = const$
friction coefficient	μ	resistance	R
friction force	$F_f = -\mu v$	voltage on the resistor	$V_R = RI$
force balance with friction	$m\dot{v} = F_e + F_f$	voltage balance with resistor	$V_L + V_C = V_R$
relaxation time	$\tau = m/\mu$	relaxation time	$\tau = L/R$
oscillation equation with friction	$\ddot{x} + \tau^{-1}\dot{x} + \omega_0^2 x = 0$	oscillation equation with resistor	$\ddot{q} + \tau^{-1}\dot{q} + \omega_0^2 q = 0$
equilibrium position	$x = 0, v = 0$	equilibrium position	$q = 0, I = 0$
external force	$F_{ext} = F_0 \sin \omega t$	external voltage	$V_{ext} = V_0 \sin \omega t$
forced frequency	ω	forced frequency	ω
force balance with external force	$m\dot{v} = F_e + F_{ext}$	voltage balance with external voltage	$V_L + V_C = V_{ext}$
oscillation equation with external force	$m\ddot{x} + kx = F_{ext}$	oscillation equation with external voltage	$L\ddot{q} + q/C = V_{ext}$

2. Determine the phase curve for the specified set of parameters.

3. Carry out the above calculations with larger and smaller values of frequency of forced oscillations. Analyze the results.

4. By changing the parameters of the system, detect the resonance phenomenon.

Oscillations of charge and current under periodic external voltage
is realized with frequency of forced oscillations.
As approaching frequency forced oscillations
to frequency of natural oscillations amplitude increases,
which corresponds to the resonance phenomenon.
Forced oscillations of circuit are electrical analogue
of forced oscillations of pendulum and spring.

Direction of further work. In this and previous chapters, we considered the processes described by differential equations. In the future, various systems are considered, also characterized by differential equations, which are usually more difficult. Therefore, in a next chapter, we consider some methods for their qualitative research, which we will use when analyzing mathematical models characterized by systems with lumped parameters.

Appendix

Since, as has been repeatedly noted, there is a clear analogy between electrical and mechanical oscillations, the results obtained above remain valid for forced mechanical oscillations. In particular, we will talk about forced oscillations of spring. We also consider examples of nonlinear electrical oscillations that are close to the nonlinear mechanical oscillations described in the Appendix to the previous chapter.

1   Forced oscillations of spring

As already noted, there is a pronounced analogy between the oscillations of spring and electrical circuit. In this regard, all the remarks made above about forced electrical oscillations are true for spring under periodic changing external forces. This process in the absence of friction is described by equation[7]

$$m\ddot{x} + kx = F_0 \sin \omega t$$

with known initial deviation of spring x_0 and initial velocity v_0, where x is the deviation of spring of the equilibrium position, $\dot{x} = v$ is the velocity of spring, m is its mass, k is an elasticity coefficient, F_0 is an amplitude of the external forced, ω is a frequency of external oscillations.

This equation is similar to (4.11) and has the corresponding properties. Figure 4.7 shows the phase curve for forced oscillations of spring for $L = 1$, $C = 1$, $q_0 = 1$, $I_0 = -1$, $a_0 = 1$, $\omega = 0.75$. This is closed, i.e., we have a periodic solution. This result corresponds to a superposition of natural and forced oscillations. For the considered case the frequencies of forced and natural oscillations are equal to $\omega = 0.75$ and $\omega_0 = 1$, respectively, which allows us to find their periods $T = 4/3 \, 2\pi$ and $T_0 = 2\pi$. Then the period of spring oscillation is the smallest common multiple of these quantities, i.e., 8π. If the frequency of the natural and forced oscillations turn out to be incommensurable, then the phase curves of the system can no longer be closed lines, and no periodic function corresponds to the solution of the problem. However, the spring movement remains fairly regular. In particular, the maxima of its deviation from the equilibrium position do not change with time. In this case, we are dealing with *quasiperiodic oscillations*[8]

The forced oscillations of spring with friction are described by the equation

$$m\ddot{x} + \mu\dot{x} + kx = F_0 \sin \omega t,$$

where μ is a friction coefficient.

Task 4.6 *Forced oscillation of spring*. Based on the numerical solving of the equation of forced spring oscillations with initial conditions make the following analysis.

1. Determine the change in time of the position and velocity of spring in the absence of friction, as well as the phase curve, using the parameter values from Figure 4.7.

2. Carry out calculations by increasing and decreasing the value of frequency induced oscillations by an integer (small) number of times. Analyze the results.

3. By setting the incommensurable frequency values of forced and intrinsic oscillations, detect quasiperiodic oscillations[9].

4. Perform calculations for the forced spring oscillations with friction.

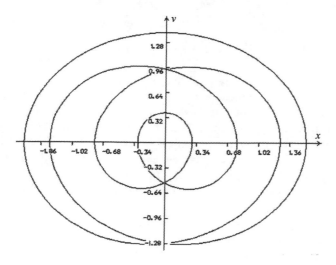

FIGURE 4.7: Phase curve for the forced oscillations of spring.

2   Circuit with nonlinear capacity

As in mechanical problems, nonlinear electrical oscillations are of interest. Consider again a circuit consisting of inductor, capacitor, and resistor. However, instead of the formula $V_C = q/C$, which characterizes the linear dependence of voltage on the capacitor from the charge, now this dependence is nonlinear[10]. The electrical circuit with nonlinear capacity is described by the equation[11]

$$L\ddot{q} + R\dot{q} + U(q) = 0,$$

where $U = U(q)$ is a monotonically increasing function, see Figure 4.8.

One can prove that this system has a unique equilibrium position that is asymptotically stable[12].

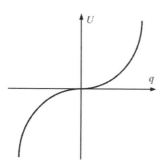

FIGURE 4.8: Nonlinear dependence of voltage on the capacitor from the charge.

3   Van der Pol circuit

Another (and more interesting) example of nonlinear electrical oscillations is given by the **Van der Pol circuit**, which includes a voltage source acting as a negative resistor[13].

The process in question is described by the nonlinear **Van der Pol equation**

$$\ddot{I} - \gamma(1 - \beta I^2) + \omega_0^2 I = 0,$$

where positive constants β and γ are parameters of the problem.

At certain values of these coefficients in the Van der Pol circuit, oscillations are observed that differ significantly from those considered previously. In this case, there is some closed phase curve in the phase plane. It would seem that we have already encountered a similar situation when considering undamped oscillations of pendulum, spring, and circuit. This is true for the equilibrium positions called the center, which we will meet more than once in subsequent chapters. However, the peculiarity of the situation is that in previous cases there is an infinite number of closed phase curves, while here we are talking about a single curve of this kind.

Other phase curves passing near the considered one are "wound" on it from the inside (see Figure 4.9) or from the outside. A phase curve with a similar property is called a *limit cycle* (more precisely, a stable limit cycle)[14]. Note that over time, regardless of the initial state, the system reaches oscillations of the same amplitude. At the same time, for oscillations corresponding to a harmonic oscillator, amplitude is determined by the initial state of the system. The described type of oscillations is also called *self-oscillations*. Therefore, phase curves with time can go both to equilibrium position and to limit cycles. Thus, we can meet two classes of limiting states of a dynamical system, called *attractors*.

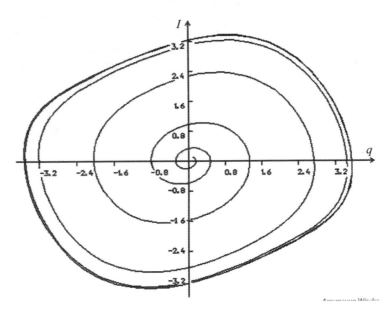

FIGURE 4.9: Phase curve for the Van der Pol equation.

Task 4.7 *Circuit Van der Pol*. Based on the numerical solving of the equation Van der Pol with initial conditions make the following analysis.

1. Make sure that for sufficiently small values of the parameter γ the behavior of the system is close enough to the properties of the harmonic oscillator.

2. By changing coefficients β and γ equation, detect self-oscillations. Determine the change over time charge and current, as well as the phase curve. Analyze the results.

3. Verify that under the conditions of the limit cycle (with the parameters already selected), the

behavior of the system over a sufficiently large time interval does not depend on the initial state of the system. Select initial conditions such that the corresponding phase curve "wound" around the closed curve from the inside, as shown in Figure 4.9. Now choose the initial conditions so as to get "winding" from the outside.

Notes

[1] Electromagnetism theory is devoted, for example, to books [204], [241], [247], [257], [278], [324], [343]. Electrical oscillations are considered also in [12], [66], [116], [179], [190], [223], [262], [268].

[2] Electrical oscillations in the wire are considered in Chapter 12. In this case, we have a distributed parameters system, since the characteristics of the wire change not only with time, but also along the length of the wire.

[3] The current satisfies analogical equation.

[4] Here we observe the ***dissipation of energy***, which consists of the transition of energy of ordered processes (in particular, the kinetic energy of a moving body and the energy of electrical current) into energy of disordered processes (for example, heat or radiation). Systems in which the energy of ordered movement decreases over time due to its dissipation, passing into other forms of energy, such as heat, are called ***dissipative***. If energy dissipation occurs in a closed system, then the entropy of the system will increase. Dissipative systems with distributed parameters will be considered in the third part of the course, in particular, in Chapter 10.

[5] From the equality we get

$$A(\omega_0^2 - \omega^2)\sin(\omega t + \varphi) + A\omega\tau^{-1}\cos(\omega t + \varphi) = a_0 \sin \omega t.$$

Using the formulas

$$\sin(\omega t + \varphi) = \sin \omega t \cos \varphi + \cos \omega t \sin \varphi, \quad \cos(\omega t + \varphi) = \cos \omega t \cos \varphi - \sin \omega t \sin \varphi,$$

we obtain

$$\left[(\omega_0^2 - \omega^2)\cos \varphi - \omega\tau^{-1}\sin \varphi\right]A \sin \omega t + \left[(\omega_0^2 - \omega^2)\sin \varphi + \omega\tau^{-1}\cos \varphi\right]A \cos \omega t = a_0 \sin \omega t.$$

Now we find

$$\tan \varphi = \frac{\omega\tau^{-1}}{\omega_0^2 - \omega^2}, \quad A = \frac{a_0}{(\omega_0^2 - \omega^2)\cos \varphi + \omega\tau^{-1}\sin \varphi}.$$

Using the relationship between trigonometric functions, determine

$$\sin \varphi = -\frac{\omega\tau^{-1}}{\sqrt{(\omega_0^2 - \omega^2)^2 + (\omega\tau^{-1})^2}}, \quad \cos \varphi = \frac{(\omega_0^2 - \omega^2)}{\sqrt{(\omega_0^2 - \omega^2)^2 + (\omega\tau^{-1})^2}}.$$

Thus, the oscillation amplitude is

$$A = \frac{a_0}{\sqrt{(\omega_0^2 - \omega^2)^2 + (\omega\tau^{-1})^2}}.$$

[6] Easy to see that the maximal deviation from the equilibrium position for the circuit with resistor is realized not at the coincidence of the frequency of the forced and natural oscillations. Indeed, formula (4.14) defines the functional dependence of amplitude on the frequency of forced oscillations. To find the maximum point of the specified function, its derivative is zero (the problems of finding the extremum are discussed in more detail in Chapters 16 and 21). We find

$$\frac{\partial A}{\partial \omega} = a_0 \frac{\partial}{\partial \omega}\left[(\omega_0^2 - \omega^2)^2 + (\omega\tau^{-1})^2\right]^{-3/2} =$$

$$a_0\omega\left[-2(\omega_0^2 - \omega^2) + \tau^{-2}\right]\left[(\omega_0^2 - \omega^2)^2 + (\omega\tau^{-1})^2\right]^{-3/2}.$$

Equating this value to zero, we find $\omega^2 = \omega_0^2 - 0.5\tau^{-2}$. Thus, the maximum amplitude is observed when the value of the frequency of forced oscillations is less than the frequency of the natural oscillations. These values practically coincide with a sufficiently long relaxation time.

[7] The oscillations of the spring without external force was be considered in Chapter 3.

[8] About quasiperiodic oscillations see [191].

[9] It is clear that any two rational numbers chosen as the frequency of the natural and forced oscillations will be comparable. Consequently, the corresponding oscillations will be periodic. However, the period of oscillations can be quite large. Therefore, to simulate the quasiperiodic of their oscillations, one should choose their own and forced oscillations so that their smallest common multiple is large enough.

[10] This is possible, in particular, for a capacitor with a ferroelectric.

[11] About electrical circuit with nonlinear capacity, see [191].

[12] Similar properties are possessed by a circuit with a nonlinear inductance, which corresponds to an inductor with a ferromagnetic core, see [191].

[13] About the Van der Pol circuit, see [12], [190], [239], [262].

[14] We consider the limit cycles in Chapter 5.

Chapter 5

Elements of dynamical system theory

Most of the mathematical models considered in this course are related to evolutionary processes. This applies in particular to the problems discussed in the previous chapters. To analyze such problems, one should get acquainted with the main characteristics of the general theory of dynamical systems[1].

The most important of the introduced concepts are be the phase space, phase curve, integral curve, equilibrium position, its stability, and limit cycle. The phase space is a set of various possible values of state functions. Any point in it characterizes a certain state of the system. During the evolution of the system, we observe the movement of a point in phase space. The corresponding movement path is called a phase curve. The graph of the dependence of state functions on time is called the integral curve.

If the state of the system does not change over time, then we are in a position of equilibrium. Thus, the equilibrium position is a stationary solution of the corresponding differential equations. If the deviation of the system from the equilibrium position certainly leads to its return to its original state, then the equilibrium position is stable. Having determined the equilibrium position of a dynamical system and analyzing its behavior in the neighborhood of these states, we can predict the evolution of the system for various process parameters. Sometimes the system eventually goes into some vibrational mode, which corresponds to a closed phase curve. In this case, we have a limit cycle. Some examples of limit cycles are given in the Appendix[2].

Lecture

1 Evolutionary processes and differential equations

In all the phenomena considered earlier (and in most of those that we will turn to later), the state of the system changes with time[3]. Thus, the objects of our study were *evolutionary processes* or *dynamic systems*. We give their simplest classification.

A process or system is called *deterministic* if all its development, both in the future and in the past, is uniquely determined by the state of the system at the initial moment of time. For example, in the problems of classical mechanics[4], knowing the initial position and initial velocity of the material point and having the equations of movement, we can predict its further movement and completely restore the background of the system. At the same time, when studying thermal conductivity or diffusion, we are able to predict the future, but not restore the past of the system[5]. These phenomena are accompanied by an increase in entropy and are not reversible in time. In problems of quantum mechanics one has to deal with processes that are not determined in principle[6]. Thus, from the Heisenberg uncertainty relation, it follows in principle that it is impossible to simultaneously find the coordinate and momentum of a particle with an arbitrarily large degree of accuracy.

A process is called *finite-dimensional* if, for a complete description of the considered phenomenon at an arbitrary moment in time, it suffices to indicate a finite number of its

DOI: 10.1201/9781003035602-5

characteristics. The term ***lumped-parameter system*** is also used here[7]. In particular, a mechanical system consisting of n material points is described by $6n$ state functions. In this case, each material point corresponds to its three coordinates in space and the three components of the velocity vector. All previously considered evolutionary processes were finite-dimensional. ***Infinite-dimensional processes*** or ***distributed-parameter systems***[8] are associated with the movement of a fluid flow[9], the propagation of electromagnetic waves[10], and heat transfer. These phenomena cannot be described with a sufficient degree of accuracy using a finite number of time-varying characteristics.

A process is called ***differentiable*** if all state functions that describe it are continuously differentiable functions of independent variables. Almost all the phenomena considered in this course possess this property. At the same time, the propagation of a shock wave[11] and the Brownian motion[12] of a particle correspond to non-differentiable processes[13].

In this chapter, we consider general principles for the study of deterministic finite-dimensional differentiable evolutionary processes described by ordinary differential equations. We define the most important concepts of this theory.

> ***Theory of differential equations is a mathematical apparatus for studying deterministic finite-dimensional differentiable evolutionary processes.***

2   General notions of dynamic systems theory

Consider some evolutionary process. Many possible values of its state functions are called ***phase space***. So, as we already know, the movement of a material point on a line is characterized by its position and velocity, i.e., two characteristics. Thus, the phase space corresponding to it is a plane, i.e., two-dimensional Euclidean space $\mathbb{R}^2$, where $\mathbb{R}$ is the numerical line that is the set of real numbers. This space is also called the ***phase plane***. When describing the movement of a system of n material points in classical mechanics, we obtain the Euclidean space $\mathbb{R}^{6n}$ of dimension $6n$. Naturally, the phase space of a finite-dimensional process (a system with lumped parameters) is finite-dimensional, and an infinite-dimensional process (a system with distributed parameters) is infinite.

Let M be a phase space of an evolutional process, and an element x of M is its initial state (the state of the system at the time $t = 0$). Denote by $\Lambda_t x$ the state of this at the time t. In particular, for the body falling equation considered in Chapter 1, an element x from M represents its initial position and initial velocity, which in the general case is characterized by a pair of numbers x_{10} and v_{10}. Then $\Lambda_t x$ means the height $x(t)$ and the velocity $v(t)$ of the body at time t, provided that at the initial time it was at point x_{10} and had the velocity v_{10}.

If this process is deterministic, then the state $\Lambda_t x$ is determined uniquely by the initial state for all positive and negative values of the parameter t. Thus, for all real t, i.e., for $t \in \mathbb{R}$ we can determine the transformation Λ_t of the set M to itself[14], which transforms any initial state x of the system to its state $\Lambda_t x$ at the time t. In particular, for the process of the body fall, Λ_t is a transformation that allows us to determine its position and velocity at the time t from the value of the initial height and initial velocity of the body. Actually, in the process of solving the equation of the body fall with the indicated initial conditions, we get these results.

Determine properties of the transformation Λ_t. It is obvious that $\Lambda_0 x$ is a state of system at the time $t = 0$, i.e., $\Lambda_0 x = x$ for any initial state x[15]. Suppose for a value s we have $y = \Lambda_s x$. Denote by z the state $\Lambda_t y$ for a value $t \in \mathbb{R}$. Then, the system goes to state z from the initial state x in time $t + s$ (see Figure 5.1). Therefore, we can determine an operation[16] denoted by $\bullet$ on the set of all such transformations $\{\Lambda_t\}$ by the equality $\Lambda_t \bullet \Lambda_s = \Lambda_{t+s}$ for

all $t, s \in \mathbb{R}$. This operation has following properties
1. $\Lambda_t \bullet (\Lambda_s \bullet \Lambda_r) = (\Lambda_t \bullet \Lambda_s) \bullet \Lambda_r$ for all $t, s, r \in \mathbb{R}$;
2. $\Lambda_0 \bullet \Lambda_t = \Lambda_t \bullet \Lambda_0 = \Lambda_t$ for all $t \in \mathbb{R}$;
3. $\Lambda_t \bullet \Lambda_{-t} = \Lambda_{-t} \bullet \Lambda_t = \Lambda_0$ for all $t \in \mathbb{R}$.

A set with these properties is called a **group**[17]. In particular, the set of all transformations $\{\Lambda_t\}$ with operation $\bullet$ is called the **dynamic transformation group**[18].

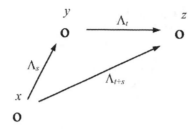

FIGURE 5.1: Evolution of a process.

The pair $(M, \{\Lambda_t\})$ with the phase space M and dynamic transformation group $\{\Lambda_t\}$ is called the **phase flow**. Determine its properties. For any value $x \in M$, the mapping φ from the set of real numbers to the phase space determined by the equality $\varphi(t) = \Lambda_t x$ for all $t \in \mathbb{R}$ is called the **movement** of the point x. In other words, the movement of a point is a collection of points of the phase space to one can get from a given point in the future, as well as those from where one can get to this point. The set of all value s of map φ, i.e., the set of all states $\varphi(t)$ for $t \in \mathbb{R}$ is called the **phase curve** and the set of all pairs[19] $(t, \varphi(t))$ is called the **integral curve**.

Obviously, over time, the evolution of the system in the phase space occurs along the phase curve. Note that exactly one phase curve passes through each point of the phase space. Indeed, if no phase curve had passed through a point at all, this point could not have corresponded to the state of the system, and therefore, could have been an element of the phase space. If two phase curves pass through the point, then this would mean that there are two versions of the evolution of the dynamical system for a given initial state, which contradicts the condition of the determinism of the process.

Consider a phase flow with a phase space M and dynamic transformation group $\{\Lambda_t\}$. If the set M is n-dimensional, then for any point $x \in M$ the movement φ is a function of n variables. This flow is differentiable if the function φ is continuously differentiable. The **phase velocity** of the flow at the point $x \in M$ is the vector of the velocity of movement change at the initial time is the value (see Figure 5.2)

$$v(x) = \frac{d}{dt}\Lambda_t x \Big|_{t=0}.$$

A **vector field** v on M maps an arbitrary point x of the phase space to the phase velocity $v(x)$. A point x is called a **singular point** of the vector field if $v(x) = 0$. It is obvious that a point $x \in M$ is a singular point of a vector field whenever this is an equilibrium position of the phase flow. Indeed, the phase velocity at a point determines the value and the direction of the change velocity of the state functions, i.e., the movement of the system.

The general problem of the differential equation theory is a determining of the movement of the phase flow by its vector field. Indeed, if the vector field $v = v(x)$ is given, then we have the differential equation

$$\dot{x} = v(x).$$

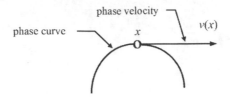

FIGURE 5.2: Phase velocity.

More exact, this is the system of ordinary differential equations if the dimension of the phase space is greater than one. This system is **autonomous**, i.e., it does not include directly the independent variable t. By solving of this system, we determine the set of movements that correspond different initial states of system.

Illustrate these concepts by simple examples.

General problem of the differential equation theory
is a determining of the movement of the phase flow by its vector field.

3   Change in species number with excess food

The simplest case corresponds to a one-dimensional phase space. Consider the simplest biological problem[20]. The change in time with the number of a biological species under favorable conditions is investigated. This means that his birth rate invariably exceeds mortality[21]. The velocity of change in the number of species x is determined by its natural positive increase α, which leads to the differential equation

$$\dot{x} = \alpha x. \tag{5.1}$$

In this case the phase space M is the set of all non-negative real numbers. Its dimension is equal to one that is the quantity of the state functions. Indeed, this evolutional process is finite dimensional. Therefore, for any state x, its phase velocity $v(x)$ is proportional to the value x, see Figure 5.3.

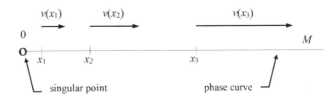

FIGURE 5.3: Phase velocities for the equation (5.1).

The vector field v is characterized by the equality $v(x) = \alpha x$, $x \in M$, see Figure 5.3. For any initial state x_0, the solution of the equation (5.1) is determined by the formula $x(t) = x_0 \exp(\alpha t)$. Therefore, the movement of the point x_0 is determined by the equality $\varphi(t) = \Lambda_t x_0 = x_0 \exp(\alpha t)$.

We can have two variants. For $x_0 \neq 0$ the phase curve, i.e., the set of all possible values $\varphi(t)$ consists of all positive numbers. Thus, all phase curves for non-zero initial states coincide and fill the entire set M without a point $x = 0$. For $x_0 = 0$ we have $\varphi(t) = 0$ for all t. Therefore, the point $x_0 = 0$ is the equilibrium position of the considered system. By the equality $v(x_0) = 0$, this is the singular point of the vector field. Note that we have only two phase curves only. Besides, exactly one phase curve passes through each point of the phase space. Integral curves, i.e., the graphs of the function φ for different initial states x_0 are shown in Figure 5.4.

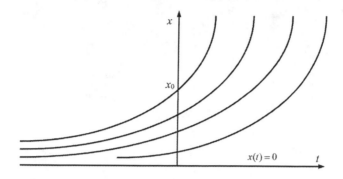

FIGURE 5.4: Family of integral curves for the equation (5.1).

We consider other interpretations of the equation (5.1) in Appendix.

Change in species number with excess food
is characterized by one-dimensional phase space
with two phase curves and unique equilibrium position.

4 Oscillations of pendulum

By Chapter 3, the small oscillations of pendulum are described by the equation

$$\ddot{\theta} + \omega^2\theta = 0, \tag{5.2}$$

where θ is the angular displacement of the pendulum from the vertical state, ω is the natural frequency of the oscillations determined by the length of the thread.

Determine the functions $x_1 = \theta$, $x_2 = \dot{\theta}$. Transform the equation (5.2) to the system

$$\begin{cases} \dot{x}_1 = x_2 \\ \dot{x}_2 = -\omega^2 x_1 \end{cases}. \tag{5.3}$$

Thus, the movement of the pendulum is described by two state functions that are the angle x_1 and angular velocity x_2. The corresponding phase space is the Euclidean plane $\mathbb{R}^2$.

The vector field v in this case is the matrix

$$\begin{pmatrix} 0 & 1 \\ -\omega^2 & 0 \end{pmatrix}.$$

Its action to the vector $x = (x_1, x_2)$ is the right-hand sides of the system (5.3), that is the phase velocity of the system

$$v(x) = \begin{pmatrix} 0 & 1 \\ -\omega^2 & 0 \end{pmatrix} \begin{pmatrix} x_1 \\ x_2 \end{pmatrix} = \begin{pmatrix} x_2 \\ -\omega^2 x_1 \end{pmatrix}, \quad x \in \mathbb{R}^2.$$

This vector field for the case $\omega = 1$ is given in Figure 5.5*b*.

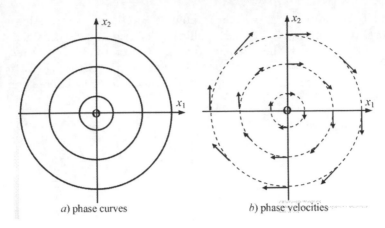

a) phase curves b) phase velocities

FIGURE 5.5: Phase curves and phase velocities for the pendulum oscillation equality with identity frequency.

If the point $x = (x_1, x_2)$ is in the first quadrant, i.e., $x_1 > 0$, $x_2 > 0$, then the first component of the vector of phase velocity is positive, and the second one is negative. Therefore, the first phase coordinate increases, and the second one decreases. If $x_2 = 0$, then the function x_1 has a maximal value; after that this function decreases. The simultaneous decrease in the variables x_1 and x_2 occurs for $x_1 > 0$, $x_2 < 0$ and continues until the coordinate x_1 becomes equal to zero. Besides, according to the second equality (5.3), the derivative of x_2 vanishes, and function x_2 itself reaches its minimum. In the future, the coordinate x_2 begins to increase, and x_1 still decreases, with both x_1 and x_2 being negative. As x_2 increases, it reaches zero. According to the first equality in (5.3), at this moment in time, the derivative of x_1 vanishes, and function x_1 reaches a minimum. In the future, both coordinates x_1 and x_2 increase, and the inequalities $x_1 < 0$, $x_2 > 0$ hold true. Sooner or later, during the increase, the coordinate x_1 will vanish. In this case, x_2 takes its maximum value. Then both coordinates turn out to be positive and us again fall into the first quadrant. A new process cycle begins.

It is obvious that the phase curve, i.e., the graph of movement is closed. There are concentric circles for $\omega = 1$ (see Figure 5.5a)[22]. This system has a unique equilibrium position determined by the equation $v(x) = 0$. This is the origin. Physically, this means the following. If the pendulum is in a strictly vertical position (the deviation angle is zero) and at the same time has zero velocity, then in the future it will remain in this position. The origin is the only singular point of the vector field. For any other initial states, the pendulum moves along closed phase curves, i.e., state functions are periodic. Indeed, it was previously noted that the general solution to system (5.3) is

$$x_i(t) = a_i \sin(\omega t + \varphi_i), \quad i = 1, 2.$$

The concrete values of amplitudes a_1 and a_2 and phases φ_1 and φ_2 are determined by initial states of the system. The integral curves of this process were given in Chapter 3.

Oscillations of the pendulum are characterized
by a two-dimensional phase space with a unique equilibrium position
and a family of concentric circles as phase curves.

5 Stability of the equilibrium position

In the study of evolutionary processes and the corresponding mathematical models, it is extremely important to know the equilibrium positions of the considered system, as well as its behavior in the vicinity of these states. Consider a differential equation (in the general case, we have a system of differential equations) defined by the vector field v

$$\dot{x} = v(x). \tag{5.4}$$

The corresponding equilibrium positions are solutions x of the algebraic equation (more precisely, systems, of nonlinear algebraic equations of order equal to the dimension of the phase space)[23]

$$v(x) = 0. \tag{5.5}$$

The equilibrium position of the system x^* is called **asymptotically stable** if, for any initial state x_0 sufficiently close to x^*, the solution of equation (5.4) with the initial state x_0 tends to x^* as $t \to \infty$. If x^* is a stable equilibrium, then all phase curves falling in some neighborhood of x^* end at this point. The equilibrium position is **Lyapunov stable** if for a neighborhood U there exists a neighborhood V such that any phase curve starting in region V lies entirely in a neighborhood of U. Thus, the equilibrium position is Lyapunov stable if any phase curve starting sufficiently close to the equilibrium position will never be far enough from her. The equilibrium position x^* is called **unstable** if an arbitrary phase curve extends beyond any of its neighborhood[24].

The simplest mechanical illustration of stable and unstable equilibrium position is a ball located in a pit and on a mountain (see Figure 5.6). In principle, both on the mountain and in the pit, the ball can be arbitrarily long, i.e., indeed we are dealing with a state of equilibrium. However, an arbitrarily small deviation of the ball from the top of the mountain leads to its rolling down under the action of gravity, i.e., to his distance from the equilibrium position. Deviation of the ball from the bottom of the pit leads to its return to its original state[25].

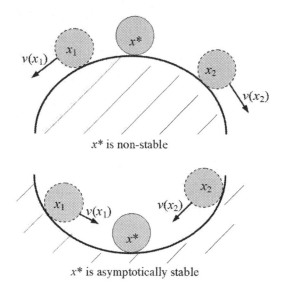

x^* is non-stable

x^* is asymptotically stable

FIGURE 5.6: Stability and non-stability of equilibrium positions.

The equilibrium positions for systems characterized by equations (5.1) and (5.2) are unstable and Lyapunov stable, respectively. An example of a system from a species under adverse conditions. This means that the mortality of a species invariably exceeds its birth rate. This process is described by the equation

$$\dot{x} = -\alpha x,$$

where a positive parameter α characterizes the extinction rate of a species. This equation has the solution $x(t) = x(0)\exp(-\alpha t)$. This tends to the equilibrium position $x^* = 0$ as $t \to \infty$ corresponding complete extinction of the species.

The general classes of equilibrium position for the two-dimensional cases origin as the equilibrium position is given in Figure 5.7[26]. The corresponding integral curves (more exact

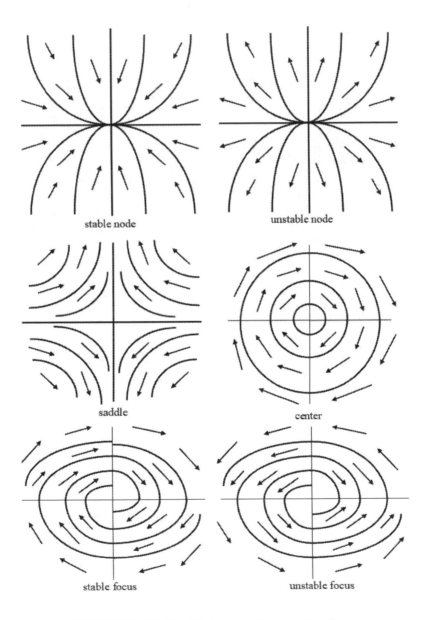

FIGURE 5.7: Equilibrium positions on the plane.

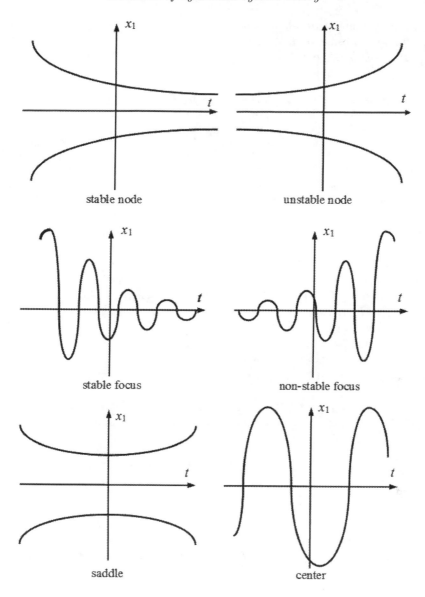

FIGURE 5.8: Integral curves in the neighborhood of the zero equilibrium position.

õl that is the first of these functions), see Figure 5.8. The **stable node** is characterized by the fact that all phase curves passing in its vicinity certainly end at this point, and the corresponding integral curves monotonously tend to this point at $t \to \infty$. For an **unstable node**, all phase curves go out of the vicinity of the equilibrium position, and the corresponding integral curves move monotonically away from the equilibrium position with time. The **saddle** is characterized by the fact that all phase curves (with the exception of two entering the equilibrium position and two coming out of it) begin far from the equilibrium position, approach it and then move away. The integral curves in this case behave in a similar way. In the case of a **stable focus**, the phase curves have the form of a twisting spiral, twisting to the equilibrium position. A stable focus is the equilibrium position for the equation of pendulum oscillation in the presence of friction. For an **unstable focus**, the phase curves have the form of a spinning spiral, and the integral curves move away from the equilibrium position in a non-monotonic manner. In the case of the ***center***, the phase curves are closed, and the integral curves are periodic functions. This applies, in particular, to the equation of oscillation of the pendulum.

Task 5.1 *Nonlinear pendulum oscillations with friction.* Consider the equation of large pendulum oscillation with friction (see Chapter 3)

$$ml\ddot{\theta} + \mu\dot{\theta} + mg\theta = 0.$$

Based on the numerical solving of this equation with initial conditions make the following analysis.

1. Choose the zero initial conditions. Make sure that the state of the system does not change over time, and the phase curve reduces to a point, i.e., we are, in reality, dealing with a position of equilibrium.

2. Define small initial conditions with non-zero friction. Make sure that over time the system goes into equilibrium, which corresponds to an asymptotically stable equilibrium position. Focusing on the behavior of integral and phase curves, we conclude that we are dealing with a stable focus.

3. Choose small initial conditions with a zero value of the coefficient of friction. Make sure that the integral curves turn out to be periodic functions, and the corresponding phase curves turn out to be closed lines. Since the phase curves do not go beyond the neighborhood of the equilibrium position, we are dealing with a Lyapunov-stable position.

4. Choose the initial deviation of the pendulum equal to π and the zero initial velocity, make sure that the state of the system does not change with time, i.e., this state is an equilibrium position. Moreover, to observe the invariance of the state of the system should be for a short time interval.

5. Choose the initial conditions are close enough to those specified in paragraph 4, but different from them. Make sure that the system moves away from the equilibrium position, i.e., this is unstable.

> *Equilibrium position is a system state that does not change with time.*
> *Phase curve corresponding to the equilibrium position is a point.*
> *System located in a small neighborhood of an equilibrium position*
> *may to a state of equilibrium,*
> *to move away from it or to remain in its neighborhood.*
> *Most important classes of equilibrium positions in the plane are*
> *stable and unstable node, stable and unstable focus, saddle and center.*

6   Limit cycle

Consider another amazing version of the evolution of a dynamic system. Let us have the differential equations

$$\begin{cases} \dot{x}_1 = x_2 + x_1(1 - x_1^2 - x_2^2) \\ \dot{x}_2 = -x_1 + x_2(1 - x_1^2 - x_2^2) \end{cases}. \tag{5.6}$$

A qualitative analysis of the studied problem is more convenient to carry out not in the Cartesian, but in the polar coordinate system. We introduce new variables that the angle φ and radius r using the equalities

$$x_1 = r\cos\varphi, \; x_2 = r\sin\varphi.$$

Then the given system is transformed to the equations

$$\dot{r} = r(1 - r), \tag{5.7}$$

$$\dot{\varphi} = -1. \tag{5.8}$$

In equations (5.7) and (5.8), state functions can be studied independently of each other, which explains the feasibility of changing variables.

Equation (5.7) has three equilibrium positions $r = -1$, $r = 0$, and $r = 1$. The value $r = -1$ is not of interest, since in polar coordinates the quantity r cannot be negative, i.e., does not belong to the phase space of the system. It is easy to verify that the equilibrium position $r = 0$ is unstable, and $r = 1$ is stable. In this case, the radial coordinate increases monotonically for $r < 1$ since, according to (5.7), the derivative of r is positive and monotonically decreases for $r > 1$ because the derivative of r is negative. By equality (5.8), the angular coordinate decreases with a constant speed equal to -1.

We now turn to system (5.6). It has a unique equilibrium position $x_1 = 0$, $x_2 = 0$ corresponding to a value of $r = 0$, which means that it is unstable. Suppose for definiteness that the initial state of the system is characterized by coordinates x_1, x_2 satisfying the inequality $x_1^2 + x_2^2 < 1$. As applied to equation (5.7), this corresponds to the inequality $r < 1$. Then, according to equation (5.7), function r will tend to unity over time. Equation (5.8) indicates the rotation of the point in the phase plane with a constant angular velocity in a clockwise direction, i.e., in the direction opposite to the increase in the angle φ. As a result, we get a spinning spiral (unstable focus), which is "wound" around a circle of unit radius centered at the origin (see Figure 5.9). Let equality $x_1^2 + x_2^2 = 1$ be true at the initial moment of time, i.e., $r = 1$. According to equality (5.7), the value of r will not change with time. However, the angular coordinate will change at a constant speed. As a result, the corresponding phase curve is a circle, and the integral curves are periodic functions. Finally, if the condition $x_1^2 + x_2^2 > 1$ is satisfied at the initial time, then $r > 1$. Then, as can be seen from Figure 5.9, the radial coordinate decreases and asymptotically approaches the equilibrium position $r = 1$. At the same time, the angular velocity remains constant. Thus, the phase curves will "wrap" around the unit circle from the outside (see Figure 5.9).

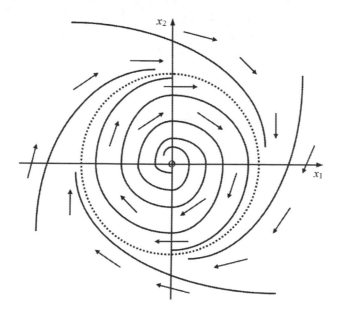

FIGURE 5.9: Phase curves for the system (5.6).

In this case, we have the ***limit cycle*** that is ***stable***[27]. If we replace the direction of the phase velocities (arrows in Figure 5.9) with the opposite, i.e., go back in past, we get an **unstable limit cycle**. Oscillations arising in a system with a limit cycle are called **self-oscillations**[28]. They differ significantly from those that arise in a system with the equilibrium center. In the latter case, the amplitude is determined by the initial state of the systems (see, for example, the oscillation of a pendulum). In self-oscillations, the initial state does not play a fundamental role. Over time, the system will go into oscillations determined solely by its internal structure. So, in the considered example of the amplitude, the oscillations of the functions x_1 and x_2 tend to unity. Stable equilibrium positions and limit cycles are attractors of the system, i.e., any phase curve starting in a certain neighborhood of it tends to this set of points in the phase space with time[29].

Task 5.2 *Limit cycle*. Based on the numerical solving of the system (5.6) make the following analysis.

1. By setting the initial state of the system to zero, make sure that this is, in reality the equilibrium position.

2. Setting the initial state of the system inside the unit circle, find that the indicated equilibrium position is not stable, and the system goes to the limit cycle with an increase in the oscillation amplitude. Make sure that the limiting amplitude of the oscillation does not depend on the initial state of the system.

3. By setting the initial state of the system on a circle of unit radius, detect a periodic change in state functions over time.

4. By asking the initial state of the system outside the unit circle, to detect the output of the system to the limit cycle with decreasing vibration amplitude. Make sure that the limiting amplitude of the oscillation does not depend on the initial state of the system.

> *Limit cycle is a closed phase curve such that any phase curve*
> *sufficiently close to it either approaches it unlimitedly*
> *(stable limit cycle), or leaves its neighborhood (unstable limit cycle).*

Direction of further work. *Having established some properties of dynamical systems, we can continue the analysis of mathematical models of dynamical systems with lumped parameters. The considered processes will be described, as a rule, by systems of nonlinear differential equations.*

Appendix

We consider various systems, which under certain assumptions can be described by the equation (5.1). In addition, two systems of differential equations with a limit cycle are given. The first of them describes a system of chemical reactions, and the second is characterized by the presence of two limit cycles that are stable and unstable.

1   Exponential growth systems

Section 3 of this Chapter considered a mathematical model for the evolution of a biological species under ideal conditions. It assumes a constant predominance of the birth rate of the species over mortality and is described by equation (5.1). As is known, its solution is an exponentially increasing function. This result has a natural interpretation. Let there be a certain population of organisms of the same species. Over time, someone dies, and someone is born, which is due to a change in the number of species. But since the birth rate exceeds mortality, the number of descendants of the population at a given time will exceed the

TABLE 5.1: Systems with exponential growth of state.

Science	Phenomenon	State function	Cause
biology	unlimited population growth	species number	excess fertility over mortality
physics	nuclear decay	number of neutrons	formation of free neutrons in the decay of heavy nuclei
chemistry	chain reaction	concentration	formation of active substances during chemical reactions
medicine	epidemic	number of infected	infecting healthy people sick
economics	inflation	price level	price increases with increasing demand for goods

number of ancestors. Over time, an increasing population will give an even greater number of descendants, and the next generation will be even larger. Thus, under the conditions of applicability of the model, we observe an exponential growth of the population.

Note that a similar behavior of the system is observed for different processes. For example, in physics this is a characteristic of a nuclear explosion. Indeed, during the decay of a nucleus of heavy elements (such as uranium) free neutrons are formed. They initiate the decay of new nuclei, as a result of which new neutrons are released. Their interaction with nuclei leads to the release of even more neutrons. Thus, the chain reaction of nuclear decay is accompanied by an exponential increase in the number of free neutrons[30].

An analogue of the described processes in chemistry are chemical chain reactions [31]. In this case, the starting materials enter into a sequence of chemical transformations involving some intermediate active substances, which themselves are formed during each such reaction. As a result, the concentration of these substances increases exponentially.

Another example of the course of events under consideration is the spread of epidemics[32]. At the same time, infected people infect healthy people. They, in turn, infect new people. Thus, the number of infected avalanche increases.

The economic analogue of the described processes is inflation[33]. In the context of the economic crisis, prices for all kinds of goods and services are rising. To avoid social upheaval, the state raises salaries for employees, as well as pensions, scholarships, etc. People have excess money for which they buy goods. In conditions of a deficit, to restore the balance between supply and demand, producers of goods raise prices, which lead to a new round of inflation.

All these processes can be described (naturally, under extremely strong assumptions) by equation (5.1). Their comparative characteristics are illustrated in Table 5.1.

2 Brussellator

Consider a system of chemical reactions described by the equations[34]

$$\dot{x} = k_1 a - k_2 b x + k_3 y x^2 - k_4 x, \quad \dot{y} = k_2 b x - k_3 y x^2,$$

where x and y are the concentrations of the corresponding substances, a and b are constant concentrations of other substances (parameters of the problem), and the positive constants k_1, k_2, k_3, k_4 are the rates of these reactions. This system is called the **brussellator**. It can be established that under the condition $k_3 k_1^2 a^2 = k_2 k_4^2 b - k_4^3$ the system has a limit circle.

Task 5.3 *Brussellator*. Based on the numerical solving of the system with initial conditions make the following analysis.

1. Choose the system parameters so that mentioned condition is satisfied.

2. At these parameter values, choose the initial states in such a way that the phase curve reaches the limit cycle from the inside. Changing the initial states of the system, make sure that the result does not depend on these states.

3. At these parameter values, choose the initial states so that the phase curve goes to the limit cycle from the outside. Make sure that the system enters the same mode as in the previous case.

4. At these parameter values, choose the initial states so that the phase curve turns out to be closed.

3   System with two limit cycle

Consider the following system of differential equation

$$\dot{x}_1 = x_2 + x_1(x_1^2 - x_2^2 - 1)(x_1^2 - x_2^2 - 4), \quad \dot{x}_2 = -x_1 + x_2(x_1^2 - x_2^2 - 1)(x_1^2 - x_2^2 - 4).$$

After transition to the polar coordinates $x_1 = r\cos\varphi$, $x_2 = r\sin\varphi$ we get

$$\dot{r} = r(r-1)(r-2), \quad \dot{\varphi} = -1.$$

It is easy to verify that the corresponding system, written in Cartesian coordinates, has two limit cycles determined by the values $r = 1$ and $r = 2$. In particular, for any initial state located on each of the indicated circles, the solution of the problem turns out to be periodic. It is characteristic that any phase curve starting in the region with $r < 1$ extends to the circle $r = 1$ from the inside, and in the region with $1 < r < 2$ it tends from the outside. At the same time, any phase curve starting in the region $r > 2$ moves away from the circle $r = 2$. With this behavior of the system, the first limit cycle is called stable, and the second is called unstable. The behavior of the phase curves for the system in question is shown in Figure 5.10.

Notes

[1] The general theory of dynamical systems is considered, for example, in [30], [127], [220], [232], [261], [341]. Among the numerous literature on the qualitative theory of differential equations, we note monographs [14], [59], [132], [168], [212], [275], [346], [383].

[2] Other aspects of the qualitative theory of differential equations (local and global solvability, uniqueness of a solution, a continuous dependence of a solution on a parameter, bifurcation) will be considered in Chapter 19.

[3] Stationary systems are considered in Chapter 13.

[4] We have already considered some problems of classical mechanics in previous chapters. We continue to study the problems of classical mechanics in Chapter 16.

[5] The thermal conductivity process is discussed in Chapter 10.

[6] The problems of quantum mechanics are considered in Chapter 15. The problems of stochastic modelling are considered in Chapter 18.

[7] It is such problems that make up the subject of this Part.

[8] Distributed-parameter systems are discussed in Part II.

[9] The problems of fluid dynamics are discussed in Chapter 15.

[10] Electromagnetic waves are described in Chapter 12.

[11] The mathematical theory of shock waves is described, for example, in [95], [292].

[12] On the mathematical problems of Brownian motion see in [138], [209], [365].

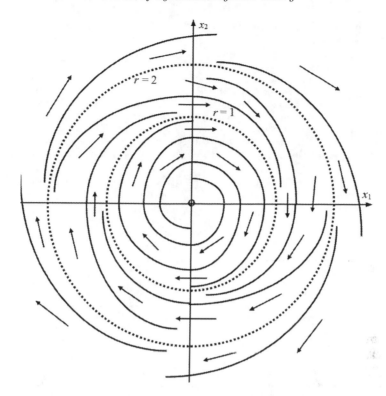

FIGURE 5.10: Phase curves for the system with two limit cycles.

[13] Naturally, discrete systems are also non-differentiable, see Chapter 17.

[14] We consider here an operator $\Lambda_t : M \to M$.

[15] Therefore, Λ_0 is the ***identity operator*** on the phase space M.

[16] An ***operation*** of order n on some set is a transformation that maps n arbitrary elements of a given set to some element of the same set. The most commonly considered operations are second-order operations such as addition, multiplication, etc. Properties of operations is a subject of ***algebra***.

[17] The considered properties are called the associativity, existence of unitary and inverse elements, respectively. The ***group*** is a fundamental mathematical concept that serves as the most important tool for the study of ***symmetry***. In this regard, groups are widely used in theoretical physics and other sciences. Group theory can be found in [3], [291], [307].

[18] In the study of transfer processes, one has to deal with processes that are not reversible in time, see Chapters 10 and 11. In this case, one can introduce the concept of a ***dynamic monoid of transformations***, for which there is no requirement of transform reversibility.

[19] Namely, the graph of the map φ.

[20] The study of the evolution of biological species is the subject of population dynamics and is considered in Chapter 7.

[21] Naturally, the constancy of the positive increase in the number of a biological species in life can be observed only in exceptional situations over a limited time interval. In practice, there exists various constraints that are the limited food and habitat required, the presence of competitors and natural opponents (predators), etc., see Chapter 7. Under these conditions, the growth will no longer be a constant value, but will decrease and may even become negative. Thus, the model under consideration is extremely idealized and has a limited scope. However, do not forget that any model is idealized and has a specific scope, see Chapter 1.

[22] In general, phase curves are ellipses here.

[23] Almost all the concepts introduced in this topic apply to systems with distributed parameters that are characterized by infinite-dimensional phase spaces, see Part II. Moreover, the equilibrium positions will

already satisfy not more algebraic, but more difficult equations that correspond to stationary systems, see Chapter 13. Equilibrium positions are also of interest for discrete dynamical systems, see Chapter 17.

24 For a more rigorous definition of the considered classes of equilibrium positions, it is required to indicate which points can be considered close enough, and what should be understood by the **neighborhood** of the point. In principle, the proximity of points can be reduced to the concept of a neighborhood, assuming that the point y is sufficiently close to the point x if it belongs to a sufficiently small neighborhood of this point. If a **norm** is defined on the phase space, then two points are considered sufficiently close if the norm of their difference is a sufficiently small number. In a more general case, proximity can be estimated using a **metric** expressing the meaning of the distance between points. This is possible if the phase space is a metric space. In an even more general case, proximity is estimated using the concept of **topology**, i.e., in the theory of topological spaces. The above concepts can be found in any course of **functional analysis**, see, for example, [153], [185], [283].

25 Naturally, the rolling ball will inertia slip through the pit, i.e., equilibrium position. However, due to the frictional forces absorbing part of the energy, it will certainly slide in the end there. But if we do not take into account the force of friction, then, having slipped the equilibrium position (pit), the ball will rise exactly to the same height at which it was initially on the other side of the pit. After that, it will roll in the opposite direction, i.e., toward the equilibrium position. This process is described by the well-known harmonic oscillator equation and is characterized by an equilibrium position that is Lyapunov stable.

26 Consider a linear homogeneous system of differential equations

$$\begin{cases} \dot{x}_1 = a_{11}x_1 + a_{12}x_2 \\ \dot{x}_2 = a_{21}x_1 + a_{22}x_2 \end{cases},$$

where the coefficients a_{ij} does not depends from t, $i, j = 1, 2$. Thus, this system is autonomous. This can be transformed to $\dot{x} = Ax$ where x is the vector with components x_1 and x_2, and A is a matrix with equation coefficients, i.e.,

$$A = \begin{pmatrix} a_{11} & a_{12} \\ a_{21} & a_{22} \end{pmatrix}.$$

The equilibrium position is determined by the equality $Ax = 0$ that is the system of linear algebraic equations. If this matrix is non-degenerate, i.e., $a_{11}a_{22} \neq a_{12}a_{21}$ the last equation has zero solution only. Therefore, the equilibrium position is the origin of the phase plan. Then determine eigenvalues of the matrix A from the characteristic equation

$$\lambda^2 - (a_{11} + a_{22})\lambda + (a_{11}a_{22} - a_{12}a_{21}) = 0.$$

If both roots of this equation are real and have the same sign, the equilibrium position is a **node**. If they are real, but have different signs, then we have a **saddle**. If the roots are complex (and therefore complex conjugate) with a nonzero real part, then the equilibrium position is a **focus**. Finally, if they are purely imaginary, then we have a **center**. The equilibrium position is asymptotically stable if both roots have a negative real part, and unstable if at least one of the eigenvalues of the matrix has a positive real part. There are also some special equilibrium points corresponding, in particular, to the case of a degenerate matrix A. In particular, for the equation of a harmonic oscillator we have a system of equations (5.3), where $a_{11} = 0$, $a_{12} = 1$, $a_{21} = -\omega^2$, $a_{22} = 0$. Thus, the characteristic equation has the form $\lambda^2 + \omega^2 = 0$. It has purely imaginary roots, i.e., equilibrium is indeed the focus. The equation of oscillation of a pendulum with friction

$$\ddot{\theta} - \tau^{-1}\dot{\theta} + \omega^2\theta = 0.$$

can be transformed to the system

$$\dot{x}_1 = x_2, \quad \dot{x}_2 = -\omega^2 x_1 - \tau^{-1}x_2.$$

Now we have $a_{11} = 0$, $a_{12} = 1$, $a_{21} = -\omega^2$, $a_{22} = -\tau^{-1}$. Then we determine the characteristic equation $\lambda^2 - \tau^{-1}\lambda + \omega^2 = 0$. This has the roots

$$\lambda_{1,2} = -\frac{1}{2\tau} \pm \sqrt{\frac{1}{4\tau^2} - \omega^2}.$$

For large enough values τ, these values are complex and have a negative real part. Therefore, the equilibrium position is a stable focus. Naturally, for systems with a higher dimension of the phase space, there are more diverse variants of the evolution of dynamic processes.

27 Numerous examples of limit cycles (including those given above) are described, for example, in [12], [132], [164], [175], [239], [245].

28 The phenomenon of self-oscillations and limit cycles are widely encountered in practice. The work of watches, musical instruments, hearts, etc., is based on this principle.

29 The equilibrium positions and limit cycles do not exhaust the set of possible attractors of a dynamical system. **Strange attractors** are also possible, representing limited areas with unstable trajectories inside.

Such systems are characterized by chaotic behavior. An example of such a system is the **Lorenz attractor**, the corresponding autonomous system of third-order nonlinear homogeneous differential equations associated with the process of convective heat transfer, see [142], [239], [333].

[30] For nuclear reactions see, for example, [34].

[31] Chapter 6 is devoted to the mathematical modelling of chemical reactions. Chain chemical reactions are considered, for example, in [194].

[32] Mathematical models in epidemiology are considered, for example, in [7], [46], [73], [244], see also Chapters 7 and 17.

[33] We will return to the mathematical model of inflation in Chapter 8.

[34] In Chapter 6, we indicate what these reactions are and explain why they are described by these equations.

Chapter 6

Mathematical models in chemistry

We have already received certain information in the field of qualitative and quantitative methods of studying differential equations and got acquainted with the simplest models of physical processes. We now turn to the consideration of mathematical models of chemical processes of mathematical models. Its subject is, first of all, the study of the structure of molecules and the change in this structure in the process of interaction of substances with each other. Naturally, we do not have the opportunity to get acquainted with the whole variety of mathematical problems that arise in such a vast and inexhaustible science[1]. Therefore, we will inevitably have to limit ourselves to considering only one of the branches of chemistry, which, however, plays a key role in it. This is chemical kinetics that is a deep scientific field related to *physical chemistry* and studying the rates of various chemical reactions.

The determination of mathematical models of the considered processes begins with the writing of a stoichiometric equation characterizing the number of molecules of the starting materials and reaction products involved in the elementary chemical interaction. On the basis of the stoichiometric equation, in accordance with the law of mass action, kinetic differential equations are established for the concentrations of all substances involved in the reaction.

The simplest linear differential equations of chemical kinetics correspond to monomolecular reactions. More difficult reactions, in which several molecules are involved at once, are described by nonlinear equations with various properties. In practice, most often it is necessary to deal not with a separate chemical reaction, but with a whole system of reactions. As an example, we consider the system of chemical reactions described by the Volterra–Lotka equations. Other reaction systems are discussed in the Appendix.

Lecture

1 Chemical kinetics equations

The central problem of chemistry is the study of chemical reactions. A *chemical reaction* consists in the conversion of one or more substances, called *starting materials*, into one or more other substances that are *reaction products*. A chemical reaction occurs as a result of the interaction of molecules, as well as individual atoms and ions. In this case, the atomic nuclei do not change, but either the starting materials merge into a new, more complex reaction product, or the molecules decompose into simpler components, or one or more atoms of the interacting molecules exchange. We study dynamic processes associated with the rates of chemical reactions and are the subject of *chemical kinetics*.

Suppose we have starting materials A_1, A_2, ..., A_m. After its interaction we get reaction products B_1, B_2, ..., B_n. Suppose, using μ_1 molecules of the chemical substances A_1, μ_2 molecules of the substances A_2, etc. we get ν_1 molecules of the reaction product B_1, ν_2 molecules of the product B_2, etc.[2] This process is schematically indicated as follows

$$\mu_1 A_1 + ... + \mu_m A_m \longrightarrow \nu_1 B_1 + ... + \nu_n B_n. \tag{6.1}$$

DOI: 10.1201/9781003035602-6

The relation (6.1), which is a formal denotation of the considered reaction, is called the **stoichiometric equation**. This includes natural numbers μ_1, ..., μ_m, ν_1, ..., ν_n, called **stoichiometric coefficients**. The sum $n = \mu_1 + ... + \mu_m$, which characterizes the number of molecules participating in a single reaction, is called the **order of the chemical reaction** (6.1).

For example, the formation of water (reaction product) as a result of the interaction of oxygen with hydrogen (starting materials) occurs as follows. As a result of the fusion of two hydrogen molecules ($\mu_1 = 2$) with one oxygen molecule ($\mu_2 = 1$), two water molecules are formed ($\nu_1 = 2$). Thus, we have a third-order reaction characterized by a stoichiometric equation $2H_2 + O_2 \rightarrow 2H_2O$.

Let us turn now to the mathematical description of the considered phenomenon. During the chemical reaction, the **concentrations** of the starting materials and reaction products change. This is a number of molecules located in a unit volume. It is natural to choose these characteristics as system state functions. The mathematical model of reaction should describe the change over time of the concentrations of all substances involved in this reaction.

Try to quantify the course of the chemical reaction. A change in the structure of molecules occurs as a result of their collision. Obviously, the higher the concentration of the starting materials, the more often they will collide with each other and react. Thus, the rate of change in the concentrations of the considered substances (both initial and reaction products) should be directly proportional to the concentration of each of the reacting substances. This value is positive for the reaction products (they appear in the process of chemical conversion) and negative for the starting materials (their concentration decreases due to the reaction). The above considerations determine the essence of the **law of mass action**.

Before describing the change in the concentration of substances involved in the general reaction (6.1), we turn to significantly simpler processes. In the same way, modelling of any phenomena of the surrounding world is carried out. At first, individual laws are revealed for extremely simple phenomena, and then the obtained results are extended to a more general case.

We begin by considering the simplest first-order reaction, also called a **monomolecular reaction**. It is characterized by the decomposition of a starting substance into two reaction products

$$A \longrightarrow B_1 + B_2. \tag{6.2}$$

An example of such a reaction is the decomposition of ethyl bromide into ethylene and hydrogen bromide $C_2H_5Br \rightarrow C_2H_4 + HBr$.

The rate of change in the concentration b_1 of the reaction product B_1 will be proportional to the concentration a of the starting material A. Here inafter, the same lower case letter is used to indicate the concentration of a substance. Indeed, the more molecules of substance A are in a unit volume, the greater their number will decay per unit time, forming the reaction product B_1. As a result, we obtain the relation

$$\dot{b}_1 = ka,$$

called the kinetic equation of reaction (6.2). A positive value of k is called the **rate constant of chemical reaction**. It is a parameter of the process under study and depends on temperature, pressure, the presence of a catalyst and other factors. Sometimes the reaction rate constant is included in the recording of the stoichiometric equation (this becomes especially important when a system of chemical reactions is considered, see below). Thus, a more complete form of the reaction (6.2) takes the form

$$A \xrightarrow{k} B_1 + B_2.$$

Determine the law of change in the concentration b_2 of the product B_2 of reaction (6.2). Obviously, as a result of the decay of one molecule of substance A, one molecule of each of the reaction products is obtained. Consequently, the same number of molecules of substances B_1 and B_2 appears per unit time. Thus, the kinetic equation for the second reaction product is written in the same way as for the first

$$\dot{b}_2 = ka.$$

The concentration of starting material A decreases. Moreover, the number of disintegrated molecules is exactly equal to the number of molecules of each of the substances resulting from the reaction. Then the rate of change in concentration a differs from the rate of change of b_1 and b_2 only by a sign, which corresponds to the kinetic equation

$$\dot{a} = -ka.$$

Based on this analysis, we conclude that the rate of change in the concentrations of all substances involved in the reaction is the same. This value is called the **reaction rate** and is determined exclusively by the constant k. For reaction products, the rate of concentration change is equal to the reaction rate, and for the starting material, it is equal to the reaction rate taken with the opposite sign.

To assess the effect of stoichiometric coefficients on the form of the kinetic equation, we consider a particular case of reaction (6.2), when the reaction products B_1 and B_2 coincide, i.e., the molecule of the starting material breaks up into two molecules of a single reaction product. The corresponding stoichiometric equation is

$$A \longrightarrow 2B. \tag{6.3}$$

It can be the decomposition of a hydrogen molecule into two atoms $H_2 \to 2H$.

It was previously noted that the change in the concentrations of reacting substances depends solely on the concentrations of the starting materials. In reactions (6.2) and (6.3), there is a unique starting substance A. This means that the law of change in concentration a is the same for both reactions. Thus, it remains to write the kinetic equation for the reaction product. Obviously, during the decay of one molecule of the starting material, two molecules of the reaction product are obtained. Thus, the rate of increase in the concentration b of substance B is twice the rate of decrease in function a. Thus, we get

$$\dot{b} = 2ka.$$

Now consider the second-order reaction, i.e., **bimolecular reaction**. It can be the **synthesis reaction** that is the combination of two substances in one reaction product

$$A_1 + A_2 \longrightarrow B. \tag{6.4}$$

An example of such a reaction is the formation of carbon dioxide due to the fusion of a carbon monoxide molecule and an oxygen atom $CO + O \to CO_2$.

The rate of change in the concentration of substance B for reaction (6.4) is directly proportional to the concentration of each of the starting materials, and hence their product. Then the kinetic equation for the reaction product takes the form

$$\dot{b} = ka_1 a_2.$$

Since the number of reaction product molecules that appear in a unit volume per unit time is exactly equal to the number of molecules of each of the starting materials that have reacted, the velocities of change in concentrations a_1 and a_2 coincide and differ from the rate of change in the concentration of the reaction product only by a sign. As a result, we obtain kinetic equations

$$\dot{a}_1 = -ka_1a_2, \ \ \dot{a}_2 = -ka_1a_2.$$

Now suppose that substances A_1 and A_2 in reaction (6.4) coincide, i.e., we have

$$2A \longrightarrow B. \tag{6.5}$$

Its example is the fusion of two oxygen atoms into one molecule $2O \to O_2$.

The reaction products in the stoichiometric equations (6.4) and (6.5) are the same. Therefore, the equation for the function b in this case is written in the same way as in the previous one. It is only necessary to keep in mind that the starting materials A_1 and A_2 for reaction (6.5) coincide. Equating to each other the concentrations a_1 and a_2 in the kinetic equation with respect to the reaction product (6.5), we obtain the relation

$$\dot{b} = ka^2.$$

Note that for each newly formed molecule of substance B, according to the stoichiometric equation (6.5), there are two molecules of the starting substance. Thus, the rate of decrease in the concentration of the latter will be twice the rate of increase in function b. Thus, the corresponding kinetic equation is

$$\dot{a} = -2ka^2.$$

Now we have already can finally consider a general chemical reaction characterized by stoichiometric equation (6.1). The rate of change in the concentration b_i of the reaction product B_i is proportional to the corresponding coefficient ν_i times the product of all concentrations a_j of the starting materials raised to the power μ_j. As a result, we obtain

$$\dot{b}_i = k\nu_i \prod_{j=1}^{n} a_j^{\mu_j}, \ i = 1, ..., m. \tag{6.6}$$

Similarly, the rate of change in concentration a_i is proportional to the stoichiometric coefficient a_i with the minus sign, multiplied by the product of the concentrations of all the starting materials raised to the corresponding degree. As a result, we obtain the equation

$$\dot{a}_i = -k\mu_i \prod_{j=1}^{n} a_j^{\mu_j}, \ i = 1, ..., m. \tag{6.7}$$

We have the following equalities

$$\frac{\dot{b}_1}{\nu_1} = ... = \frac{\dot{b}_n}{\nu_n} = -\frac{\dot{a}_1}{\mu_1} = ... = -\frac{\dot{a}_m}{\mu_m} = k \prod_{j=1}^{n} a_j^{\mu_j}.$$

The value on the right-hand side of these equalities is called the **rate of chemical reaction** (6.1). It shows how quickly the concentration of substances involved in this reaction changes.

Add initial conditions for the equations (6.6), (6.7)

$$a_i(0) = a_{i0}, \ i = 1, ..., m, \ \ b_j(0) = b_{j0}, \ j = 1, ..., n, \tag{6.8}$$

where a_{i0}, b_{j0} are the initial concentrations of the corresponding substances. The problem (6.6)–(6.8) is the mathematical model of the studied process[3]. Solving it, one can determine the law of change with time of the concentrations of all substances involved in the reaction, depending on the reaction rate constant k and all initial concentrations.

Note that the system of equations (6.7) can be solved without using the concentration of reaction products. Having determined the concentration values a_i, we can proceed to solving equations (6.6) with the already known right-hand sides, which is a fairly simple problem. Below we will consider some special cases of reaction (6.1), which allow a complete qualitative analysis of the corresponding system of kinetic equations.

Chemical reaction is characterized by concentrations
of starting materials and reaction products.
rate of change in the concentrations is positive for reaction products
and negative for starting materials, and this is proportional
to the concentration product of starting materials
in the degree equal to the corresponding stoichiometric coefficient.

2 Monomolecular reaction

Determine the law of concentration change for the monomolecular reaction (6.2). As already noted, the mathematical model of this process includes the differential equations

$$\dot{a} = -ka, \ \dot{b}_i = ka, \ i = 1, 2$$

with initial conditions

$$a(0) = a_0, \ b_i(0) = b_{i0}, \ i = 1, 2.$$

Define the law of change in the concentration of the starting material from the problem[4]

$$\dot{a} = -ka, \ a(0) = a_0.$$

Its solution is $a(t) = a_0 \exp(-kt)$. Thus, the function a decreases exponentially and tends to zero at $t \to \infty$ for any initial state a_0. The corresponding equation has a unique equilibrium position, which turns out to be asymptotically stable.

Using the change law of the starting material, from the equalities

$$\dot{b} = ka, \ b_i(0) = b_{i0}, \ i = 1, 2$$

we find the functions

$$b_i(t) = b_{i0} + k \int_0^t a(\tau)d\tau = b_{i0} + ka_0 \int_0^t e^{-k\tau}d\tau = b_{i0} + a_0(1 - e^{-kt}), \ i = 1, 2.$$

The functions b_i increase to the sum $b_{i0} + a_0$, $i = 1, 2$. The results are shown in Figure 6.1. Thus, the second stage of the study is completed. We found the solutions of the considered system of differential equations with the corresponding initial conditions. Now it remains only to interpret the results.

We see that the concentration of the starting material decreases with time, while the amount of reaction products monotonically increases, which naturally agrees with natural sense. Note that the rate of change of all quantities tends to zero. As the concentration of the starting material decreases, an ever smaller number of its molecules decompose, and over time, the entire starting material will react successfully. In turn, the concentration of reaction products is steadily increasing as the concentration of the starting material decreases. For each decaying molecule of the starting material, as is known, one molecule of reaction products is formed. Thus, over time, the concentration of the reaction products tends to the sum of their initial concentration and initial concentration of the starting material.

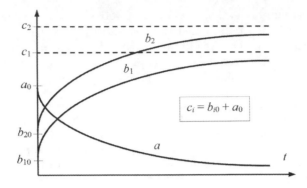

FIGURE 6.1: Concentration change for a monomolecular reaction.

Let us evaluate the effect of process parameters on the course of the reaction. The initial concentration of the starting material does not qualitatively affect the results. The larger it is, the larger the supply of molecules is, which means that the longer the decomposition process will take place (see Figure 6.2a). However, all the substance will react sooner or later, i.e., the limiting value of the concentration of the starting substance does not depend on its initial concentration. But the final value of the reaction products depends on this value. The initial values of the reaction products do not have any effect on the concentration of the starting material and cause a proportional change in the concentration of the reaction products, since the starting material (cause) affects the result, but not the reaction products (effect). The higher the reaction rate constant, the more intense the reaction, and therefore, the more molecules of the starting material will decompose per unit time. As the reaction rate increases, decomposition processes occur more intensively (see Figure 6.2b). It is clear that the reaction rate has no effect on the limit values of all substances, but determines the steepness of the corresponding curves, i.e., the rate tending to the equilibrium.

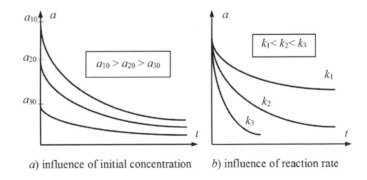

a) influence of initial concentration b) influence of reaction rate

FIGURE 6.2: Influence of parameters for the monomolecular reaction.

Concentration of the starting material of the monomolecular reaction exponentially decreases to zero, and the concentration of the reaction products monotonically increases to a value equal to the sum of the initial concentration of this and the starting substance.

3 Bimolecular reaction

Consider more difficult synthesis reaction $A_1 + A_2 \rightarrow B$. This describes by the nonlinear differential equations[5]

$$\dot{a}_i = -ka_1 a_2, \; i = 1,2, \; \dot{b} = ka_1 a_2$$

with initial conditions

$$a(0) = a_{i0}, \; i = 1,2, \; b(0) = b_0.$$

The equations for the concentration of the starting materials do not include function b. Then the first two equations of the system under consideration can be solved independently of the third. It follows from the equality $\dot{a}_1 = \dot{a}_2$ that the functions a_1 and a_2 can only differ by a constant, i.e.,

$$a_2(t) = a_1(t) + c, \; t \geq 0.$$

From the initial conditions, we get $c = a_{20} - a_{10}$. Excluding the function a_2, from the first kinetic equation, we obtain

$$\dot{a}_1 = -ka_1(a_1 + c).$$

This is the nonlinear differential equation

$$\dot{a}_1 + ka_1^2 + kca_1 = 0, \tag{6.9}$$

called **Bernoulli equation**[6].

By simple transformations[7], it can be established that the solution of this equation with the available initial conditions is determined by the formula

$$a_1(t) = a_{10}\frac{a_{20} - a_{10}}{a_{20}\exp(\lambda t) - a_{10}},$$

where $\lambda = k(a_{20} - a_{10})$.

By the relation between the concentrations of starting materials, we find

$$a_2(t) = a_{20}\left[1 + a_{10}\frac{1 - \exp(\lambda t)}{a_{20}\exp(\lambda t) - a_{10}}\right].$$

From the equality $\dot{a}_1 + \dot{b} = 0$ and the initial conditions, it follows that

$$b(t) = b_0 + a_{10}a_{20}\frac{\exp(\lambda t) - 1}{a_{20}\exp(\lambda t) - a_{10}}.$$

Thus, for the considered bimolecular reaction, the corresponding system of nonlinear differential equations has an analytical solution[8]. This circumstance allows a complete qualitative analysis of the system directly without the use of a special mathematical apparatus.

Suppose $a_{20} > a_{10}$, which implies that $\lambda > 0$. This assumption does not violate the generality of the study, since as A_2 we have the right to choose one of the starting materials for which the initial concentration is higher. Find

$$\lim_{t \to \infty} a_1(t) = a_{10}(a_{20} - a_{10})\lim_{t \to \infty}\left[\frac{\exp(-\lambda t)}{a_{20} - a_{10}\exp(-\lambda t)}\right] = 0,$$

Then the concentration of the starting material with a lower initial concentration decreases exponentially and tends to zero with time. Therefore, the substance A_1 will eventually react (see Figure 6.3).

Find the limit

$$\lim_{t \to \infty} a_2(t) = a_{20}\left\{1 + a_{10}\lim_{t \to \infty}\left[\frac{\exp(-\lambda t) - 1}{a_{20} - a_{10}\exp(-\lambda t)}\right]\right\} = a_{20}\left(1 - \frac{a_{10}}{a_{20}}\right) = a_{20} - a_{10}.$$

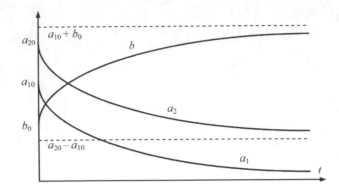

FIGURE 6.3: Concentration change for a bimolecular reaction.

Thus, the concentration of the second starting material decreases and tends to the difference in the initial concentrations of the starting materials (see Figure 6.3). Since for each molecule of substance A_1 that has reacted, there is one molecule of substance A_2, the presence of an excess of the second substance at the initial time leads to the fact that, when almost all of the first substance enters the reaction, there remains a certain amount of substance A_2, which does not with what to react. This amount is natural and equal to the difference in initial concentrations. In particular, if $a_{20} = a_{10}$, then the concentrations of both substances decrease according to the same law and tend to zero.

Let us evaluate the change in the reaction product. Find the limit

$$\lim_{t\to\infty} b(t) = b_0 + a_{10}a_{20} \lim_{t\to\infty} \left[\frac{1 - \exp(-\lambda t)}{a_{20} - a_{10}\exp(-\lambda t)} \right] = b_0 + a_{10}.$$

Thus, the concentration of reaction product increases over time and tends to the sum of its initial concentration and the initial concentration of substance A_1. This is due to the fact that the total number of molecules of substance B formed during the reaction is exactly equal to the number of reacted molecules of each of the starting materials. We know that all molecules of substance A_1 will sooner or later react. Therefore, over time, the concentration of the reaction product increase just by the value of initial concentration of substance A_1.

By bimolecular synthesis reaction, the concentration of the starting material
with a lower initial concentration decreases monotonically to zero;
the concentration of substance with higher initial concentration decreases
to the difference between initial concentrations of the starting materials.
The concentration of reaction product monotonically increases to the sum
of the initial concentrations of the reaction product
and the starting material with a lower initial concentration.

4   Lotka reaction system

In practice, more often than not one reaction occurs, but a whole quantity of reactions. In this case, a particular chemical substance can participate in several reactions simultaneously, both as a starting substance and as a reaction product. As an example, we consider a system of chemical reactions characterized by Lotka stoichiometric equations

$$A+X \xrightarrow{k_1} 2X, \quad X+Y \xrightarrow{k_2} 2Y, \quad Y \xrightarrow{k_3} B. \tag{6.10}$$

The first two reactions are ***autocatalytic***, i.e., reaction products (substances X and Y) are at the same time the starting materials.

When compiling the corresponding kinetic equations, all reactions should be taken into account. The equations for the starting material A and the reaction product B involved in a single reaction are written naturally

$$\dot{a} = -k_1 ax, \quad \dot{b} = k_3 x.$$

The substance X is the starting material for the first and second reactions and the product in the first reaction. Then, when assessing the rate of change in its concentration, one should take into account its total change in each of the reactions

$$\dot{x} = -k_1 ax + 2k_1 ax - k_2 xy.$$

Then we have the equation

$$\dot{x} = k_1 ax - k_2 xy.$$

Determine the analogical equation for the concentration of Y

$$\dot{y} = k_2 xy - k_3 y.$$

Since substance B only forms during the reaction, its concentration steadily increases and is not included in the right-hand sides of the obtained equations. Thus, in the study of the established ratios, it cannot be considered, considering that this substance is discharged outside the reactor. The starting material A, on the contrary, is constantly consumed, and its concentration decreases, which inevitably should slow down the whole process. To maintain the reaction, we assume that a constant supply of substance A is observed in the system, which compensates for its consumption. Under these conditions, its concentration remains unchanged, being a parameter of the task. Under the assumptions made, the process is described by the following system of kinetic equations

$$\begin{cases} \dot{x} = k_1 ax - k_2 xy \\ \dot{y} = k_2 xy - k_3 y \end{cases} \tag{6.11}$$

called the ***Volterra–Lotka equations***[9].

We could prove that their solutions are periodic functions (see Figure 6.4)[10]. Suppose, for example, the initial concentrations of both considered substances are such small that the following inequalities hold $k_2 < k_1 a$, $k_2 < k_3$. Therefore, the first and third reactions (6.10) are more intensive than the second one. Then, by equalities (6.10), the derivative of the first state function is positive, and the derivative of the second one is negative, i.e., the formation of the first substance in the first reaction prevails over its decomposition in the second reaction, and the decomposition of the second substance in the second reaction prevails over its formation in the third reaction. As a result of this, we observe an increase in the concentration of the first substance and a decrease in the concentration of the second substance (see Figure 6.4, stage I). Over time, there are already so many first substances that the inequality $k_2 x > k_3$ holds. This means that the synthesis of the second substance in the second reaction begins to prevail over its decomposition in the third reaction, as a result of which we observe an increase in the concentration of both substances (see Figure 6.4, stage II). Inevitably, there comes a point in time when the second substance is already enough to satisfy the inequality $k_2 > k_1 a$, i.e., for the predominance of decay over the synthesis of the first substance. At the third stage of the process, we observe a decrease in the concentration of the first substance with an increase in the concentration of the second substance (see Figure 6.4, stage III). As soon as the first substance becomes sufficiently small, the inequality $k_2 < k_3$ will be fulfilled again, which means that the decay of the second substance in the third reaction will prevail over its synthesis in the second reaction. Thus, a decrease in both concentrations is observed (see Figure 6.4, step IV). Sooner or later, the concentration of the second substance will decrease so much that the inequality $k_2 < k_1 a$ will be satisfied, and we return to the first stage of the process.

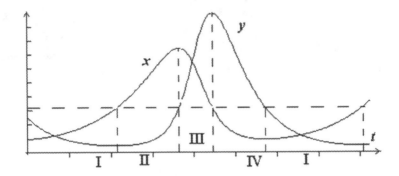

FIGURE 6.4: Solutions of the Volterra–Lotka equations.

Determine the equilibrium positions of the system. Equal to zero the value sat the right-hand sides of equations (6.11). We obtain the system of algebraic equations

$$\begin{cases} k_1 ax - k_2 xy = 0 \\ k_2 xy - k_3 y = 0 \end{cases} \qquad (6.12)$$

This system has two solutions. The zero solution is trivial and does not have any interest, because this is the case of absence of considered substances. Of course, we have no reaction in this situation. These second solution is more interesting. For this case, the concentrations are positive, and this equilibrium position is a center. In this case, the formation of each of the substances in one of the reactions is balanced by its decays in the other reaction. It is under these conditions that we observe a periodic change in the concentrations of both substances under consideration.

Task 6.1 *Volterra–Lotka system*. Based on the numerical solving of problem (6.11) make the following analysis.

1. Choose zero value as the initial concentrations for system (6.11). Make sure that this is, in reality, the equilibrium position.

2. Choosing small enough values of the initial concentrations make sure that the system does not tend to the zero equilibrium position. Therefore, this is not stable[11].

3. Choosing small arbitrary values of the initial concentrations make sure that the state functions are periodic functions.

4. Changing the initial states of the system, determine increasing and decreasing of the oscillation amplitudes of the concentrations. Explain this result.

5. Find the non-trivial equilibrium position by system (6.12). Choose this value as the initial concentrations. Solving the corresponding system, make sure that this the equilibrium position.

6. Deviating the values of the initial states of the system from those selected at the previous stage, make sure that the state of the system remains in a certain neighborhood of the equilibrium position, but tends to zero. This effect corresponds to Lyapunov stability.

Task 6.2 *System of chemical reactions.* Describe a mathematical model for the following abstract chemical reaction system

$$A+B \xrightarrow{k_1} 2C, \quad C+2D \xrightarrow{k_2} 3A, \quad 3B+D \xrightarrow{k_3} A+E, \quad 3C \xrightarrow{k_2} 2B.$$

When modeling a system of chemical reactions should write a system of kinetic equations that take into account each of the available reactions. For the system of Volterra–Lotka equations, there are two equilibrium positions, one of which is not stable, and the second is Lyapunov stable.

Direction of further work. In the next chapter, we consider mathematical models of biology problems. It turns out that the equations characterizing some chemical processes have a biological interpretation.

Appendix

Some systems of chemical reactions described by nonlinear differential equations are considered. One of them corresponds to a mathematical model called the "brusselator", and the second to the "oregonator", and the third to the "chemical niche". It is completely unexpected that the kinetic equations in the latest model also describe some processes of laser healing.

1 Brusselator

Consider the following system of chemical reactions

$$A \xrightarrow{k_1} X, \quad B+X \xrightarrow{k_2} 2Y+D, \quad 2X+Y \xrightarrow{k_3} 3X, \quad X \xrightarrow{k_4} E.$$

With a constant supply of starting materials A and B and removal of reaction products D and E, the concentrations of the corresponding substances will be known, and it is only necessary to establish the change in the concentrations of intermediate substances X and Y over time. They are described by the equations[12]

$$\dot{x} = k_1 a - k_2 bx + k_3 yx^2 - k_4 x, \quad \dot{y} = k_2 bx - k_3 yx^2.$$

Add initial conditions for the considered system

$$x(0) = x_0, \quad y(0) = y_0.$$

Depending on the combination of parameters, an equilibrium position is possible here, which is the focus (stable or unstable), as well as a limit cycle. It is realized when the following equality holds $k_3 k_1^2 a^2 = k_2 k_4^2 b - k_4^3$. In this case, the solution of the equations eventually tends to the same periodic mode, regardless of the choice of the initial state, which are sufficiently close to the limit cycle[13].

Task 6.3 *Brusselator.* Based on the numerical solving of the system make the following analysis.

1. Carry out calculations with the following parameters: $x_0 = 0.5$, $y_0 = 0.5$, $a = 3$, $b = 5$, $k_1 = 1.2$, $k_2 = 0.7$, $k_3 = 0.9$, $k_4 = 1$. Make sure that over time, the concentrations of the intermediate substances become stationary.

2. Calculate the system with $x_0 = 5$ and unchanged values of all other parameters. Make sure that the state of the system increases unlimitedly.

3. Carry out calculations with the following parameter values: $x_0 = 3.4$, $y_0 = 1.1$, $a = 3$, $b = 5$, $k_1 = 0.5$, $k_2 = 0.7$, $k_3 = 0.9$, $k_4 = 1.2$. Make sure that the state of the system changes over time periodically.

4. Calculate the system with initial conditions $x_0 = 1.7$, $y_0 = 3$, and unchanged values of all other parameters. Detect a tendency to a periodic mode with increasing amplitude. Pay attention that the process goes to the same oscillatory mode as in the previous case.

5. Calculate the system with initial conditions $x_0 = 0.3$, $y_0 = 0.3$, and unchanged values of all other parameters. Detect a tendency to a periodic mode with decreasing amplitude. Pay attention that the process goes to the same oscillatory mode as in the previous cases.

6. Give an interpretation of all the results obtained in the previous steps.

2   Oregonator

The brussellator describes a system of chemical reactions, one of which is realized as a result of the collision of three molecules at once. From a practical point of view, this situation is rare enough. We give an example of a reaction system that does not have this drawback. It assumes the presence of three reactive substances[14]. Consider the following system of reactions.

$$A+Y \xrightarrow{k_1} X, \quad X+Y \xrightarrow{k_2} P, \quad B+X \xrightarrow{k_3} 2X+Z, \quad 2X \xrightarrow{k_4} Q, \quad Z \xrightarrow{k_5} fY,$$

where the stoichiometric coefficient f is a parameter of the problem. Here, the starting materials A and B are constantly supplied to the system, and the reaction products P and Q are constantly removed from it. Their concentrations are considered known and are the parameters of the problem. The corresponding system of kinetic equations has the form

$$\dot{x} = k_1 ay - k_2 xy + k_3 bx - k_4 x^2, \quad \dot{y} = -k_1 ax - k_2 xy + fk_5 z, \quad \dot{z} = k_3 bx - k_5 z.$$

This model is called the **oregonator** and has a limit cycle with a combination of parameters[15].

Task 6.4 *Oregonator*. Make the following analysis of the system.

1. Equating to zero the right-hand sides of the equations, find the equilibrium position.

2. Choosing the states of the system found at the previous step as initial conditions, make sure that these are indeed equilibrium positions.

3. Choosing values that are close enough to the equilibrium positions as the initial states of the system, check whether these states are stable.

4. By changing the system parameters, using a computer experiment to detect the limit cycle.

5. Select the points on the phase curve corresponding to the limit cycle as the initial states of the system. Verify that the system has a periodic solution.

6. Verify experimentally whether this limit cycle is stable.

7. Give a chemical interpretation of all the results from the previous steps.

3   Chemical niche

Consider the following system of chemical reactions

$$A+X \xrightarrow{k_1} 2X, \quad A+Y \xrightarrow{k_2} 2Y, \quad 2X \xrightarrow{k_3} B,$$

$$2Y \xrightarrow{k_4} C, \quad X+Y \xrightarrow{k_5} D.$$

TABLE 6.1: Analogy between processes.

Characteristic	Chemistry	Physics
object	reactor	laser
state function	substance concentration	photon number
state increase	substance synthesis	photon generation
state decrease	substance decay	photon leaving

It is assumed that the starting material A is constantly supplied to the reactor from the outside so that the concentration a remains unchanged. The reaction products B, C, and D are removed from the reactor and are not considered. The equations for intermediate substances X and Y are as follows:

$$\dot{x} = \left(k_1 a - 2k_3 x - k_5 y\right)x, \quad \dot{y} = \left(k_2 a - k_5 x - 2k_4 y\right)y.$$

The resulting model is called the "***chemical niche***".

Of particular interest is the special case of the presented model, in which the reaction rates are related by the following equality $k_5 = 2\sqrt{k_3 k_4}$. Now we determine the kinetic equations

$$\dot{x} = \left[\varepsilon_1 - \alpha_1\left(\beta_1 x + \beta_2 y\right)\right]x, \quad \dot{y} = \left[\varepsilon_2 - \alpha_2\left(\beta_1 x + \beta_2 y\right)\right]y,$$

where $\varepsilon_i = k_i a$, $\alpha_i = 2\sqrt{k_{i+2}}$. This system corresponds to the model of "***chemical competition***"[16]. We will meet with such equations more than once. In particular, in physics they describe the process of laser healing.

4 Laser healing model

Consider a model of "physical competition" related to the ***laser*** healing[17]. In the process of external atomic excitation (pumping), the laser emits photons. The ***photon number*** x is chosen as a function of the state of the system. The rate of change of this quantity is the sum of the increase in photons due to stimulating radiation associated with the excitation of atoms and losses due to the departure of photons outside the system. Thus, the following relation holds

$$\dot{x} = \alpha X x - x/T,$$

where α is a parameter called the amplification factor, T is the time the photon spends in the system, X is the number of excited atoms[18].

In the absence of photon emission, the number of excited atoms remains unchanged due to the constant external pumping, and is equal to a value of X_0. When a photon is emitted, an excited atom returns to a more stable state. Thus, the number of excited atoms decreases by an amount proportional to the number of photons, i.e., $X = X_0 \beta$, where β is a positive coefficient. Then we have the equality

$$\dot{x} = (\varepsilon - \gamma x)x,$$

where $\gamma = \alpha\beta$, $\varepsilon = \alpha X_0 - 1/T$. We can determine the same equation for modelling of the chemical reactions system A+X $\longrightarrow$ 2X, X $\longrightarrow$ B under the continuous supply of substance A. Thus, we have an analogy between the processes in chemical reactor and laser (see Table 6.1)[19].

The above equation is characteristic of a single-mode laser emitting photons of the same wavelength. For a multimode laser, there are several types of photons that differ in their properties. In the presence of two modes, the state of the system is described by the photon number x and y of the first and second types. Repeating the previous arguments, we obtain the system

$$\dot{x} = \alpha_1 X x - x/T_1, \quad \dot{y} = \alpha_2 X y - x/T_2.$$

where all parameters have the same physical meaning as in the previous case. The number of excited atoms in this case is $X = X_0 - \beta_1 x - \beta_2 y$. Thus, we get the equations

$$\dot{x} = \Big[\varepsilon_1 - \alpha_1\big(\beta_1 x + \beta_2 y\big)\Big]x, \quad \dot{y} = \Big[\varepsilon_2 - \alpha_2\big(\beta_1 x + \beta_2 y\big)\Big]y.$$

where $\varepsilon_i = \alpha_i X_0 - 1/T_i$, $i = 1, 2$. This corresponds to the "chemical competition" model.

Notes

[1] Mathematical models of chemical kinetic are considered, for example, in [27], [56], [84], [90], [91], [115], [128], [171], [176], [194], [213], [245], [256], [277], [279], [282], [374].

[2] Naturally, the simultaneous participation in a single reaction of a significant amount is extremely unlikely. Usually no more than two molecules of the starting materials interact. However, now it is of interest to us to derive a mathematical model of the process for the general case.

[3] In Chapter 11, it will be shown that the description of chemical reactions taking into account the diffusion of reacting substances in a certain spatial region is carried out using partial differential equations. In this case, the course of chemical reactions is described in a known manner, and the concentration change due to diffusion, as is done in the study of transfer processes (see Chapter 10). Chapter 18 discusses a mathematical model of a chemical reaction with random parameters.

[4] We have now the linear homogeneous first order equation. The equation is *homogeneous* if, after transferring all terms, including unknown values, to its left-hand side, one gets zero at its right-hand side. In particular, the equation for the concentration a of a monomolecular reaction can be written in the form $\dot{a} + ka = 0$. Let us give an example of a linear non-uniform first order equation $c\dot{T} = \alpha(T_0 - T)S$. This describes a *heat transfer* with the environment and characterizes the heat flux through the surface area S, where T is the temperature, c is the heat capacity coefficient, T_0 is the ambient temperature, and α is the heat transfer coefficient. Indeed, transferring an expression including an unknown state function T to the left side of the equation, we obtain the equality $c\dot{T} + \alpha ST = \alpha ST_0$, in the right side of which there is a nonzero value. By the way, we will consider a heat transfer phenomenon in Chapter 10.

[5] In Chapter 11, the synthesis reaction will be described for the case when the considered substances are unevenly distributed over a certain region. This phenomenon is described by a system of nonlinear partial differential equations. Chapter 18 provides a mathematical model of the described reaction for the case when the reaction rate constant is a random variable. Chapter 21 will discuss the problem of determining the reaction rate constant for a synthesis reaction from the results of measuring the concentrations of reagents.

[6] The general form of the *Bernoulli equation* is $\dot{x} + a(t)x = b(t)x^n$, where the degree n is not equal to 0 or 1. The Bernoulli equation has an analytic solution for the general case even. This is the partial case of the *Riccati equation* for $n = 2$. This is true, for example, to the Verhulst equation considered in subsequent chapters. The properties of the Bernoulli equation are considered, for example, in [132].

[7] In order to simplify equality (6.9), we introduce a new unknown function $x(t) = a_1(t)^{-1}$. We have $a_1 = x^{-1}$, $\dot{a}_1 = -x^{-2}\dot{x}$. Putting these values to equality (6.9), we obtain the linear differential equation $\dot{x} - kcx = k$. Find its solution $x(t) = c' \exp(kct) - c^{-1}$, where c' is a constant. Thus, we find the general solution of equation (6.9)

$$a_1(t) = \frac{c}{cc'\exp(kct) - 1}.$$

Determine $t = 0$. Using the initial conditions and the value of the constant c, we get

$$c' = \frac{a_{20}}{a_{10}}\frac{1}{a_{20} - a_{10}}.$$

Then we determine the change law of the substance A_1

$$a_1(t) = a_{10}\frac{a_{20} - a_{10}}{a_{20}\exp(\lambda t) - a_{10}},$$

where $\lambda = k(a_{20} - a_{10})$.

[8] Note that obtaining an analytical solution of a nonlinear system by reducing it to a linear equation by successfully changing variables should be recognized as exceptional success, indicating the relative simplicity of the problem being studied, in particular, its solution is monotonic, which is completely not typical for general systems. Typical nonlinear equations are in principle not reducible to linear, which does not exclude the possibility of their qualitative, and even more so, quantitative analysis.

[9] In a subsequent chapter, it will be shown that the popular biological predator–prey model is associated with the Volterra–Lotka equations. We will determine that the considered systems of equations have a biological, economic and even socio-political interpretation.

[10] The completely analysis of the Volterra–Lotka equations is given in Chapter 7.

[11] One can sure that this is a saddle.

[12] This system was be considered in the previous chapter.

[13] Oscillation chemical reactions are indeed observed experimentally. This is characteristic, for example, for the Belousov–Zhabotinsky reaction, see, for example, [100], [150]. At the same time, some process parameters (concentration of components, color, temperature, etc.) change periodically, forming a complex spatio-temporal structure. Naturally, for a more complete mathematical description of these processes, it is necessary to take into account the diffusion of reacting substances, the temperature change in the considered area, etc. (see Chapter 11), associated with the change in the state functions of the system not only with time, but also in space. As a result, the mathematical model of the process will be a system of partial differential equations, see [373].

[14] One can prove that the two substances chemical systems with bimolecular reactions do not have any limit cycles.

[15] About the oregonator see [100].

[16] The meaning of the names "chemical niche" and "chemical competition" will be clarified in subsequent chapters, where these equations will receive a different interpretation and the complete analysis.

[17] Laser healing phenomenon is considered, for example, in [222], [323].

[18] Mathematical models of the laser healing are considered, for example, in [128].

[19] In a subsequent chapter, it will be shown that the model of a single-mode laser and the indicated system of reactions corresponds to the Verhulst equation, which also describes the change in the abundance of the biological species under conditions of limited food.

Chapter 7

Mathematical model in biology

In the previous chapter, we were acquainted with mathematical models of chemical reactions. It may seem surprising that the equations obtained in this case also describe some biological systems. Naturally, we are not even try to cover all of biology within a single chapter, confining ourselves to a cursory acquaintance with one of its sections. This is the ***population dynamics***[1]. We have to witness the dramatic events associated with the coexistence of various species in a limited area. The considered processes are described by ordinary differential equations[2], which are, as a rule, nonlinear.

At first, consider the evolution of one species. If its birthrate prevails over mortality, then the number of this species is steadily increasing. Under adverse conditions with a predominance of mortality over birthrate, the species dies out. The presence of restrictions on the amount of incoming food with a positive natural increase in the number of species leads to stabilization of the system, i.e., to establish such a value of the number of the species for which the available amount of food is just enough.

More interesting and diverse processes occur when there are two biological species. Their coexistence is possible in various forms. In one of the presented models, the competition of species in the struggle for common food is studied, during which a less adapted species is replaced by a stronger competitor. In the second model, one of the species is food for the other. As a result, we consider the famous predator–prey model described by the Volterra–Lotka equations. Finally, in the third case, each of the species has a positive effect on the other species, which corresponds to the phenomenon of symbiosis.

The Appendix notes the analogy between models of biological, chemical, and physical competition. It turns out that the phenomenon of fluctuation in the number of predators and preys is similar to the change in soil fertility and crop productivity. The coexistence of two species is considered, provided that each of them has a certain preference in food, which corresponds to the ecological niche model and is a generalization of the competition model. Another generalization of this model is the antibiotic resistance model, which characterizes the microorganisms getting used to the action of antibiotics.

Lecture

1 One species evolution

We first consider the simplest biological system represented by one species. The studied processes characterized by an ***number of the species*** x, which varies with time. Obviously, the rate of change in the number of species is determined by the ratio between its birthrate and mortality. Thus, we obtain the equation

$$\dot{x} = A - B,$$

where A and B characterize, respectively, the number of births and deaths of individuals in a unit of time. Their values are apparently proportional to the number of the species.

DOI: 10.1201/9781003035602-7

Indeed, the more individuals there are, the more they will be born and die in a fixed time interval. As a result, we obtain the relations $A = ax$, $B = bx$, where the positive constants a and b are process parameters and characterize the birthrate and mortality of the species.

Thus, we considered process described by the differential equation

$$\dot{x} = kx, \tag{7.1}$$

where the coefficient $k = a - b$ is called the **growth** in the number of species, determines the change in the number of species per unit time and can take both positive and negative values[3]. Equation (7.1) is considered with the initial condition

$$x(0) = x_0, \tag{7.2}$$

where the initial number x_0 is the parameter of the problem, taking any positive values[4]. This mathematical model is called the **Malthus model**[5].

The solution of problem (7.1), (7.2) is $x(t) = x_0 \exp(kt)$. This function has qualitatively different properties depending on the sign of the parameter k. With its positive values, we observe an exponential growth of the function x. At $k = 0$, the solution of the problem does not change with time. Finally, for negative values of this parameter, the function x monotonically decreases and tends to zero over time. Variants of behavior for solving the problem are shown in Figure 7.1. We also note that an increase and decrease in the number k leads, respectively, to a sharper or smoother change in the solution of the problem, and a change in the initial state of the system does not introduce qualitative changes in the behavior of the studied process.

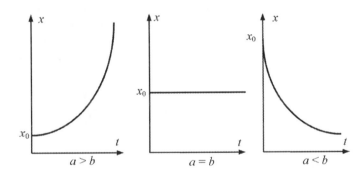

FIGURE 7.1: Variants of evolution of the species number.

Now we try to interpret the results. Negative values of the coefficient k correspond to the unfortunate case when the mortality of a species prevails over its birthrate. Therefore, in a unit of time, the number of species decreases by a certain value. Next time, more individuals will die than be born. Since in this model the increase in the number of species is considered unchanged, we are observing its gradual extinction. This sad outcome is characteristic of a species found in extremely unfavorable living conditions caused by an acute shortage of food, a sharp deterioration of the habitat, the appearance of natural opponents, the spread of epidemics and other troubles.

Positive values of the increase in the species number correspond to the excess of birthrate over mortality. This means that at any time interval more individuals are born than they die. Then, in the next interval, with the same increase, an even greater number of individuals will born. Thus, an unlimited increase in the number of species is observed, which is

TABLE 7.1: Interpretation of the Verhulst equation.

Characteristic	Physics	Chemistry	Biology
object	single-mode laser	reaction system A+X ⟶ 2X, X ⟶ B	species evolution with bounded food
state function	photon number	concentration of substance Õ	species number
restriction	number of excited atoms	substance quantity À	quantity of incoming food
state function growth	atomic emission of photons	substance Õ synthesis	birthrate of the species
state function decrease	leaving of photons from the system	substance Õ decay	species mortality

characteristic of a species under ideal conditions, i.e., in the absence of opponents and competitors, with an unlimited supply of food and a favorable environment. A similar situation is possible, for example, for microorganisms in a nutrient medium[6].

The trivial case of zero increase in numbers means that birthrate and mortality are balanced. The number of dead individuals here is compensated by the newly born, and the species number remains unchanged.

From a mathematical point of view, the zero state is the equilibrium position of a dynamical system characterized by equation (7.1). For $k < 0$, the state of the system monotonically decreases and tends to an equilibrium position that is asymptotically stable. For $k > 0$, a steady removal of the system from the equilibrium position is observed[7].

The described model has a fairly obvious flaw. With a positive increase in the species, an exponential increase in its number is observed here, which is in poor agreement with reality. In nature, the increase in the species number is constrained by restrictions on the amount of available food and free territory, the presence of natural opponents and competitors, etc. We restrict ourselves to considering the evolution of the species in conditions of limited food intake. Then, the increase in the species number in relation (7.1) will already substantially depend on the function x. It can be assumed that this dependence has the following form $k(x) = a(D - qx) - b$, where D is the amount of incoming food, q is the coefficient characterizing the consumption of food, b is the natural mortality of the species (not related to lack of food), and is the specific increase in the number of species (growth corresponding to the unit of incoming food). Under the assumptions made, the birthrate of the species is directly proportional to the excess of the amount of incoming food over the amount of necessary food (with a proportionality coefficient a). In case of food shortage, a corresponding increase in the mortality rate of the species is observed.

$$\dot{x} = \big[(aD - b) - aqx\big]x. \tag{7.3}$$

This is called the **Verhulst equation** or the **logistic equation**[8]. It also describes (up to the form of the coefficients) the photon emission process considered in the previous chapter in a single-mode laser and the change in the concentration of substance X in the system of reactions A+X⟶ 2X, X⟶B with a constant supply of substance A (see Table 7.1)[9].

The obtained relation is a special case of the Bernoulli equation. We considered the Bernoulli equation when describing the change in concentration in a second-order reaction, see Chapter 6. Repeating the transformation from the previous chapter, we find the solution to problem (7.3), (7.2)

$$x(t) = \frac{(D - b/a)x_0}{qx_0 + D - b/a - qx_0 \exp[(b - aD)t]}, \quad aD \neq b,$$

$$x(t) = \frac{x_0}{1 + x_0 qat}, \qquad aD = b.$$

By this result, the solution of the problem tends to zero if $aD \leq b$. This tends to $x^* = (D - b/a)/q$ if $aD > b$. For this case the function x increases if $x_0 < x^*$ and decreases if $x_0 > x^*$ (see Figure 7.2).

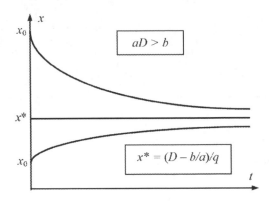

FIGURE 7.2: Species number tends to x^* if the quantity of food is bounded.

In the future, we consider more difficult equations that do not allow an analytical solution. In this regard, it would be desirable to establish the properties of equation (7.3) without resorting to a direct solution to the problem.

The sign of the derivative of the function x is determined by the term in square brackets on the right-hand side of formula (7.3). Under the inequality $aD \leq b$, this derivative is obviously negative. Thus, the function x monotonically decreases. However, as it approaches zero, its derivative steadily tends to zero. Thus, the rate of decrease of the state function gradually slows down, and its value tends to zero. For $aD > b$, the sign of the derivative depends on the current state of the system. If the initial number x_0 is less than the value of x^*, then the derivative $\dot{x}$ at the initial time is positive, which means that the function x increases. This situation is observed all the time, while the inequality $x < x^*$ is true. However, as x approaches the critical value of x^*, according to equality (7.3), the derivative $\dot{x}$ tends to zero. This suggests that as t grows, the value of $x(t)$, increasing, tends to x^*.

For $x_0 > x^*$, the derivative $\dot{x}$ at the initial instant of time is negative, and therefore, the function x decreases. As it decreases, the value of x approaches x^*, and the derivative $\dot{x}$ tends to zero. Thus, at $t \to \infty$, the function x monotonically decreases and tends to x^*. Finally, for $x_0 = x^*$ the derivative $\dot{x}$ is equal to zero, which means that the quantity x does not change, remaining equal to x^*. Therefore, regardless of the initial state of the system, the solution of equation (7.3) tends to the value x^*, see Figure 7.2[10].

The obtained results have a natural interpretation. If the initial number of a species is small enough, then the boundedness on food supplies is not so important, and the number of individuals is growing steadily. However, at the same time, the quantity of food consumed (with its constant supply) will certainly increase. A steadily increasing shortage of food restrains the growth of the species number. As a result, a certain equilibrium value of the number of species x^* is gradually established, which can be maintained at a given ratio between the amount of food available and its consumption. If initially the number of the species is too large, then there is a shortage of food, which leads to a reduction in the number of the species. However, as the number of the species decreases, food consumption decreases. Thus, the lack of food is gradually reduced, which means that the role of negative factors affecting the behavior of the system is gradually disappearing. One way or another, over time, one and the same equilibrium value x^* of the number of species is established. Variants of evolution of the considered system are given in Figure 7.3[11].

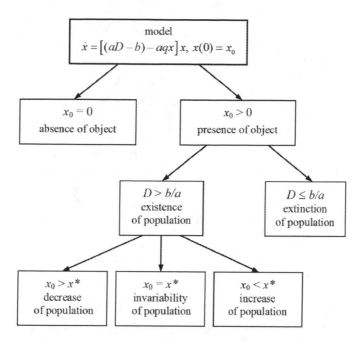

FIGURE 7.3: Variants of the system evolution for the Verhulst model.

Note that the factors restraining an increase in the species number can also be the presence of a competing species consuming the same food, or natural opponents for which this species is itself food. Thus, we come to the study of the coexistence of several biological species. In this case, the decisive role is played by the principle of interaction between species.

Malthus model describes exponential population growth under unlimited resources.
Evolution of a biologic species with limited resources
is described by the Verhulst equation that has an equilibrium position.

2   Biological competition model

Consider two biological species located in a same area. We assume that none of them directly affects the other species. However, consuming the same food, they certainly enter into intense competition between themselves in view of the limited amount of food[12].

Determine a mathematical model of the considered process. Let x_i be an number of i-th species. Obviously, the rate of its change is proportional to the number of this species

$$\dot{x}_i = k_i x_i,$$

where k_i is the growth of the number of i-th species, $i = 1, 2$. Under the unbounded quantity of food, there is no competition between species, and the increase in number and species is equal to its natural increase. By the boundedness of food, the growth of the species decreases. Repeating the arguments of the previous model, we obtain the equations

$$\dot{x}_i = (a_i N - b_i)x_i,$$

where $N = D - (q_1 x_1 + q_2 x_2)$, D is a quantity of food, a_i is a specific increase i-th species, b_i is a mortality of i-th species, q_i is a food consumption of i-th species, $i = 1, 2$.

Thus, the considered phenomenon is described by the following system of differential equations

$$\begin{cases} \dot{x}_1 = \Big[d_1 - a_1\big(q_1 x_1 + q_2 x_2\big)\Big]x_1 \\ \dot{x}_2 = \Big[d_2 - a_2\big(q_1 x_1 + q_2 x_2\big)\Big]x_2 \end{cases}, \tag{7.4}$$

where $d_i = a_i D - b_i$ is the special growth of i-th species. Formulas (7.4) are called the competition equations. This is strikingly reminiscent of the "chemical competition" model and process equations for a two-mode laser (see Appendix), although this is hardly surprising since an analogy between different classes of phenomena was discovered earlier for the corresponding processes with a single state function (see Table 7.1).

Suppose the number x_{i0} of i-th species at the initial time $t = 0$ is known, i.e., we have

$$x_i(0) = x_{i0}, \quad i = 1, 2. \tag{7.5}$$

System of differential equations (7.4) with initial conditions (7.5) is the mathematical model of the process under consideration. In view of the nonlinearity of equations (7.4), a direct solution to the problem is difficult. However, surprisingly, a direct qualitative study of the model can be carried out in the most general way. Particularly, we can determine the equality[13].

$$\frac{x_1(t)^{1/a_1}}{x_2(t)^{1/a_2}} = \frac{(x_{10})^{1/a_1}}{(x_{20})^{1/a_2}} \exp(\theta t), \tag{7.6}$$

where $\theta = d_1/a_1 - d_2/a_2 = b_2/a_2 - b_1/a_1$.

For definiteness, suppose that the inequality $\theta > 0$ holds. This assumption does not violate the generality of the situation, since for the first of the species we have the right to choose the one with a lower b_i/a_i ratio. The coincidence of these relations for different species is unlikely from a practical point of view of particular interest (as can be seen from equality (7.6), in this case its left side does not change over time). Passing to the limit in relation (7.5), we find

$$\lim_{t \to \infty} \frac{x_2(t)^{1/a_2}}{x_1(t)^{1/a_1}} = 0. \tag{7.7}$$

Condition (7.7) is realized either with an unlimited increase in the number of the first species, or when the number of the second one tends to zero. Note, however, that with the growth of the function x_1, starting from a certain moment in time, the inequality $q_1 x_1 + q_2 x_2 > d_1/a_1$. Then, in accordance with equality (7.4), we obtain a negative increase in number of the first species, i.e., its decrease. Thus, in this model, the unlimited increase in the function x_1 is obviously not realized. Consequently, condition (7.7) can be satisfied only when the number of the second form tends to zero.

Let us evaluate the change with time of the number of the first species. Since the function x_2 tends to zero at $t \to \infty$, after a sufficiently large amount of time elapses in this territory, only the first form will remain. Then, with a sufficiently high degree of accuracy, the equation

$$\dot{x}_1 = \big(d_1 - a_1 q_1 x_1\big)x_1,$$

which is equivalent to formula (7.3). Repeating the above reasoning, we establish that over

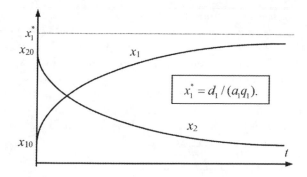

FIGURE 7.4: Change of species number of biological competition model.

time the state of this system will tend to the equilibrium position. The qualitative behavior of the considered system for relatively small initial values of species number is shown in Figure 7.4.

Obviously, the ratio $\theta_i = b_i/a_i$ characterizes the vitality of the species. The results show that in the process of struggle for existence, a less resilient species (one with a high mortality and low growth) dies out, and the number of a more resilient species stabilizes over time at a certain level, providing a balance between natural growth its number and the amount of food available.

Of course, nature does not allow playing the dramatic situation described above. If, in fact, two species risked entering into direct competition between themselves, then over time a more adapted species would certainly supersede the other. A similar result is observed in the case of competition of three or more types[14]. Given the significant duration of the existence of life, we would have long ago come to an outcome at which a more fit look remains. We note, however, that a pronounced specialization is observed in the biological world, which is characteristic of the "ecological niche" model considered in the Appendix[15].

One should consider also the degenerate case corresponding to the equality $\theta_1 = \theta_2$, when both species have equal viability. Moreover, relation (7.6) takes the form

$$\frac{x_1(t)^{1/a_1}}{x_2(t)^{1/a_2}} = \frac{(x_{10})^{1/a_1}}{(x_{20})^{1/a_2}},$$

where, after exponentiation, we can find the function $x_1 = c(x_2)^{a_1/a_2}$, where the constant c is uniquely determined by the initial states of the system. As a result, second equation (7.4) is written as follows

$$\dot{x}_2 = \left\{d_2 - a_2\left[q_1 c(x_2)^{a_1/a_2} + q_2 x_2\right]\right\}x_2.$$

The term in parentheses here is a strictly increasing function of x_2. If initially it is less than the ratio d_2/a_2, then the right-hand side of the last equality is positive. Then the function x_2 increases and will increase as long as the value in square brackets remains positive. Therefore, over time, the second type of number tends to a unique solution of the algebraic equation $q_1 c(x_2)^{a_1/a_2} + q_2 x_2 = d_2/a_2$. A similar result (with a change in the type of monotonicity) is obtained in the case when the term mentioned above initially turns out to be greater than or equal to the value of d_2/a_2. To find the limiting number of the first species, it is sufficient to use the relation established above between the number of species.

Thus, in the case of the same viability of the species, none of them dies out, and the number of each of them eventually takes on a certain value that corresponds to the possibilities of maintaining this species. The crowding out of one species by another does not occur, since in fact in this case the entire population is homogeneous, i.e., two species behave as a whole. If initially the number of one of the species prevails, then its prevalence will continue in the future. Thus, in this case, there is not one outcome of the development of events, but a whole multitude of equilibrium positions. The implementation of a specific equilibrium is determined by the initial state of the system. Variants for the system evolution in the competition model are presented in Figure 7.5.

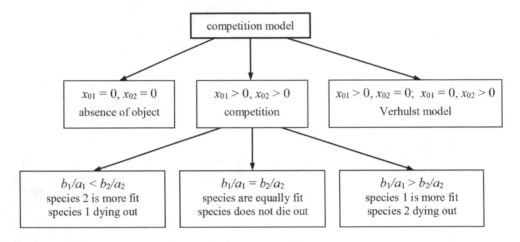

FIGURE 7.5: Variants of system evolution for the competition model.

Task 7.1 *Biological competition*. Based on the numerical solving of system (7.4) with initial conditions (7.5) make the following analysis.

1. Equating to zero the right-hand sides of equations (7.4), determine all equilibrium positions of the system.

2. Calculate the system solution with an arbitrary set of coefficients of equations and relatively small initial species number. Establish the extinction of the weakest species.

3. Leaving the coefficients of the equations unchanged, resume calculations with large initial states of the system. Make sure that the same species dies out again, but the course of the process changes. Interpret the results.

4. Change coefficients of the equations so that the second species dies out.

5. Carry out calculations for the case when the species are equal to each other in vitality. To detect an increase in the population at small initial values of the number of species and a decrease in the population when the initial number of both species is large enough.

Competition model describes the coexistence
of two biological species, consuming the same food.
According to the competition model, the strongest species survives only.

3   Predator–prey model

Consider now a coexistence of two species when the first of them is a food for the second. If only the first species (prey) lived in this environment, then it would have a natural increase of ε_1, which is constant and positive if is assumed that the prey do not lack food. Then, in the absence of predators, an exponential increase in the number of prey would be observed, as was the case in the previously considered model of evolution of the species with an unlimited amount of food. If the second species (predators) exists in isolation, then due to lack of food (i.e., prey), it would have a negative population growth of $-\varepsilon_2$ that is the coefficient of extinction of predators in the absence of prey. The outcome here would be the complete extinction of predators.

These species have serious influence on each other under their coexistence in a limited area. Obviously, the increase in the number of prey should decrease, and the more, the higher the number of predators. This is due to the fact that a greater number of predators need an appropriate amount of food, i.e., the prey. On the other hand, the growth of predators should increase the stronger, the higher the prey number, since in these conditions a greater predators number will be provided with nutritious food. As a result, we get the differential equations

$$\begin{cases} \dot{x}_1 = \left(\varepsilon_1 - \gamma_1 x_2\right)x_1 \\ \dot{x}_2 = \left(\gamma_2 x_1 - \varepsilon_2\right)x_2 \end{cases}, \tag{7.8}$$

where the coefficients γ_1 and γ_2 characterize the changes in the growth of prey and predators due to their natural interaction with each other[16]. Equations (7.8) with the corresponding initial conditions form the well-known **predator–prey** mathematical model.

To study the resulting system, we use a fairly common technique related to the **replacement of variables**[17]. Define the variables $\tau = at$, $u(\tau) = bx_1(t)$, $v(\tau) = cx_2(t)$, where the constants a, b, and c are chosen such that the resulting equations were as simple as possible. As a result, we establish the relations

$$\begin{cases} u' = \left(\varepsilon_1/a - v\gamma_1/ac\right)u \\ v' = \left(u\gamma_2/ab - \varepsilon_2/a\right)v \end{cases},$$

where u' and v' are derivatives of these functions with respect to τ. Determining the parameters $a = \varepsilon_1$, $b = \gamma_2/\varepsilon_1$, $c = \gamma_1/\varepsilon_1$, $m = \varepsilon_2/\varepsilon_1$, we get

$$\begin{cases} u' = u(1 - v) \\ v' = mv(u - 1) \end{cases}. \tag{7.9}$$

There are **Volterra–Lotka equations**. The systems (7.9) and (7.8) have the same meaning. Indeed, the state functions u and v, as well as the independent variable τ differ from the functions x_1, x_2, and t, respectively, by exclusively constant factors. Thus, we are still dealing with the number of species and time, but considered on a different scale.

Obviously, equations (7.8) have two equilibrium positions $u_1 = 0$, $v_1 = 0$, and $u_2 = 1$, $v_2 = 1$. The first of them is trivial, implemented in the absence of both species and does not represent any practical interest. The second position of the equilibrium of the system is much more interesting. Here, the number of newly born prey for some time is compensated by their number eaten by predators at the same time. In turn, birthrate and mortality among predators also coincide. As a result of this, the number of both species does not change over time, i.e., the system is in a state of dynamic equilibrium.

Consider the system out of equilibrium. Note that the number of species is obviously not negative. Equal to zero the initial number of predators with a positive value of the initial number of victims leads to the equation $u' = u$, whose solution increases exponentially. Thus, in the absence of natural adversaries and food restrictions prey multiply unlimitedly. The absence of prey at the initial time allows us to obtain the following equation with respect to the number of predators $v' = -mv$. It follows that in the absence of food, predators gradually die out. For further research, it is sufficient to consider a system with positive initial states.

As can be seen from equations (7.9), there are four different regions in the phase plane, see Figure 7.6. If at a time t the following inequalities hold

$$0 < u(t) < 1, \quad 0 < v(t) < 1, \tag{7.10}$$

then the term at the right-hand side of first equality (7.9) is positive, and the second one is negative. We get the inequalities $u'(t) > 0$, $v'(t) < 0$. Then the function u increases and the function v decreases, see Figure 7.6.

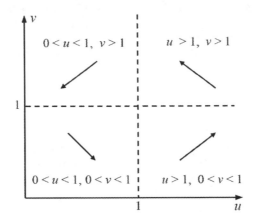

FIGURE 7.6: Directions of system evolution for the predator–pray model.

This property of the system persists all the time until relations (7.10) are satisfied. These conditions can be violated either in the case when during the increase the value of the function u exceeds unity, or when the function v reaches zero as it decreases. However, the approach of v to zero is accompanied by the tendency of its derivative to zero in accordance with second equation (7.9). Thus, the achievement of the function v by zero could only be asymptotic. However, with increasing function u and decreasing v, the derivative u' not only remains positive, but even grows. Therefore, sooner or later, the value of u will exceed unity, and the phase curve enters in a region characterized by inequalities

$$u(t) > 1, \quad 0 < v(t) < 1. \tag{7.11}$$

Under these conditions, the derivatives of both functions are positive, which means that their values will increase with time. This property is observed all the time while inequalities (7.11) are true. Obviously, with the growth of the function v, it will ever exceed the value of unity, and as a result, the following conditions hold

$$u(t) > 1, \quad v(t) > 1, \tag{7.12}$$

In the future, a decrease in the function u is observed with a simultaneous increase in v. The situation will change only if one of relations (7.12) is violated. Naturally, as the function v grows, second of conditions (7.12) cannot be violated. However, as u decreases in the end, it becomes less than unity, and we get the inequalities

$$0 < u(t) < 1, \quad v(t) > 1. \tag{7.13}$$

This means that both given functions decrease. Considering that as the function u tends to zero, its derivative tends to zero as well. We conclude that sooner or later the moment of

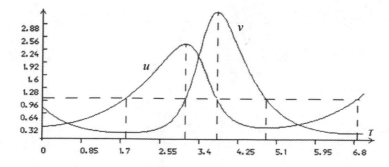

FIGURE 7.7: Solutions of system (7.9) for $u_0 = 0.3$, $v_0 = 0.8$, $m = 2$.

time will come when the function v becomes less than unity for a positive value of u. Thus, relations (7.10) will be fulfilled again, which means that the described process is repeated, see Figure 7.7.

It can be prove that the solutions of equations (7.9) turn out to be periodic functions, see Figure 7.8. In the phase plane, we observe closed curves. We have already encountered a similar type of equilibrium (center) in the study of free undamped mechanical and electrical oscillation. At the same time, the trivial equilibrium position (the origin in the phase plane) possess fundamentally different properties. Obviously, only those states that lie on the coordinate axis v in the phase plane tend to it. Other phase curves can approach the origin for some time, but subsequently inevitably move away from it, see Figure 7.8. Thus, we have an unstable equilibrium, which is a saddle.

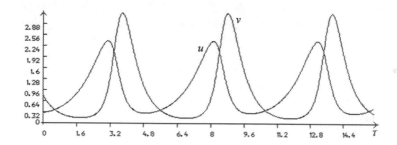

FIGURE 7.8: State functions of the predator–pray model are periodic.

Give the interpretation of the results. Suppose that at the initial time, the numbers of prey and predators are small enough. With a small number of prey, the number of predators is decreasing, and the pray are in a better position due to the small number of natural enemies. As a result, their quantity is increasing. The onset of the second stage is due to the fact that a small number of predators with an increased number of prey will no longer lack food. The number of predators begins to grow, and faster and faster due to the still observed increase in the number of prey. The third stage begins at that moment in time when the natural increase in the number of preys compensated by their destruction by predators, whose number is steadily growing. With a further increase in the predator population, the number of prey begins to gradually decrease, and the rate of increase in the number of predators slows with a decrease in the number of prey. In the end, there comes a time when there is already not enough food for the increased number of predators, as a result of which their number begins to decline. However, the number of prey continues to

TABLE 7.2: Analogy between models of the predator–prey type.

Science	Biology	Chemistry
object of study	predator–prey system	Lotka reaction system
first system element	prey	substance X
second system element	predator	substance Y
state functions	number of species	substance concentrations
first hypothesis	natural increase in prey is positive	reaction A+X→2X
second hypothesis	natural growth in predators is negative	reaction Y→B
third hypothesis	predators feed on prey	reaction X+Y→2Y
influence of the first element on the second one	few preys, few food for predators	little substance X, little formed Y
influence of the second element on the first one	few predators, few enemies for prey	little substance Y, little consumed X
trivial equilibrium	no animals, no population	no substances, no reactions
non-trivial equilibrium	there are as many prey as needed to predators	substance X and Y formed as much as consumed

decrease, as there are still too many predators. The rate of decrease in prey decreases due to a decrease in the number of predators. Then the number of prey is reduced to a certain minimum value, after which, due to the continuing decrease in the number of predators, the number of prey begins to increase. This indicates the onset of a new cycle of the considered process with a return to the initial stage[18].

It was noted earlier that the predator prey model and the Lotka chemical reaction are described by the same equations. In this regard, the above results can be given a chemical interpretation, see Table 7.2.

Task 7.2 *Predator–prey model.* Based on the numerical solving of the system (7.8) with corresponding initial conditions make the following analysis.

1. Make calculations with an arbitrary set of parameters. Establish a periodic change in the number of both species.

2. Set the following values of the following system parameters: $x_{10} = 2$, $x_{20} = 3$, $\varepsilon_1 = 3$, $\gamma_1 = 1$, $\varepsilon_2 = 2$, $\gamma_2 = 1$. Make sure that the system is in equilibrium.

3. Analyze the impact of the initial number of prey on the results, choosing the values of the remaining parameters the same as in the previous version of the account. Set successively the following values of the parameter x_{10}: 0.3, 0.8, 1.3, 2, 2.5, 3, 4, 6. Draw a graph of the dependence of the oscillation amplitude and the oscillation period on the initial state of the system. Explain the results.

Predator–prey system is described by the Volterra–Lotka equations.
number of predators and prey are periodical functions here.

4   Symbiosis model

In principle, three general forms of coexistence of two biological species are possible, unless, of course, the trivial case is excluded when the species do not affect each other. First, they can have a negative effect on each other, which corresponds to competition. Secondly, one of the species can have a positive effect on the second, and the second one has a negative effect on the first. It is this case that is realized in the predator–prey model. Finally, a third variant of the coexistence of species is also possible, which corresponds to the positive influence of both species on each other, which corresponds to the phenomenon of *symbiosis*[19].

Consider the interaction of two species in symbiotic conditions, when each of the species has a beneficial effect on the second species. It is assumed that the i-th species dies out in the absence of another species with a certain rate ε_i, $i = 1, 2$. At the same time, the greater the number of another species, the higher the growth of this species. Denoting by γ_i the coefficient of influence of the j-th type on the i-th one, we define the following type of growth number $k_i = -\varepsilon_i + \gamma_i x_j$, $i \neq j$. As a result, the mathematical model of the system under consideration is characterized by differential equations

$$\begin{cases} \dot{x}_1 = (\gamma_1 x_2 - \varepsilon_1)x_1 \\ \dot{x}_2 = (\gamma_2 x_1 - \varepsilon_2)x_2 \end{cases}, \tag{7.14}$$

which differ from relations (7.8) only in the sign at first equation. However, these models have significantly different properties.

Equating the right-hand sides of equations (7.14) to zero, we establish two equilibrium positions of the system, one of which is zero, and the second corresponds to positive values of the state functions. This is also characteristic of system (7.8). As in the case considered earlier, the phase plane (more precisely, the first quadrant corresponding to the positive values of both coordinates) is divided into four regions (see Figure 7.9).

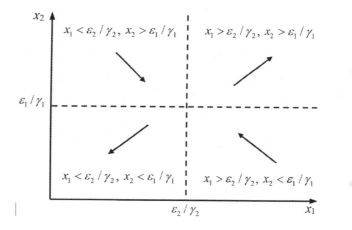

FIGURE 7.9: Directions of the system evolution in the symbiosis model.

Suppose both initial states are small enough such that

$$x_1 < \frac{\varepsilon_2}{\gamma_2}, \ x_2 < \frac{\varepsilon_1}{\gamma_1}. \tag{7.15}$$

Then, in accordance with equalities (7.14), the derivatives of both functions are negative, which means that the functions themselves decrease. Thus, in subsequent time instants, inequalities (7.15) will be fulfilled with even greater justification. Consequently, derivatives will remain negative. Thus, over time, the functions will steadily decrease and tend to zero, i.e., to a stable equilibrium position (see Figure 7.10, curve 1).

Let, on the contrary, both initial values be so large that the following conditions hold

$$x_1 > \frac{\varepsilon_2}{\gamma_2}, \ x_2 > \frac{\varepsilon_1}{\gamma_1}. \tag{7.16}$$

According to equalities (7.14), the derivatives of both state functions are positive, which means that the functions themselves increase (see Figure 7.8). Consequently, conditions

(7.16) will also be satisfied at subsequent instants of time, which will lead to an increase in the derivatives of the function under consideration. We observe their exponential growth (see Figure 7.10, curve 2).

Finally, one of the functions may turn out to be large enough and the other small enough so that the inequalities

$$x_1 > \frac{\varepsilon_2}{\gamma_2}, \ x_2 < \frac{\varepsilon_1}{\gamma_1}. \tag{7.17}$$

Then the derivative of the first of the state functions is negative, and the derivative of the second of them is positive. Thus, the first function decreases, and the second increases. This happens as long as conditions (7.17) are satisfied (see Figure 7.9). Further, two possible events are possible. Perhaps, the function x_1, decreasing, becomes smaller than the ratio ε_2/γ_2, while the function x_2, increasing, has not yet reached a value ε_1/γ_1. Thus, conditions (7.15) will be satisfied, which means that both functions will tend to zero (see Figure 7.10, curve 3). However, another situation is also possible when the function x_2, increasing, exceeds the value ε_1/γ_1 earlier than the function x_1, decreasing, reaches a value ε_2/γ_2. Thus, conditions (7.16) are valid, which means that both functions increase unlimitedly (see Figure 7.10, curve 4).

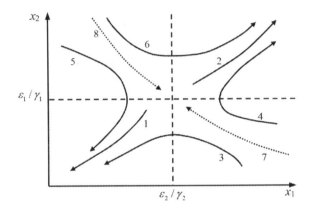

FIGURE 7.10: Phase curves for the symbiosis model.

If the first of the considered functions is sufficiently small, and the second is sufficiently large, i.e., since both inequalities (7.17) are replaced by opposite ones, a picture similar to the previous one is observed. In this case, the first and second functions are interchanged (see Figure 7.10, curves 5,6). Based on the results obtained, we see that the non-trivial position of the equilibrium is not stable, in particular, this is a saddle[20].

The obtained results can be given the following interpretation. The small initial number values of both species correspond to the extinction of each of them, since the available number of individuals in this population is not enough to support another population. In the presence of sufficiently large values of species number, both of them are in favorable conditions. Therefore, in the absence of any limiting factors, the number of both species grows unlimitedly. If, initially, the number of one of the species is relatively large, and the number of another species is quite small, then the first of the species is in unfavorable conditions, and the second in favorable. As a result, the number of the first type is decreasing, and the second is growing. Two possible out-comes. If the number of the first species decreases earlier than the number of the second species, then over time both species will find themselves in unfavorable conditions, and the entire population will die out. Otherwise, the population is growing unlimitedly[21].

Task 7.3 *Symbiosis model.* Based on the numerical solving of problem (7.14) with corresponding initial conditions, make the following analysis.

1. Find the non-trivial equilibrium position of system (7.14). Choosing it as the initial state of the system, make sure that the state functions do not change with time.

2. Choose the initial state of the system close enough to the non-trivial equilibrium. Make sure that this position is not stable.

3. Calculate the system for small initial states of the system. Make sure that both species die out.

4. Carry out calculations for large initial states of the system. Ensure that the number of both species increases indefinitely.

5. By choosing a large number of the first species and a small number of the second one, find the case when the number of both ultimately increases unlimitedly, although one of the state functions does not change monotonously.

6. Choose the initial states of the system so that both species die out, but the number of one of them did not change monotonously.

Task 7.4 *System of biological species.* Describe a mathematical model of a system of biological species in which the first three species compete with each other, the fourth and fifth are in a state of symbiosis, and the fifth species feeds on the third.

> *The symbiosis model describes the coexistence of two species,*
> *having a positive effect on each other.*
> *According to the symbiosis model, species either die out or multiply indefinitely.*

Direction of further work. In the next chapter, we will make sure that the mathematical models of population dynamics discussed above also describe some economic systems.

Appendix

The biological competition model has natural physical and chemical analogues, and the Volterra–Lotka equations underlying the predator–prey model also describe changes in the concentrations of the corresponding system of autocatalytic reactions and changes in crop productivity and soil fertility[22]. If in the model of biological competition it is assumed that each species, while consuming general food, gives some preference to its type of food, then the model is an "ecological niche", which allows the coexistence of both species[23]. Then a model of epidemiology is considered. Another generalization of the competition model is the antibiotic resistance model, which describes the addiction of microorganisms to the action of the antibiotic.

1 Models of chemical and physical competition

As we already know from the previous chapter, the system of chemical reactions

$$A+X \xrightarrow{k_1} 2X, \quad A+Y \xrightarrow{k_2} 2Y, \quad 2X \xrightarrow{k_3} B,$$

$$2Y \xrightarrow{k_4} C, \quad X+Y \xrightarrow{k_5} D.$$

under the condition of constant supply of the starting substance A is described by the equations

$$\dot{x} = \left(k_1 a - 2k_3 x - k_5 y\right)x, \quad \dot{y} = \left(k_2 a - k_5 x - 2k_4 y\right)y.$$

TABLE 7.3: Characterization of predator–prey models.

Denotation	Chemistry	Biology	Agriculture
x_1	concentration of substance 1	prey number	soilfertility
x_2	concentration of substance 2	predator number	productivity of land
ε_1	formation rate of substance 1	prey growth	fertility restoration
ε_2	formation rate of substance 2	predator decrease	productivity decrease
γ_1	conversion rate of substance 1 to substance 2	loss of prey due to being eaten by predators	fertility depletion with increasing productivity
γ_2	conversion rate of substance 2 to substance 1	predator growth due to consumption of prey	productivity increase due to minerals in the soil

If $k_5 = 2\sqrt{k_3 k_4}$, we get the system that is analogical to (7.4). Based on the above analysis, we conclude that over time one of the intermediate substances (which one, depends on a combination of system parameters) will fully react, and the second will reach a certain limit value.

In the previous chapter, it was noted that an equation of type (7.4) also describes the emission of photons in a two-mode laser. The results obtained indicate that over time one of the laser modes will certainly suppress the other. Thus, over time, the laser will emit photons of the same wave-length. Thus, we are convinced that the previously established analogy between phenomena of a physical, chemical and biological nature turns out to be quite deep[24].

2   Fluctuations in yield and fertility

Consider some agricultural crops. Its yield largely depends on the content of minerals in the soil. In turn, soil fertility is associated with crop yields, because plants consume nutrients in the soil. We assume that in the absence of minerals, the crop yield x_2 decreases with a speed of ε_2 and in the absence of plants, the field restores its fertility 1 with a speed of ε_1. The higher the crop yield, the more minerals are consumed, which leads to a proportional (with a proportionality coefficient γ_1) decrease in soil fertility. On the other hand, the increase in crop yield is directly proportional (with a proportionality coefficient of γ_2) to soil fertility. As a result, we again come to the well-known equations (7.8)

$$\dot{x}_1 = (\varepsilon_1 - \gamma_1 x_2)x_1, \quad \dot{x}_2 = (\gamma_2 x_1 - \varepsilon_2)x_2.$$

Knowing the properties of this system, we conclude that crop yields and soil fertility periodically change over time. The increase in yield over time leads to depletion of the soil, which in turn leads to a decrease in yield. As the number of plants growing on the field decreases, a gradual restoration of the mineral content in the soil occurs, which over time leads to an increase in yield. A comparative analysis of the chemical, biological, and agricultural models described by equations (7.8) is given in Table 7.3.

3   Ecological niche model

As we know, the coexistence in a limited territory of two different species that consume the same food will inevitably lead to the extinction of a less adapted species. Nevertheless, in nature there is an innumerable variety of living organisms. This indicates the presence of some mechanism, as a result of which the coexistence of several species is possible. In particular, the survival of both species could be expected in the case when the species have some preference for any type of food, which corresponds to the ecological niche model. It is characterized by the following system of differential equations

$$\dot{x}_1 = \big(\varepsilon_1 - \alpha_{11}x_1 - \alpha_{12}x_2\big)x_1, \quad \dot{x}_2 = \big(\varepsilon_2 - \alpha_{21}x_1 - \alpha_{22}x_2\big)x_2,$$

where x_i is an abundance of i-th species, $\varepsilon_i = \big(a_{i1}D_1 + a_{i2}D_2 - b_i\big)$ is a specific abundance of i-th species, $\alpha_{ij} = \big(a_{i1}q_{j1} + a_{i2}q_{j2}\big)$ is a specific coefficient of consumption of j-th type by i-th species, D_j is a quantity of j-th food, a_{ij} is a specific growth of i-th species determined by the consumption of j-th food, q_{ij} is a consumption of j-th food by i-th species, $i, j = 1, 2$. According to these results, each species has a positive natural increase in abundance, but it decreases due to limited food. There are two types of food that are consumed by both species, but to a different extent, determined by the coefficients α_{ij}. The amount of food consumed is considered directly proportional to the abundance of the species.

The analysis of this model is carried out in a subsequent chapter in economic interpretation. In particular, it will be shown (in relation to the biological case) that, under conditions

$$\frac{\varepsilon_1}{\alpha_{11}} > \frac{\varepsilon_2}{\alpha_{21}}, \quad \frac{\varepsilon_1}{\alpha_{12}} < \frac{\varepsilon_2}{\alpha_{22}},$$

that mean that each species prefers its type of food, both competitors can coexist without harming each other. This explains the use of the name "ecological niche" for this model. Such results shed some light on the extremely high degree of specialization observed in wildlife[25].

Task 7.5 *Ecological niche model.* Based on the numerical solving of the problem make the following analysis.

1. Choose the parameters of the system so that the first species dies out for any initial states of the system.

2. Choose the system parameters so that the first species dies out for any initial states of the system.

3. Choose the system parameters in such a way that, in some initial states of the system, the first species dies out, and for others, the second.

4. Choose the system parameters so that both species do not die out.

4   SIR model for spread of disease

Consider a mathematical model of epidemiology[26]. In this model, a certain population is considered, which is divided into three groups of individuals. There are susceptible, i.e., healthy people who can get sick, infected, i.e., sick, and recovered, who are considered immunized, which means that they can no longer get sick[27]. The **SIR model** is characterized by the following system of differential equations

$$\dot{S} = -\beta N^{-1}IS, \quad \dot{I} = \beta N^{-1}IS - \gamma I, \quad \dot{I} = \gamma I,$$

where S, I, and R are the numbers of susceptible, infected, and recovered, N is the total population, i.e., the sum of individuals of all three of the above groups, β is a parameter characterizing the intensity of contacts between individuals, and γ is a parameter characterizing the intensity of recovery of infected individuals[28].

By the first of the considered equations, the number of healthy individuals decreases in proportion to both the number of healthy individuals and the number of infected, since the sick infect susceptible individuals. According to the second equation, the number of infected increases due to infected of healthy people and decreases due to recovery of patients. Finally, the number of those who have recovered increases in proportion to the number of infected. The above equations are supplemented with initial conditions.

Task 7.6 *SIR model*. Make the following analysis.

1. Adding all three equations, make sure that the sum of all three groups of individuals, i.e., the size of the population N does not change over time.

2. Establish the equilibrium position of the system. Analyze the results obtained.

3. Solving the numerically reduced equations with initial conditions, make sure of the validity of the conclusions made in the previous paragraphs.

4. Using a computer experiment, establish the dependence of the maximum value of infected on the initial number of infected, as well as the intensity of contacts and the intensity of recovery.

5. Using a computer experiment, establish the dependence of the time to reach the maximum number of infected (the peak of the epidemic) on the above parameters.

5   Antibiotic resistance model

Consider a mathematical model of population dynamics associated with medicine[29]. There is an organism susceptible to infection by certain microorganisms. Their population, once in a favorable environment, begins to grow sharply. After some time, an antibiotic is used to treat the infected body. As a result, the number of microorganisms is sharply reduced. However, over time, microorganisms adapt to the action of the antibiotic. This is due to the appearance of mutants that are not sensitive to this antibiotic. As a result, the growth of the mutant population begins. This phenomenon is called *antibiotic resistance*.

In the considered system we have two types of bacteria that are the initial population and mutants. Since both of them develop in the same organism, it can be assumed that competition is at the core of their coexistence. However, unlike the system (7.4), it is necessary to take into account mutations, i.e., the descendants of one type of bacteria may be bacteria of another type. Finally, the effect of the antibiotic should be taken into account, which has a sharply negative effect on the main population and does not affect the mutants.

Try to determine the mathematical model of the considered phenomenon. We will base on the competition model, which is described by the equations

$$\dot{x}_1 = \Big[d_1 - a_1\big(q_1 x_1 + q_2 x_2\big)\Big]x_1, \quad \dot{x}_2 = \Big[d_2 - a_2\big(q_1 x_1 + q_2 x_2\big)\Big]x_2,$$

where x_1 is an number of sensitive bacteria, x_2 is an number of resistant bacteria, and all coefficients have the same meaning as before.

Further, it should be noted that resistant bacteria are obtained as a result of mutations in the main (first) population of the bacteria[30]. Obviously, the more bacteria of the first type there are, the more bacteria of the second type will appear due to mutations. As a result, the second of the considered equations must be replaced by the following relation

$$\dot{x}_2 = \Big[d_2 - a_2\big(q_1 x_1 + q_2 x_2\big)\Big]x_2 + b_1 x_1, \tag{7.18}$$

where parameter b_1 characterizes the frequency of emergence of resistant bacteria due to mutations of sensitive bacteria. If we assume that in the process of mutations resistant bacteria can lose the property of resistance, then the first equation should be written as

$$\dot{x}_1 = \Big[d_1 - a_1\big(q_1 x_1 + q_2 x_2\big)\Big]x_1 + b_2 x_2.$$

We have yet to consider the effect of the antibiotic. We assume that the antibiotic destroys the first type of bacteria and does not have any effect on the second type[31]. In this regard, we add the negative term to the last equation. As a result, we get

$$\dot{x}_1 = \left[d_1 - a_1 \left(q_1 x_1 + q_2 x_2 \right) \right] x_1 + b_2 x_2 - c(x_1)^\theta, \tag{7.19}$$

where c and θ are positive parameters.

Thus, the considered process is characterized by equations (7.18), (7.19) with natural initial conditions. For a more accurate description of the process, you can make some restrictions on the parameters of the system.

1) The initial time corresponds to the initial stage of infection. In this regard, the initial number of the main bacterial population is quite small, while the number of the second type of bacteria (mutants) can be considered equal to zero.

2) At the initial stage of infection, the disease has not yet been identified, and treatment has not begun. In this regard, the value of c can be considered time-dependent (it can be interpreted as the accepted dose of the antibiotic), which is equal to zero at the early stage of infection.

3) The coefficients b_1 and b_2 are quite small, since mutations are quite rare. Their influence becomes noticeable when the number of the main population of bacteria is relatively large. Then, at the early stage of infection, in the absence of the initially second type and for small values of number and the first type of bacteria, the second equation of the system can be neglected. Under these conditions, the first equation of the system actually reduces to the Malthus equation, see Section 1. Consequently, the number of the bacterial population exponentially grows, i.e., the disease worsens.

4) At the stage of exacerbation of the disease, the number of the main population of bacteria is large enough. There is already a shortage of food, i.e., the first equation is close to the Verhulst equation. In addition, the influence of the last term in equation (7.18) becomes significant, i.e., due to mutations, bacteria of the second type appear. Their effect on the system as a whole is insignificant not only because their number is much smaller than the number of the first type of bacteria, but also because under natural conditions (in the absence of an antibiotic) the main population of bacteria is more viable. In the context of the competition model (see Section 2), this corresponds to the inequality $d_1/a_1 > d_2/a_2$.

5) With a large number and a population of bacteria, antibiotic treatment begins. Therefore, the quantity c is already non-zero. On the right-hand side of equation (7.19), the last term is non-zero. The term characterizing the birthrate in the equation is linear with respect to x_1. Lack of food is characterized by a quadratic term with respect to x_1. In order for the influence of the antibiotic to become decisive, the exponent θ is chosen more than two. As a result, the value on the right-hand side of equation (7.19) becomes negative, and is sufficiently large in magnitude, since the number of this type of bacteria is very large. Then the number of bacteria is sharply reduced.

Now we can make sure that the above model really describes the phenomenon of antibiotic resistance. Perhaps the bacterial number due to the action of the antibiotic fell to a fairly small value so that the body can be considered cured. However, in conditions of a significant reduction in the number of bacteria in the main population (result of treatment), an already insignificant population of mutant bacteria is found to be in favorable conditions. Their population is growing, since the antibiotic has no effect on it, and the first type of bacteria (competing species) is suppressed by the antibiotic. Number of resistant mutants is growing, and over time and over time, it is this type of bacteria that becomes dominant. Thus, the effectiveness of antibiotic gradually decreases, and over time its further use does not make sense[32].

Notes

[1] The problems of population dynamics are considered, for example, in [7], [45], [85], [244], [287], [342], [355], [362], [372]. About other mathematical models of biology, see [7], [50], [72], [73], [82], [134], [244], [245], [289], [326], [355], [348], [374].

[2] In Chapter 11, it will be shown that the tasks of the struggle for the existence of biological species, taking into account their migration over a certain territory, are characterized by partial differential equations. Mathematical models of various biological systems described by partial differential equations are considered in [85], [245], [244]. We also note discrete models (see Chapter 17, and also [45], [85], [244], [287], [310]) and stochastic models (see Chapter 18, and also [7], [57], [244], [287], [342], [355] of biological systems).

[3] This equation, both for positive and negative values of the parameter k, was considered in Chapter 5.

[4] The number of the species cannot be negative, and the case $x_0 = 0$ is not of special interest due to its triviality.

[5] Chapter 17 discusses a discrete analogue of the Malthus model. Chapter 18 explores stochastic analogs of the Malthus model, where birth and death of an individual are random events. In addition, the value of the increase in the number of a species may turn out to be a random variable.

[6] In fact, we will encounter the Malthus model when describing the phenomenon of antibiotic resistance at the stage of infection of the body, see Appendix. Chapter 17 will discuss a discrete analogue of the Malthus model.

[7] We have already noted these properties in Chapter 5.

[8] Similar to the bimolecular reaction equation considered in the previous chapter, we have the equation with quadratic nonlinearity, i.e., with an equation like Riccati. Another version of this equation arises when describing the ***body falling*** taking into account air resistance, where the resistance force is proportional to the square of the velocity. Similarly, we described the movement of the probe in Chapter 1. Indeed, according to Newton second law, we have $m\ddot{x} = mg - k\dot{x}^2$. Then the velocity of the body $v = \dot{x}$ satisfy the equation with quadratic nonlinearity $m\dot{v} = mg - kv^2$. A stochastic model of the evolution of a biological species with quadratic nonlinearity is considered, for example, in [288].

[9] In the next chapter, we will be convinced that the Verhulst equation can also be given an economic interpretation. Chapter 17 will discuss the discrete counterpart of the Verhulst model. It is interesting that under certain conditions the discrete Verhulst model has qualitatively different properties, in particular, to describe the chaotic behavior of the system.

[10] Note that we were able to establish the most important qualitative properties of the considered without using its analytical solution. We will use this technique in the future, since, unfortunately, the fact of finding an analytical solution of nonlinear differential equations should be recognized as an extremely rare event and is true evidence of the relative simplicity of the system under consideration. It remains, however, the possibility of an approximate solution to the problem. It should be noted that almost all quite serious mathematical models are studied just numerically, and with the use of computer technology. However, the possibilities of numerical methods for analyzing the general properties of the mathematical model and studying the effect on the system of the various values included in it are clearly limited. In the best case, with some degree of accuracy we find a solution to the problem on a fixed time interval for a specific set of input parameters. At the same time, we do not receive any information about what will happen to the system subsequently or at other values of the parameters. One can, of course, carry out the calculations repeatedly. This is usually done for lack of something more acceptable. However, there is absolutely no guarantee that we will not miss the most interesting events, especially since many nonlinear systems can qualitatively change the properties of the solution with a slight change in the parameters. It should also be borne in mind that the observed non-standard behavior of the solution is often very easy to confuse with all kinds of malfunctions of the numerical algorithm. In this regard, when analyzing models, it is advisable to use a reasonable combination of qualitative and quantitative methods, if, of course, this is possible, but this is not always possible.

[11] The results obtained have a natural meaning from an environmental point of view. Each territory is able to feed a very specific number of individuals of this species. This is especially true, for example, for nature reserves focused on the preservation of a particular species. Mathematical models of ecology are considered, for example, in [68], [173], [225], [244], [266], [267], [272], [327], [355].

[12] Competition of species can occur not only for food, but also for the habitat. However, in the mathematical analysis of the system, the subject of biological competition does not play a fundamental role.

[13] From the equalities (7.4) it follows

$$d_i - a_i(q_1 x_1 + q_2 x_2) = \frac{\dot{x}_i}{x_i} = \frac{d}{dt} \ln x_i, \ i = 1, 2.$$

Therefore, we get

$$\frac{1}{a_1}\frac{d}{dt}\ln x_1 = \frac{d_1}{a_1} - (q_1 x_1 + q_2 x_2), \quad \frac{1}{a_2}\frac{d}{dt}\ln x_2 = \frac{d_2}{a_2} - (q_1 x_1 + q_2 x_2).$$

Subtracting the second from the first equality, we have

$$\frac{d}{dt}\left(\ln x_1^{1/a_1}\right) - \frac{d}{dt}\left(\ln x_2^{1/a_2}\right) = \theta,$$

where $\theta = d_1/a_1 - d_2/a_2 = b_2/a_2 - b_1/a_1$. Thus, we exclude the nonlinear term, which is the general difficulty of the system (7.4). Transform the previous formula to the form

$$\frac{d}{dt}\left[\ln \frac{(x_1)^{1/a_1}}{(x_2)^{1/a_2}}\right] = \theta.$$

Integrate this equality, using initial conditions (7.5). We get

$$\frac{x_1(t)^{1/a_1}}{x_2(t)^{1/a_2}} = \frac{(x_{10})^{1/a_1}}{(x_{20})^{1/a_2}}\exp(\theta t).$$

In Chapter 8, we give a qualitative analysis of the system of equations (7.4) (naturally, with a different interpretation of the model) without using their analytical transformation.

[14] Competition of an arbitrary number of biological species, by analogy with system (7.4), is described by the equations

$$\dot{x}_i = \left(d_i - a_i \sum_{j=1}^{m} q_j x_j\right)x_i, \quad i = 1,...,m.$$

Obviously, we have

$$\frac{d_i}{a_i} - \frac{1}{a_i}\frac{d}{dt}\ln x_i = \sum_{j=1}^{m} q_j x_j, \quad i = 1,...,m.$$

Suppose the index j corresponds to the species in which the ratio d_i/a_i takes on the greatest value. Then, by analogy with relation (7.6), the equality

$$\frac{(x_j)^{1/a_j}}{(x_i)^{1/a_i}} = \frac{(x_{j0})^{1/a_j}}{(x_{i0})^{1/a_i}}\exp\left[\left(\frac{d_j}{a_j} - \frac{d_i}{a_i}\right)t\right], \quad i = 1,...,m, \ i \neq j.$$

It follows that the i-th species dies out over time. Thus, the coexistence of many species consuming the same food leads to the survival of only one of the species that has proven to be the most viable. Similar results can be obtained for analogues of the model under consideration, which relate to other subject areas. In particular, for a multi-mode laser, one of the modes of all the others is suppressed regardless of their number, see Appendix.

[15] Unfortunately, the conditions for applicability of the biological competition model can be artificially created by a "Master of nature", having decided by frivolity to move species from one place to another. A similar situation developed, for example, in Australia, where rabbits were brought, which became rivals to Australian sheep. Soon it became clear that rabbits are significantly superior to sheep in fertility. The consequences were not slow to affect.

[16] We have already considered similar equations in Chapter 5 when describing a specific system of chemical reactions. In particular, equations (7.8) with $\gamma_1 = \gamma_2$ (a decrease in the number of prey and an increase in the number of predators by eating the first by the second equally), up to the physical meaning of the quantities included in them, describes the reaction system

$$A{+}X \xrightarrow{k_1} 2X, \quad X{+}Y \xrightarrow{k_2} 2Y, \quad Y \xrightarrow{k_3} B.$$

These equations has also other interpretations, see Appendix and next chapters. Chapter 21 discuss the problem of determining the coefficients of the Volterra–Lotka equations from the results of measuring the number of species.

[17] The ***transition to dimensionless variables*** made it possible to significantly reduce the number of system parameters and, therefore, simplify the procedure for a qualitative study of the problem under consideration. If the initial formulation of the equation of state included four coefficients, then in the transformed form the model contains a single parameter m, which does not have a significant effect on the qualitative behavior of the system. A serious simplification of the equations of state and a reduction in the number of system parameters explains the rather high popularity of this technique. This is the basis if the similarity theory, see Chapter 11.

[18] The results obtained have a certain meaning from the standpoint of ecology. Predators in the absence of victims cannot exist at all. On the other hand, the absence of predators leads to a sharp increase in the prey

population, which also has negative consequences, causing a disturbance in the ecological balance. However, the coexistence of predators and victims gives this system a certain stability. The presence of predators prevents the unwanted growth of the prey population, and the presence of prey provides the possibility of the existence of a predator population.

[19] The phenomenon of symbiosis is widespread in wildlife, see [77]. For example, insects, pollinating plants, feed on their nectar. Sea anemone protects hermit crab, which moves sea anemone from place to place. Ants protect aphids by feeding on its secretions.

[20] In the last two cases, in principle, a situation is possible when, for example, when condition (7.17) is fulfilled, as the value of the function x_1 approaches to the value ε_1/γ_1 the function x_2, increasing, approaches the ratio ε_2/γ_2. Thus, the expressions on the right-hand sides of both equalities (7.14) tend to zero, which means that the derivatives of both functions also tend to zero. As a result, the system tends to a non-trivial equilibrium position (see Figure 7.10, curve 7). This situation is characteristic of the saddle: there are two phase curves (the second corresponds to the case when both conditions (7.17) are replaced by opposite inequalities, see Figure 7.10, curve 8), tending to this equilibrium position.

[21] In the considered model, with a large number of both species, their steady growth occurs, which is not in good agreement with the real situation. Inhibition of growth in the number of species can occur due to limited food. In particular, it can be assumed that the decrease in growth depends on the number of species quadratically, and each species eats its own food, we come to the following equations

$$\dot{x}_1 = \left(\gamma_1 x_2 - \varepsilon_1 - \beta_1 x_1^2\right)x_1, \quad \dot{x}_2 = \left(\gamma_2 x_1 - \varepsilon_2 - \beta_2 x_2^2\right)x_2,$$

where the coefficients β_1 and β_1 characterize food intake. Choosing the quadratic nature of the dependence in the expression describing the decrease in the increase in the number of species due to lack of food with a linear dependence in the term characterizing symbiosis, we emphasize that with a sufficiently large number of species, the negative effect of lack of food will prevail over the positive effect of symbiosis. If the numbers of both species are quite small, then the restrictions on the amount of food can be neglected, and the behavior of the system is similar to the previous case, i.e., the entire population is dying. If the number of at least one of the species is relatively large, then the number of the second species begins, and over time, food shortages begin to affect. In this case, either a lack of food and a low number of both species will lead to their complete extinction, or stabilization of their number is possible.

[22] Mathematical models in agriculture are considered in [104].

[23] A niche model is considered in more detail in a subsequent chapter in its economic interpretation.

[24] In subsequent chapters, we will be convinced that the equations considered in this chapter can be given a completely different interpretation.

[25] Naturally, the specialization of species can be carried out not only in food, but also in the habitat.

[26] Mathematical models of epidemiology are discussed, for example, in [28], [46], [74].

[27] Among other models of epidemiology, we note the **SEIR model**, in which the group of exposed individuals is additionally distinguished. It takes into account that the disease has a certain incubation period, i.e., a person who was in contact with a sick person, not immediately getting sick. In addition, some of the sick die. Chapter 17 will describe one discrete model of epidemiology, which, in addition, considers different categories of patients.

[28] Chapter 18 discusses the case when the parameters β and γ included in the SIR model are random values.

[29] Mathematical models of medicine are considered, for example, in [11], [23], [32], [72], [226], [281], [321], [310]. The antibiotic resistance considered model is provided at [317].

[30] In principle, various mutations occur in nature. However, we are only interested in the sensitivity of bacteria to the action of an antibiotic. Thus, we attribute all bacteria with these properties to the first type, and not those with the second type.

[31] Thus, we are considering a bactericidal antibiotic. There exists also bacteriostatic antibiotics that do not kill bacteria, but inhibit their birthrate. To describe such antibiotics, one cannot add an additional negative term to the equation, but modify the parameter d_1, which is responsible for the birth rate of the first type of bacteria. In particular, if we divide it into a certain power x_1, more than unity, then the model will describe the decrease in the birthrate of sensitive bacteria under the influence of an antibiotic. You can also consider antibiotics of a mixed (both bactericidal and bacteriostatic) action, introducing both modifications at the same time into the first equation, see [317].

[32] If the body has not recovered before the moment when the number of resistant bacteria became noticeable, then the tactics of treatment should be changed. You can try to choose another antibiotic that acts on the currently prevailing type of bacteria. In this way, the patient can be cured, although it is possible that over time mutants will appear that are not sensitive to the action of a new antibiotic. By the way, if you stop treatment at a stage when the initial bacterial population has not declined enough (the disease is

not cured), then we return to the action of the competition model. Then the bacterial population increases in number and crowds out mutants, a less viable lack of antibiotic.

Chapter 8

Mathematical model of economics

Almost all the equations obtained in the previous chapter can be interpreted economically. In this chapter, we study simple mathematical models of economic systems[1] described by ordinary differential equations[2]. Mainly considered are two classes of simple economic processes associated with the relationships of various economic entities and with some forms of market relations. The analysis of the first class of models begins with the study of one firm that produces goods in the presence of limited demand. The competition model already considers two competing firms in conditions of limited demand for manufactured products. If they produce the same product, then the weaker firm goes broke over time. If each of them focuses on the release of its goods, then firms can coexist. This case is described by the niche model. The Appendix also considers models of economic cooperation and the "racketeer–entrepreneur", which are economic analogues of the symbiosis and predator–prey models discussed in the previous chapter.

As state functions of the second class of models, the price of the goods, incomes of the population and the volume of output are selected. We study the change in these characteristics over time depending on the form of economic management. The models of the free and monopolized market, as well as (in the Appendix) the inflation model, which is characteristic for the total state regulation of the economy, are considered. It turns out that fluctuations in prices, population incomes, and output can be described by the Volterra–Lotka equations that we have already encountered. In addition, the Appendix presents an extremely important Solow model of economic growth.

Lecture

1 One company evolution

Let us first consider the simplest economic system represented by unique firm. The process under study is characterized by the **working capital** of firm x, which varies with time. It is assumed that the entire amount of money received from the sale of the manufactured goods is invested in production. The rate of capital change is determined by the relation between the income and expenses of the company. Thus, we obtain the relation

$$\dot{x} = A - B,$$

where A and B characterize, respectively, the income and expenses of the company per unit time. Their values, apparently, can be considered proportional to the values of capitals. The larger the capital of the company, the greater changes can occur over a fixed period of time. As a result, we obtain the relations $A = ax$, $B = bx$, where the positive constants a and b are process parameters and characterize the growth of the firm's income and expenses per unit time. Therefore, the considered process is described by the differential equation

$$\dot{x} = kx, \tag{8.1}$$

where the coefficient $k = a - b$ is called **capital gains**, it characterizes the change in the capital of the company per unit time and can take both positive and negative values.

DOI: 10.1201/9781003035602-8

TABLE 8.1:   Interpretation of the Verhulst equation.

Characteristic	**Physics**	**Chemistry**	**Biology**	**Economics**
object	single-mode laser	reaction system A+X ⟶ 2X X ⟶ B	species evolution with bounded food	firm with bounded demand
state function	photon number	concentration of substance X	species abundance	capital of firm
restriction	number of excited atoms	substance quantity A	quantity of incoming food	level of demand
state function growth	atomic emission of photons	substance X synthesis	birthrate of the species	expansion of production
state function decrease	leaving of photons from the system	substance X decay	species mortality	production decline

Equation (8.1) corresponding to the **Malthus model** is considered with the initial condition

$$x(0) = x_0, \tag{8.2}$$

where the start-up capital x_0 is the parameter of the problem, taking any positive values.

The solution to problem (8.1), (8.2) with constant value of capital gains is $x(t) = x_0 \exp(kt)$. With positive capital gains, the firm's income exceeds its costs, and we get an exponential growth of the function x. When $k = 0$, income coincide with expenses, and the solution to the problem does not change over time. Finally, for negative values of this parameter, the function x decreases monotonically, which corresponds to the case of the ruin of the company.

For a more accurate description of the process under study, it is necessary to take into account restrictions on the consumption of manufactured goods. In this case, the capital gain is likely to decrease with increasing function x. Indeed, the greater the capital of the company, the greater the volume of output, and therefore, the greater the decline in demand for goods due to the inevitable saturation of the market. We accept the following dependence of capital gains $k(x) = a(D - qx) - b$, where D is the demand for manufactured products (an unchanged quantity of goods required per unit of time), q is the quantity of goods produced per unit of invested capital, b is the expenses of the company associated with the production of goods and not depending on the quantity of goods sold, and a is profit, received from the sale of a unit of goods. Thus, the income of the company is determined by the ratio between supply and demand.

Thus, the state equation is

$$\dot{x} = \big[(aD - b) - aqx\big]x. \tag{8.3}$$

We obtain the well-known **Verhulst equation**. If the inequality $D > b$ is true, then the income received from the sale of the goods exceeds the expenses of the company, for any initial state of the system (start-up capital) the capital of the company tends to $x^* = (D - b/a)/q$ over time. If the starting capital of the company is small enough, i.e., the inequality $x < x^*$ holds, then demand exceeds supply, the firms income grows, and it expands its production. However, as the market is saturated with goods, capital gains decrease. If the initial capital of the company is so large that the inequality $x > x^*$ is satisfied, then there is an overproduction of goods (recall that in the framework of this model all money is invested in the production of a single product). Since not all manufactured products are bought up, the company suffers losses and gradually turns production down. In the course of restoring the balance between supply and demand, the decline in the capital of the company is reduced. A comparative analysis of the various interpretations of the Verhulst equation is given in Table 8.1[3].

Now we have to be acquainted with the various forms of relations between the two economic entities.

Evolution of a company under limited demand is described by the Verhulst model. By the Verhulst model, the firm capital is stabilizing.

2 Economic competition model

Two firms are considered that produce the same product and are focused on the same consumer. It is assumed that in the absence of sales problems, both firms have some profit, i.e., the costs of manufacturing or purchasing products pay off as a result of their sale. All available capital is invested in production. Thus, having received additional funds from the sale of goods, firms expand production, and with a decrease in profits, the output of goods decreases accordingly.

The state functions x_1 and x_2 are the capitals of the firms. Obviously, the velocities of its change are proportional to the amount of capital, i.e., the more money a company has, the more it produces products, and therefore, the more profit it will receive per unit of time. Then the process under consideration is described by the equations

$$\dot{x}_i = k_i x_i,$$

where k_i is the capital gain of the corresponding firm.

If all the products of the company are sold out, then the capital gain k_i is considered equal to a positive value ε_i, which characterizes the production efficiency. As is known, the corresponding equations have exponential solutions, i.e., with unlimited consumption of goods, firms' incomes increase exponentially over time to the great joy of their owners.

Naturally, such a situation is not realized in practice, since the number of potential buyers and their need for a product is obviously limited. Obviously, the more goods are produced, the less likely they are to ensure their full sale in conditions of limited consumption. In this regard, we assume that the decrease in the growth of firms income will be proportional to the sum of the capital of both firms. We get the equality $k_i = \varepsilon_i - (x_1 + x_2)/\beta_i$, where the coefficient β_i here characterizes the sales efficiency of manufactured products and is associated with the organization of advertising, service culture, etc. Thus, the higher the marketing efficiency of products, the less the problems arising with the sale of manufactured goods affect the rate of change of capital of the company. As a result, we obtain the equations

$$\dot{x}_i = \left[\varepsilon_i - \frac{1}{\beta_i}(x_1 + x_2)\right]x_i, \quad i = 1, 2, \tag{8.4}$$

which are the basis of the *economic competition model*[4].

These equations are close to those that were considered in the previous chapter when describing the process of coexistence of two biological species consuming the same food and have the corresponding properties. For the qualitative behavior of the system, the decisive role here is the value $\varepsilon_i \beta_i$, which characterizes the effectiveness production and marketing of goods (see Figure 8.1). A company whose production and marketing of goods is less efficient is inevitably ruined[5].

For $\varepsilon_1 \beta_1 > \varepsilon_2 \beta_2$, the weaker second firm is ruined (see Figure 8.1 a). In the case when the starting capitals of both are small enough, at first both firms completely sell their products, grow richer and expand their production. However, as the market is saturated with goods, the first firm gradually displaces its weaker competitor (curve 1). Its capital eventually stabilizes at a value equal to $\varepsilon_1 \beta_1$, corresponding to the level of demand for

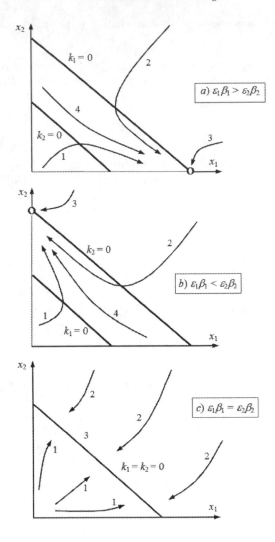

FIGURE 8.1: Phase curves for the competition model.

manufactured products. If the starting capitals of both firms are too large, then the market is oversaturated with goods. Firms curtail production until the moment when the balance between supply and demand is restored (curve 2). After that, in a competitive environment, the stronger first firm wins. If initially the start-up capital of a more competitive company is too large and far exceeds the start-up capital of the second company, then in the process of reducing production a weak company goes bankrupt before there is a balance between supply and demand. In this case, the phase curve does not fall into the competition zone, which is a trapezoid bounded by the straight lines $k_1 = 0$ and $k_2 = 0$. Both state functions will monotonically decrease, the second of them tending to zero, and the first to a non-zero equilibrium position (curve 3). If the total starting capital of firms is not very small and not very large, then the initial state of the system is in the zone of competition. In this case, both state functions change monotonously, and the first of them increases and tends to its equilibrium position, and the second decreases to zero (curve 4).

TABLE 8.2: Evolution the outcome (a) for the competition model. First company goes bankrupt.

x_{10}	x_{20}	Curve	x_1	x_2
small	small	1	increase	at first increase then decrease
large	large	2	at first decrease then increase	decrease
large	small	3	decrease	decrease
reasonably	reasonably	4	increase	decrease

TABLE 8.3: Evolution the outcome (b) for the competition model. Second company goes bankrupt.

x_{10}	x_{20}	Curve	x_1	x_2
small	small	1	at first increase then decrease	increase
large	large	2	decrease	at first decrease then increase
small	large	3	decrease	decrease
reasonably	reasonably	4	decrease	increase

If the inequality $\varepsilon_1\beta_1 < \varepsilon_2\beta_2$ is true, then the second company wins in the process of competition. The behavior of the system in this case is similar to the previous one with the only difference that firms change places (see Figure 8.1 b).

Equality $\varepsilon_1\beta_1 < \varepsilon_2\beta_2$ means that firms are equally competitive (see Figure 8.1 c). None of them are able to supplant their competitor. If the starting capitals of firms are small, then both of them expand their production until the market is saturated with goods (curves 1). If the goods on the market are in abundance, then both firms curtail production (curves 2). A feature of this option is the presence of an infinite number of equilibrium positions (segment 3). Moreover, the exit to a specific equilibrium position is determined by the initial state of the system.

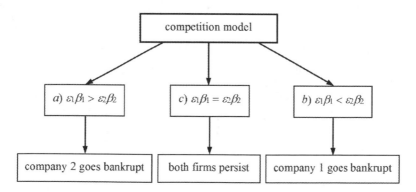

FIGURE 8.2: Possible outcomes for the competition model.

Possible outcomes in the competition model are presented in Figure 8.2, and the evolution of the system for different outcomes in Tables 8.2, 8.3, and 8.4.

TABLE 8.4: Evolution the outcome (c) for the competition model. Both firms persist.

x_{10}	x_{20}	**Curve**	x_1	x_2
small	small	1	increase	increase
large	large	2	decrease	decrease
projection of a point on the interval 3	projection of a point on the interval 3	equilibrium point	constant	constant

Task 8.1 *Economic competition.* Based on the numerical solving of system (8.4) with corresponding initial conditions make the following analysis.

1. Find all possible equilibrium positions of system (8.4). By choosing each of them as the initial state of the system, make sure that this is indeed an equilibrium state. Experimentally check the equilibrium position for stability by choosing the initial state of the system from its neighborhood.

2. Make the calculations under the condition $\varepsilon_1\beta_1 < \varepsilon_2\beta_2$. Make sure that the second company goes broke at any initial state of the system. Choose the system parameters in such a way that the system evolution would be similar to curves 1, 2, 3, and 4 in Figure 8.1 *a*.

3. Make the calculations under the condition $\varepsilon_1\beta_1 > \varepsilon_2\beta_2$. Make sure that the second company goes broke at any initial state of the system. Choose the system parameters in such a way that the system evolution would be similar to curves 1, 2, 3, and 4 in Figure 8.1 *b*.

4. Make the calculations under the equality $\varepsilon_1\beta_1 = \varepsilon_2\beta_2$. Make sure that in this case, none of the firms goes bankrupt. Choose the system parameters so that the capitals of both firms monotonously increase, monotonically decrease and do not change over time.

Task 8.2 *Competition between three firms.* Give a model of competition between three firms and prove that in the process of competition, two weaker firms go bankrupt.

Evolution of two firms producing identical products
under limited demand is described by the competition equations.
By the competition model, the weakest firm is going a bankrupt.

3 Economic niche model

In the model considered above, for two firms producing the same product, long-term coexistence is impossible, since a weaker firm will inevitably go bankrupt. An exception here is the unlikely degenerate case of the equality $\varepsilon_1\beta_1 = \varepsilon_2\beta_2$, when firms practically do not differ in properties. The situation will decisively change if each company focuses on its own customer, preferring the production of any specific product. At the same time, we get a model similar to the previously considered ecological niche model

$$\dot{x}_1 = \left(\varepsilon_1 - \frac{1}{\beta_{11}}x_1 - \frac{1}{\beta_{12}}x_2\right)x_1, \quad \dot{x}_2 = \left(\varepsilon_2 - \frac{1}{\beta_{21}}x_1 - \frac{1}{\beta_{22}}x_2\right)x_2, \qquad (8.5)$$

where β_{ij} is the sales efficiency of the i-th product by the j-th company. To bring the resulting model, which is naturally called the "economic niche", to the biological analogue considered in the previous chapter, it suffices to introduce the parameters $\alpha_{ij} = 1/\beta_{ij}$.

To analysis of the system, we replace the variables

$$\tau = at, \quad u(\tau) = bx_1(t), \quad v(\tau) = cx_2(t),$$

where the constants a, b, c will be chosen so that in the new variables equations (8.5) would be as simple as possible. The functions u and v and the variable τ differ, respectively, from

x_1, x_2 and t only by constant factors. In this regard, they also characterize the capital of firms and time, expressed in a different coordinate system.

After the transformations made, we obtain the relations

$$u' = \frac{1}{\beta_{12}ac}\left(\varepsilon_1 c\beta_{12} - \frac{c\beta_{12}}{b\beta_{11}}u - v\right)u, \quad v' = \frac{1}{\beta_{21}ab}\left(\varepsilon_2 b\beta_{21} - u - \frac{\beta_{21}b}{\beta_{22}c}v\right)v,$$

where u', v' are the derivative of the functions u, v by τ. Determine the parameters

$$a = \frac{\beta_{11}\varepsilon_1}{\beta_{21}}, \quad b = \frac{1}{\varepsilon_1\beta_{11}}, \quad c = \frac{1}{\varepsilon_1\beta_{12}}, \quad A = \frac{\beta_{21}}{\beta_{12}}, \quad B = \frac{\beta_{21}\varepsilon_2}{\beta_{11}\varepsilon_1}, \quad C = \frac{\beta_{12}\beta_{21}}{\beta_{11}\beta_{22}}.$$

We get the equations

$$u' = A(1 - u - v)u, \quad v' = (B - u - Cv)v. \tag{8.6}$$

Let us consider the variants of evolution of this system[6]. Different situations are possible depending on the sign of the values located on the right-hand sides of relations (8.6).

Outcome a)[7]. Under the conditions $B \le 1$, $B/C \le 1$ (except $B = C = 1$), the equalities (8.6) three variants of the system evolution are possible (see Figure 8.3 a[8]):

1) if $u + v < 1$, $u + Cv < B$, then the function u increases to unity, and v increase, and then v decreases to zero;

2) if $u + v < 1$, $u + Cv > B$, then the function u increases to unity, and v decreases to zero;

3) if $u + v > 1$, $u + Cv > B$, then the function u decreases, and then u increases to unity (maybe only decreases to unity), and v decreases to zero.

Thus, for all initial states the solutions of the system tends to the values $u = 1$, $v = 0$.

Outcome b)[9]. Under the conditions $1 \le B$, $C \le B$ (except $B = C = 1$) three variants of the system evolution are possible (see Figure 8.3 b[10]):

1) if $u + v < 1$, $u + Cv < B$, then the function u increases, and then decreases to zero, and v increases to B/C;

2) if $u + v > 1$, $u + Cv < B$, then the function u decreases to zero, and v increases to B/C;

3) if $u + v > 1$, $u + Cv > B$, then the function u decreases to zero, and v decreases, and then increases to B/C (maybe only decreases to B/C).

Thus, for all initial states the solutions of the system tends to the values $u = 0$, $v = B/C$.

Outcome c)[11]. Under the conditions $C < B < 1$ four variants of the system evolution are possible (see Figure 8.3 c):

1) if $u + v < 1$, $u + Cv < B$, then the functions u and v increase;

2) if $u + v < 1$, $u + Cv > B$, then the function u increases, and v decreases;

3) if $u + v > 1$, $u + Cv < B$, then the function v increases, and u decreases;

4) if $u + v > 1$, $u + Cv > B$, then the functions both u and v decrease.

Depending on the initial state of the system, the solutions to the problem eventually tend to $u = 1$, $v = 0$ or to $u = 0$, $v = B/C$.

Outcome d)[12]. Under the conditions $1 < B < C$ four variants of the system evolution are possible (see Figure 8.3 d):

1) if $u + v < 1$, $u + Cv < B$, then the functions u and v increase;

2) if $u + v < 1$, $u + Cv > B$, then the function v increases, and u decreases;

3) if $u + v > 1$, $u + Cv < B$, then the function u increases, and v decreases;

4) if $u + v > 1$, $u + Cv > B$, then the functions both u and v decrease.

For all initial states the solutions of the system tends to the values $u = (CB)/(C + 1)$, $v = (B + 1)/(C + 1)$.

Outcome e^{13}. The case $B = C = 1$ is degenerate. The states of the system tend some values lying on the straight line $u + v = 1$ (to which it depends on the initial conditions, see Figure 8.3 e). From a mathematical point of view, the degenerate case is interesting in that the limiting states of the system (attractor) do not form a finite set of points on the phase plane (equilibrium positions), but a line that is a segment of the straight line $u + v = 1$ with non-negative values of the state functions. In this case, the state of the system tends to some values lying on the line $u + v = 1$ (to which it depends on the initial conditions, see Figure 8.3 e).

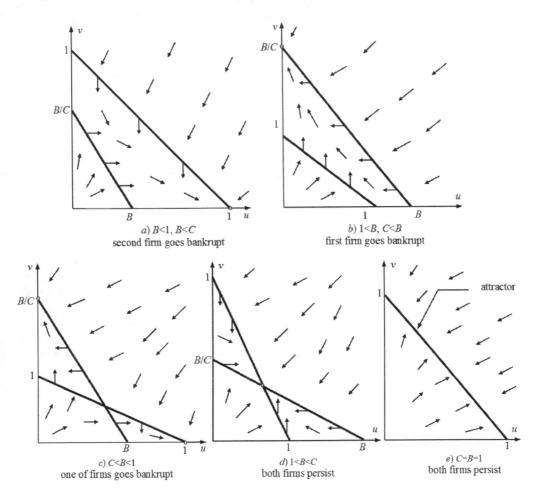

FIGURE 8.3: Directions of the system evolution for the economic niche model.

Analyze these results. The outcome $a)$ corresponds to the inequalities

$$\varepsilon_1 \beta_{11} > \varepsilon_2 \beta_{21}, \quad \varepsilon_1 \beta_{12} > \varepsilon_2 \beta_{22}.$$

Here, the first firm is superior to the second in both types of goods, since the product of the production and marketing efficiency of both types of goods is higher. As a result, the second firm will inevitably go bankrupt, i.e., its capital tends to zero. On outcome $b)$, the opposite inequalities are realized

$$\varepsilon_1 \beta_{11} < \varepsilon_2 \beta_{21}, \quad \varepsilon_1 \beta_{12} < \varepsilon_2 \beta_{22}.$$

They correspond to the ruin of the first firm, which is weaker. Both of these variants actually return us to the previously considered model of economic competition.

At the outcome c), the following relations hold

$$\varepsilon_1\beta_{11} < \varepsilon_2\beta_{21}, \quad \varepsilon_1\beta_{12} > \varepsilon_2\beta_{22}.$$

Here one of the firms survives, depending on the value of the initial state of the system. This situation corresponds to the case when the first company works more effectively with the second type of product, and the second with the first.

The opposite case d), characterized by the inequalities, will be more interesting.

$$\varepsilon_1\beta_{11} > \varepsilon_2\beta_{21}, \quad \varepsilon_1\beta_{12} < \varepsilon_2\beta_{22}.$$

Here, any of the companies produce both types of goods, but give preference to their own. As a result, they are able to coexist peacefully, with each company filling its own economic niche. The results show that each company must certainly strive to find its consumer, otherwise it will be replaced by a stronger competitor. Only powerful firms, giant enterprises can afford the luxury of not being afraid of competition, confidently crowding out rivals from their sphere of production.

The degenerate case e) corresponds to the equalities

$$\varepsilon_1\beta_{11} = \varepsilon_2\beta_{21}, \quad \varepsilon_1\beta_{12} = \varepsilon_2\beta_{22}.$$

This situation means that firms actually have the same properties, with the possible exception of the initial states. Under these conditions, none of them can displace the other, as a result of which both firms coexist. A similar situation was also observed in the competition model.

Task 8.3 *Economic niche.* Based on the numerical solving of system (8.6) with corresponding initial conditions make the following analysis.

1. Find all possible equilibrium positions of system (8.6). By choosing each of them as the initial state of the system, make sure that this is indeed an equilibrium state. Experimentally check the equilibrium position for stability by choosing the initial state of the system from its neighborhood.

2. Make the calculations under the condition $\varepsilon_1\beta_{11} < \varepsilon_2\beta_{21}$, $\varepsilon_1\beta_{12} < \varepsilon_2\beta_{22}$. Make sure that the firm goes broke at any initial state of the system.

3. Make the calculations under the condition $\varepsilon_1\beta_{11} > \varepsilon_2\beta_{21}$, $\varepsilon_1\beta_{12} > \varepsilon_2\beta_{22}$. Make sure that the second firm goes broke at any initial state of the system.

4. Make the calculations under the condition $\varepsilon_1\beta_{11} < \varepsilon_2\beta_{21}$, $\varepsilon_1\beta_{12} > \varepsilon_2\beta_{22}$. Choosing the initial conditions, detect the ruin of both the first or second firms.

5. Make the calculations under the condition $\varepsilon_1\beta_{11} > \varepsilon_2\beta_{21}$, $\varepsilon_1\beta_{12} < \varepsilon_2\beta_{22}$. Make sure that both firms coexist peacefully regardless of the choice of initial conditions.

6. Make the calculations under the condition $\varepsilon_1\beta_{11} = \varepsilon_2\beta_{21}$, $\varepsilon_1\beta_{12} = \varepsilon_2\beta_{22}$. Carrying out calculations with different initial conditions, make sure that the system goes to different equilibrium positions.

> *Evolution of two firms producing two different products*
> *under limited demand is described by niche equations.*
> *According to the niche model, both firms can coexist*
> *in the case of the preference of each company for a own product.*

4 Free market model

Let us turn now to a fundamentally different economic model. Consider the mechanism of changing prices for goods and population incomes in the free market. It is assumed that there are a large number of independent producers of goods and sellers that are keenly competing with each other. For simplicity, we assume that the volume of products does not change. The rate of change in population incomes is considered to be proportional to the value of income x_1. The more funds people have, the greater the changes are possible. It is believed that the population has a constant source of income, and the money received is spent exclusively on the purchase of these goods. Naturally, in the absence of consumption of goods, no funds are spent, which means that population incomes are growing at a constant rate ε_1. Incomes are reduced due to the acquisition of goods in proportion to the price level x_2 with some proportionality coefficient γ_1. Thus, the rate of change in population income is considered equal to $k_1 = (\varepsilon_1 - \gamma_1 x_2)$.

The rate of change in prices is considered proportional to the price level, and in the absence of population incomes, goods are not bought, and entrepreneurs are forced to lower prices with a rate ε_2. Price increases are carried out with increasing incomes of the population, and hence the demand for goods, with a proportionality coefficient of γ_2. As a result, we obtain the following formula for determining the rate of change in prices $k_2 = (\varepsilon_2 - \gamma_1 x_2)$.

Under the considered assumptions, the change in population incomes and commodity prices will be described by the following system of differential equations

$$\dot{x}_1 = (\varepsilon_1 - \gamma_1 x_2)x_1, \quad \dot{x}_2 = (\gamma_2 x_1 - \varepsilon_2)x_2. \tag{8.7}$$

It exactly coincides with the equations characterizing the predator prey model. As already noted in previous chapters, the solution to the problem will change periodically (see Figure 8.4).

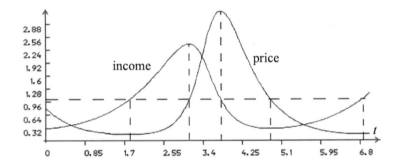

FIGURE 8.4: Change in income and prices in the free market with parameter values $x_{10} = 0.3$, $x_{20} = 0.8$, $\varepsilon_1 = 1$, $\varepsilon_2 = 1$, $\gamma_1 = 1$, $\gamma_2 = 1$.

Give an interpretation of these results. Suppose that at the initial time, population incomes and price levels are relatively low, i.e., the following inequalities hold

$$x_1 < \varepsilon_2/\gamma_2, \quad x_2 < \varepsilon_1/\gamma_1.$$

Then the derivative of x_1 is positive, and the derivative of x_2 is negative. At low incomes of the population, its purchasing power is small, which means that not all manufactured goods are sold out (recall that the volume of output in this model remains unchanged). In the conditions of intense competition in the free market, being under constant threat of ruin, the entrepreneur is forced to lower prices for goods. As a result, at the first stage of the studied process, we observe a decrease in the price level and an increase in population incomes (see Figure 8.4). Sooner or later, such a moment of time t_1 comes when the incomes of the population become high enough and the equality $x_1(t_1) = \varepsilon_2/\gamma_2$ is satisfied. In this case, the right-hand side of the second equation (8.7) will turn out to be zero, which corresponds to the extremum (in this case, the minimum) of the function x_2.

Since the price level is still quite low, there is a further increase in population incomes. Thus, we have the relations

$$x_1 > \varepsilon_2/\gamma_2, \quad x_2 < \varepsilon_1/\gamma_1.$$

At the second stage of the studied process, population incomes are quite high, and the price level is low. Under these conditions, according to equations (8.7), both derivatives are positive, i.e., we observe an increase in both incomes of the population (since prices are still relatively low) and prices (purchasing power of the population is already quite high). The time t_2 arrives when the increasing function x_2 reaches the ratio ε_1/γ_1. After that, the relations

$$x_1 > \varepsilon_2/\gamma_2, \quad x_2 > \varepsilon_1/\gamma_1.$$

Now consider the third stage of the process, characterized by a decrease in household incomes (prices are quite high) and price increases (the purchasing power of the population is high). As the populations income decreases, the moment t_3 comes when equality $x_1(t_3) = \varepsilon_2/\gamma_2$ holds. Then we get

$$x_1 < \varepsilon_2/\gamma_2, \quad x_2 > \varepsilon_1/\gamma_1.$$

There is a further decline in incomes (prices are still quite high) and prices fall due to a decrease in the purchasing power of the population. This process continues until the time t_4, when prices fall to ε_1/γ_1, after which we return to the first stage, characterized by an increase in income due to lower prices and a further decrease in the price level due to the low purchasing power of the population. A new cycle of the process under consideration begins.

The results obtained confirm that economic laws are as objective as physical, chemical, etc. A businessman in a free market cannot but lower prices when the purchasing power of the population falls, because otherwise his competitors will get ahead of him and easily sell all his goods. In the conditions of intense competition, it is necessary to sell goods as quickly as possible, even at reduced prices, than to stubbornly wait for a completely crazy customer who for some reason will purchase goods at a high price, if similar things can be bought nearby at least a little cheaper. On the other hand, with high incomes, the population quickly buys all the goods available for sale, and maintaining the changed balance between supply and demand in the conditions of constant production volume laid down in this model can be done exclusively by raising prices. And how can one not raise prices if the people directly sweeps goods from the counter.

Task 8.4 *Free market.* Based on the numerical solving of problem (8.7) with corresponding initial conditions make the following analysis.

1. By choosing all parameters of the system in an arbitrary way, make sure that the prices of goods and incomes of the population change periodically.

2. Determine the system parameters so that the state of the system does not change over time. Explain the results. Indicate the type of equilibrium position, based on the behavior of the system in its neighborhood.

3. Find the dependence of the period of oscillation of the state functions on the initial income level, as well as on parameter ε_1. Give an economic interpretation of these results.

Another variant of the relationship between the level of prices and incomes of the population is presented in the Appendix. Below we consider a different way of regulating the balance between supply and demand.

> *Free market model is based on the Volterra–Lotka equations.*
> *By the free market model, the population incomes*
> *and prices change periodically over time.*

5   Monopolized market model

The preceding model is characteristic of the free market, when there are relatively many independent producers of goods and there is intense competition between them. Only the threat of competition can force manufacturers of goods and trading organizations to take such a desperate step as lowering prices for goods and services if supply exceeds demand. One of the businessman who earlier feels a change in market conditions and manages to timely reduce prices and quickly sell their goods, will be able to make significant profits. If some stubborn man did not want or did not manage to carry out a timely reduction in prices, then in principle he will not be able to sell his products and will inevitably go bankrupt to the joy of more quick competitors.

In conditions of highly monopolized production and marketing of products, entrepreneurs have the opportunity in the same situation to avoid undesirable price reductions without the threat of complete ruin. The monopolist is not afraid of competition and can restore the imbalance between demand and supply not by lowering prices and thereby increasing demand for goods, but by decreasing production, i.e., supply reduction. A monopolistic entrepreneur can reduce production, considering it more profitable to sell less goods, but at a higher price. He is sure that the population experiencing a certain need for this product will simply have no other choice but to purchase it at a fairly high fixed price, even if in a smaller quantity.

By this model, the price of goods remains unchanged. This is not state function of the system. However, along with the income of the population x_1, the volume of output x_3 is now change. In the absence of consumption of goods, the entrepreneur does not reduce prices, but the volume of production with a rate of ε_3. An increase in the volume of production occurs with increasing incomes of the population (with a proportionality coefficient of γ_3), i.e., its purchasing power. In the absence of consumption of goods, i.e., when $x_3 = 0$, population incomes grow with rate ε_1 since unspent funds will accumulate in people. When goods are consumed, expenses increase in direct proportion (with a coefficient of γ_1) to the number of goods available for sale, i.e., volume of production. Based on these considerations, we have the following equations

$$\dot{x}_1 = (\varepsilon_1 - \gamma_1 x_3)x_1, \quad \dot{x}_3 = (\gamma_3 x_1 - \varepsilon_3)x_3. \tag{8.8}$$

We again get a predator–prey model with a well-known periodic solution. The appearance of this solution is characterized by Figure 8.4 with the replacement of the index "2" by "3" and a new interpretation of the results[14].

At low initial values of income and output according to equations (7.8), a decrease in production is observed. Lacking sufficient funds, people are forced to reduce the consumption of goods, as a result of which their costs are reduced. Since people who consume little, accumulate money, their incomes increase. Over time, too few goods will remain on the market, not enough to sustain even reduced demand, and people will gradually accumulate a certain amount of money that they could spend on the purchase of goods. In these conditions, the entrepreneur increases the output of goods. However, as long as there is still a relatively small quantity of goods, the needs of the population remain unsatisfied, and people continue to accumulate funds. As more and more goods are produced, demand begins to be satisfied and, in the end, supply becomes smaller. The expenditures of the population for the purchase of goods increase, and the purchasing power decreases. There comes a time when a significant amount of goods again remains unsold. To save the entrepreneur from imminent ruin can decrease either prices or production, i.e., restoration of the balance between supply and demand. In a monopolistic market, he can, without fear of competition, leave prices unchanged. Thus, a reduction in production is inevitable. The release of goods will be resumed only when the population manages to accumulate a sufficient amount of funds for the purchase of goods. Thus, the entrepreneur produces goods in some not very large batches, and this process is periodically repeated.

Task 8.5 *Monopolized market.* Based on the numerical solving of problem (8.8) with corresponding initial conditions make the following analysis.

1. By choosing all parameters of the system in an arbitrary way, make sure that the volume of output and incomes of the population change periodically.

2. Choose the system parameters so that the state of the system does not change over time. Evaluate the behavior of the system in a neighborhood of this equilibrium position. Give an economic interpretation of this result.

3. Determine the dependence of the amplitude of oscillations in the volume of production of goods on the initial level of income, as well as on parameter γ_2. Explain the results.

> *Monopolized market model is based on the Volterra–Lotka equations.*
> *By the monopolized market model, the population incomes*
> *and the volume of output change periodically in time.*

Direction of further work. *In the next chapter, we will make sure that the mathematical models discussed above also describe some processes of sociology, political science, and psychology.*

Appendix

The economic niche model is similar to the biological model "ecological niche" described in the previous chapter, in which two species are considered, consuming, generally speaking, the same food, but giving preference to their own. Under these conditions, there is a possibility of peaceful coexistence of species.

We considered the processes taking place in a free and monopolized market. In conditions of completely state regulation of the economy, the imbalance between supply and demand is restored by fundamentally different means. With a deterioration in the standard of living of the population and a drop in its purchasing power, there is no decrease in prices or a decrease in production now, but an increase in official salaries. This, in turn, leads to higher prices, which corresponds to the inflation model.

Mutually beneficial cooperation of firms is described by a model of the "symbiosis" type. If two friendly firms that produce components of the same products initially have small means, then they will inevitably go bankrupt. Sufficiently large starting capital firms provide them with complete prosperity. With the gradual saturation of the market with goods, either the ruin of firms or the stabilization of the system in question occurs. The interaction of an entrepreneur and a racketeer is described by equations of the "predator-prey" type with a periodic change in the income of both actors.

The final part presents the macroeconomic Solow model of economic growth[15].

1   Ecological niche model

We considered in the previous chapter the ecological niche model described by the equations

$$\dot{x}_1 = \left(\varepsilon_1 - \alpha_{11}x_1 - \alpha_{12}x_2\right)x_1, \quad \dot{x}_2 = \left(\varepsilon_2 - \alpha_{21}x_1 - \alpha_{22}x_2\right)x_2,$$

where x_i is the abundance of i-th species, α_{ij} is a specific coefficient of consumption of j-th type by i-th species, ε_i is a specific growth abundance of i-th species, $i, j = 1, 2$. Denoting $\beta_{ij} = 1/\alpha_{ij}$, we get the economic niche model considered above.

The properties of this model, up to the meaning of the characteristics included in it, are described in Figure 8.3. Repeating the previous arguments, we come to the following scenarios. The first case, when the second form dies out for any initial states, corresponds to the inequalities

$$\frac{\varepsilon_1}{\alpha_{11}} > \frac{\varepsilon_2}{\alpha_{21}}, \quad \frac{\varepsilon_1}{\alpha_{12}} > \frac{\varepsilon_2}{\alpha_{22}}.$$

The first species here has the ratio of population growth to consumption of both types of food greater that the second one. Thus, the first species in all respects is more resilient, and the extinction of the second species is inevitable. Similarly, opposing inequalities inevitably entail the extinction of the first species. In fact, in these two variants, the presence of two different types of food does not appear, and we essentially return to the biological model of competition.

Under the inequalities

$$\frac{\varepsilon_1}{\alpha_{11}} < \frac{\varepsilon_2}{\alpha_{21}}, \quad \frac{\varepsilon_1}{\alpha_{12}} > \frac{\varepsilon_2}{\alpha_{22}},$$

only one species can survive. The initial state of the system determines which particular species is dying. More interesting is the opposite situation, characterized by inequalities

$$\frac{\varepsilon_1}{\alpha_{11}} > \frac{\varepsilon_2}{\alpha_{21}}, \quad \frac{\varepsilon_1}{\alpha_{12}} < \frac{\varepsilon_2}{\alpha_{22}}.$$

Here, both species are, in principle, capable of consuming any food, but they have their preference. As a result, they are able to coexist peacefully. Thus, each species fills its ecological niche, which is in good agreement with the real situation and corresponds to the name of the model under consideration.

2 Inflation model

Another variant of market relations is associated with strong state regulation of the economy. The state is a commodity producer and seller, as well as an employer of the population, determining its income. For some reason, let the incomes of the population turn out to be too low and the prices prohibitively high, which leads to a disruption in the balance between supply and demand. When a similar situation is observed in a free market, the restoration of balance occurs due to lower prices, and therefore, stimulation of increased demand. In the case of a monopolized market, production can be reduced, thereby bringing the supply level closer to the declining demand level. In this case, the third scenario is being developed.

Can the state reduce the price of goods? An entrepreneur in a free market is forced to take such an action. He prefers to sell as many products as possible, at least at a reduced price, and thereby be enriched by less efficient competitors. The state has no competitors, and therefore it does not need to take such actions. In addition, price reductions will inevitably turn existing enterprises into unprofitable ones. The state is also not very interested in reducing the volume of production. This private monopolist can go on temporary conservation of his enterprise. He can make staff reductions, thereby reducing the cost items, and channel the proceeds to more profitable areas of activity. It is much more difficult for a state institution to quickly and toughly implement such unpopular decisions. In addition, the director of such an enterprise receives a fixed official salary, and not profit from the production and sale of goods, which does not very stimulate his operational action.

The state has powerful mechanism, obviously absent from a private entrepreneur. It defines and controls the entire financial policy. To increase the purchasing power of the population and reduce social tension, it can go on increasing the salaries of public servants, as well as pensions, benefits, scholarships, etc. Where does the state find considerable means to increase the incomes of the population? Can it by the power of any effective action increase the well-being of its long-suffering citizens? Apparently not. But it is all about an increase in the number of banknotes in circulation. The implementation of this event is quite within the power of even a not very developed state, which is in the midst of a severe economic crisis.

In this model, the change in the price of goods x_2 is still characterized by the income of the population (more precisely, by the amount of money supply available to it), and the income of the population x_1 is determined by the constantly indexed wage calculated by the state based on the current price level. Naturally, the higher the price of goods, the more often the state is forced to raise wages for its employees. Thus, the change in household income will be proportional to the price of the goods. As a result, we have the system of equations

$$\dot{x}_1 = \alpha x_2, \quad \dot{x}_2 = \beta x_1, \tag{8.9}$$

where the positive constants α and β are system parameters.

Differentiating the first equality (8.9) and using the second one, we get the equation

$$\ddot{x}_1 = \alpha \beta x_1.$$

Analogically, after differentiation the second equality (8.9) and substitution the result to the second one, we obtain

$$\ddot{x}_2 = \alpha \beta x_2.$$

These equations have the solutions[16]

$$x_1(t) = a_1 \exp(-\lambda t) + b_1 \exp(\lambda t), \quad x_2(t) = a_2 \exp(-\lambda t) + b_2 \exp(\lambda t),$$

where $\lambda = \sqrt{\alpha \beta}$ and the constants a_1, b_1, a_2, b_2 can be found from the initial conditions. By these results, we observe an exponential increase in prices and incomes of the population.

Give an interpretation of these results. Because of the low demand for goods, the state increases the wages of its employees, since the machine that prints banknotes is in good hands. As the money supply available to the population increases, its purchasing power increases, which leads to an increase in demand for goods. With a constant volume of production and increasing demand, the goods will inevitably be in short supply. In principle, it would be possible to eliminate it by increasing the volume of production, which is not so prosperous state is clearly beyond the power. It would be possible, in principle, to lower the level of wages. However, this is already facing serious social upheaval. Thus, the only reliable way out of this situation is associated with higher prices, which inevitably leads to lower demand with constant supply. However, rising prices will certainly lead to the impoverishment of the masses and lower levels of consumption. To restore the balance between supply and demand in order to avoid a socio-political crisis, the state again raises wages, which will certainly cause another round of prices, etc. The considered process is the *inflation*[17].

3 Model of economic cooperation

Coexistence of firms does not necessarily follow the path of competition. The opposite situation is also possible, when firms mutually beneficial cooperate. Suppose that two firms produce components so that the products of one of the firms cannot be sold without the corresponding products of the other (roughly speaking, one firm produces bolts and the

TABLE 8.5:   Analogy between models of the symbiosis type

Science	Biology	Economics
phenomenon	symbiosis	cooperation
x_i	species abundance	firm income
ε_i	species extinction rate in the absence of another species	firm ruin rate in the absence of another firm due food limitations
γ_i	species abundance increase in the presence of friendly species	capital firm increase in the presence of friendly company
β_i	growth decrease of species abundance because of food limitations	growth decrease of company income because of market limitations

other produces nuts for these bolts). Under these conditions, firms simply cannot exist without each other. At the same time, the success of one of them favorably affects the financial situation of another company. As a result, we determine the equations

$$\dot{x}_1 = \big(\gamma_1 x_2 - \varepsilon_1\big)x_1, \quad \dot{x}_2 = \big(\gamma_2 x_1 - \varepsilon_2\big)x_2,$$

where x_i is the capital of the i-th company, ε_i is its ruin rate in the absence of another company, γ_i is a parameter of the influence of the j-th company on the i-th, $i = 1, 2, j \neq i$.

The obtained relations, up to the notation, coincide with the biological model of symbiosis (see the previous chapter). Knowing the properties of the latter model, we can conclude that with small initial capital, both firms go bankrupt. Sufficiently large starting capitals of firms lead to complete prosperity, which is expressed in the unlimited increase in their capitals. If one of the companion firms is poor and the other rich, then for some time the rich company will gradually become poorer, supporting its weaker companion. At the same time, a poor company gets richer, having a sufficiently strong company as a companion. Then there are two possible scenarios. Perhaps the poor company will get stronger before the rich company is significantly weakened, and then both firms will grow rich indefinitely. However, the opposite situation is also possible when a stronger firm weakens before a weaker firm grows stronger. In this case, both firms will eventually go broke. A comparative analysis of the economic and biological model of the symbiosis type is given in Table 8.5.

Task 8.6 *Economical cooperation model.* Based on the numerical solving of the problem make the following analysis.

1. Make the calculations for small initial states. Make sure that both firms go bankrupt.

2. Make the calculations for large initial states of the system. Make sure that the capital of both firms is growing unlimitedly.

3. Changing the initial state of the system, detect non-monotonous evolution of capital of firms.

4. Find a nontrivial equilibrium position of the system. Make sure that it is not stable.

The unlimited expansion of production in the model of economic cooperation, of course, is not implemented in practice. To clarify the model, one can take into account the limited consumption of manufactured products, i.e., market saturation gradually as production increases. Suppose that a decrease in a company's capital growth due to market saturation depends quadratically on its capital value. The result is equations

$$\dot{x}_1 = \big(\gamma_1 x_2 - \varepsilon_1 - \beta_1 x_1^2\big)x_1, \quad \dot{x}_2 = \big(\gamma_2 x_1 - \varepsilon_2 - \beta_2 x_2^2\big)x_2,$$

where the coefficients β_1 and β_2 characterize the saturation of the market with goods of the firms. These relations exactly coincide with the equations of the "symbiosis with limited food" model (see Chapter 7). Depending on the combination of parameters, stabilization or ruin of both firms is possible here.

TABLE 8.6: Analogy between predator–prey type models.

Characteristic	Biological model	Economic model
x_1	abundance of prey	income of entrepreneur
x_2	abundance of predator	income of racketeer
ε_1	increase of prey in the absence of predators	natural profit of entrepreneur
ε_2	extinction of predators in the absence of prey	natural expenses of racketeer
γ_1	extermination of prey by predators	decrease of entrepreneurial income by the action of the racketeer
γ_2	predator increase in the presence of prey	increase of racketeer's income by the robbery of entrepreneur

Task 8.7 *Ecological niche model.* Based on the numerical solving of the given problem make the following analysis.

1. Calculate the equations with small initial states. Make sure that both firms go broke.

2. Calculate the equations with large initial states. Choose the system parameters so that the capitals of both firms increase unlimitedly over time.

3. Calculate the equations with large initial states. Choose the system parameters so that the capitals of both firms stabilize over time.

4 Racketeer—entrepreneur model

Consider another specific type of economic "cooperation". This is the relationship between the entrepreneur and the racketeer. The state functions here are their incomes x_1 and x_2, respectively. It is assumed that the entrepreneur has his own company with a constant source of income, and in the absence of a racketeer, his income increases with a rate ε_1. The racketeer lives solely at the expense of the entrepreneur, and in the absence of the latter spends the available funds with a rate ε_2. An increase in the entrepreneur's income leads to a proportional enrichment of the racketeer, which, in turn, is carried out through the withdrawal of funds from the entrepreneur. This process is described by the equations

$$\dot{x}_1 = (\varepsilon_1 - \gamma_1 x_2)x_1, \quad \dot{x}_2 = (\gamma_2 x_1 - \varepsilon_2)x_2,$$

where the parameters γ_1 and γ_2 characterize the influence of the actors on each other.

These equations correspond to the predator prey biological model. Naturally, the solutions of the problem are periodic functions. The analogy between biological and economic objects is presented in Table 8.6.

Give an interpretation of the results. With increasing entrepreneurial income, the racketeer has the opportunity to collect more and more tribute. However, as the "appetite" of the racketeer grows, the entrepreneur begins to go broke, which eventually affects the racketeer's income. Since the complete ruin of the entrepreneur is detrimental to the racketeer, he is forced to reduce the required sum of money. As a result, over time, the entrepreneur restores the financial position of his company, shaken due to exorbitant exactions. Then the racketeer gets the opportunity to catch up. A new process cycle begins.

We have already considered the Volterra–Lotka equations in describing the processes of chemistry (the Lotka reaction system A+X→2X, X+Y→2Y, Y→B), biology (predator–prey system), and agriculture (fertility–yield system) and economics (models of a free and monopolized market, as well as a racketeer–entrepreneur model). Their comparative characteristics are presented in Table 8.7[18].

TABLE 8.7:   Characteristic of "predator–prey" models.

Application	Model	"Predator"	"Prey"
chemistry	Lotka system	X concentration	Y concentration
biology	predator–prey	predator	prey
agriculture	yield–fertility	yield	fertility
economics	free market	price	income
economics	monopolized market	volume of goods	income
economics	racketeer–entrepreneur	racketeer	entrepreneur

5   Solow model of economic growth

An economic system is given in which the whole economy is considered as a whole and is closed from outside influence. All firms exist in a competitive environment and seek to maximize their profits. The system produces the only product that is used for consumption and investment. The process conditions do not change over time. Government expenses and taxes are not taken into account.

For this system, some macroeconomic indicators are considered. There are **gross domestic product** Y, **gross investment** I, **consumption fund** C, **fixed assets** K, **number of people employed** in the production sector L. Gross domestic product is determined by the production fund and employment, i.e., $Y = F(K, L)$, where F is called the **production function**. As such, the **Cobb–Douglas function** can be chosen $F(K, L) = AK^\alpha L^\beta$, where A, α, and β are positive constants. Then gross domestic product is consisted of investment and consumption, i.e., $Y = I + C$.

The change in production assets is determined by their constant decrease (depreciation) and investments

$$\dot{K} = -\mu K + I,$$

where μ is wear factor. Finally, the number of employees is increasing at a constant rate, i.e.,

$$\dot{L} = \nu L,$$

where ν is a positive constant. Besides, initial values of production assets and employment are known

$$K(0) = K_0, \quad L(0) = L_0.$$

The above relations characterize the **Solow model** of economic growth. It underlies the analysis of the rate of change of capital and the economic effect of economic progress and is the starting point for many macroeconomic theories[19].

Notes

[1] Various mathematical problems and models of economic systems are described, for example, in books [2], [11], [18], [21], [50], [51], [60], [65], [75], [78], [82], [87], [92], [108], [111], [155], [160], [173], [184], [196], [238], [335], [363], [382].

[2] Lecture 11 will discuss some models of economic processes described by systems with distributed parameters, see also [92]. Discrete economic models are considered in Chapter 17, see also [184], see also Lecture 17. The use of game theory to describe economic systems is given in Lecture 17, see also [21], [87], [111], [363]. Stochastic models of economics see Chapter 18 and [78], [92], [173], [335]. Optimization methods in economic problems are described, for example, in [75], [155], [184].

[3] Table 8.1 is Table 7.1 from the previous chapter, supplemented by an economic interpretation of the Verhulst equation.

[4] Chapter 21 discuss the problem of determining the coefficients of the equations of economic competition from the results of measuring the capital of firms.

[5] The presence of three or more companies producing the same product also leads to the gradual absorption of the most powerful company of its unlucky competitors (see the Appendix to the previous chapter). This inevitably leads to a monopolization of the market.

[6] One can immediately determine the equilibrium position of the considered system. Equating to zero the right-hand sides of equations (8.6), we obtain

$$A(1 - u - v)u = 0, \quad (B - u - Cv)v = 0.$$

This system of algebraic equations has four solutions

$$u = 0, \; v = 0; \quad u = 0, \; v = B/C; \quad u = 1, \; v = 0; \quad u = \frac{C - B}{C - 1}, \; v = \frac{B - 1}{C - 1}.$$

The first of them corresponds to the ruin of both firms and is always unstable (see all variants of Figure 8.3). The second solution is a stable equilibrium (node) at $C < B$ (see Figure 8.3 a and c) and an unstable equilibrium $B < C$ (see Figure 8.3 b and d). The third equilibrium position is stable (also a node) at $B < 1$ (see Figure 8.3 b and c) and unstable at $B > 1$ (see Figure 8.3 a and d). Finally, the fourth equilibrium exists either for $B < 1$ and $C < B$ (and then it is not stable, in particular, the saddle, see Figure 8.3 a and c), or for $B > 1$ and $B < C$ (and then it is stable in particular a node, see Figure 8.3 d). This is because the functions u and v must be positive.

[7] The outcome a is an analogue of the case a of the competition model.

[8] Figure 8.3 a corresponds to the case when both inequalities realizing the outcome a are strict. If one of these relations is realized in the form of equality, then the middle region in the figure is a triangle, not a quadrangle. However, this does not affect the outcome.

[9] The outcome b is an analogue of the case b of the competition model.

[10] Figure 8.3 b corresponds to the case when both inequalities realizing the outcome of b, are strict. If one of these relations is realized in the form of equality, then the middle region in the figure is a triangle, not a quadrangle.

[11] The outcome c has no analogue in the competition model.

[12] The outcome d has no analogue in the competition model.

[13] The outcome e is an analogue of the case c of the competition model.

[14] In the models of both the free and the monopolized market, the incomes of the population act as "prey". However, while in the free market model "prices" appear to be "predators", then in the monopolized market model this is the volume of goods produced.

[15] **Macroeconomics** is a part of economic theory that studies economics as a whole, see, for example, [302].

[16] Indeed, consider the corresponding characteristic equation $\lambda^2 = \alpha\beta$. Its solutions $\lambda_{1,2} = \pm\sqrt{\alpha\beta}$ are real, what determines the form of the given generalized solutions.

[17] Mathematical inflation models are considered, for example, in [184].

[18] In the next chapter, we consider another interpretation of the predator–prey model.

[19] In the given Solow model, there is an uncertainty associated with the distribution of the gross domestic product between investment and consumption. This circumstance allows us to pose optimal control problems on the basis of this model. On the Solow model, see [2], [51], [184].

Chapter 9

Mathematical models in social sciences

This chapter discusses some dynamic models of social life, mostly characterized by the previously described equations. In particular, the processes related to political science[1], sociology[2], and psychology[3] are analyzed.

First, simple mathematical models of political science are considered. The studied systems can be interpreted as different stages of the transition from totalitarianism to democracy. At the initial stage, it is assumed that after the collapse of the totalitarian system in the absence of democratic traditions in the country, a constitution is adopted that allows a multi-party system. Under these conditions, some political associations are formed that do not have a clear socio-political platform and are grouped, as a rule, around energetic and ambitious leaders who want to come to power and keep it. They compete among themselves for the votes, which corresponds to the model of political competition described by the well-known equations of competition. In these conditions, the strongest of competitors inevitably wins. Over time, complete domination of one party is established in the country. The transition from formal to real multi-party system is possible in the case when a political association (naturally, not a party of power) begins to focus on a specific socio-political layer of population. At the same time, various parties begin to express the interests of not only the people as a whole, but individual social strata. We come to a political niche model described by the corresponding equations.

At the next stage, the opposition can expand its political niche and come to power. After that, it does not have the opportunity to transform into a new party in power and appeal to all the people at once. It has certain obligations to those social groups that brought it to power. Under the same conditions, the former party in power can no longer remain the party of the whole people. It is impossible for it to achieve the support of those sections of the population on which the new government relies. Thus, we get two parties with clearly defined political platforms that are capable of succeeding each other in power. The result is a third model of political struggle.

In addition, this chapter discusses the model of trade union activity described by the Volterra–Lotka equations, as well as the model of allied relations, characterized by the equations of symbiosis.

The Appendix describes some models of social processes, which are new interpretations of models of competition, niche, and predator–prey.

Lecture

1 Political competition

Suppose that in some country a democratic constitution was adopted for the first time, allowing for a multi-party system. Then political groups appear that claim power and compete with each other. In view of the absence of established democratic traditions in the country, these groups do not yet have a more or less clear socio-political program. In the absence of explicit support from the specific social strata of the population, they seek to enlist the support of the entire people at once and go to the polls under popular slogans.

For simplicity, we restrict ourselves to considering the political struggle of two parties for influence among voters. We describe the system by the influence functions x_1 and x_2, which represent the number of voters who intend to vote for this party. Each of the parties is actively fighting to expand its influence and, in the absence of restrictions on the number of voters, would have had a constant positive increase in influence. However, in view of the limited number of voters, the growth of their influence sooner or later is reduced and may even become negative. Naturally, the larger the number of voters who have already made their choice at a given moment in time, the fewer potential voters who still can give preference to any of the parties. Thus, the increase in the influence of the i-th party is determined by the formula

$$k_i = \varepsilon_i - \frac{1}{\beta_i}(x_1 + x_2), \quad i = 1, 2,$$

where ε_i is the natural increase of the parties influence, determined by its political activity, and β_i is the process parameter characterizing the decrease in the increase in the influence of the party due to the fact that at the given time, voters in the amount of $x_1 + x_2$ people have already given their preference to one of the parties. The more attractive the party for the voter, the less it is threatened by a decrease in influence. As a result, we obtain

$$\dot{x}_i = \left[\varepsilon_i - \frac{1}{\beta_i}(x_1 + x_2)\right]x_i, \quad i = 1, 2. \tag{9.1}$$

These relations determine the model of political competition, which actually coincides with the models of competition discussed in previous chapters.

According to a previous study, the output depends from the product $\varepsilon_i\beta_i$, which characterizes the party. That party, which has a stronger political program and more actively influences voters (ideological, economic, and psychological advantage in the fight against political opponents) is gradually displacing its competitors from the political scene[4]. Thus, when parties come up with a nationwide program, one has the monopolization of power.

The results obtained indicate that an authoritarian regime is being established in the country. Its fundamental difference from the previous totalitarian regime is the presence of a relatively democratic constitution. The opposition has a formal legal right to exist, but does not have a real opportunity to exercise this right and gain a foothold in the government. It will certainly be forced out over time from the political arena. Opposition parties exist, but do not exert any influence on the political life of the country. Their formal existence is extremely beneficial to the party in power, which has the opportunity to declare its clear commitment to a democratic form of government, demonstrating its high civilization to the world community and receiving certain dividends for it.

Rivalry of parties focused on all voters is described by the competition equations.
By the competition model, the result of the political struggle
is the monopolization of power by the strongest of the parties.

2   Political niche

In the above model of political competition, a weaker party is gradually being squeezed out of the political arena. Under these conditions, the state has a tendency to transition to a one-party system, which in fact leads to the monopolization of the entire political system. An obstacle to such monopolization may be the creation of parties that were initially oriented not at the abstract voter, but at specific sections of the population with more or less clear group interests.

Consider the political struggle of two parties expressing the interests of certain segments of the population. The dynamics of this political competition is described by the equations

$$\dot{x}_1 = \left(\varepsilon_1 - \beta_{11}^{-1}x_1 - \beta_{12}^{-1}x_2\right)x_1, \quad \dot{x}_2 = \left(\varepsilon_2 - \beta_{21}^{-1}x_1 - \beta_{22}^{-1}x_2\right)x_2, \quad (9.2)$$

where the coefficient β_{ij} characterizes the effectiveness of the political program of the i-th party based on the j-th group of the population. The obtained ratios actually coincide with the previously considered economic niche model. This is the *political niche model*.

By the previous chapter, under the conditions $\varepsilon_1\beta_{11} > \varepsilon_2\beta_{21}$, $\varepsilon_1\beta_{12} < \varepsilon_2\beta_{22}$, each party works more effectively with its own group of voters. Then both parties have a chance to survive and be represented in parliament and local authorities. In principle, the fulfillment of these ratios can be achieved by one party, which interacts quite effectively with a specific social layer of the population. A relatively weak opposition can exist only if it has a sufficiently clear political orientation.

The obtained results indicate the futility of the existence of parties with a vague social program. Among such parties, only a stronger party of power survives. Recall also that the term "party" itself has the meaning of a "part" rather than a "whole". If in the first model the party in power will surely oust all its competitors from the political arena, then in the model of the political niche, the government is forced to endure the opposition, which holds a number of deputies in parliament and is able to influence political life in the country to a certain extent. However, the parliamentary majority is still in the party in power[5].

Task 9.1 *Political niche model. Qualitative analysis.* Perform a qualitative analysis of the model of the political niche and provide an interpretation of the results obtained.

Task 9.2 *Political niche model. Quantitative analysis.* Based on the numerical solving of problem (9.2) with corresponding initial conditions make the following analysis.

1. Find the equilibrium positions of system (9.2). Explain their classification.

2. Make the calculations under the condition $\varepsilon_1\beta_{11} < \varepsilon_2\beta_{21}$, $\varepsilon_1\beta_{12} < \varepsilon_2\beta_{22}$. Make sure that the second party wins at any initial state of the system.

3. Make the calculations under the condition $\varepsilon_1\beta_{11} > \varepsilon_2\beta_{21}$, $\varepsilon_1\beta_{12} > \varepsilon_2\beta_{22}$. Make sure that the first party wins at any initial state of the system.

4. Make the calculations under the condition $\varepsilon_1\beta_{11} < \varepsilon_2\beta_{21}$, $\varepsilon_1\beta_{12} > \varepsilon_2\beta_{22}$. Choosing the initial conditions, detect the victory of both the first and second party.

5. Make the calculations under the condition $\varepsilon_1\beta_{11} > \varepsilon_2\beta_{21}$, $\varepsilon_1\beta_{12} < \varepsilon_2\beta_{22}$. Make sure that both parties retain a certain effect for any initial states of the system.

> *Competition of political parties expressing the interests*
> *of different sectors of the population is described by niche equations.*
> *According to the political niche model, in the conditions of orientation*
> *of parties to different strata of the population, their long coexistence is possible.*

3 Bipartisan system

The transition to the third stage of the political process of democratization is due to the gradual expansion of the opposition's political niche. The opposition does not bear any responsibility for the socio-political and economic life in the country, confining itself to sharp criticism of the party in power. The latter is forced to act actively. Therefore, it inevitably make all kinds of mistakes from time to time, offend someone's interests, and cause some discontent among fellow citizens. Over time, an increasing number of offended voters cast their votes to the opposition, which does not skimp on promises. Sooner or later, the political niche of the opposition is expanding to the size of a parliamentary majority, and a change of power is taking place in the country.

The victorious opposition, being bound by the promises made earlier and widely publicized by the electoral program, has no opportunity to speak on behalf of the whole people. It must defend the direct interests of those sections of the population that brought her to power. On the other hand, the losing party in power is no longer able to remain a nationwide party. Those sections of the population who are keenly interested in carrying out deep reforms of the new government will not go after it. However, all dissatisfied with these reforms can unite around the new opposition. As a result, we get two parties with a fairly clear socio-political platform, more or less consistently expressing the direct interests of specific social strata of the population.

Note that in most countries with established democratic traditions, political life is largely determined by the activities of two parties or party blocs. One of which is called left, and the second one is right. The first of them adheres to social democratic traditions and prefers the principle of equality, and the second one is a liberal party and professes the principle of freedom.

The goal of the left party is to reduce social inequality between different sectors of the population. It protects, first of all, the interests of the most needy poor citizens, rightly believing that the wealthier citizens themselves are able to protect their interests. In this regard, the left-wing party is fighting to raise pensions and benefits, to reduce unemployment, and to increase benefits for the least protected socially.

However, the question arises, where can one find the means to implement the noble social program of the left party? The desired goal can be achieved by implementing a fairly strict tax policy and increasing the influence of the state in managing the economy. By increasing taxes, the left-wing government acquires additional funds from wealthier strata of the population and directs the proceeds to social needs, thereby smoothing out the contradictions between the rich and the poor[6]. Carrying out the nationalization of industry, the left government creates new jobs and protects unprofitable enterprises from the inevitable financial collapse, which threatens the loss of work to many workers.

In turn, the right-wing party advocates for the freedom of private enterprise. By its opinion, the living standards of the population is a direct consequence of the high efficiency of production. In this regard, the right-wingers consider it necessary to protect the interests of a relatively small, but significantly more active group of citizens, those whose successful professional activity determines the high level of well-being of the country. To this end, the right-wing government seeks to increase the flow of capital into the country, both domestic and foreign. It liberalizes tax policy, closes unprofitable state enterprises, and reduces various kinds of government spending. The right-wingers strive to reduce the role of the state in managing the economy and are conducting the denationalization of enterprises, realizing that private enterprise in a free market allows for more efficient management of production compared to clumsy state regulation.

We would like to model the process of coexistence of two parties with such different political platforms. As state functions, we choose the popularity of the left and right parties, which we will denote by x_1 and x_2, respectively. Party popularity at a given point in time is recorded using opinion polls. These results are fixed in the parliamentary, presidential, municipal elections. The winning party gets the opportunity to form its own government and try to implement its political and economic program. If a party justifies the confidence of its voters, then it will be able to hold out until the next election and remain in power for a new term. With unsuccessful government policies, the opposition comes to power.

All activities of the left government are aimed at improving the living standards of the population, increasing the incomes of the broad masses. The right-wing party seeks to increase production efficiency and achieve a recovery in the country economy. In this regard, we also include the average income of the population y_1 and production efficiency y_2 among the functions of the state. Thus, we need to establish how the popularity of the left and right, as well as population incomes and production efficiency, will change over time.

The velocity of change in the influence of any of the parties will still be considered proportional to the value of this influence. Moreover, an increase in the influence of the left-wing party is possible with a relatively low standard of living, when an increasing number of citizens will need social protection. At the same time, this growth will occur only with a sufficiently high production efficiency, otherwise a significant number of voters will give preference to the right-wing party. Under the same conditions, the influence of conservatives decreases. The result is equations

$$\dot{x}_1 = (\alpha_2 y_2 - \alpha_1 y_1)x_1, \quad \dot{x}_2 = (\alpha_1 y_1 - \alpha_2 y_2)x_2, \tag{9.3}$$

where α_1, α_2 are process parameters.

With an increase in the influence of the left party and a decrease in the influence of the right one, an increase in the incomes of the broad sections of the population is observed, since with the coming to power the left begins to fight unemployment, raise pensions, social benefits for the poor, etc. At the same time, as it is not sad, there is a drop in production efficiency. Indeed, as a result of creating new jobs, increasing deductions for social needs and other noble deeds, incomes in enterprises are inevitably reduced. In addition, an increase in taxes inevitably leads to a steadily increasing desire among entrepreneurs to evade their payment and to the outflow of capital abroad. On the contrary, with the right-wing party coming to power (that is, in the context of their influence and the popularity of the left-wing party growing), a desperate struggle begins to improve the country's economy, but with an inevitable reduction in social spending and rising unemployment. As a result, we obtain

$$\dot{y}_1 = (\beta_1 x_1 - \beta_2 x_2)y_1, \quad \dot{y}_2 = (\beta_2 y_2 - \beta_1 y_1)y_2, \tag{9.4}$$

where β_1, β_2 are process parameters.

The equalities (9.3), (9.4) with the corresponding boundary conditions make up a mathematical model of a two-party system. We carry out a mathematical analysis of this problem. Equations (9.3), (9.4) imply the relations

$$\frac{d \ln x_1}{dt} + \frac{d \ln x_2}{dt} = 0, \quad \frac{d \ln y_1}{dt} + \frac{d \ln y_2}{dt} = 0.$$

After integration, we have the formulas

$$x_1(t)x_2(t) = c_1, \quad y_1(t)y_2(t) = c_2,$$

where the constants c_1, c_2 are determined by the initial conditions of the system.

Determining the functions x_2 and y_2 from the previous equalities, transform the system (9.3), (9.4) to the equations

$$\dot{x}_1 = \left(\alpha_2 c_2 y_1^{-1} - \alpha_1 y_1\right)x_1, \quad \dot{y}_1 = \left(\beta_1 x_1 - \beta_2 c_1 x_1^{-1}\right)y_1.$$

Replace the variables

$$\tau = at, \quad u(\tau) = bx_1(t), \quad v(\tau) = cy_1(t),$$

where the parameters a, b, c are chosen so that the preceding equations are as simple as possible. As a result, we obtain the following equations

$$u' = \frac{\alpha_2 c_2}{ac}\frac{u}{v}\left(1 - \frac{\alpha_1}{\alpha_2 c_2 c^2}v^2\right), \quad v' = \frac{\beta_2 c_1}{ab}\frac{v}{u}\left(\frac{\beta_1}{\beta_2 c_1 b^2}u^2 - 1\right).$$

Determine the values

$$a = \sqrt{\alpha_1 \alpha_2 c_1}, \ b = \sqrt{\frac{\beta_1}{\beta_2 c_1}}, \ b = \sqrt{\frac{\alpha_1}{\alpha_2 c_2}}, \ m = \sqrt{\frac{\beta_2}{\alpha_1 \alpha_2 \beta_1}}.$$

We get the system

$$u' = uv^{-1}(1 - v^2), \quad v' = mu^{-1}v(u^2 - 1). \tag{9.5}$$

Determine properties of this system. Suppose the following inequalities hold

$$0 < u(t) < 1, \ \ 0 < v(t) < 1 \tag{9.6}$$

at the initial time. By the equalities (9.5) the function u increases, and the function v decreases. When the function v approaches zero from the first equation it follows that the derivative u' tends to infinity. Thus, at some point in time, the function u will exceed the value of unity, and we get the relations

$$u(t) > 1, \ \ 0 < v(t) < 1.$$

As a result, an increase in both functions occurs, which continues until the second of them becomes more than one. After the inequalities

$$u(t) > 1, \ \ v(t) > 1,$$

then the function u decreases, and the function v increases until the time, in which the following equality holds $u = 1$. By the conditions

$$0 < u(t) < 1, \ \ v(t) > 1$$

both functions decrease. If the function u approaches zero, then according to the second equation (9.5), the derivative v' decreases monotonically, which means that over time the function v will reach unity. As a result, inequalities (9.6) will be satisfied, and then the process resumes. One can determine that the solutions of the system are periodic functions.

Let us analyze the obtained results from the point of view of political science. Suppose that at the initial moment of time the influence of the right party prevails, the incomes of the population are relatively low, and the production efficiency is quite high, which corresponds to the relations (9.6). Under these conditions, there is a serious increase in the popularity of the left party, because their potential voters are quite active in view of the low incomes of the broad masses of the people. Besides, the influence of the right one decrease, because the more affluent sections of the population, traditionally oriented toward them, remain relatively passive with high production efficiency. If parliamentary elections take place at this time, then the good left party will come to replace the government of the right one to the great joy of the broad masses.

In order to increase incomes of the population, honestly implementing its election program, the left-wing government increases the size of pensions and social benefits for the poor. This is achieved by increasing taxes (and where else?), which are paid mainly by the more affluent segments of the population. To combat unemployment, new jobs are being created. To this result, enterprises are nationalized, since it is precisely in state institutions under the rule of the left-wing party that a relatively high level of employment and relatively high salaries can be achieved for numerous workers. The consequence of the government policy is a gradual increase in the income of the broad masses and a clear decrease in social tension.

However, the glorious political course of the compassionate left party has, unfortunately, also negative consequences. Rigid tax policy of the government leads to the fact that a significant number of entrepreneurs become not interested in the development of domestic industry. They prefer to export their capital abroad, where the tax burden is not so severe. State-owned enterprises, which employ too many low-skilled and relatively well-paid workers and employees, become unprofitable and operate mainly through state subsidies. Funds for such subsidies and for the implementation of a broad social program go from the state treasury due to increasing taxes. However, in conditions of the export of a significant part of capital abroad and the massive tax evasion, the state receives less and less money from its irresponsible fellow citizens. With increasing costs and lowering state revenues, the government is forced to increase inflation, which leads to destabilization of production and a decrease in its efficiency, as well as an inevitable increase in prices[7].

The onset of the economic crisis, characterized by low production efficiency. This leads to an inevitable decline in the popularity of the left government and a significant increase in the influence of the right party. The poor, who have achieved a certain increase in their well-being, are no longer distinguished by high political activity and even dare to express obvious dissatisfaction with their benefactors in connection with a steady rise in prices. On the other hand, the political activity of wealthier citizens who are vitally interested in stabilizing the economy and improving production efficiency is increasing. In these conditions, the right-wing opposition wins in the next parliamentary elections to the general satisfaction of the ungrateful fellow citizens.

The new government begins to actively fight for the recovery of the economy. It drastically reduces government spending, significantly reduces taxes, and denationalizes production. Due to tax cuts, significant capital, including foreign capital, is sent to domestic enterprises. Unprofitable enterprises go bankrupt due to a serious reduction in government subsidies. In order to increase production efficiency, unprofitable enterprises are transferred to private hands, and the greedy owner, striving to achieve maximum profit, dismisses unnecessary employees who have forgotten how to work properly during the inglorious rule of the left party. One begin a long-awaited financial stabilization and a marked increase in production efficiency.

Unfortunately, the policy of liberalizing the economy has also undesirable consequences. The reduction of government spending leads to a decrease in the size of pensions, benefits, scholarships. Due to the transfer of state organizations to private hands and the ruin of unprofitable enterprises, unemployment is growing rapidly. The stabilization of the economy is achieved to a large extent due to the least protected sections of the population. As a result, in the country, social tension deepens. The onset of the socio-political crisis is accompanied by a sharply increased political activity of the broad masses. At the same time, the activity of the more affluent segments of the population receiving relatively large incomes is noticeably reduced. As a result, the popularity of the left opposition, heroically defending the interests of workers, is growing, and the influence of the right government is decreasing. Therefore, in the next parliamentary elections, a left-wing party comes to power. A new cycle of the process under consideration[8]. In practice, we really observe the alternate coming to power of parties with different political platforms[9].

Task 9.3 *Bipartisan system.* Based on the numerical solution of system (9.3), (9.4) with corresponding initial conditions make the following analysis.

1. Determining the system parameters in an arbitrary way, make sure that all the functions of the state of the system change periodically.

2. By changing the coefficients of the equations for fixed initial states of the system, find its equilibrium position.

3. By changing the initial states with constant coefficients of the equations, find the equilibrium position of the system.

4. Establish the dependence of the oscillation period on the parameter β_1.

5. To establish the dependence of the amplitude of the oscillation of the function y_2 on the parameter α_1.

Bipartisan system is described by the system of four nonlinear differential equations.
According to the bipartisan system model,
parties alternately succeed each other in power.

4   Trade union activity

Consider a simple dynamic model of trade union activity. The state functions here are the degree of influence of the trade union x_1 and social positions x_2 in the enterprise. It is assumed that with a significant deterioration in social positions, the influence of the trade union grows at a certain velocity ε_1, and in the complete absence of union influence, social positions deteriorate at a constant velocity ε_2. An increase in union influence is accompanied by an improvement in social positions, which in turn leads to a drop in union influence. Thus, the mathematical model of this process is the system

$$\dot{x}_1 = \big(\varepsilon_1 - \gamma_1 x_2\big)x_1, \ \ \dot{x}_2 = \big(\gamma_2 x_1 - \varepsilon_2\big)x_2,$$

where the constants γ_1 and γ_2 are parameters of the model.

The above equations have already been encountered previously in the description of chemical, biological, and economic processes. We are again dealing with the Volterra–Lotka equations, i.e., with the predator–prey model. As you know, they have a periodic solution (see Figure 9.1). Give an interpretation of these results.

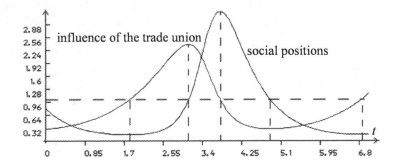

FIGURE 9.1: Oscillations of trade union influence and social positions.

If at the initial time the influence of the trade union is weak enough, and social positions are low, then the value on the right-hand side of the first equation is positive, and for the second one it is negative. Thus, at the first stage of the process, there is an increase in the influence of the trade union, because more and more workers, in poor social conditions at work, turn to the union for help. Besides, we observe a further deterioration of social positions, because with weak influence of the trade union, the greedy entrepreneur seeks to maximize profits due to the brutal exploitation of disenfranchised workers. With the growing influence of the trade union, the moment comes to the transition to the second stage, when the entrepreneur is already forced to reckon with the trade union and, under the threat of a strike, goes to a certain satisfaction of the fair demands of his employees. The improvement of social positions is taking place with the continuous growth of the influence of the trade union, which enjoys growing support from workers and employees. However, as living standards rise for ungrateful people, the need to support the union gradually disappears. In the third stage, the influence of the trade union begins to decline, while social positions in the workplace are still growing. Indeed, the union is still strong enough, and therefore the entrepreneur does not risk holding any bad events. However, with the decline in the influence of the trade union, the greedy owner begins to struggle to increase his income, naturally, at the expense of innocent workers. At the fourth stage, a further decrease in the influence of the union is observed, since social positions are still relatively high, and the union did not manage to gain strength. One observes the deterioration of living conditions. However, with the worsening living standards of workers, the union regains its influence again and the process repeats.

New interpretations of the predator–prey model are provided in the Appendix.

The change in the influence of the trade union and
social positions at work is described by the Volterra–Lotka equations.
According to this model, system characteristics are periodic functions.

5 Allied relations

Consider two countries fighting a common adversary. It is assumed that the enemy is much stronger and able to defeat each of the allies individually. However, with their active interaction, the adversary may be defeated. The stronger one ally, the easier it is to wage war on another ally. As a result, we get the equations

$$\dot{x}_1 = (\gamma_1 x_2 - \varepsilon_1)x_1, \quad \dot{x}_2 = (\gamma_2 x_1 - \varepsilon_2)x_2,$$

where x_i is the force of i-th ally, ε_i is the velocity of its defeat in the absence of another ally, and γ_i is the influence parameter of j-th ally to the i-th one.

This **allied relations model**, up to the notation, coincide with the biological model of symbiosis and the model of economic cooperation. Knowing the properties of the latter, we have the following results. If both allies are relatively weak, then the enemy wins completely. Otherwise, the forces of the allies will increase unlimitedly. In the case when one of the allies is weak enough and the other is strong, the outcome of hostilities is determined by the ratio between all the parameters of the system. Here, victory and defeat of the allies are possible.

Task 9.4 *Allied relations.* Based on the numerical solution of the system make the following analysis.

1. Find the equilibrium position of the system and analyze the behavior of the system in their neighborhoods.

2. Make sure that with relatively small values of the initial states of the system, the allies are defeated.

3. Make sure that with relatively large values of the initial states of the system, the allies win.

4. For the case when one of the allies is strong enough and the other is comparatively weak, establish both victory and defeat of the allies. Provide an interpretation of current events.

Allied relations are described by the symbiosis model.
According to the model, both victory and defeat of allies are possible
depending on the initial states of the system.

Direction of further work. *After consideration the simplest models characterized by systems with lumped parameters, we move on to systems with distributed parameters. The corresponding state functions depend on several variables. We have to work with partial differential equations.*

Appendix

The following are new interpretations of models that we are familiar with such as competition (the struggle for leadership in a group and tribes over territory), niches (split of a group, separation of tribes, and family relations) and predator–prey (effectiveness of the state apparatus, relations between the metropolis and the colony).

1   Competition models

Let us consider new variants of competition models. The first of them is the struggle for group leadership. There is an isolated group of people who, for one reason or another, find themselves quite closely related. This may be a newly formed professional team, a group of random companions, suddenly found themselves in long-term isolation from the outside world, a group of tourists, a gang of bandits, etc. In such a community, there is usually an official leader initially recognized by all. However, a situation is possible where either such a leader was not originally, or he dropped out of the group for any reason, or completely discredited himself. In such circumstances, as a rule, someone from the group tries to take the whole initiative on themselves. Moreover, often several people claim leadership. We consider a situation when there are two potential leaders in a group, between which there is a struggle for influence in the group.

Determine functions x_1 and x_2, which characterize the degree of influence of each of the leaders. Each of the leaders has a certain activity, as a result of which the influence of the i-th leader increases with velocity ε_i in the case of unlimited group. However, in reality, the growth in popularity of leaders decreases the more, the more members of the group have already given preference to one of the leaders at the moment. As a result, we arrive at the equations

$$\dot{x}_i = \left[\varepsilon_i - \frac{1}{\beta_i}(x_1 + x_2)\right]x_i, \quad i = 1, 2, \tag{9.7}$$

where the coefficients β_i are the parameters of the problem and characterize the degree of attractiveness of the plans proposed by this leader.

Once again we get a competition model, see, in particular, equations (9.1). As you know, the leader who has higher activity and more attractive plans, or rather, the one whose product $\varepsilon_i \beta_i$ takes on the greatest value, wins here. Thus, over time, the whole group recognizes one leader, who turned out to be the most authoritative.

Another version of the competition type model is associated with the **struggle of tribes** for territory. It is assumed that the tribes have the same way of life and are has the same habitat. In the case of small numbers of people and unlimited free territory, tribes are constantly developing new lands. However, in conditions of limited uninhabited territory, they enter into competition among themselves. It is necessary to establish the change over time of the territory controlled by each of the tribes, provided that the initial values of these quantities are known. The considered process is described by the same following system of nonlinear differential equations (9.7), where x_i is the size of the controlled territory of the i-th tribe, ε_i is the activity of the tribe, and β_i is the development efficiency of the territory. We get another version of the competition model. Its outcome is predetermined. A stronger tribe will gradually take over the entire territory, absorbing or destroying all its competitors.

A comparative analysis of competition models in various interpretations is given in Table 9.1.

TABLE 9.1: Characterization of competition models.

Science	model	state function	competition object
physics	two-mode laser	numbers of different photon classes	excited atoms
chemistry	"competition" of reacting substances	concentration of "competing" substances	reactant
biology	competition of biologic species	abundance of species	food
economics	economic competition	funds of firms	consumers of goods
political science	competition of parties	influence of parties	votes
sociology	struggle of tribes	occupied territory	free territory
psychology	struggle for group leadership	influence of leaders	group recognition

2 Niche models

Let us return to the consideration of the problem of competition of leaders in a group. If each of the potential leaders does not seek to enlist the support of the whole team, but focuses mainly on some grouping united by common interests, then we again come to the niche model. With some combination of parameters (see above), one leader is not wined by another, and both of them retain a certain influence. Each of them eventually becomes the head of a part of the collective. Thus, there is a split of the group into two more or less isolated and significantly more stable (compared to the original team) groups with their recognized authorities[10]. This process of the group split is described by equations (9.2)

$$\dot{x}_1 = \left(\varepsilon_1 - \beta_{11}^{-1}x_1 - \beta_{12}^{-1}x_2\right)x_1, \quad \dot{x}_2 = \left(\varepsilon_2 - \beta_{21}^{-1}x_1 - \beta_{22}^{-1}x_2\right)x_2. \tag{9.8}$$

We now turn to the problem of the struggle of tribes for territory. Suppose that each tribe prefers a special habitat, for example, some prefer the steppes or river valleys, while others feel more comfortable in the mountains or forests. Then events may develop in a different scenario. In this case, the process is described by equations (9.8), where x_i is the territory controlled by the parameters of the i-th tribe, ε_i is the activity of the tribe, and β_{ij} characterize the effectiveness of the development of the i-th tribe of the j-th territory. With a certain combination of parameters (see previous chapters), it is possible a ***demarcation of tribes***. Each of them gradually captures the territory to which it gives greater preference. It is no accident that the borders of states most often have a natural character corresponding to the line of demarcation of the peoples inhabiting them.

Consider another example of a niche model, now from psychology. We will analyze family relationships. There is a family consisting of a husband, wife, and his mother. It is assumed that the mother-in-law and daughter-in-law are in a warm family relationship, struggling to assert their influence on the only man in the family. And he, the poor fellow, is torn between an ardent love for his mother and wife. The state functions here are the degrees of influence x_1 and x_2, respectively, of the mother-in-law and the daughter-in-law on the beloved son and husband.

It is assumed that the rate of change in the influence of each of the women is proportional to the value of this value. Both the mother-in-law and the daughter-in-law have an excellent relationship with the "head of the family". The increase in their influence includes a certain positive value ε_i, determined by the merits of this woman. If the increase in influence were limited only by this factor, then we would observe an unlimited increase in the influence of each of the women. In practice, the growth of influence is naturally held back. In particular, there is also a drop in influence, characterized by the expression $-\alpha_{ii}x_i$, where the parameter

α_{ii} characterizes the decrease in the influence of the i-th woman, determined by her personal imperfections according to the man. Finally, the mother-in-law and the daughter-in-law strive to reduce the influence of the rival on the man, as a result of which the increase in the influence of the i-th woman also includes the negative expression $-\alpha_{ij}x_j$, where the constant α_{ij} characterizes the degree of antipathy to this woman by her opponent.

Using these suppositions, we describe the considered phenomenon by the system

$$\dot{x}_1 = (\varepsilon_1 - \alpha_{11}x_1 - \alpha_{12}x_2)x_1, \quad \dot{x}_2 = (\varepsilon_2 - \alpha_{21}x_1 - \alpha_{22}x_2)x_2,$$

which is equivalent to (9.8). Depending on the combination of parameters, events can develop in different scenarios.

Under the inequalities

$$\varepsilon_1/\alpha_{11} > \varepsilon_2/\alpha_{21}, \quad \varepsilon_1/\alpha_{12} > \varepsilon_2/\alpha_{22},$$

daughter-in-law finally loses mother-in-law. As a result of this, her influence on her spouse gradually disappears, which inevitably entails a divorce or unconditional submission of the daughter-in-law to a decisive and imperious mother-in-law. Opposite relations on the contrary mean a complete defeat of the mother-in-law. She either obeys a more energetic daughter-in-law, or is forced to live separately from ungrateful spouses. The conditions

$$\varepsilon_1/\alpha_{11} = \varepsilon_2/\alpha_{21}, \quad \varepsilon_1/\alpha_{12} = \varepsilon_2/\alpha_{22},$$

mean the equality of forces of the warring parties. Not one of them is able to finally win. Inequalities

$$\varepsilon_1/\alpha_{11} > \varepsilon_2/\alpha_{21}, \quad \varepsilon_1/\alpha_{12} < \varepsilon_2/\alpha_{22}$$

correspond to the case when the decisive influence is not the personal qualities of women, i.e., the relationship of advantages to disadvantages, but by the degree of their mutual antipathy. The outcome here also leaves much to be desired. Depending on the initial state of the system, i.e., the relations of the warring parties with the man toward the beginning of their joint residence, either the daughter-in-law or the mother-in-law will win. Finally, the following case is characterized by the conditions

$$\varepsilon_1/\alpha_{11} < \varepsilon_2/\alpha_{21}, \quad \varepsilon_1/\alpha_{12} > \varepsilon_2/\alpha_{22}$$

Here, the personal qualities of women prevail over their mutual antipathy. Each of them finds its own niche, and peace and tranquility in the family come over time[11].

3   Predator–prey models

Consider the functioning of the state apparatus during the implementation of large-scale economic and political reforms. We will be interested in, first of all, the effectiveness of the reforms, which can be conditionally characterized by the function x_1 changing over time. To implement reforms, an appropriate state apparatus is needed, the number of which x_2 is chosen as the second state function. We will assume that under natural conditions, in the absence of opposition to economic transformation on the part of the state apparatus, reforms develop at a velocity ε_1. However, as the number of government officials grows, there is a proportional decline in the pace of reform. In turn, in the event of a failure of economic reforms, the state apparatus is reduced at a given speed of ε_2. With the deepening of reforms, it is necessary to attract an increasing number of government officials to implement the government program. As a result, we get the equations

$$\dot{x}_1 = (\varepsilon_1 - \gamma_1 x_2)x_1, \quad \dot{x}_2 = (\gamma_2 x_1 - \varepsilon_2)x_2,$$

TABLE 9.2: Predator–prey models[12].

Science	model	predator	prey
chemistry	Lotka system	first product	second product
biology	predator–prey	predator	prey
agriculture	yield–fertility	yield	fertility
economics	free market	price	income
economics	monopolized market	volume of goods	income
economics	racketeer–entrepreneur	racketeer	entrepreneur
sociology	trade union activity	social positions	trade union
sociology	state apparatus	state apparatus	reforms
sociology	metropolis–colony	colony	metropolis

where γ_1 and γ_2 are system parameters. The resulting model of the state apparatus is similar to the predator–prey model and has periodic solutions.

As the pace of reform increases, there is a need to increase the number of employees involved in government administration who are called upon to directly implement government decisions. However, the expanded state apparatus is gradually becoming a brake on reforms, and the pace of their implementation is steadily decreasing. The slowdown in the pace of reform indicates a low efficiency of the state apparatus. To overcome the difficulties that have arisen, a reorganization of the apparatus is being carried out with a constant reduction in the number of state officials. The state apparatus is becoming more mobile, and the pace of reform is accelerating. However, for their maintenance, the available number of civil servants is for some reason not enough. The apparatus is gradually growing, and the course of reforms, on the contrary, is slowing down. We come to a new stage in the process under consideration. The results obtained confirm the amazing ability of bureaucrats to survive in any state upheaval.

The relationship between the colony and the metropolis is now being considered. It is assumed that in the absence of a metropolis the colony becomes richer, and in the absence of a colony the metropolis becomes poorer. The richer the colony, the more the metropolis receives from it. The stronger the metropolis, the more it exploits the colony. As a result, we conclude that the ***metropolis–colony model*** is yet another version of the predator–prey model. The outcome of the events is known to us. The metropolis seeks to enrich itself as much as possible at the expense of the colony. However, with the gradual weakening of the colony, the metropolis is forced to reduce the amount of tribute levied. As a result, over time, the colony regains its wealth, after which the metropolis for some time gets the opportunity to more intensively exploit the colony. After this, the colony is poorer, and after it the metropolis and events begin to repeat itself.

Task 9.5 *Models of the relationship of two subjects.*
1. Give a new interpretation of the competition model.
2. Give a new interpretation of the niche model.
3. Give a new interpretation of the symbiosis model.
4. Give a new interpretation of the predator–prey model.

The characteristics of the models of the main classes of the relationship between the two subjects are given in Table 9.2[13].

TABLE 9.3:   Characteristics of two-subject models.

Science	Predator–prey	Competition	Niche	Symbiosis
physics		two-mode laser		
chemistry	Lotka system	chemical "competition"	chemical "niche"	
biology	predator–prey	biological competition	ecological niche	symbiosis
agriculture	yield–fertility			
economics	free market monopolized market racketeer–entrepreneur	economic competition	economic niche	economic cooperation
sociology	trade union activity state apparatus metropolis–colony	struggle of tribes	demarcation of tribes	allied relationships
political science	political competition	political niche		
psychology	struggle for group leadership	group split family relationships		

Notes

[1] Mathematical models of political science are considered, for example, in [5], [166], [224], [271].

[2] Mathematical models of sociology are considered in [40], [48], [50], [70], [130], [136], [149], [152], [173], [208], [219], [339], [358]. In addition to differential equations, the methods of probability theory and mathematical statistics is used to describe the phenomena of sociology, see [173], [70]; and game theory, see [48], [70]. Chapter 18 is devoted to stochastic models, and Chapter 17 is devoted to the use of game theory.

[3] Mathematical models of psychology are considered, in [5], [166], [224], [271].

[4] As we already know, the presence of three or more competing parties leads to similar results, see, in particular, Chapter 7.

[5] Naturally, it is assumed here that the party in power complies with the rules of the game, allowing the opposition to openly uphold their political platform.

[6] The redistribution of something from an area where there is a lot of it to an area where it is not enough is characteristic of transfer processes (heat conduction, diffusion, etc.), see Chapters 10 and 11.

[7] See the inflation model in a previous chapter.

[8] We have already considered the systems with periodic solutions. They were described by qualitatively different equations. In particular, the oscillation equation of the pendulum is linear, in the predator–prey model we had a system of two nonlinear equations, and in this case the process is described by four nonlinear equations. However, in all these cases, we observe a balance between two objects. There are kinetic and potential energies, predators and preys, left and right parties. Everywhere the presence of this balance makes the system relatively stable. In all cases, the corresponding equilibrium position is the center.

[9] It is characteristic that the higher the economic level in the country, the longer the period of fluctuation. We also note that in this case, as in all others, we did not describe the real situation, but the result of modelling the process under study. Modern left and right parties in democracies have long learned to pursue a more flexible policy and are not so different in their socio-economic programs. The political life of society is much more complex and only to a small extent lends itself to clear mathematical modelling. However, the same should be said about physical, chemical, and other processes. It hardly makes sense to sharply contrast the natural sciences and the humanities.

[10] Similar results can have a slightly different interpretation. Perhaps leaders will seek their niche not in gaining popularity among part of the group members, but in securing for themselves a certain area of activity in which each of them is recognized as an unconditional authority. Therefore, when forming a new creative team, one of the leaders can take on the solution of professional issues, and the other prefers of organizational matters. In a tourist group, leaders can divide economic and tactical functions among themselves. In such a situation the group continues to exist as a whole under the indispensable condition that each of the leaders recognizes the primacy of the leadership in its field of activity.

[11] The described model does not take into account the reaction to the events of the culprit and the subject of contention. However, is it really necessary to ask the opinion of an authoritative head of the family when

a serious dispute is between loving women? By the way, the relationship between the mother-in-law and the son-in-law is characterized by no less exciting events.

[12] Table 9.3 is an extension of Table 8.7.

[13] Some of the models listed here should not be taken too seriously. In a way, this is just a game. In a sense, modeling is always a game. However, the game is not useless.

Part II

Systems with distributed parameters

Part I of the course examined the various processes characterized by lumped parameter systems. The state functions of these systems depend on a single variable that is the time. Their mathematical models are based on ordinary differential equations. In this case, the corresponding phase space is finite dimensional, i.e., the state of the system at each moment of time can be characterized by a finite set of numbers.

However, in practice, the state of the system often changes not only over time, but also from point to point of the considered object. The corresponding state functions already depend on several arguments, among which there are one or more spatial variables. Here, partial differential equations are already used as state equations. They correspond to an infinite dimensional phase space. This set of properties corresponds to the distributed parameter systems that are the subject of Part II of this course.

This Part consists of six chapters. Chapters 10 and 11 are devoted to the transfer processes. They are characterized by the fact that in some environment some substance is unevenly distributed (heat, mass, population of a biological species, goods, information, etc.), then due to some circumstances a redistribution of this substance is observed from an area where it is relatively abundant, into an area where it is significantly less. The typical example of these processes is described by the heat equation that is a parabolic partial differential equation. In these lectures, deal with transfer processes related to different subject areas. Qualitative and quantitative methods of their analysis are described.

In Chapter 12, wave processes are considered. They are distributed analogues of the mechanical and electrical oscillations described in the first part of the course. Examples of such processes are vibrations of a string or membrane and the propagation of waves of various nature, for example, electromagnetic, sound, etc. Such processes are described by equations of a hyperbolic type, a typical example of which is the equation of string vibrations.

A feature of the simplest transfer processes is the tendency of the system to a equilibrium position, and the simplest wave processes are the fluctuations around the equilibrium position. If these equilibrium positions are set as the initial state of the system, then the state functions are not change with time, and the system turns out to be stationary. In Chapter 13, stationary systems are considered. They are described by partial differential equations of elliptic type, examples of which are the Poisson equation and its homogeneous analogue Laplace equation. Such equations describe, in particular, electrostatic, gravitational, and some other types of physical fields.

In practice, it is often done to deal not with one, but with several interconnected state functions of a system of different nature. In this case, the mathematical model of this process is based on a system of heterogeneous equations. Vivid examples of such problems, which are of extremely important theoretical and practical importance, are given by the mechanics of fluid and gas, see Chapter 14. Various systems of equations relating the velocity vector, pressure, and density of a fluid or gas are considered here.

Final Chapter 15 is devoted to the problems of quantum mechanics. The considered mathematical models are based on the amazing Schrödinger equation. This is a partial differential equation with respect to a wave function that takes complex values and has a probabilistic interpretation.

Chapter 10

Mathematical models of transfer processes. 1

We begin the analysis of mathematical models characterized by systems with distributed parameters, by considering the transfer processes. They are characterized by the transfer of a certain substance (heat, material, charge, etc.) from the area where its concentration (the amount of substance per unit length, area or volume) is high enough to the area where the value of this concentration is relatively small. Under these conditions, for certain reasons, a redistribution of the considered substance is observed, as a result of which the system tends to some equilibrium state over time. A similar class of phenomena includes thermal conductivity, diffusion, electrical conductivity, etc. All of them under certain conditions are described by identical partial differential equations of the type of heat transfer.

In the first chapter on transfer processes, we mainly consider the heat equation, which describes the temperature redistribution in a certain region under the influence of heat fluxes. It is the basis of mathematical models of thermal physics[1]. The chapter gives the derivation of the heat equation and the formulation of the most important boundary value problems for it. Using the method of separation of variables, this equation is solved with a known temperature at the initial moment of time and at the boundary of the considered region.

The Appendix describes some generalizations of the heat equation. The heat equation with a known heat flux at the boundary of the given interval is solved. We consider also the mathematical model of the diffusion process that describes the distribution of the concentration of a liquid or gas in a certain volume and is also characterized by the heat equation.

Lecture

1 Heat equation

A non-uniformly heated body is considered. Some of its parts are warmer than others. Over time, for some reason, the cooling of the warmer areas and the heating of the colder occur[2]. Thus, heat transfer is observed in this body. We construct a mathematical model of the described phenomenon.

The state function of the studied process is the **temperature** u. Obviously, the object in question is distributed, as long as the temperature changes from point to point of a given region. For simplicity, we assume that this body is quite long and thin. In this case, we can assume that all the characteristics of the process are averaged over the cross section and depend on the only spatial coordinate x directed along its axis[3]. It is necessary to obtain a relation to determine the dependence of temperature on time and spatial variable. Thus, the problem consists in determining the function of two variables $u = u(x, t)$.

At first, determine the change of heat quantity in a certain part of the body from the point x to $x + \Delta x$ over time from t to $t + \Delta t$. In order to heat a body of mass m with

DOI: 10.1201/9781003035602-10

temperature u_1 to temperature u_2, it is necessary to expend heat

$$Q = cm(u_2 - u_1), \tag{10.1}$$

where the proportionality coefficient c is called the **heat capacity** and is a characteristic of the material of which the body consists.

The mass in the last equality is not a property of the material, i.e., system parameter. This is because the mass largely depends on the size of the object. The real parameter of the system that determines the mass of an object is the **density** characterizing the mass of a unit volume. Since in this case the body is one-dimensional, its linear density ρ is considered, i.e., mass per unit length. Therefore, the mass of a homogeneous body of length Δx is $m = \rho \Delta x$. Thus, equality (10.1) is reduced to the following form

$$Q = c\rho(u_2 - u_1)\Delta x.$$

The disadvantage of this formula is the assumption that all the characteristics of the process remain unchanged in the area $[x, x + \Delta x]$. In reality, at least the temperature of the body, and in the case of an inhomogeneous body, i.e., consisting of an inhomogeneous material, and the parameters c and ρ characterizing the properties of the material vary from point to point. Thus, this formula makes sense only on an arbitrarily small segment[4] $[x, x + dx]$. Denoting it by the length dx, and the corresponding value of the amount of heat by dQ, we obtain the equality[5]

$$dQ = c\rho(u_2 - u_1)dx.$$

Suppose u_1 is a temperature of the body at the time t, and u_2 is its temperature at the next time $t + \Delta t$. Then the change of heat quantity over time from t to $t + \Delta t$ in the small considered area is

$$dQ = c\rho\big[u(x, t + \Delta t) - u(x, t)\big]dx.$$

However, we consider not an idealized arbitrarily small interval $[x, x + dx]$, but an extended segment $[x, x + \Delta x]$. Obviously, the value

$$Q_1 = \int\limits_{x}^{x+\Delta x} c\rho\big[u(x, t + \Delta t) - u(x, t)\big]dx \tag{10.2}$$

characterizes the heat quantity that must be expended to heat[6] the considered part of the body at a given time interval[7].

Thus, we have set the value of the change in the heat quantity. Now it remains to be determined by what means it could have happened. There are many reasons that can cause a change in body temperature. This is the mechanical movement of a heated body (convective heat transfer), and heat exchange with the environment, and thermal radiation, and the action of external heat sources (furnace, chemical reaction, laser, etc.)[8]. However, in this case, we restrict ourselves to considering changes in body temperature solely due to the phenomenon of **heat conduction**, in which there is a transition of heat from the warmer part of the body to the colder.

This process can be characterized by the density of the heat flux or, briefly, the **heat flux** q, which is the heat quantity passing per unit time through a given point[9]. If the heat flux remains unchanged, then during the time from t to $t + \Delta t$ the following heat quantity pass through this point $Q = q\Delta t$. In the case of a variable heat flux, this equality only makes sense on an arbitrarily small time interval dt

$$dQ = qdt.$$

Then the heat quantity passing on the time interval $[t, t + \Delta t]$ is equal to the integral

$$Q = \int_t^{t+\Delta t} q\, dt.$$

We are interested in the change of heat quantity over a given time interval in the $[x, x + \Delta x]$ interval. For definiteness, we choose the direction of the coordinate coinciding with the direction of the heat flux, see Figure 10.1. Then, at the point x, the flux enters the region under consideration, and at the point $x + \Delta x$ it exits[10]. Thus, the change of heat quantity in this area over the studied time interval is equal to

$$Q_2 = \int_t^{t+\Delta t} \left[q(x, t) - q(x + \Delta x, t) \right] dt. \tag{10.3}$$

Determine the relation between the heat flux and the temperature for transforming this formula.

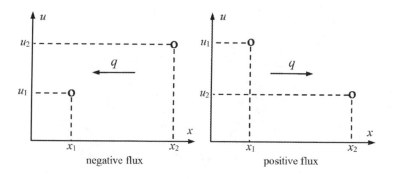

FIGURE 10.1: Directions of the heat flux.

Consider two points with coordinates x_1 and x_2, with $x_1 < x_2$. Let u_1 and u_2 denote the temperatures of these points. We estimate the heat flux from one point to another. Obviously, it will be the larger, the greater the temperature difference and the smaller the distance between the points. Note that heat spreads from a hot body to a cold one. Thus, for $u_1 < u_2$, the direction of the heat flux is positive, and for $u_1 > u_2$ it is negative (see Figure 10.1). As a result, we get the formula

$$q = -\lambda \frac{u_2 - u_1}{x_2 - x_1},$$

where the proportionality coefficient λ is called the thermal conductivity coefficient or briefly **thermal conductivity** and is a process parameter characterizing the body's ability to conduct heat.

When considering not two separate points, but a continuous body, the last equality retains meaning with arbitrarily large proximity of the considered points. By passing the point x_2 to x_1, we obtain on the right-hand side of the last equality the derivative of temperature with respect to the spatial variable. Thus, we get the equality

$$q = -\lambda \frac{\partial u}{\partial x}$$

that is called the **Fourier law**. Putting the value of the heat flux to the equality (10.3), we obtain

$$Q_2 = \int\limits_t^{t+\Delta t} \left(\lambda \frac{\partial u}{\partial x}\Big|_{x+\Delta x} - \lambda \frac{\partial u}{\partial x}\Big|_x \right) dt. \tag{10.4}$$

We assume that the heat balance in this area in the studied time interval is carried out exclusively due to the phenomenon of heat conduction. Thus, the equality $Q_1 = Q_2$. Using the formulas (10.2), (10.4), we get

$$\int\limits_x^{x+\Delta x} c\rho[u(x,t+\Delta t) - u(x,t)]dx = \int\limits_t^{t+\Delta t} \left(\lambda \frac{\partial u}{\partial x}\Big|_{x+\Delta x} - \lambda \frac{\partial u}{\partial x}\Big|_x \right) dt. \tag{10.5}$$

To transform formula (10.5), the **mean value theorem** is used. According to this statement, if the function $f = f(y)$ is continuous on some interval $[a,b]$, then on this interval there exists a point c such that the equality[11]

$$\int\limits_a^b f(y)dy = f(c)(b-a).$$

Using this theorem for both integrals of the equality (10.5), we get

$$c(\xi)\rho(\xi)\big[u(\xi,t+\Delta t) - u(\xi,t)\big]\Delta x = \left[\lambda(x+\Delta x)\frac{\partial u(x+\Delta x, \tau)}{\partial x} - \lambda(x)\frac{\partial u(x,\tau)}{\partial x}\right]\Delta t,$$

where $\xi \in [x, x+\Delta x]$, $\tau \in [t, t+\Delta t]$. Dividing by $\Delta x \Delta t$, we obtain

$$c(\xi)\rho(\xi)\frac{u(\xi,t+\Delta t) - u(\xi,t)}{\Delta t} = \frac{\lambda(x+\Delta x)u_x(x+\Delta x, \tau) - \lambda(x)u_x(x,\tau)}{\Delta x}.$$

Passing to the limit[12] as $\Delta t \to 0$, $\Delta x \to 0$, determine the equality

$$c\rho\frac{\partial u}{\partial t} = \frac{\partial}{\partial x}\left(\lambda\frac{\partial u}{\partial x}\right) \tag{10.6}$$

that is true for all arguments x and t. This relation includes the partial derivatives of the unknown function u with respect to the variables x and t[13]. As a result of this, we are dealing with a **partial differential equation**[14]. This is called the **heat equation** and is the basis of the mathematical model of the studied process. If the considered body is homogeneous, then the parameters c, ρ, and λ are constant, and the heat equation takes the form[15]

$$u_t = a^2 u_{xx}, \tag{10.7}$$

where u_t and u_{xx} are the first derivative with respect to t the second derivative with respect to x of the function u, and the parameter $a = \sqrt{\lambda/c\rho}$ is called **thermal diffusivity** coefficient.

As we know, the mathematical model of the system is not only a differential equation, but also the boundary conditions that supplement it. Previously, state functions depended on a single variable, i.e., time. In this case, a spatial variable was also added. In this regard, the mathematical model of the considered process includes conditions both in time and in spatial variable.

Consider the heat equation for a body of length L. We assume that the left end is located at the origin, and the right has the coordinate $x = L$. We will study the process starting at time $t = 0$. The heat equation is first order in time and second in spatial coordinate[16]. For the uniqueness of the solution of the problem, it is required to set an additional corresponding number of boundary conditions.

The time condition is quite natural. We know the temperature at the initial time[17]. It is characteristic that the initial body temperature can vary from point to point. Then the initial condition has the form

$$u(x, 0) = \varphi(x),\tag{10.8}$$

where φ is a known temperature distribution. In addition to this initial condition, two boundary conditions must be specified, since the second derivative of the desired function enters the heat equation. We can have here different cases.

On the left end of the body, i.e., at the point $x = 0$, the law of temperature change α can be given

$$u(0, t) = \alpha(t).$$

This is the ***first type boundary condition***. An alternative here may be to set the heat flux β on the left boundary, which corresponds to the ***second type boundary condition***.

$$\lambda u_x(0, t) = \beta(t).$$

In particular, if $\beta = 0$, then we have the ***thermal insulation*** condition. Sometimes, one considers the ***third type boundary condition***

$$\lambda u_x(0, t) = k\big[u(0, t) - u_0(t)\big],$$

which describes the heat exchange with the environment, where the parameter k is the ***heat transfer coefficient***, and the function u_0 is the ambient temperature. Similar conditions can be set on the right end of the body.

We restrict ourselves to the case when the mathematical model includes one boundary condition at each end of the body[18]. Since the behavior of the system at both boundaries is independent of each other, various combinations of boundary conditions are permissible. In the future, we restrict ourselves to the consideration of the most important problems when the same conditions are set at the ends of the body. In particular, the ***first boundary value problem*** includes the heat equation in the form of (10.6) or (10.7) with the initial condition (10.8) and two first type boundary conditions

$$u(0, t) = \alpha_1(t), \quad u(L, t) = \alpha_2(t),\tag{10.9}$$

where the functions α_1 and α_2 are given. There are replaced by the second type boundary conditions

$$\lambda u_x(0, t) = \beta_1(t), \quad \lambda u_x(L, t) = \beta_2(t)\tag{10.10}$$

for the ***second boundary value problem***, where the functions β_1 and β_w are known.

Below we find a solution to the first boundary value problem for the heat equation. The corresponding second boundary value problem is considered in the Appendix[19].

The process of heat transfer due to the thermal conductivity phenomenon
is described by a partial differentiation equation.
The mathematical model of this process
is a boundary value problem for the heat equation.

2 First boundary value problem for the homogeneous heat equation

Consider the heat equation

$$u_t(x,t) = a^2 u_{xx}(x,t), \; 0 < x < L, \, t > 0 \qquad (10.11)$$

with initial condition

$$u(x,0) = \varphi(x), \; 0 < x < L \qquad (10.12)$$

and first type boundary conditions

$$u(0,t) = 0, \; u(L,t) = 0, \, t > 0, \qquad (10.13)$$

where the function φ and parameters a and L are known[20]. This first boundary value problem is called **homogeneous**, because of the property of the boundary conditions[21].

Try to find the solution of the problem by the formula

$$u(x,t) = X(x)T(t), \qquad (10.14)$$

where the functions X and T are chosen so as to achieve the desired result. In particular, substituting this value in the boundary conditions (10.13), we have

$$X(0)T(t) = 0, \; X(L)T(t) = 0, \, t > 0.$$

These equalities can be satisfied in two cases. Particularly, the function T is identically equal to zero, or the function X vanishes on the boundary. In the first case, according to the formula (10.14), the search function u vanishes everywhere. However, this contradicts the initial condition (10.12) if the trivial case $\varphi = 0$ is excluded from consideration. Thus, come to the equalities

$$X(0) = 0, \; X(L) = 0. \qquad (10.15)$$

Put the function u from the equality (10.14) to the formula (10.11); we get

$$X(x)T'(t) = a^2 X''(x)T(t).$$

Then we have

$$\frac{T'(t)}{a^2 T(t)} = \frac{X''(x)}{X(x)}.$$

Here on the left-hand side is a term that depends only on t, and the value at the right-hand side depends only on x. Such equality is possible only if these values are constants. Denoting the corresponding constant by λ, we get two equalities

$$T'(t) - \lambda a^2 T(t) = 0, \, t > 0, \qquad (10.16)$$

$$X''(x) - \lambda X(x) = 0, \, 0 < x < L. \qquad (10.17)$$

We have obtained two ordinary differential equations characterizing the temporal and spatial parts of a hypothetical solution to the problem and connected by a common constant λ. Due to the possibility of studying equations (10.16) and (10.17) independently of each other, the method used to analyze the considered boundary value problem is called the **method of separation of variables**.

Note that for the ordinary second order differential equation (10.17) we previously obtained the boundary conditions (10.15). Thus, we have the **boundary value problem** for the considered equation[22]. First, we note that it has an identically zero solution. It may seem that, as a result, this task is meaningless, since after substituting such a solution in formula (10.14), we obtain the function u identically equal to zero, which contradicts the initial condition (10.12). However, in problem (10.17), (10.15) not only the function X is unknown, but also the constant λ. Therefore, one can expect that for some values of λ the boundary value problem will have a non-trivial, i.e., non-zero solutions. The problem of determining non-trivial functions X satisfying equation (10.17) with boundary conditions (10.15) and the corresponding values of λ is called the **Sturm–Liouville problem**. Such solutions of the boundary value problem are called the **eigenfunctions** of the problem, and the corresponding values of λ are called the **eigenvalues**[23].

The characteristic equation $z^2 - \lambda = 0$ for the problem (10.17) has the solutions $z_{1,2} = \pm\sqrt{\lambda}$. The general solution of the equation (10.17) for the positive λ is

$$X(x) = c_1 e^{\sqrt{\lambda}x} + c_2 e^{-\sqrt{\lambda}x}.$$

Putting it to the boundary conditions (10.15), we conclude that the constants c_1 and c_2 and the function X too are equal to zero. For $\lambda = 0$, the general solution of the equation is

$$X(x) = c_1 + c_2 x.$$

Using (10.15), we get $X = 0$. Finally, for $\lambda < 0$ we find

$$X(x) = c_1 \sin \sqrt{-\lambda}x + c_2 \cos \sqrt{-\lambda}x.$$

By the first equality (10.15), we obtain $X(0) = c_2 = 0$. Determine the value

$$X(L) = c_1 \sin \sqrt{-\lambda}L = 0.$$

The constant c_1 cannot be zero, otherwise the solution will be trivial. So, $\sin \sqrt{-\lambda}L = 0$. Now we find the numbers

$$\lambda_k = -\left(\frac{k\pi}{L}\right), \ k = 1, 2, \dots.$$

There are the eigenvalues. The corresponding eigenfunctions are

$$X_k(x) = c_k \sin \frac{k\pi}{L}x,$$

where the constants c_k are arbitrary, $k = 1, 2, \dots$.

For $\lambda = \lambda_k$ the equation (10.16) has the general solution

$$T_k(t) = b_k \exp \frac{ak\pi}{L}t,$$

where the constants b_k are arbitrary. Putting the solutions of the ordinary differential equations (10.16), (10.17) to the formula (10.14), we get

$$u_k(t) = X_k(x)T_k(t) = \varphi_k \sin \frac{k\pi}{L}x \exp \frac{ak\pi}{L}t, \ k = 1, 2, \dots,$$

where $\varphi_k = b_k c_k$. Obviously, for all numbers k and constants φ_k this function u_k satisfies the equality (10.11) and boundary conditions (10.13). The sum of all these functions

$$u(x,t) = \sum_{k=1}^{\infty} \varphi_k \sin \frac{k\pi}{L}x \exp \frac{ak\pi}{L}t \tag{10.18}$$

has the same properties[24].

The last thing to be done is to choose the numbers φ_k so that the function u_k determined by formula (10.18) satisfies the initial condition (10.12). After the corresponding substitution, we have

$$u(x,0) = \sum_{k=1}^{\infty} \varphi_k \sin \frac{k\pi}{L} x = \varphi(x).$$

Multiply this equality by the function $\sin \frac{n\pi}{L} x$ for any number n and integrate the result by x from 0 to L. We get[25]

$$\sum_{k=1}^{\infty} \varphi_k \int_0^L \sin \frac{k\pi}{L} x \sin \frac{n\pi}{L} x dx = \int_0^L \varphi(x) \sin \frac{n\pi}{L} x dx. \qquad (10.19)$$

Using the sine product formula, we have

$$\int_0^L \sin \frac{k\pi}{L} x \sin \frac{n\pi}{L} x dx = \frac{1}{2} \int_0^L \left[\cos \frac{(k-n)\pi}{L} x + \cos \frac{(k+n)\pi}{L} x \right] dx.$$

For $k \neq n$ we find

$$\int_0^L \sin \frac{k\pi}{L} x \sin \frac{n\pi}{L} x dx = \frac{L}{2(k-n)\pi} \sin \frac{(k-n)\pi}{L} x \Big|_0^L + \frac{L}{2(k+n)\pi} \sin \frac{(k+n)\pi}{L} x \Big|_0^L = 0.$$

If $k = n$, we get

$$\int_0^L \left(\sin \frac{k\pi}{L} x \right)^2 dx = \frac{1}{2} \int_0^L \left(1 - \cos \frac{2k\pi}{L} x \right) dx = \frac{L}{2} - \frac{L}{4k\pi} \sin \frac{2k\pi}{L} x \Big|_0^L = \frac{L}{2}.$$

Substituting the found values of the integrals in the formula (10.19), we find the values[26]

$$\varphi_k = \frac{2}{L} \int_0^L \varphi(x) \sin \frac{k\pi}{L} x dx, \ k = 1, 2, \dots. \qquad (10.20)$$

Thus, the solution of the problem (10.11)–(10.13) is determined as the ***Fourier series*** (10.18) with the coefficients φ_k, calculated by the formula (10.20).

To clarify the results, we consider a particular case with $X = \pi$, and $\varphi(x) = \sin x$. Thus, we consider a body of length π with zero temperature at the ends and the initial body temperature distributed according to a sinusoidal law. Substituting the given function φ in equalities (10.20), we find the coefficients

$$\varphi_1 = 1, \ \varphi_k = 0, \ k = 2, 3, \dots.$$

From the formula (10.18), it follows that the solution of the problem is

$$u(x,t) = e^{-a^2 t} \sin x.$$

By this formula, the temperature distribution along the body at any time is sinusoidal. Besides, the body temperature at any of its point exponentially decreases and tends to zero at $t \to \infty$. Thus, we observe a gradual cooling of the body, and the intensity of cooling is the highest in its center, see Figure 10.2.

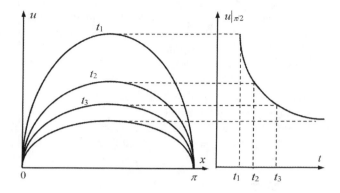

FIGURE 10.2: The body temperature for the first boundary value problem.

Let us analyze the obtained results. At the initial time, the temperature is not evenly distributed. According to the Fourier law, the heat fluxes occur, directed from the center to the ends of the body. Because heat leaves the central part of the body, the temperature decreases there. If the body was thermally insulated, then we would observe the heating of its ends due to the influx of heat from the center. However, at the ends of the body, zero temperature is constantly maintaincd. This means that the heat input from the center is compensated by the removal through the border into the environment.

By the Fourier law, the value of the heat flux is proportional to the temperature difference. As already noted, over time, the temperature in the center decreases, but at the ends it remains unchanged. Thus, the temperature difference between the center and the boundaries (more precisely, the derivative u_x) decreases with time. Because of this, the heat flux also decreases. Then in a unit of time less and less heat leaves the center. Consequently, the cooling of the body continues with a monotonously decreasing speed. Over time, all the heat from the body will go into the environment. Therefore, a uniform (zero) temperature distribution will be established inside the body[27].

One can also evaluate the effect of thermal diffusivity on the process. Obviously, the larger the value of the parameter a, the faster the cooling of the body occurs (see Figure 10.3). Thus, this coefficient characterizes the intensity of heat transfer for this material.

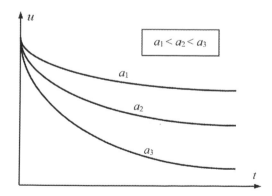

FIGURE 10.3: The body cools the faster, the greater the coefficient a.

Task 10.1 *First boundary value problem for the heat equation*. Consider the heat equation $c\rho u_t = \lambda u_{xx}$ on the interval $[0, L]$ with initial condition $u(x, 0) = b \sin \pi x / L$ and homogeneous first type boundary conditions. Make the following analysis.

1. Give a physical interpretation of the problem.

2. Find its solution by the method of separation of variables.

3. Determine the effect of body length L, density ρ, and maximum initial temperature b on the result. Explain the physical meaning of the results.

4. Find the law of change with time of the heat flux λu_x at the right end of the body.

5. Determine the law of change with time of temperature in the middle of the body. Assess the effect of the thermal conductivity coefficient λ on this result. Explain the results.

> *The solution of the heat equation with homogeneous boundary conditions*
> *is determined as a Fourier series by the method of separation of variables.*
> *Its coefficients are determined by the initial temperature distribution.*

3   Non-homogeneous heat equation

The body can also be affected by some external heat source associated with the action of the furnace, chemical reaction, laser radiation, etc.[28] These phenomena can be characterized using the density of the heat source F that is the heat quantity received or given away by the body per unit time. The value of F is a function of time and a spatial variable for the general case. In the presence of external heat sources, the process is described by the **non-homogeneous heat equation**

$$c\rho \frac{\partial u}{\partial t} = \frac{\partial}{\partial x} \left(\lambda \frac{\partial u}{\partial x} \right) + F.$$

Consider the homogeneous body of length L. We get the equation

$$u_t(x, t) = a^2 u_{xx}(x, t) + f(x, t), \ 0 < x < L, \ t > 0, \tag{10.21}$$

where $a = \sqrt{\lambda/c\rho}$, $f = F/c\rho$. The equation (10.21) is completed by the initial condition

$$u(x, 0) = \varphi(x), \ 0 < x < L \tag{10.22}$$

and homogeneous first type boundary conditions[29]

$$u(0, t) = 0, \ u(L, t) = 0, \ t > 0. \tag{10.23}$$

By the **Fourier method**, determine the solution of the problem (10.21)–(10.23) as the Fourier series

$$u(x, t) = \sum_{k=1}^{\infty} u_k(t) \sin \frac{k\pi}{L} x \tag{10.24}$$

that is the analogue of (10.18). The function u_k is unknown here. We can find it after putting this function u to the equalities (10.21) and (10.22). Find the derivatives[30]

$$u_t = \sum_{k=1}^{\infty} u_k' \sin \frac{k\pi}{L} x, \ \ u_{xx} = -\sum_{k=1}^{\infty} \left(\frac{k\pi}{L} \right)^2 u_k \sin \frac{k\pi}{L} x.$$

Putting it to the equality (10.21), we get

$$\sum_{k=1}^{\infty} u_k' \sin \frac{k\pi}{L} x = -\sum_{k=1}^{\infty} \left(\frac{ak\pi}{L} \right)^2 u_k \sin \frac{k\pi}{L} x + f.$$

Multiply this equality by $\sin\frac{n\pi}{L}x$. After integration we have

$$\sum_{k=1}^{\infty} u_k' \int_0^L \sin\frac{k\pi}{L}x \sin\frac{n\pi}{L}x dx = -\sum_{k=1}^{\infty}\left(\frac{ak\pi}{L}\right)^2 u_k \int_0^L \sin\frac{k\pi}{L}x \sin\frac{n\pi}{L}x dx + \int_0^L f \sin\frac{n\pi}{L}x dx.$$

Calculating the integrals, as was done in the derivation of formula (10.18), we obtain

$$u_k'(t) = \left(\frac{ak\pi}{L}\right)^2 u_k(t) + f_k(t), \ k = 1, 2, \ldots, \tag{10.25}$$

where

$$f_k(t) = \frac{2}{L}\int_0^L f(x,t) \sin\frac{k\pi}{L}x dx. \tag{10.26}$$

Multiplying the equality (10.22) by the function $\sin\frac{n\pi x}{L}$ after integration with using the formula (10.24), we find

$$u_k(0) = \varphi_k(t), \ k = 1, 2, \ldots, \tag{10.27}$$

where

$$\varphi_k = \frac{2}{L}\int_0^L \varphi_k(x) \sin\frac{k\pi}{L}x dx. \tag{10.28}$$

Find the solution of the equation (10.25) with initial condition (10.27). The equality (10.25) can be transformed to

$$e^{-\frac{ak\pi}{L}t}\frac{d}{dt}\left[u_k(t)e^{\frac{ak\pi}{L}t}\right] = f_k(t).$$

Multiply this equality by $\exp\frac{ak\pi}{L}t$ and integrate the result by t from zero to an arbitrary value t. Using the initial condition (10.27), we get

$$u_k(t)e^{\frac{ak\pi}{L}t} - \varphi_k = \int_0^t f_k(\tau)e^{\frac{ak\pi}{L}\tau}d\tau.$$

Then we find

$$u_k(t) = \varphi_k e^{-\frac{ak\pi}{L}t} + \int_0^t f_k(\tau)e^{\frac{ak\pi}{L}(\tau-t)}d\tau. \tag{10.29}$$

Thus, the solution of the problem (10.21) – (10.23) is determined by the formula (10.24), where the function u_k is found from the equality (10.29), and the Fourier coefficients of the functions f and φ are defined by the equalities (10.26) and (10.28).

Consider an example of the problem (10.21) – (10.23) with parameters

$$L = \pi, \ a = 1, \ \varphi(x) = 0, \ f(x,t) = \sin x.$$

We analyze the heat transfer phenomenon for the body of unit length with unit thermal diffusivity. The initial and boundary temperature of the body is zero. Finally, there exists a heat source with sinusoidal distribution such that its action is maximal in the center of the body and decreases to zero as it approaches the boundaries.

We turn now to the solving of the problem. Obviously, all the Fourier coefficients of the function φ and all the Fourier coefficients of the function f, starting from the second, are equal to zero, and $f_1 = 1$. As a result, from formulas (10.29) we find

$$u_1(t) = 1 - e^{-t}, \ u_k(t) = 0, \ k = 2, 3, \dots .$$

Using the formula (10.24), we find the solution of the problem

$$u(x, t) = (1 - e^{-t}) \sin x.$$

Let us analyze the results. Obviously, at any non-zero point in time, the temperature is distributed according to a sinusoidal law along the length of the body (see Figure 10.4). Moreover, the temperature at any point in the region rises more at the center, less at a distance from it. Note that the velocity of temperature increase at any point x gradually decreases. With an unlimited increase in time, the temperature at this point tends to the value $\sin \pi x$. Thus, over time, the temperature distribution is characterized by the function $u_\infty(x) = \sin x$.

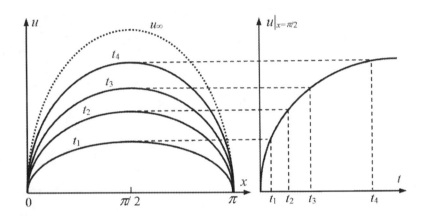

FIGURE 10.4: The temperature of the body under the heat source influence.

Try to explain the obtained result. Thus, initially zero temperature is maintained in the body. However, a heat source acts on the body, as a result of which the body heats up more strongly in its center with a decrease to the right and left of it. As a result, the body heats up, and the more, the closer to the center. Due to the uneven distribution of temperature, heat flows arise from a more hot region to a colder one, i.e., from the center to the borders. Since the temperature at the borders is invariably zero, and heating occurs in the center, heat exits through the borders into the environment. Thus, the body, on the one hand, heats up under the action of an existing heat source, and, on the other hand, cools due to heat removal through the boundaries.

Note that the action of the heat source is constant. However, the higher the body temperature in the center, the greater the temperature difference between the center and the boundaries, since the temperature there is invariably zero. Therefore, over time, the heat flux across the boundary increases in accordance with the Fourier law. Over time, an ever-increasing heat removal across the border compensates for the influx of heat due to the action of the source. As a result, the system goes into some equilibrium state characterized by the function u_∞[31].

Task 10.2 *Non-homogeneous heat equation*. Consider the heat equation $u_t = \lambda u_{xx} + e^{-t} \sin x$ on the interval $[0, \pi]$ with initial condition $u(x, 0) = \sin x$ and homogeneous first type boundary conditions. Make the following analysis.

1. Give a physical interpretation of the problem.

2. Using the Fourier method, find a solution of the problem and give its physical interpretation.

3. Determine the law of change with time of temperature in the middle of the body. Explain the result.

4. Set the behavior of the system with an unlimited increase in time.

5. Determine the influence of the thermal conductivity coefficient λ on the solution. Explain the physical meaning of the result.

6. Find the law of change with time of the heat flux λu_x on the left end of the body. Explain the result.

> *The solution of the non-homogeneous heat equation*
> *with homogeneous boundary conditions is found by Fourier method.*
> *The system tends to an equilibrium position for a constant heat source.*

Direction of further work. We got acquainted with the heat equation and some methods for solving them. Further study of transfer processes will continue in a subsequent chapter[32]. In particular, we will be convinced that in addition to the heat transfer process and the diffusion phenomenon considered in the Appendix, there are other processes described by the heat equation. We will also consider approximate methods for solving the corresponding problems. In addition, some special mathematical models related to transfer processes will be described.

Appendix

We analyzed the process of heat transfer due to the phenomenon of heat conduction. However, the body temperature can also change under the influence of other factors. Below we consider some generalizations of the heat equation.

Earlier, the first boundary value problem was solved for the heat equation when the temperature was given at the boundary of the body. Now we consider the second boundary value problem with a known heat flux at the boundary.

Finally, we will see that the phenomenon of diffusion[33] is also described by the heat equation.

1   Generalizations of the heat equation

Earlier, we considered the case when the heat transfer was caused solely by the phenomenon of heat conduction due to heat fluxes in an unevenly heated body. In addition, the effect of heat sources was investigated. However, there may be other reasons for the change in body temperature. One such reason is convection associated with the movement of the body in question, for example, the flow of a liquid or gas. If a heated body moves with velocity v, then instead of relation (10.6) a more general equation is obtained

$$c\rho \frac{\partial u}{\partial t} = \frac{\partial}{\partial x}\left(\lambda \frac{\partial u}{\partial x}\right) - v\frac{\partial u}{\partial x}.$$

In the presence of heat transfer between the considered body and the environment, the heat equation takes the form

$$c\rho \frac{\partial u}{\partial t} = \frac{\partial}{\partial x}\left(\lambda \frac{\partial u}{\partial x}\right) - k(u_0 - u),$$

where u_0 is the ambient temperature, and k is the heat transfer coefficient characterizing the intensity of thermal interaction between the body and the environment. We can consider the general case when there are simultaneously all of the above factors, and the system is described by the equation[34]

$$c\rho\frac{\partial u}{\partial t} = \frac{\partial}{\partial x}\left(\lambda\frac{\partial u}{\partial x}\right) - v\frac{\partial u}{\partial x} - k(u_0 - u) + F.$$

All considered models belong to a one-dimensional body. However, the assumption that the body can be considered as a one-dimensional object is far from always justified. If we are dealing with a flat body of sufficiently small thickness, then we can neglect the change in its characteristics in thickness and consider the two-dimensional object. In this case, the body temperature will already depend on time t and two spatial variables x and y. The heat equation for a non-homogeneous two-dimensional body has the form

$$c\rho\frac{\partial u}{\partial t} = \frac{\partial}{\partial x}\left(\lambda\frac{\partial u}{\partial x}\right) + \frac{\partial}{\partial y}\left(\lambda\frac{\partial u}{\partial y}\right).$$

For the homogeneous body, we can write

$$u_t = a^2(u_{xx} + u_{yy}).$$

If we cannot interpret this object as a one-dimensional or flat body, then it is necessary to consider heat transfer in space. In this case, the temperature will already depend on three spatial variables x, y, z. The corresponding heat equation in the simplest case has the form

$$u_t = a^2(u_{xx} + u_{yy} + u_{zz}).$$

Note that in mathematical physics the **Laplace operator** Δ is widely used, which is the sum of the second derivatives with respect to all spatial variables[35]. If we ignore the number of spatial variables, the heat equation of a homogeneous object, regardless of its dimension, can be written uniformly in the form

$$u_t = a^2\Delta u.$$

Naturally, in the multidimensional case, one can consider a body that is not homogeneous, and also take into account the influence of various types of heat transfer mentioned above.

The mathematical model of heat transfer in all cases, along with the heat equation in one form or another, includes boundary conditions. Moreover, the initial condition invariably involves setting the body temperature at the initial time at all points of the object. However, the boundary conditions depend on the features of the considered physical process. For the considered above generalizations of the heat equation in the one-dimensional case, the same versions of the boundary conditions can be set as for the simplest heat equation (10.6). But the multidimensional case has its own peculiarities.

First of all, we note that in the one-dimensional case, the ranges of changes in the spatial and temporal coordinates were independent of each other. As a result, the temperature as a function of two variables is determined in a rectangular region. However, in the multi-dimensional case, both two-dimensional and three-dimensional, spatial coordinates, cannot be considered independently of each other, determining the shape of the studied object[36].

In particular, let us consider the heat transfer in a two-dimensional or three-dimensional set Ω with boundary[37] S, see Figure 10.5. The **first boundary value problem** for the multidimensional heat equation involves setting the temperature on the entire boundary S at any time[38].

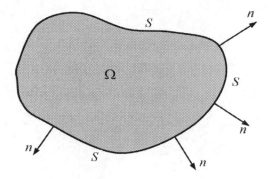

FIGURE 10.5: The set Ω with boundary S and directions of external normal n.

The heat flux is defined at an arbitrary point on the boundary for the **second boundary value problem** by the equality

$$q = \lambda \frac{\partial u}{\partial n},$$

where n is the external **normal** to the boundary of S at a given point, see Figure 10.5. Naturally, mixed boundary conditions are also possible when different conditions are set on different parts of the border.

2 Second boundary value problem for the heat equation

We considered before the first boundary value problem for the homogeneous and non-homogeneous heat equation. Extend these results to the second boundary value problem. Consider the non-homogeneous heat equation

$$u_t = a^2 u_{xx} + \cos x$$

on the interval $(0, \pi)$ with initial and boundary conditions

$$u(x, 0) = 0,$$

$$u_x(0, t) = 0, \ u_x(\pi, t) = 0.$$

Thus, we study the process of heat transfer in the presence of a heat source, the influence of which is characterized by the cosine entering the right side of the equation. Given the properties of this function, we conclude that heat is supplied to the left side of the body, and it is removed from its right side. In this case, the influence of the heat source increases as one approaches the boundary of the body. At the initial time, the body temperature is zero. The ends of the body are thermally insulated, i.e., interaction with the environment at the boundary of the body does not occur.

Using the known technique, one can determine the problem solution[39]

$$u(x, t) = a^{-2} \left[1 - \exp(-a^{-2}t) \right] \cos x.$$

Thus, at any moment of time, the temperature distribution along the body has a form of cosine, besides, temperature on the left side of the body rises and decreases on the right side (see Figure 10.6). As a result, the temperature distribution $u_\infty(x) = \cos x$ is obtained over time. Establish the physical meaning of the results.

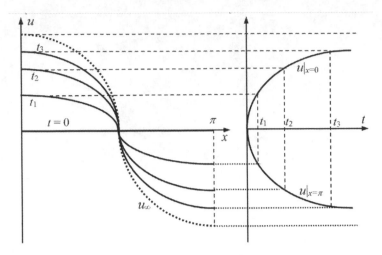

FIGURE 10.6: The body temperature for the second boundary value problem.

Thus, at the initial time, the temperature is evenly distributed over the body. The temperature changes under the influence of an external heat source. In this case, the left half of the body heats up, and the right half cools down, and the closer to the border, the stronger the influence of the source. Note that the action of the heat source does not change over time. However, as the temperature rises to the left and decreases to the right, a heat flux appears, acting from left to right. Due to the thermal insulation of the ends of the body, its interaction with the environment is absent. The heating of the body in its left part and cooling in the right part are gradually compensated by the heat flux from the hot to the cold region, i.e., thermal conductivity phenomenon. Thus, a thermodynamic equilibrium is established inside the body, characterized by the limiting value u_∞. Note that with an increase in the thermal diffusivity a, the system quickly enters an equilibrium state.

Task 10.3 *Second boundary value problem for the homogeneous heat equation.* Consider the homogeneous heat equation on the interval $(0, 1)$ with initial condition $u(x, 0) = \cos \pi x$ and homogeneous second type boundary conditions. Make the following analysis.

1. Give a physical interpretation of the problem statement.
2. Using the method of separation of variables, find a problem solution.
3. Determine the equilibrium position of the system.
4. Determine the law of change with time of body temperature at its ends.
5. Determine the influence of parameter a on the obtained results.
6. Explain the physical sense of all previous results.

3 Diffusion equation

In Part I, we repeatedly met with a situation where qualitatively different processes related to completely different subject areas are described by the same equations. Thus, equations can have fundamentally different interpretations. This property holds not only for ordinary differential equations, but also for partial differential equations. After analysis of the problems of the heat transfer phenomenon, we now turn to the diffusion process.

Consider a volume filled with gas, and the distribution of gas over the volume is assumed uneven. Under these conditions, we observe the flow of substance from the area where there is a lot of it to a less filled area. This process is called **diffusion** and is characterized by the **concentration**[40]. This is the number of gas molecules per unit volume or per unit length in the one-dimensional case. For simplicity, suppose that the volume in question is a sufficiently long and thin cylinder. Under these conditions, we can assume that the concentration u depends only on time t and one spatial coordinate x directed along the axis of the cylinder. To determine a mathematical model of this process, we establish the changes in the substance quantity in an interval $[x, x + \Delta x]$ over time from t to $t + \Delta t$.

If the gas concentration is constant, then the substance quantity in an interval with length Δx is $Q = u\Delta x$. Since in reality the concentration of a substance can vary from point to point, we can use the previous equality only in an arbitrarily small area. As a result, we establish $dQ = u\,dx$. Then the substance quantity in the interval $[x, x + \Delta x]$ is

$$Q = \int\limits_{x}^{x+\Delta x} u\,dx.$$

Now the change of the substance quantity for the time from t to $t + \Delta t$ is

$$Q_1 = \int\limits_{x}^{x+\Delta x} \big[u(x, t + \Delta t) - u(x, t)\big]dx \tag{10.30}$$

that is an analogue of the formula (10.2).

There are various reasons that can cause a redistribution of a substance within the body. This can be diffusion proper (transfer of a substance from a region with a high concentration to an area where it is small), mechanical movement of a substance, supply or removal of a substance through a side surface of a cylinder, appearance or disappearance of a substance due to chemical reactions[41], mass transfer due to an uneven temperature distribution, or pressure inside the body (thermodiffusion and barodiffusion, respectively), etc. We restrict ourselves to considering only the mass transfer due to diffusion.

The diffusion process can be described by the **diffusion flux** q. This is a substance quantity passing per unit of time through a given point[42]. If the diffusion flux does not change, then in the time Δt we obtain the following substance quantity $Q = q\Delta t$. For the variable diffusion flux, we get $dQ = q\,dt$. Then for the time from t to $t + \Delta t$ we obtain

$$Q = \int\limits_{t}^{t+\Delta t} q\,dt.$$

Thus, the change of the substance quantity on the interval $[x, x + \Delta x]$ is

$$Q_2 = \int\limits_{t}^{t+\Delta t} \big[q(x, t) - q(x + \Delta x, t)\big]dt$$

that is an analogue of the formula (10.2).

Let us determine the relationship between the diffusion flux and concentration. We consider some points x_1 and x_2 for $x_1 < x_2$, at which the concentration of the substance is equal to u_1 and u_2, respectively. Due to the difference in concentrations, a diffusion flux arises, the value of which is the higher, the greater the difference in concentrations and the smaller the distance between the points. Considering that the flow is directed from the region with a high concentration to the region where it is relatively small, we obtain the equality

$$q = -D\frac{u_2 - u_1}{x_2 - x_1}.$$

TABLE 10.1: Analogy between thermal conductivity and diffusion.

Physics direction	Thermal physics	Molecular physics
phenomenon	thermal conductivity	diffusion
local characteristic	temperature	concentration
global characteristic	heat quantity	substance quantity
flux	heat flux	diffusion flux
law	Fourier law	Ficks law
transfer coefficient	thermal conductivity	diffusion coefficient
state equation	heat equation	diffusion equation

Here, the parameter D is called the **diffusion coefficient** and characterizes the intensity of the diffusion process for a given substance. When deriving the last relation, only two isolated points were considered. When considering a continuous medium, this equality only makes sense if the distance between the points is arbitrarily small. Passing to the limit at $x_2 \to x_1$, we get

$$q = -D\frac{\partial u}{\partial x}$$

called the **Fick's law**. As a result, we obtain the equality

$$Q_2 = \int\limits_t^{t+\Delta t} \left(D\frac{\partial u}{\partial x}\Big|_{x+\Delta x} - D\frac{\partial u}{\partial x}\Big|_x\right)dt \qquad (10.31)$$

that is an analogue of the formula (10.4). The material balance on the segment $[x, x + \Delta x]$ for the time from t to $t + \Delta t$ is characterized by the equality $Q_1 = Q_2$, which takes the form

$$\int\limits_x^{x+\Delta x} [u(x, t + \Delta t) - u(x,t)]\,dx = \int\limits_t^{t+\Delta t} \left(D\frac{\partial u}{\partial x}\Big|_{x+\Delta x} - D\frac{\partial u}{\partial x}\Big|_x\right)dt,$$

because of conditions (10.30) and (10.31). Using the mean theorem after division by the product $\Delta x \Delta t$ and passing to the limit as $\Delta x \to 0$, $\Delta t \to 0$, determine the equality

$$\frac{\partial u}{\partial t} = \frac{\partial}{\partial x}\left(D\frac{\partial u}{\partial x}\right). \qquad (10.32)$$

This is called the **diffusion equation**. For the partial case with constant parameter D we get

$$u_t = a^2 u_{xx}, \qquad (10.33)$$

where $a = \sqrt{D}$.

The resulting relation differs from the heat equation (10.7) only in the physical meaning of the quantities included in it. For the diffusion equation, boundary value problems similar to those considered above. Since, from a mathematical point of view, the heat conduction and diffusion equations coincide, their solutions have the same properties. An analogy between the phenomena of thermal conductivity and diffusion is presented in Table 10.1.

Task 10.4 *First boundary value problem for the non-homogeneous diffusion equation.* Consider the diffusion equation with substance source $u_t = Du_{xx} + \sin \pi x/L$ on the interval $(0, L)$ with homogeneous initial condition and homogeneous first type boundary conditions. Make the following analysis.

  1. Give a physical interpretation of the problem statement.

2. Using the Fourier method, find a solution of this problem.

3. Find the equilibrium position of the system.

4. Determine the law of change in the concentration of a substance in the middle of the body.

5. Determine the law of change of the diffusion flow at the ends.

6. Determine the effect of parameters D and L on the results.

7. Explain all previous results.

Notes

[1] The problems of thermal physics are considered, for example, in [22], [24], [80], [94], [154], [202], [259], [284], [294].

[2] We will return to this issue in a subsequent chapter.

[3] In the Appendix, the heat equation for the multidimensional case will be given.

[4] The characteristics do not have time to change on an arbitrarily small interval $[x, x + dx]$. Then the temperature and process parameters at each point here can be considered equal to their values at point x. Moreover, the corresponding equality will be fulfilled with an arbitrarily high degree of accuracy.

[5] The terms dx and dQ are called the **differentials** of the argument x and the function $Q = Q(x)$. The differentials are considered in mathematical analysis, see, for example, [13], [61], [79], [148], [206], [344], [353].

[6] Naturally, in reality, in the considered time interval, the body may not heat up, but cool. In this case, the determined quantity Q_1 will be negative, i.e., heat is not supplied, but removed.

[7] The appearance of the integration procedure in formula (10.2) can be explained as follows. Suppose that the considered interval $[x, x + dx]$ is divided into a number of parts, on each of which the characteristics of the system are constant. Then the heat quantity falling over the entire considered interval will be summed up from its values determined on each of the obtained subinterval in accordance with the above formula for the heat quantity if the parameter is constant. The resulting value is actually the corresponding integral sum. If we now consider the limit of the expression obtained, when the length of the maximum of the subintervals tends to zero, then the result obtained exactly corresponds to the classical definition of the **integral**, see, for example, [13], [61], [79], [148], [206], [344], [353].

[8] We will return to this problem in the Appendix.

[9] In the study of heat transfer in space, heat flux is understood as the heat quantity passing per unit time through a unit surface of the body.

[10] Naturally, the value of the heat flux itself can also be negative, i.e., the movement of heat can be either one way or the other.

[11] The mean theorem has a natural geometric meaning. The value of the integral of the function f on the segment $[a, b]$ is equal to the area of the curved trapezoid bounded by the corresponding curve, the straight lines $y = a$, $y = b$ and the y axis, see Figure 10.7. Then, according to this statement, on the considered interval under there is a point c such that the area of this trapezoid is equal to the area of the rectangle with sides equal to the length of the segment and the value of the function f at the specified point.

[12] The problem of justifying passage to the limit in the derivation of equations of mathematical physics is discussed in Chapter 19.

[13] The determination of the heat equation for the stationary case is considered in Chapter 12.

[14] The **theory of partial differential equations** is an independent branch of mathematics, see, for example, [96], [141], [147], [169], [192], [215], [235], [269], [274], [349], [361].

[15] All considered in this and subsequent chapters are of parabolic type. In particular, we consider a second order partial differential equation for a function of two variables $u = u(x, y)$, solvable with respect to the highest derivatives

$$a_{11}u_{xx} + 2a_{12}u_{xy} + a_{22}u_{yy} = F\big(x, y, u, u_x, u_y\big), \tag{10.34}$$

where u_x, u_y, etc. are the corresponding partial derivatives. The properties of this equation depend from the sign of the value

$$D = a_{12}^2 - a_{11}a_{22}$$

called the discriminant of the equation. For $D > 0$, the equation is called **hyperbolic**, for $D < 0$, this is

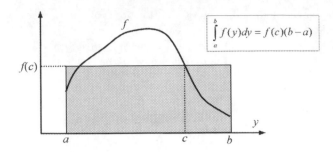

FIGURE 10.7: Geometric sense of the mean theorem.

elliptic, and for $D = 0$, this is *parabolic*. In particular, for the heat equation $u_t = a^2 u_{xx}$ with the variable $y = t$, we have $a_{11} = a^2$, $a_{22} = 0$, $a_{12} = 0$. Thus, $D = 0$, and therefore, the heat equation is parabolic. Using a special transformation of independent variables $x = x(\xi, \eta)$, $y = y(\xi, \eta)$, the equation (10.34) can always be reduced to the canonical form

$$u_{\eta\eta} = \Phi(\xi, \eta, u, u_\xi, u_\eta).$$

Hyperbolic equations are associated with wave processes (see Chapter 12), and elliptic equations are associated with stationary systems (see Chapter 13). In principle, the coefficients of the highest derivatives can depend on the variables x, y. In this case, the type of equation can vary from point to point. This applies, for example, to the **Tricomi equation** (see [274]) $u_{xx} = xu_{yy}$, which is hyperbolic for positive values of x, parabolic if $x = 0$, and hyperbolic for negative x. This equation has applications in *transonic gas dynamics*, where a change in the type of equation corresponds to the transition of the gas flow velocity through the speed of sound, see [33].

[16] In the multidimensional case, the heat equation has a second order in any spatial variable, see Appendix.

[17] In principle, sometimes it makes sense to solve a problem with a known temperature value not at the initial, but at the final time. In this case, we do not predict the further course of events from the known temperature at a given time, chosen as the reference point, but try to restore the backstory of the system, i.e., find out how the system came to this known state. Unfortunately, the heat equation with known data at a finite moment in time forms an ill-posed problem. The problems of the well-posedness of mathematical physics problems are discussed in Chapter 19.

[18] There are practical situations when one end of the body is inaccessible, and the missing information can be obtained at the other end. In this case, we do not have any boundary conditions at the first end, and two boundary conditions are set at the second end, for example, both the temperature and the heat flux are known. However, the corresponding boundary value problem is ill-posed. Nevertheless, it can be solved using the theory of inverse problems of mathematical physics, see Chapter 21.

[19] The determining of boundary conditions in the multidimensional case has its own peculiarities, see Appendix. When considering heat transfer in a body of infinite length, the **Cauchy problem** makes sense, which includes the heat equation and the initial condition for $x \in (-\infty, \infty)$. The missing boundary conditions here are compensated by the assumption that as we approach infinity, all the characteristics of the system decay. In the next chapter, we will also consider the **Stefan problem** for the heat equation, in which the boundary of the object in question changes with time. This situation arises, for example, when describing melting ice, when the studied ice body decreases in volume over time.

[20] To agreement the initial state of the system with the boundary conditions, it is assumed here that $\varphi(0) = 0$ and $\varphi(L) = 0$.

[21] Suppose we consider a non-homogeneous first boundary problem, i.e., we have the boundary conditions (10.9). Find its solution by the formula

$$u(x, t) = v(x, t) + w(x, t), \qquad (10.35)$$

where the function w is chosen such that it satisfies the boundary conditions (10.9). It is easiest to consider it linear with respect to the spatial variable x, i.e., define $w(x, t) = a(t) + b(t)x$. To find the functions a and b, it suffices to set the function w equal to $\alpha_1(t)$ for $x = 0$ and $\alpha_2(t)$ for $x = L$. It is easy to verify that

$$w(x, t) = \alpha_2(t) + (L - x)[\alpha_1(t) - \alpha_2(t)]/L. \qquad (10.36)$$

In this case, the function v is equal to zero at both boundaries, being the difference between the functions

u and w, which take the same value there. Substituting the function u defined by formula (10.35) into the heat equation (10.11), we have the equality

$$v_t - a^2 v_{xx} = (u_t - a^2 u_{xx}) - (w_t - a^2 w_{xx}) = -w_t,$$

where the value at its right-hand side is known by the formula (10.36). Substituting the function u in the initial condition (10.12), we have the equality

$$v(x,0) = u(x,0) - w(x,0) = \varphi(x) - w(x,0)$$

with known value at right-hand side. Thus, the function v satisfies the above non-homogeneous heat equation with the above initial condition and homogeneous boundary conditions. The technique for solving this problem is carried out in Section 3 of this chapter. After that, the solution of the non-homogeneous boundary value problem for the heat equation will be determined by formula (10.35) with the function w defined by equality (10.36).

[22] We will also meet boundary value problems in Chapter 16 when solving problems of minimizing functionals. Additional properties of boundary value problems for differential equations, including nonlinear ones, will be described in Chapter 19.

[23] The analogues of the Sturm–Liouville problems are also encountered for other types of equations. Consider, for example, a system of linear algebraic equations

$$\begin{cases} (\lambda - 1)x_1 + 2x_2 = 0 \\ 2x_1 + (2+\lambda)x_2 = 0 \end{cases}$$

with parameter λ. Obviously, for any λ, there exists the trivial solution $x_1 = 0$, $x_2 = 0$. However, for $\lambda_1 = 2$ and $\lambda_2 = -3$ we have the systems

$$\begin{cases} x_1 + 2x_2 = 0 \\ 2x_1 + x_2 = 0 \end{cases}, \quad \begin{cases} -4x_1 + 2x_2 = 0 \\ 2x_1 - x_2 = 0 \end{cases}$$

that have non-unique non-trivial solutions. The considered system of linear algebraic equations can be described in the matrix form $Ax = \lambda x$ with unknown vector $x = (x_1, x_2)$ and matrix

$$A = \begin{pmatrix} 1 & -2 \\ -2 & -2 \end{pmatrix}.$$

[24] Here we actually use the linearity property of the heat equation. Naturally, when working with an infinite sum, i.e., with extreme caution should be exercised. Here the questions of series convergence arise, i.e., the existence of an infinite sum, see, for example, [13], [61], [79], [148], [206], [344], [353]. We also note that in this case we are talking about the **Fourier series**.

[25] These transformations need rigorous justification.

[26] Equalities (10.20) determine the **Fourier coefficients** of the function φ.

[27] The phenomenon considered is associated with the **second law of thermodynamics**, which characterizes the tendency of the system to an equilibrium state, which is associated with an increase in **entropy**. The considered equations are not reversible in time, i.e., according to this state of the system, it is impossible to restore its background. Mathematically, this means that the boundary value problem for the heat equation with a known final rather than initial state is not correct, see Chapter 20. Actually, the presence of irreversible processes in time determines the very concept of time, for which there is a fundamental difference between the past and the future. Transfer processes are not strictly determined in the sense that only the future, but not the past can be determined from a given state of the system, see Chapter 5. In this regard, the dynamic group of transformations will no longer correspond to them, since the reversibility condition of the corresponding transformation will be violated. However, the concept of a **dynamic transformation monoid**, for which there is no corresponding reversibility requirement, makes sense.

[28] The source may be negative, for example, in the case of a refrigerator.

[29] It was noted earlier that in order to solve the problem under non-homogeneous boundary conditions, the solution to the problem is sought in the form $u = v + w$, where the function w is selected so that it satisfies the same boundary conditions as the function u. Then the function v will satisfy the inhomogeneous heat equation with homogeneous boundary conditions.

[30] Differentiation of the series is not obvious and needs justification.

[31] We are convinced that the concept of equilibrium also makes sense for partial differential equations. However, unlike systems with lumped parameters, the equilibrium position here is a function of spatial variables. We also note that, as can be seen from the obtained results, the equilibrium state of the system is not necessarily uniform. These two concepts coincide when the coefficients of the equation are constant and there is no influence of sources. In the multidimensional case, equilibrium states, i.e., the limiting states of a dynamical system at $t \to \infty$ is described by partial differential equations, see Chapter 13.

[32] Stationary heat transfer are considered in Chapter 13, and the discrete heat transfer model is discussed in Chapter 17. Chapter 21 discuss some inverse heat transfer problems.

[33] Mass transfer problems relate to molecular physics, see [94], [145], [154], [211], [304].

[34] Another cause of changes in body temperature may be thermal radiation.

[35] We will also meet with the Laplace operator in the future when describing wave processes and stationary system.

[36] In areas of a special form due to a special change of variables, the analysis of the equation can be simplified. For example, in the two-dimensional case, if the studied region Ω is a circle of radius R, then we can use the polar coordinates by the equalities $x = r \cos \varphi$, $y = r \sin \varphi$. The Laplace operator in the polar coordinate system is

$$\Delta u = u_{xx} + u_{yy} = \frac{1}{r} \frac{\partial}{\partial r} \left(r \frac{\partial u}{\partial r} \right) + \frac{1}{r^2} \frac{\partial^2 u}{\partial \varphi^2}.$$

Although the transition to polar coordinates is characterized by the terms with variable coefficients, in this representation the parameters r and φ are independently of each other, in particular, $0 < r < R$ and $0 < \varphi < 2\pi$. If, now the system has radial symmetry, then the derivative of the desired function with respect to the angular variable will be zero. In this case, the heat equation in the circle will take the form

$$\frac{\partial u}{\partial t} = \frac{a^2}{r} \frac{\partial}{\partial r} \left(r \frac{\partial u}{\partial x} \right).$$

In addition to transforming from the function of three variables to the function of two variables, there is another important circumstance. In the original statement of the problem in Cartesian coordinates, the method of separation of variables is not applicable, since one variable cannot be separated from another at the boundary of the circle characterized by the equality $x^2 + y^2 = R^2$. However, for the above form of equation, the variable separation method works successfully. In the three-dimensional case, spherical, cylindrical, and some other coordinate systems are also used. We will explicitly use the transition to other coordinate systems in the description of stationary systems, see Chapter 13.

[37] In the two-dimensional case, the boundary of some bounded plane object is a closed curve that bounds this object. The boundary of a three-dimensional body is its bounding surface. In the general case, the dimension of the boundary is one less than the dimension of the region itself.

[38] One may get the impression that when formulating the first boundary value problem for the heat equation in the multidimensional case, the above rule was violated, consisting in the fact that the number of additional conditions should be equal to the maximum order of the derivative included in the equation. In our case, the equation has a second order in spatial variables. However, we have only one boundary condition, setting the temperature at the boundary. In fact, in a spatially one-dimensional case, the boundary consisted of two points, i.e., the left and right boundaries. Thus, the previously set boundary conditions for the heat equation also imply determining exactly one condition in each boundary. If, for example, in the two-dimensional case, the considered region is a rectangle characterized by the inequalities $0 < x < L$, $0 < y < M$, then the boundary S consists of four sides of the rectangle. Thus, the first boundary-value problem requires knowledge of the temperature at $x = 0$, $x = L$, $y = 0$, and $y = M$. Thus, we obtain two conditions with respect to each spatial variable, which fully corresponds to the rule used previously.

[39] If the solution of the second boundary value problem for the homogeneous heat equation in accordance with the *method of separation of variables* is sought in the form $u(x,t) = X(x)T(t)$, then with respect to the function X we get the boundary value problem

$$X''(x) - \lambda X(x) = 0, \ 0 < x < L; \ X'(0) = 0, \ X'(L) = 0.$$

It is easy to see that its solutions are cosines, not sines. As a result, the solution of the second boundary value problem for the non-homogeneous heat equation according to the Fourier method is sought in the form of a series in cosines, the coefficients of which are determined by the heat source and the initial temperature of the system. The result is a solution to this problem in the specified form.

[40] We have already encountered concentration in the description of chemical processes, see Chapter 6. In a subsequent chapter, we will consider the diffusion of substances in the presence of chemical reactions.

[41] We will consider this case in a subsequent chapter.

[42] In the two-dimensional case, the flux is the substance quantity passing per unit time through a unit length, and in the three-dimensional case, through the surface of a unit area.

Chapter 11

Mathematical models of transfer processes. 2

In the previous chapter, we considered the processes of heat and mass transfer. Both of these phenomena are described by the same equation, called the heat equation in thermal physics and diffusion one in molecular physics. These processes are described by the same equations, however, the quantities included in them have different physical meanings and different physical sections. However, by replacing the variables, it is possible to obtain a mathematical model in dimensionless characteristics that can be given a different interpretation. This idea underlies the similarity theory, the application of which is devoted to Section 1 of this chapter.

In Section 2, we continue the study of mathematical models of transfer processes and make sure that the product distribution process, which would seem to be extremely distant from physics, is also described by this equation. For systems with lumped parameters, we have repeatedly encountered a situation where the same mathematical relationships can have different interpretations, being mathematical models of completely different phenomena. A similar situation is observed for systems with distributed parameters.

When analyzing systems with lumped parameters, we have already noted that finding an analytical solution to the problem is possible only in exceptional cases. Moreover, this is realized for more difficult systems with distributed parameters. In Section 3, we consider the finite difference method for solving the heat equation, which is a generalization of the Euler method described in Chapter 2. Using it, we can find an approximate solution to a large class of mathematical physics problems.

In Chapter 6, we considered mathematical models of chemical reactions, and in Chapter 10, the diffusion process. Both processes are described by the concentrations of the considered substances. In practice, a situation is possible when some substances diffuse under the conditions of chemical reactions. For this phenomenon, we can use a mathematical model that combines both of these effects. It is a system of diffusion equations for reacting substances, the right parts of which include terms describing these reactions, see Section 4.

Another mathematical model presented in this chapter is associated with the process of heat conduction under conditions of a change in the state of aggregation of a substance. In this case, due to phase transitions (from solid to liquid, from liquid to gaseous, or vice versa), the boundary of the considered object change with time. The process under consideration is characterized by a moving boundary, on which it is necessary to set specific Stefan condition, see Section 5.

Each direction of the chapter is further developed in the Appendix. In particular, it is shown that transfer processes are not limited to the phenomena of heat conduction, diffusion and distribution of goods. Some additional results are given in the field of an approximate solution of partial differential equations. One biological model of a distributed system is described, as well as some analogue of the Stefan problem with a medical interpretation.

Lecture

1 Heat equation and similarity theory

In the analysis of mathematical models, a transition is often made from natural physical quantities (independent variables, state functions, system parameters) to some complex quantities, composed in a certain way from physical quantities, based on the nature of the studied phenomenon. Let us explain this approach, which is the subject of the *similarity theory*[1], as an example of a boundary value problem for the heat equation.

The heat transfer is considered in a homogeneous thin long body uniformly heated at the initial moment of time, one end of which is thermally insulated, and on the other there is heat exchange with the environment. The mathematical model of this process is the boundary value problem

$$c\rho u_t = \lambda u_x x, \ \ 0 < x < L, \ t > 0, \tag{11.1}$$

$$u(x, 0) = \varphi, \ 0 < x < L, \tag{11.2}$$

$$u_x(0, t) = 0, \ \lambda u_x(L, t) = k[u(L, t) - u_0], \ t > 0, \tag{11.3}$$

where $u = u(x, t)$ is the body temperature at point x and time t, c is the heat capacity, ρ is the density, λ is the thermal conductivity, L is the body length, φ is the initial temperature, k is the heat transfer coefficient, u_0 is the temperature the environment.

This problem statement contains many parameters. We try to replace the variables in such a way as to simplify the problem statement. First, passing from the temperature u to the difference $u - u_0$, we achieve that the second boundary condition (11.3) becomes homogeneous, i.e., in its right-hand side, the term ku_0, which is independent of the state function of the system, disappears. In this case, the structure of the remaining relationships that make up the mathematical model remain unchanged. Indeed, the new state function also satisfies the heat equation, take a constant value $u - u_0$ at the initial time, and at the left end of the body, we again obtain a homogeneous condition of the second type. Secondly, if we divide the new state function by the difference $\varphi - u_0$, then the right-hand side of the initial condition will be equal to unity, and not to some common value while maintaining the structure of the problem. Thirdly, if we pass from the spatial variable x to the ratio x/L, then from an arbitrary length of the body we pass to the body of unit length.

Determine the variables

$$\tau = t/\theta, \ \xi = x/L, \ v = (u - u_0)/(\varphi - u_0),$$

where the transition from time t to the new variable τ due to the choice of the constant θ will be made so that the resulting relations have the simplest possible form. We have

$$v_\tau = (\varphi - u_0)^{-1} u_\tau = (\varphi - u_0)^{-1} u_t t_\tau = (\varphi - u_0)^{-1} \theta u_t,$$

$$v_\xi = (\varphi - u_0)^{-1} u_\xi = (\varphi - u_0)^{-1} u_x x_\xi = (\varphi - u_0)^{-1} L u_x,$$

$$v_{\xi\xi} = (v_\xi)_\xi = (\varphi - u_0)^{-1} L (u_x)_\xi = (\varphi - u_0)^{-1} L u_{xx} x_\xi = (\varphi - u_0)^{-1} L^2 u_{xx}.$$

Substituting the obtained values into equation (11.1), we have

$$\theta c \rho v_\tau = L^2 \lambda v_{\xi\xi}.$$

Choosing the constant $\theta = L^2 \lambda / c\rho$, we get the equation

$$v_\tau = v_{\xi\xi}, \ \ 0 < \xi < 1, \ \tau > 0. \tag{11.4}$$

The initial condition is

$$v(\xi, 0) = 1, \ 0 < \xi < 1. \tag{11.5}$$

Determine the second boundary condition

$$L\lambda v_\xi(1, \tau) = kv(1, \tau).$$

Introduce the constant Bi$= k/L\lambda$, called the **Biot number**. As a result, we get the boundary conditions

$$v_\xi(0, \tau) = 0, \ v_\xi(1, \tau) = \text{Bi}\, v(0, \tau), \ \tau > 0. \tag{11.6}$$

Comparing problems (11.1) – (11.3) and (11.4) – (11.6), it can be noted that when their common structure coincides, the second problem is much simpler. Instead of seven numeric parameters, it includes a single parameter that is the Bio number. In addition, both the state function and the spatial variable turn out to be dimensionless quantities, being the ratios of two values of the same type, respectively, of temperatures and distances. It is easy to verify that the variable τ is also a dimensionless quantity called the **Fourier number**. However, since problem (11.4) – (11.6) is connected exclusively with dimensionless quantities, its properties can be used not only to analyze the initial system with a different set of system parameters, but also to study other transfer processes characterized by models of the type (11.1) – (11.3). This is the meaning of the similarity theory[2].

Task 11.1 *Diffusion equation and the theory of similarity.* Give a statement of the mathematical model of the mass transfer process similar to system (11.1) – (11.3). Using the method described above, bring it to problem (11.4) – (11.6), paying attention to the physical meaning of the Biot and Fourier numbers.

> *By the similarity theory, the model can be simplified*
> *with reducing the number of system parameters.*

2   Goods transfer equation

Comparing the derivation of the equations of heat conduction and diffusion, as well as the final result obtained, we draw attention to the undoubted proximity of the mathematical description of these phenomena. Consider another process related to the economics, and, it would seem, not having any relation to those considered earlier.

We consider some goods that are in demand in a certain area and are unevenly distributed there. Over time, one can observe the redistribution of goods from the area where it was in excess, and therefore did not enjoy special demand, to the area where it is relatively small, which means that it is in significant demand. The distribution of goods can be characterized using the **density of goods** that is the quantity of goods located in a unit of territory. For simplicity, we restrict ourselves to considering the one-dimensional case, for example, the events occurring along some sufficiently large isolated railway line. Then the density of the goods u depends only on the spatial coordinate x and the time t.

Determine the change of the goods quantity on the interval $[x, x + \Delta x]$ during the time from t to $t + \Delta t$. If the goods density in the considered area is constant, then the quantity of goods on an interval of Δx is $Q = u\Delta x$. Then we have the equality $dQ = udx$ for the variable goods density. Now we get the goods quantity

$$Q = \int_x^{x+\Delta x} udx$$

on the interval $[x, x + \Delta x]$. Thus, the change in the quantity of goods over time from t to $t + \Delta t$ in the indicated section is equal to

$$Q_1 = \int\limits_{x}^{x+\Delta x} \big[u(x, t + \Delta t) - u(x, t)\big]\,dx.$$

We restrict ourselves to considering changes in the goods density in the region solely for their transportation from the region with a high goods density to the region where it is low, which is in full accordance with the principles of commercial activity. This phenomenon is described by the goods flux q that is the goods passing in a unit of time through a given point.

If the goods flux is a constant, then we get the quantity of goods $Q = q\Delta t$ in the time Δt. For the variable goods flux we obtain the formula $dQ = qdt$. Thus, we get

$$Q = \int\limits_{t}^{t+\Delta t} qdt$$

for time interval $[t, t+\Delta t]$. Thus, the change in the goods quantity of the segment $[x, x+\Delta x]$ for a given time is equal to

$$Q_2 = \int\limits_{t}^{t+\Delta t} \big[q(x, t) - q(x + \Delta x, t)\big]\,dt.$$

We establish the relationship between the flux and the density of goods. Consider some points x_1 and x_2 for $x_1 < x_2$ proportional to the distance between these points. Indeed, the goods flux will be greater in the event of a greater imbalance of goods available in a small area, since it is in these conditions that it will be advantageous to deal with its deliveries. Finally, the product will be distributed from an area where there is a lot of it, to an area where it is not enough. As a result, we arrive at the equality

$$q = -D\frac{u_2 - u_1}{x_2 - x_1},$$

where the coefficient D is a process parameter that characterizes the possibility of transporting goods in this area. Passing to the limit at $x_2 \to x_1$, we determine the formula

$$q = -D\frac{\partial u}{\partial x}.$$

Then the change in the quantity of goods is

$$Q_2 = \int\limits_{t}^{t+\Delta t} \left(D\frac{\partial u}{\partial x}\Big|_{x+\Delta x} - D\frac{\partial u}{\partial x}\Big|_{x}\right)dt.$$

The law of conservation of goods quantity on the interval $[x, x + \Delta x]$ for the time from t to $t + \Delta t$ under the above assumptions is characterized by the equality $Q_1 = Q_2$. As a result, we obtain the formula

$$\int\limits_{x}^{x+\Delta x} \big[u(x, t + \Delta t) - u(x, t)\big]\,dx = \int\limits_{t}^{t+\Delta t} \left(D\frac{\partial u}{\partial x}\Big|_{x+\Delta x} - D\frac{\partial u}{\partial x}\Big|_{x}\right)dt.$$

After applying the mean value theorem, dividing by $\Delta x \Delta t$ and passing to the limit, we get

$$\frac{\partial u}{\partial t} = \frac{\partial}{\partial x}\left(D\frac{\partial u}{\partial x}\right).$$

This is the **goods transfer equation**. In the homogeneous case, the goods transportation coefficient D is constant; and the last equality is reduced to

$$u_t = Du_{xx}. \tag{11.7}$$

The obtained relation, up to the notation, coincides with the heat and diffusion equations considered in the previous chapter. Similar boundary value problems can be posed for it. The Appendix will show that mathematical models of other processes are also reduced to similar equations.

Task 11.2 *First boundary problem for the goods transfer equation.* Consider equation (11.7) on the interval $[0, \pi]$ with initial condition $u(x,0) = \sin 2x$ and the homogeneous first type boundary conditions. Make the following analysis.

1. Give an economic interpretation of the problem. Moreover, the zero value of the function u can be understood as the middle value of the goods density in the given area.

2. Using the method of separation of variables, find the solution of the problem.

3. Find the law of change with time of the goods flux at the ends of this area. Explain the result.

4. Determine the equilibrium position of the system. Explain the result.

5. Determine the influence of the coefficient D on all previous results. Explain the economic meaning of the results.

The goods transfer phenomenon is described by the heat equation.

3 Finite difference method for the heat equation

When analyzing systems with lumped parameters, we already noted that finding an analytical solution to the differential equation is possible only in exceptional cases. Moreover, this statement remains valid for partial differential equations. However, in the absence of an analytical solution, it remains possible to find an approximate solution to the problem.

In Chapter 2, the Euler method for the approximate solution of differential equations was described. Its idea is to partition the domain of definition of the unknown function into parts and approximate its derivative at the boundaries of these intervals with the corresponding difference relation. Using the same idea as applied to partial differential equations leads to the **finite difference method**[3]. Let us describe this method for the first boundary value problem of the heat equation. Consider the non-homogeneous heat equation

$$u_t(x,t) = a^2 u_{xx}(x,t) + f(x,t), \quad 0 < x < L, \ t > 0 \tag{11.8}$$

with the initial condition

$$u(x,0) = \varphi(x), \quad 0 < x < L \tag{11.9}$$

and the boundary conditions

$$u(0,t) = \alpha(t), \ u(L,t) = \beta(t), \ t > 0, \tag{11.10}$$

where the parameters a, L and the functions φ, α, β are known.

Consider the boundary value problem (11.8) – (11.10) on the interval $[0, T]$. Using the ideas of the Euler method described earlier, we divide it into parts with a step $\tau = T/N$ by the points $t_j = \tau j$, $j = 0, ..., N$. Since in this case we are dealing with functions of two variables, we also divide the spatial interval $[0, L]$ with the step $h = L/M$ by the points $x_i = hi$, $i = 0, ..., M$. The solution of equation (11.8) will be sought exclusively at points (x_i, t_j), called the **grid nodes**. In this case, the derivatives contained in this equation should be replaced by some approximate equalities relating the values of the function in question at some nodes of the grid.

To approximate the time derivative entering the left side of equality (11.8), we use the same formula as in the Euler method. In particular, the value of the derivative of the function u at the arbitrary point t_j are approximately equal to the difference between its values at the subsequent point t_{j+1} and the given point t_j divided by the distance between these points, i.e., the step τ. Moreover, the arbitrary value x_i is chosen as the spatial argument of both the derivative and its approximation. Thus, we use the following relation

$$\frac{\partial u(x_i, t_j)}{\partial t} \approx \frac{u(x_i, t_{j+1}) - u(x_i, t_j)}{\tau}.$$

It can be shown that for the approximate calculation of the second derivative of a function $g = g(x)$ the following formula can be used[4]

$$\frac{\partial^2 u(x_i, t_j)}{\partial x^2} \approx \frac{u(x_{i-1}, t_j) - 2u(x_i, t_j) + u(x_{i+1}, t_j)}{h^2}.$$

Put these formulas to equality (11.8) for the arbitrary point (x_i, t_j). We get[5]

$$\frac{u_i^{j+1} - u_i^j}{\tau} = a^2 \frac{u_{i-1}^j - 2u_i^j + u_{i+1}^j}{h^2} + f_i^j, \ i = 1, ..., M-1, \ j = 0, ..., N-1, \qquad (11.11)$$

where the values $u_i^j = u(x_i, t_j)$, $f_i^j = f(x_i, t_j)$ are called the **grid functions**. Formula (11.11) is called the **explicit difference scheme** for the heat equation[6]. In addition to it, we must add initial condition (11.8) at the point x_i and boundary conditions (11.9) at the point t_j. We get

$$u_i^0 = \varphi_i, \ i = 0, ..., M, \qquad (11.12)$$

$$u_0^j = \alpha^j, \ u_M^j = \beta^j, \ j = 1, ..., N, \qquad (11.13)$$

where $\varphi_i = \varphi(x_i)$, $\alpha^j = \alpha(t_j)$, $\beta^j = \beta(t_j)$. Equalities (11.11) – (11.13) are used for finding an approximate solution of boundary problem (11.8) – (11.10).

Let us describe an algorithm for solving the problem based on an explicit difference scheme. First of all, on the basis of formula (11.12), all values of the desired grid function u_i^j are found for $j = 0$, i.e., on the zero time layer. Then, from formulas (11.13), all its boundary values are determined, i.e., corresponding to the indices $i = 0$ and $i = M$. Further, the main relation (11.11) is to be used. Suppose that for an arbitrary fixed number j, all values on a given time layer are known. Then its value on the subsequent time layer is determined by the formula

$$u_i^{j+1} = u_i^j + \tau a^2 \frac{u_{i-1}^j - 2u_i^j + u_{i+1}^j}{h^2} + \tau f_i^j, \ i = 1, ..., M-1. \qquad (11.14)$$

Thus, we find all values u_i^j from the previous time layer to the next layer, i.e., from $j = 0$ to the final number $j = N - 1$.

The finite difference method can be used for all known functions. It can be extended to the generalizations of the heat equation from the previous chapter. We can use it for the second boundary too[7].

Note that the above explicit difference scheme has one significant drawback. It is effectively if the following **stability condition** holds

$$\frac{a^2\tau}{h^2} \leq \frac{1}{2}. \tag{11.15}$$

This is also called the **Courant condition** or, more fully, the **Courant–Friedrichs–Levy condition**[8]. If this relation is violated, the errors that appear during the approximation of derivatives accumulate from layer to layer with time. Thus, taking a number of time steps, we get a significant distortion of the results. An implicit difference scheme, which is more difficult but certainly stable, will be described in the Appendix.

Task 11.3 Finite difference method for the heat equation. Consider equation (11.8) on the unit interval for $a = 1$ with initial condition (11.9) and boundary conditions (11.10). Make the following analysis.

1. Choose a function as the solution of the considered problem. Putting this value to equalities (11.8) – (11.10), find the corresponding functions f, φ, α, and β. Now we know that for these functions this problem has the solution v. It can be used for checking the exactness of the approximate solution of the problem determined by the finite difference method.

2. Solve problem (11.8) – (11.10) with known functions from the previous step of the task by the explicit scheme of the difference scheme. Moreover, to select a step in a spatial variable, a given unit interval is divided into 10 parts, and the value of the time step is selected based on the stability condition (11.15). 10 time steps are taken. The results obtained are compared with the exact solution of the problem.

3. Solve the problem on this interval by choosing the step h twice as much, still choosing the time step based on the stability condition and taking 10 time steps. Compare the resulting approximate solution with the exact solution.

4. Solve the problem by choosing a step h half as much as in step 2, choosing, as before, a time step based on the stability condition and making 10 time steps. Compare the resulting approximate solution with the exact solution.

5. Solve the problem with the steps $h = 0.1$, $\tau = 0.1$. Estimate the change from layer to layer over time of the error

$$\delta_j = \max_{i=1,2,\ldots,M-1} |u_i^j - v_i^j|,$$

where the approximate solution u_i^j is found by formulas (11.11) – (11.13), and $v_i^j = v(x_i, t_j)$ with given function v.

The finite difference method is used for the approximate solving
of boundary value problems for the heat equation.
The explicit difference scheme for the heat equation
requires the fulfillment of the stability condition.

4   Diffusion of chemical reactants

Chapter 6 considered the equations of chemical kinetics that describe the change in the concentration of reacting substances over time. In Chapter 10, a mathematical model of the diffusion process was presented, which allows one to establish the concentration distribution of a certain substance in a certain area. It is natural to assume that when considering chemical reactions occurring in a certain area in which reacting substances are distributed unevenly, a mathematical model should be used that takes into account both the chemical reaction and the diffusion of the starting materials and reaction products.

Consider, for example, a synthesis reaction A+B→C. By Chapter 6, this process is described by the equations

$$\dot{a} = -kab, \tag{11.16}$$

$$\dot{b} = -kab, \tag{11.17}$$

$$\dot{c} = kab \tag{11.18}$$

with corresponding initial conditions, where a, b, and c are the concentrations of the substances A, B, and C, respectively, k is a rate constant of reaction.

Suppose now that the reaction occurs in a certain region, which for simplicity is one-dimensional, and all the considered substances are distributed unevenly over it. Then the concentrations of substances a, b, and c depend both on time and on the spatial variable x. The mathematical model of this process consists of diffusion equations for each reagent, including terms describing the chemical reaction and, as a source of substance (positive or negative) contained in the right-hand sides of equations (11.16) – (11.18). As a result, we obtain the equations

$$a_t = D_a a_{xx} - kab, \tag{11.19}$$

$$b_t = D_b b_{xx} - kab, \tag{11.20}$$

$$c_t = D_c c_{xx} + kab, \tag{11.21}$$

where D_a, D_b, and D_c are diffusion coefficients of the corresponding substances. It is necessary add the boundary conditions here.

To solve the obtained problem, one must first solve the boundary value problem for the system of nonlinear equations (11.19), (11.20), and then the linear inhomogeneous equation (11.21) with the corresponding initial and boundary conditions. If the last problem, in principle, can be investigated using the methods described in the previous chapter, then for the analysis of the system of nonlinear equations (11.19), (11.20), approximate solution methods should be applied.

Similar results can be obtained for various chemical reactions, in particular those discussed in Chapter 6. In the Appendix, we analyze other problems that are generalizations of some previously considered systems with concentrated parameters to similar events occurring in space.

Task 11.4 *Approximate solving of diffusion equations for the synthesis reaction.* The first boundary value problem for equations (11.19) – (11.21) is considered. Make the following actions.

1. Describe the finite difference method for solving the first boundary value problem for the equations (11.19) – (11.21).

2. Create a computer program in accordance with the developed algorithm.

3. Carry out the calculations of the investigated system using the developed program.

4. Give an interpretation of the results.

Diffusion of reacting chemicals is described by a system of partial differential equations for the concentrations of all the substances in question containing diffusion terms and terms describing the chemical reaction.

5 Stefan problem for the heat equation

In the process of changing body temperature, a change in its state of aggregation can occur. In particular, when the temperature passes through the melting point, the substance transfers from the solid phase to the liquid phase upon melting or the reverse transitions from the liquid phase to the solid phase upon solidification[9]. In this case, the phase boundary changes over time, moreover, the melting (solidification) temperature is invariably maintained at the boundary itself, and the latent heat of melting (solidification) is also released. The process under consideration is described by the Stefan problem for the heat equation[10]. We restrict ourselves to the study of a spatially one-dimensional case.

Let there be a sufficiently long piece of ice. From the left end $x = 0$, a certain heat flux q is introduced, which can be considered constant. As a result of this, the ice melts so that in a certain neighborhood of the left border, water forms, and the area occupied by the water changes with time. The position of the interface between the liquid and solid phases at time t is denoted by $\xi(t)$. The purpose of the study is to determine the law of change of the moving boundary, i.e., function ξ.

Thus, at time t, water is located in the region $0 < x < \xi(t)$. The distribution of the water temperature there is characterized by the heat equation

$$c\rho u_t(x,t) = \lambda u_{xx}(x,t), \ 0 < x < \xi(t), \ t > 0, \tag{11.22}$$

where u is the water temperature, c is its heat capacity, ρ is the density, and λ is the heat conductivity.

As already noted, the heat flux q is given at the left end. In principle, the heat flux is defined as the product of the coefficient λ by the derivative of the function u with respect to the spatial variable. As a result, we obtain the boundary condition

$$-\lambda u_x(0,t) = q, \ t > 0. \tag{11.23}$$

The minus sign in the left-hand side of the equality is explained by the fact that the water at the left end is heated, which means its temperature is a decreasing function of the spatial variable.

At the right end of the considered area, i.e., at the interface, a constant ice melting temperature is set, which is zero. Thus, we obtain the boundary condition

$$u(\xi(t), t) = 0, \ t > 0. \tag{11.24}$$

The state of the system at the initial time is known. Thus, the following initial condition holds

$$u(x, 0) = u_0(x), \ t > 0, \tag{11.25}$$

where u_0 is initial temperature distribution. Based on the physical meaning of the problem, it follows that u_0 is a decreasing function, and $u_0(\xi(0)) = 0$, because of equality (11.24). Using condition (11.25), one assumes that the initial position ξ_0 of the moving boundary is known. Thus, there is still an additional initial condition

$$\xi(0) = \xi_0. \tag{11.26}$$

If the movement law of the moving boundary were known, then the existing relations would be sufficient to find the temperature distribution. However, in the absence of information on the function ξ, an additional condition is required to solve the problem.

Suppose at time t, the boundary is at the point $\xi(t)$, and at time $t + \Delta t$, this is at the point $\xi(t + \Delta t)$. If during this time interval the quantity of substance m melted, then this required an of heat quantity $Q_1 = \chi m$, where χ is latent heat of melting ice. Over the period from t to $t + \Delta t$, the region filled with water increased by $\xi(t + \Delta t) - \xi(t)$. Given that the density ρ is the mass per unit length, we conclude that this mass is $m = \rho\big[\xi(t + \Delta t) - \xi(t)\big]$. Thus, in order to melt the ice over a period from t to $t + \Delta t$, it is necessary to expend the heat quantity

$$Q_1 = \chi\rho\big[\xi(t + \Delta t) - \xi(t)\big]. \tag{11.27}$$

The resulting heat was generated due to the incoming heat flux. As you know, heat flux is the amount of heat passing per unit time through a given point. Then, if the heat flux q arrives at this point, then in the time interval Δt, then we have the heat quantity $Q_2 = q\Delta t$. The heat flux arriving at the point $\xi(t)$ on the left during the time interval Δt is

$$Q_2 = -\lambda u_x(\xi(t), t)\Delta t \tag{11.28}$$

The heat balance in the region of melting ice in the considered time interval is characterized by the equality $Q_1 = Q_2$. Then from relations (11.27), (11.28) it follows

$$-\lambda u_x(\xi(t), t)\Delta t = \chi\rho\big[\xi(t + \Delta t) - \xi(t)\big].$$

Dividing this equality by Δt and passing to the limit as $\Delta t \to 0$, we get

$$\chi\rho\dot\xi(t) = -\lambda u_x(\xi(t), t), \; t > 0. \tag{11.29}$$

Equality (11.23), called the **Stefan condition**, together with the initial condition (11.26) characterizes the desired function ξ.

Thus, to find the law of movement of the moving boundary, and at the same time also the unknown temperature distribution in the liquid phase, we obtain relations (11.22) – (11.26), (11.29) that make up the **Stefan problem** for the heat conduction equation or the more fully **one-phase Stefan problem**[11].

The heat transfer process under the condition of a phase transition
is characterized by the Stefan problem for the heat equation,
which includes the Stefan condition on the moving boundary.

Direction of further work. Having familiarized ourselves with the mathematical models of transfer processes, we turn in the next chapter to wave processes that describe mechanical, electrical, and other types of oscillations of extensional objects. The resulting systems with distributed parameters are described by another type of partial differential equations and are generalizations of mechanical and electrical oscillations considered in Chapters 3 and 4 for systems with lumped parameters.

Appendix

The Appendix develops the ideas described in the lecture. In particular, in the previous chapter it was established that qualitatively different processes of heat and mass transfer are described by the same partial differential equation. In this chapter, we made sure that the distribution process of goods is also characterized by this equation. Below we will present a general scheme for deriving the transport equation and consider other phenomena of nature and society, also described by this equation.

We have already noted that the exact solution of the partial differential equation can only be found in exceptional cases. The finite difference method for solving the boundary value problem for the heat equation was previously described. However, the explicit difference scheme used in this case is applicable under rather restrictive conditions on the algorithm parameters. An implicit scheme for solving the same problem, which is absolutely stable, will be described below. Above, we considered a mathematical model of a chemical reaction under conditions when the reacting substances are unevenly distributed over a given area. It combines expressions describing a chemical reaction and the diffusion of substances involved in the reaction. Similarly, the migration process of two competing biological species over a given territory can be considered. It is based on the equation of species migration, described by an equation of the type of thermal conductivity, supplemented by terms characterizing the phenomenon of biological competition.

A similar idea is used in the final section of the Appendix, which describes a model of tumor development in conditions of hormone resistance. It describes the change over time of tumor tissue under the influence of hormone therapy in the case of the tumor cells getting used to the action of an antibiotic. Moreover, on the one hand, a change in the size of the tumor is in a sense analogous to the volume of ice in the process of melting and is characterized by a condition like Stefan. On the other hand, the phenomenon of hormone resistance is to some extent analogous to the antibiotic resistance considered in Chapter 7.

1 Overview of transfer processes

From a mathematical point of view, the phenomena of heat conduction, diffusion, and transportation of goods turn out to be extremely close, although their nature is significantly different. In all three cases, the process of transferring a certain substance (heat, mass, goods), which is called a **measure**, is investigated. The measure is a global characteristic of the object of study and expresses the quantity of the substance in question contained in a certain area[12].

The local characteristic of the object here is a **state function** u (temperature, concentration, density of goods), which actually determines the measure of the unit length of a given area[13]. Then the formula used earlier

$$Q_1 = \int\limits_{x}^{x+\Delta x} \left[u(x, t + \Delta t) - u(x, t) \right] dx$$

defines a change in the measure of the segment $[x, x + \Delta x]$ over time from t to $t + \Delta t$. Note that the measure is determined by the spatial integral[14] of the state function.

The state function is initially unevenly distributed in this area. For some reasons (for each specific case, see below), there is a tendency to equalize the state function, which determines the course of dynamic processes. In this regard, a **measure flux** (heat, mass, goods) arises, characterizing the quantity of the considered substance, passing per unit time, through a given point[15]. The flux of the measure is proportional to the velocity of change of the state function in the x axis direction, i.e., derivative of the state function[16]. The proportionality coefficient here is a certain **equilibrium coefficient** D (thermal conductivity, diffusion, goods transfer), which is a process parameter and characterizes the transfer intensity of the considered substance (heat, mass, goods) in a given environment. In this case, a certain **equilibrium law** is obtained (Fourier, Fick, etc.), which establishes a connection between the flux and the state function and expresses the tendency of the given system to the equilibrium state. The result is a change in measure for time interval $[t, t + \Delta t]$. Thus, the change in the goods quantity of the segment $[x, x + \Delta x]$ for a given time is equal to

$$Q_2 = \int\limits_{t}^{t+\Delta t} \left(D\frac{\partial u}{\partial x}\Big|_{x+\Delta x} - D\frac{\partial u}{\partial x}\Big|_{x} \right) dt$$

in this space-time region under the influence of the corresponding fluxes.

TABLE 11.1:   Comparative characteristic of transfer processes.

Process	Heat transfer	Diffusion	Goods transfer
measure	heat quantity	mass	goods quantity
state function	temperature	concentration	goods density
measure flux	heat flux	diffusion flux	goods flux
equilibrium law	Fourier	Fick	
equilibrium coefficient	thermal conductivity	diffusion coefficient	goods transfer coefficient
law of measure conservation	law of heat conservation	law of mass conservation	law of goods conservation
transfer equation	heat equation	diffusion equation	goods transfer equation

Now we use the ***law of the measure conservation*** on the segment[17] $[x, x + \Delta x]$ for the time from t to $t + \Delta t$, characterized by the equality $Q_1 = Q_2$. The last relation is transformed using the mean value theorem and the limit transition at which the indicated region is compressed to a point. The result is the ***transfer equation*** (heat, diffusion, transfer of goods)

$$u_t = Du_{xx}.$$

A comparative description of the various transfer processes is given in Table 11.1.

The considered processes are characterized by the fact that due to the uneven distribution of the state function in this region, fluxes arise, due to which the system gradually tends to an equilibrium state. Let us try to evaluate the mechanism of this phenomenon in various interpretations.

The meaning of the phenomenon of thermal conductivity, roughly speaking, is as follows. Body temperature is associated with the kinetic energy of the molecules, and hence with their velocities. In the process of movement, the molecules collide and exchange energy. In this case, the situation in which a fast molecule transfers part of the energy to a slower molecule upon their collision seems to be much more likely than vice versa. As a result, fast molecules slow down, and slow ones accelerate. Thus, a gradual equalization of the velocities of molecules, and hence their energy, is observed. At the macroscopic level, we observe the cooling of warmer parts of the body and the heating of its colder parts. Thus, heat fluxes occur, leading to a redistribution of temperature, see Figure 11.1.

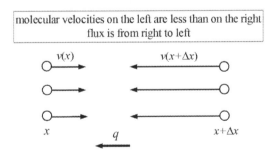

FIGURE 11.1: Heat flux occurrence.

In studying the diffusion phenomenon at the microscopic level, we again consider the movement of molecules. In the area with a high concentration contains a larger number of molecules. Naturally, the ***mean free path*** of the molecule, i.e., the distance traveled by a molecule between two collisions will be the greater, the less molecules can meet in its path. Thus, in a unit of time, from a region with a high concentration of a substance to a region where the concentration is relatively low, more molecules are likely to transfer than in the opposite direction. Thus, a diffusion flux is observed, due to which the gas concentration in the given volume is equalized, see Figure 11.2.

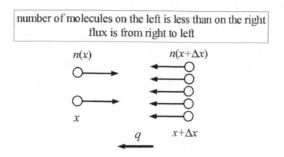

FIGURE 11.2: Diffusion flux occurrence.

Product transfer has a natural explanation. If in a certain area there is a lot of goods, then there it is unlikely to be in great demand. But where there is little of it, one can expect more demand for the product. In this regard, merchants are most likely to take the goods from the area where it is in abundance to the area where there is a shortage of it.

The described phenomena do not exhaust the class of transfer processes. In particular, the processes of liquid or gas *filtration* through a porous body[18] and *neutron diffusion* in a nuclear reactor are quite close to the diffusion phenomenon[19].

Consider now some electrically conductive medium in which some charges are unevenly distributed. In the region with a high charge density, in accordance with the **Coulomb law**, significant repulsive forces will act, tending to separate the charges in different directions. Naturally, the greatest influence of these forces is in the direction where the charge density is minimal, since relatively small repulsive forces act there. Thus, a flow of charges is observed, which determines the phenomenon of **electrical conductivity**[20]. Similar phenomena are associated with the movement of electrons and holes in the crystal lattice of a semiconductor[21].

Among the transfer processes is the movement of the liquid[22]. Molecules of a fluid move at different velocity. In the process of their collision, an exchange of impulses occurs. As a result, faster fluid layers slow down, and slower ones accelerate, due to the **viscosity** of the fluid. At the macroscopic level, a gradual equalization of the velocities of its various layers is observed.

A peculiar transfer process is the **migration of a biological species**. Suppose that a species is distributed unevenly over any territory. Then its number is not the same in different areas. Due to limited food, animals are moving from an area with a high abundance of a species, where there is intense competition for food, to an area where the abundance of a species is relatively small, and more likely to find food. As a result, a "species flux" is observed, as a result of which the species gradually spreads throughout the territory. This distribution will be uniform if the entire area under consideration has the same food reserves and other living conditions of the species. Below we will consider the process of migration of two competing species, unevenly distributed over the territory[23].

The process of **population migration** is close to the phenomenon described above. Naturally, the reason for the dispersal of the population will be not only a shortage of food in a limited area, but also various social factors.

This class of processes includes also the **epidemic propagation** from an area where there is a large number of infected people to an area where there are relatively few of them.

When studying the process of disseminating information, the role of "interacting particles" is people who differ significantly in terms of the information level. This is an information density that is the amount of information that a person has. In the process of human interaction, information is exchanged, as a result of which the level of awareness is aligned. Here, however, it should be borne in mind that at the same time, the awareness of the information source is not reduced. However, due to an increase in the level of awareness of poorly informed people, a more uniform distribution of information density is gradually being established in the considered system. The educational system, the media, scientific conferences, espionage, rumors, etc. in fact, they are certain forms of information transfer.

To a certain extent, the social struggle of the population should also be attributed to transfer phenomena. Different segments of the population can vary significantly in their income. There is a temptation to redistribute these incomes, i.e., to seize surplus funds from the richer and transfer them to the poor. At a milder level, this problem is solved by state social policy (progressive tax, social assistance to the poor, unemployment benefits, etc.). Sometimes equalization of incomes of the population is carried out by force. This is the social revolution. Note that in the model of social struggle, not a spatial variable, but layers of the population are used as a distributed coordinate. In particular, the function of the state of the system here can be considered population incomes, which vary from person to person and over time.

The transfer processes are associated with the functioning of the capital market. Various entrepreneurs can invest their available funds in certain enterprises and services. If, for example, it turns out that a large number of entrepreneurs produce refrigerators, and a small number produce washing machines, then it will become clear over time that refrigerators will be in abundance on the market, and washing machines will be in short supply. Under these conditions, many entrepreneurs will find it profitable to redesign their enterprises for the production of washing machines in the hope of obtaining higher incomes from the sale of goods in high demand. Thus, there is a gradual equalization of the concentration of capital.

Similar phenomena are observed in the labor market. If, for example, the janitors specialty is in short supply, while the need for businessmen remains relatively low, janitors are given fairly high salaries, while among businessmen there is significant unemployment. As a result, there are more and more people who want to become janitors and less and less who want to become businessmen. Thus, the system under consideration tends to an equilibrium state.

Let us now turn to the pricing process. Suppose that different entrepreneurs charge different prices for a product. Naturally, those from whom the price was too high will not be able to sell their products. In order to avoid ruin, they are forced to lower the price. On the other hand, the entrepreneur, who has not appointed a too high price, quickly sells the goods. Then, in order to obtain higher incomes, a cunning entrepreneur raises the price. Thus, there is a gradual equalization of prices for this product from various entrepreneurs[24].

Naturally, the listed examples do not exhaust the extensive list of transfer processes[25].

2   Finite difference method: Implicit scheme

For the approximate solving of boundary value problems for the heat equation, we have already used the finite difference method. However, the explicit difference scheme presented earlier is conditionally stable. As a result, for its practical application it is necessary to choose an extremely small time step, which often leads to an excessively large amount of computation. In this regard, in practice, another algorithm is often used here.

Let us return to the study of the first boundary value problem for the heat equation. Thus, the boundary value problem is given

$$u_t(x,t) = a^2 u_{xx}(x,t) + f(x,t),\ 0 < x < L,\ t > 0,$$

$$u(x,0) = \varphi(x),\ 0 < x < L,$$

$$u(0,t) = \alpha(t),\ u(L,t) = \beta(t),\ t > 0,$$

where the parameters a and L and the functions f, φ, α are β are known.

As in the case of an explicit difference scheme, a grid is introduced in the considered set, characterized by points (x_i, t_j), where $x_i = hi$, $i = 0, ..., M$, $t_j = \tau j$, $j = 0, ..., N$. Here, as before, the steps in time and spatial variable are determined by the equalities $\tau = T/N$, $h = L/M$, where T is the final time. The heat equation is used to approximate the derivatives entering into it using the well-known formulas of numerical differentiation. The only difference is that the expression approximating the second derivative is not chosen on the previous, but on the next time layer. As a result, the following relationships are obtained

$$\frac{u_i^{j+1} - u_i^j}{\tau} = a^2 \frac{u_{i-1}^{j+1} - 2u_i^{j+1} + u_{i+1}^{j+1}}{h^2} + f_i^j,\ i = 1, ..., M-1,\ j = 0, ..., N-1, \qquad (11.30)$$

$$u_i^0 = \varphi_i,\ i = 0, ..., M, \qquad (11.31)$$

$$u_0^j = \alpha^j,\ u_M^j = \beta^j,\ j = 1, ..., N, \qquad (11.32)$$

where all previously accepted designations remain valid. Equality (11.30) is called the *implicit difference scheme* for the heat equation[26]. You can make sure that this scheme is *absolutely stable*, i.e., in the process of counting from layer to layer over time, errors here do not accumulate. Thus, the values of the step of the algorithm can be selected solely for reasons of the desired accuracy, without taking into account the stability condition (11.15), which is used for an explicit difference scheme.

It would be surprising if this advantage were not accompanied by any disadvantage. Indeed, if recurrence formula (11.14) follows from the explicit scheme (11.11), which allows one to explicitly calculate the values of the desired grid function from layer to layer, then in relation (11.30) on the unknown subsequent time layer, three values of the unknown function corresponding to spatial indices $i - 1$, i and $i + 1$. Thus, at each time step, we obtain a system of linear algebraic equations. Fortunately, each equation in this system, except for the boundary ones corresponding to condition (11.32), includes only three unknown values[27]. This circumstance allows us to find its solution without using general methods for solving systems of linear algebraic equations that require a large amount of computation for sufficiently large values of the number of points M.

Denote $y_i = u_i^j$, $i = 0, ..., M$. Then we have the system

$$y_{i-1} + Ay_i + y_{i+1} = b_i,\ i = 1, ..., M-1, \qquad (11.33)$$

$$y_0 = b_0,\ y_{M+1} = b_{M+1} \qquad (11.34)$$

for the $j + 1$ step of time, where

$$A = -\left(2 + \frac{h^2}{a^2\tau}\right),\ b_i = \frac{h^2}{a^2}\left(\frac{u_i^j}{\tau} + f_i^j\right),\ i = 1, ..., M-1,\ b_0 = \alpha^j,\ b_M = \beta^j.$$

According to the *tridiagonal matrix algorithm*, it assumes that two adjacent unknown values are related by the equality[28]

$$y_i = p_{i+1}y_{i+1} + q_{i+1},\ i = M-1, M-2, ..., 0. \qquad (11.35)$$

To find the unknown parameters p_{i+1} and q_{i+1} from equality (11.35), we determine the value y_{i-1} and substitute the result in equality (11.33). We get

$$p_i y_i + q_i + A y_i + y_{i+1} = b_i.$$

This can be transformed to

$$y_i = -\frac{1}{A + p_i} y_{i+1} + \frac{b_i - q_i}{A + p_i}.$$

Comparing the result with formula (11.35), we obtain recurrence formulas for finding the unknown coefficients

$$p_{i+1} = -\frac{1}{A + p_i}, \quad q_{i+1} = \frac{b_i - q_i}{A + p_i}, \quad i = 1, 2, ..., M - 1. \tag{11.36}$$

For using these formulas it should know these parameters for $i = 1$. From formula (11.35) for $i = 0$, it follows that

$$y_0 = p_1 y_1 + q_1 = b_0,$$

because of the first equality (11.34). This is true whenever

$$p_1 = 0, \quad q_1 = b_0. \tag{11.37}$$

Thus, we have the following algorithm of solving system (11.33), (11.34).

i) By formulas (11.37), the first values of the unknown coefficients are determined.

ii) From formulas (11.36), all values of the unknown coefficients are found.

iii) From the second equality (11.34), the last value of the desired quantity is found.

iv) By formula (11.35), the solutions to the problem (11.33), (11.34) are determined, from the penultimate to the first.

Now we have an algorithm for finding the desired grid function on the subsequent time layer, and, therefore, an approximate solution of the first boundary value problem for the heat equation based on the implicit difference scheme[29]. Other boundary value problems for partial differential equations can be solved in a similar way[30].

Task 11.5 *Implicit difference scheme for the heat equation.* We consider the same for the first boundary value problem of heat equation (11.8) – (11.10), as in Task 11.3. Make the following analysis.

1. Choose the exact solution to the problem, as was done in Task 11.3, with the corresponding values of the functions included in the right-hand sides of the equation and boundary conditions.

2. Write down an algorithm for solving this problem based on an implicit difference scheme and a tridiagonal matrix algorithm, as well as an explicit difference scheme, see Task 11.3.

3. Create a program for solving the problem with the indicated algorithms.

4. To carry out calculations under conditions of stability violation of the explicit difference scheme.

5. Compare the calculation results by both methods with each other and with the exact solution to the problem.

3   Competitive species migration

In Chapter 7, the problem of the competition of two species consuming the same food was analyzed. In this case, competitors are unevenly distributed over some territory, for simplicity characterized by a one-dimensional region. The state of the system here is characterized by the densities of the considered species, i.e., their number per unit length. Then, the change in the density of species occurs both due to their migration from the region where the density of the species is high, to a relatively uninhabited region, and due to interspecific competition[31].

The resulting mathematical model combines the features of diffusion type equations characterizing the migration of species over a given territory and competition equations describing interspecific interactions. As a result, we obtain the following system of equations

$$\frac{\partial u_i}{\partial t} = \frac{\partial}{\partial x}\left(D_i \frac{\partial u_i}{\partial x}\right) + \left[d_i - a_i(q_1 u_1 + q_2 u_2)\right] u_i,$$

where u_i is the density of i-th species, D_i is its migration coefficient, a_i is the specific growth of the i-th species, q_i is the food consumption of the i-th species, d_i is the effective growth of the i-th species $i = 1, 2$, see Chapter 7. Distributions of the density of species over a given territory at the initial moment of time, as well as at the boundary of a given region, are known[32].

Task 11.6 *Approximate solving of migration equations of competing species.* The first boundary value problem for the above equations is considered. Make the following actions.

1. Approximate the task in accordance with the finite difference method using an implicit difference scheme.

2. Develop an algorithm for solving the problem using the tridiagonal matrix algorithm method.

3. Create a computer program in accordance with the developed algorithm.

4. Carry out the calculations of the studied system using the developed program.

5. Give an interpretation of the results.

4   Hormone treatment of the tumor with hormone resistance

We have already considered mathematical models of diffusion during the synthesis and migration of competing biological species, which are distributed analogs of the previously studied models of a specific chemical reaction and the interaction of two biological species. In both cases, when constructing the equations of state, we took as a basis a model of the corresponding transfer process, supplemented by terms describing the interaction of the objects in question. We give an additional example of a mathematical model with similar structure. We are talking about the hormonal treatment of certain forms of **cancer** and the resulting phenomenon of **hormone resistance**[33].

We consider the development of a tumor, which gradually increases in volume under the influence of certain factors. These may be some types of hormones present in the body. There is a certain analogue with the gradual freezing of water at low temperatures.

For the treatment of these types of cancer, hormone therapy is used, in which, under the influence of drugs, the production of the corresponding hormones is reduced. In this way, it is possible to slow down the growth of the tumor or even achieve its reduction in volume. Similarly, by heating, it is possible to slow down the freezing of water or even to melt the ice formed. The process of tumor growth under the influence of a stimulating factor and its reduction due to the suppression of this factor can in some sense be described using the Stefan problem considered earlier.

It is known that over time, the effectiveness of hormonal treatment may decrease due to an increase in the number of cancer cells resistant to the action of the used drugs. This phenomenon, called hormone resistance, in a certain sense is similar to the phenomenon of bacteria getting used to the action of antibiotics, described in Chapter 7. Based on Stefan problem to describe the size of a tumor and the antibiotic resistance model to describe a decrease in the effectiveness of hormonal treatment, one can try to construct a mathematical model of the considered process.

Two types of cancer cells that are sensitive and resistant to hormone treatment are considered. The state of the system will be characterized by the functions $u_s = u_s(x, t)$ and $u_r = u_r(x, t)$, describing the concentration of these cell types at the corresponding point at a specified moment time. The distribution of these cells in space will be characterized by the corresponding fluxes, and therefore, by diffusion terms. Both types of cells have some natural growth, which is constrained by the limited living space common to all cells. As a consequence, terms corresponding to the competition model appear in these equations. In addition, we take into account the possibility of transitions from one cell type to another. Finally, there is also the effect of a drug that inhibits the reproduction of sensitive cells and does not affect resistant cells[34].

For simplicity, we restrict ourselves to considering the spatially one-dimensional case[35]. Suppose the left end of the segment is fixed. Here is the focus of the onset of the tumor and fluxes are set that are similar to the influx or removal of heat from outside in the Stefan problem, see condition (11.23). The right end corresponds to the outer border of the tumor, which means that it is characterized by a zero concentration of cancer cells. Since the state of the system changes with time, the position of this boundary changes with time. To determine the law of movement of the boundary $\xi = \xi(t)$, a condition of Stefan type is written. The initial distribution of cancer cells, and hence the initial position of the moving border, are considered known.

Thus, we get the following mathematical model

$$\frac{\partial u_s}{\partial t} = D_s \frac{\partial^2 u_s}{\partial x^2} + \left[\frac{a_s}{1 + f(t)(u_s)^\theta} - b_s(u_s + u_r) \right] u_s + c_{rs}, \ 0 < x < \xi(t), \ t > 0,$$

$$\frac{\partial u_r}{\partial t} = D_r \frac{\partial^2 u_r}{\partial x^2} + \left[a_r b_r(u_s + u_r) \right] u_r + c_{sr}, \ 0 < x < \xi(t), \ t > 0,$$

$$u_i(x, 0) = u_{i0}(x), \ 0 < x < \xi_0, \ i = r, s,$$

$$-D_i \frac{\partial u_i}{\partial x} = q, \ t > 0, \ i = r, s,$$

$$u_i(\xi(t), t) = 0, \ t > 0, \ i = r, s,$$

$$\frac{d\xi(t)}{dt} = -D_s \frac{\partial u_s(\xi(t), t)}{\partial x} - D_r \frac{\partial u_r(\xi(t), t)}{\partial x}, \ t > 0,$$

where D_s and D_r are diffusion coefficients of sensitive and resistant cancer cells, a_s and a_r are their growths, b_s and b_r are their growth reduction due to limited living space, and c_{rs} and c_{sr} are the frequencies of intertype transitions of cells. In this case, $f(t)$ determines the dose of the hormonal drug at a given time, and θ is its intensity[36].

Note some additional restrictions for the considered model. The equalities $u_{s0}(\xi_0) = 0$, $u_{r0}(\xi_0) = 0$ mean that the initial concentration of both types of cells on the right border is zero, which corresponds to the initial border of the tumor. Inequality $u_{s0}(x) \gg u_{r0}(x)$ means that sensitive cells predominate in the system initially. The condition $a_s/b_s > a_r/b_r$ suggests that in the absence of treatment, sensitive cells are more viable. In addition, the coefficients D_s and D_r are sufficiently small, i.e., tumor growth is quite slow. The parameters c_{rs} and c_{sr} are also quite small, i.e., intertype transitions are relatively rare. By the calculations, with a certain combination of parameters, the presented model describes tumor growth during the natural course of the process, reduction in growth and even decrease in size of the tumor under the influence of hormonal treatment by reducing the concentration of sensitive cells, reducing the effectiveness of hormonal treatment due to the increase in the concentration of resistant cells, restoration of concentration sensitive cells and a decrease in the concentration of resistant cells in case of discontinuation of treatment[37].

Notes

[1] For similarity theory, see [118], [182].

[2] In fact, we have already used the theory of similarity when analyzing the predator–prey model, see Chapter 7.

[3] About the finite difference method, see, [280], [301], [325], [340], [347].

[4] The definition of approximate differentiation formulas is based on the expansion of the function in a *Taylor series*. Let there be some function $g = g(x)$ with the proper number of derivatives. Then we have its representation in the form

$$g(x + h) = g(x) + g'(x)h + \frac{1}{2}g''(x)h^2 + \frac{1}{6}g'''(x)h^3 + \dots + \frac{1}{n!}g^{(n)}(x)h^n + \dots . \tag{11.38}$$

Then we get

$$g(x + h) = g(x) + g'(x)h + o(h),$$

where the value $o(h)$ has a higher (second) order of smallness than h, i.e., characterized in that $o(h)/h \to 0$ as $h \to 0$. Now we get

$$g'(x) = \frac{g(x + h) - g(x)}{h} - \frac{o(h)}{h} \approx \frac{g(x + h) - g(x)}{h}$$

for sufficiently small values of h. As a result, we obtained the formula used earlier for the approximate calculation of the first derivative. Since the discarded quantity has a first order of smallness with respect to h, it is said that the above formula has a ***first order of approximation***. To obtain the formula for calculating the second derivative, we take into account the four terms on the right-hand side of equality (11.38). We get

$$g(x + h) = g(x) + g'(x)h + \frac{1}{2}g''(x)h^2 + \frac{1}{6}g'''(x)h^3 + o(h^3),$$

where the last term on the right-hand side has a higher (fourth) order of smallness than h^3. What interests us is not its specific meaning, but only its order of smallness. In the following formulas, by will be understood different quantities $o(h^3)$ having the same order of smallness. Replacing here h by $-h$, we have

$$g(x - h) = g(x) - g'(x)h + \frac{1}{2}g''(x)h^2 - \frac{1}{6}g'''(x)h^3 + o(h^3).$$

Moreover, the last term in the right-hand side is not the same here, but of the same order as the corresponding term in the right-hand side of the previous equality. As a result of the addition of these relations, we have

$$g(x + h) + g(x - h) = 2g(x) + g''(x)h^2 + o(h^3).$$

Then

$$g''(x) = \frac{g(x - h) - 2g(x) + g(x + h)}{h^2} + \frac{o(h^3)}{h^2} \approx \frac{g(x - h) - 2g(x) + g(x + h)}{h^2}.$$

As a result, we obtain the formula used in the finite difference method for the approximate calculation of the second derivative. Moreover, since the discarded quantity here has a ***second order of smallness*** with respect to h, they say that this formula has a second order of approximation. Naturally, the higher the approximation order, the more accurate the formula for calculating the derivative.

[5] In Chapter 17 it will be shown that the discrete analogue of the heat equation in the stationary case has a direct physical meaning, being a discrete model of the system under consideration. In particular, for the numerical analysis of this system, you can use the tridiagonal matrix algorithm described in the Appendix.

[6] The explicit difference scheme has a first order of approximation with respect to time and a second order of approximation with respect to a spatial variable due to the used approximation orders of derivatives included in the heat equation.

[7] Suppose that we consider heat equation (11.8) with initial condition (11.9) and the second type boundary conditions of the kind

$$u_x(0, t) = \alpha(t), \ u_x(L, t) = \beta(t), \ t > 0.$$

To solve the obtained boundary value problem, we can use the explicit difference scheme (11.11) with formula (11.12) to find the grid function on the zero time layer. In the above boundary conditions, we replace the first derivative at the boundaries with its corresponding approximations. We get

$$u_x(0, t_j) \approx \frac{u(h, t_j) - u(0, t_j)}{h}, \ u_x(L, t_j) \approx \frac{u(L, t_j) - u(L - h, t_j)}{h}.$$

Then we obtain the formulas

$$\frac{u_1^j - u_0^j}{h} = \alpha^j, \ \frac{u_M^j - u_{M-1}^j}{h} = \beta^j, \ j = 1, ..., N.$$

Now we obtain the following algorithm for the approximate solution of the second boundary value problem. First, by formula (11.12), the grid function is found on the zero time layer. Suppose now that its value on the j-th time layer is known. Then, from the basic calculation formula (11.14), all values are found on the subsequent time layer, starting with the value $i = 1$ and ending with $i = M - 1$. After that, the boundary values are determined from the above equalities

$$u_0^{j+1} = u_1^{j+1} - h\alpha^{j+1}, \ u_M^{j+1} = u_{M-1}^{j+1} + h\beta^{j+1}.$$

As a result, all the values on the next time layer will be known, and one can find a solution to the problem on a next layer in time. Note that, since we used first order accuracy formulas to approximate the boundary conditions, the described scheme for solving the second boundary value problem for the heat equation has a first order approximation both in time and in spatial variable.

[8] To assess the lack of the explicit difference scheme in connection with the presence of the stability condition, we consider a particular case of the problem being solved. Suppose that the problem is solved for $a = 1$, $L = 1$, $T = 1$. Let us be completely satisfied with the solution of the problem with a step in the spatial variable $h = 0.1$. Then, for reasons of stability, the value of the time step $\tau \leq h^2/2$ should be chosen. Thus, the maximum allowable time step will be equal to 0.005. Therefore, when choosing 10 points in the spatial variable in a unit square, we need to perform 200 time steps. If we want more accurate results by choosing the value of step $h = 0.01$, then we have to choose a time step of 0.00005, i.e., perform 20000 time steps. Thus, the use of an explicit difference scheme in the region of relatively large sizes leads to an extremely large amount of computation. In this regard, instead of an explicit scheme, more difficult but more effective implicit schemes are often used, see Appendix.

[9] Similarly, when the temperature passes through the boiling point, the substance transfers from the liquid phase to the gaseous phase during boiling or the reverse transitions from the gaseous phase to the liquid phase during condensation.

[10] Stefan problem is considered in [10], [119], [231], [293].

[11] The ***two-phase Stefan problem*** involves the consideration of both phases of the substance in question, i.e., in this case, water and ice. The second phase is characterized by the heat equation in the region to the right of the moving boundary.

[12] In analysis, the ***measure*** is the set function with the following properties. The measure is always non-negative. The measure of the empty set is zero. The measure is additive, i.e., the measure of combining two disjoint sets is the sum of their measures. Naturally, the notion of measure used above corresponds to all these properties. We also note that the concept of measure is invariably associated with the integration operation. A general theory of measure is given, for example, in [129].

[13] In the two-dimensional case, the state function (e.g., concentration) characterizes the measure (e.g., mass) of a unit area, and in the three-dimensional case, units of volume.

[14] In the two-dimensional case, the measure is determined by a surface integral, and in the three-dimensional case, by a volume integral.

[15] In the two-dimensional case, the flux determines the measure passing in a given time through the elementary segment, and in the three-dimensional and in the three-dimensional case, this determines the measure passing through the elementary section.

[16] In the multidimensional case, the measure flux is proportional to the gradient of the state function.

[17] In the two-dimensional case, the law of conservation of the measure characterizes the change in the measure on the elementary part of the surface over the considered time interval, and in the three-dimensional - in the elementary volume.

[18] About filtration, see [228].

[19] About neutron diffusion in a nuclear reactor, see [334].

[20] About electrical conductivity, see [204], [278], [324], [343].

[21] About semiconductor physics, see [309].

[22] The problems of hydrodynamics are considered in Chapter 14.

[23] Other distributed biology models are discussed, for example, in [85], [244], [245].

[24] Other distributed economics models are considered in [92]. Note also the Black–Scholes model in financial mathematics, see [38].

[25] The heat equation is also used in **differential geometry**, see [379] and in the theory of **random processes**, see [161]. We also note that the Schrödinger equation considered in Chapter 15 formally can be written in the form of the heat equation.

[26] Chapter 17 shows that the difference scheme itself can have a direct physical meaning.

[27] In this case, the matrix of the corresponding system of linear algebraic equations is tridiagonal.

[28] Since the considered system is linear, it is natural to assume that any pair of unknowns will also be connected by some linear relation. The tridiagonal matrix algorithm can be found in any literature on the finite difference method and calculation methods in general, see, for example, [243], [280], [301], [325], [340], [347]. This is also known as the **Thomas algorithm**.

[29] The implicit difference scheme, like the explicit one, has the first order of approximation in time and the second in spatial variable. To obtain a second order of approximation in both spatial and temporal variables, you can use, for example, the **Crank–Nicholson scheme**, see [347]. In it, the second derivative with respect to the spatial variable is approximated in the form of a half-sum of approximations used in explicit and implicit schemes. Thus, the relation

$$\frac{u_i^{j+1} - u_i^j}{\tau} = \frac{a^2}{2} \left(\frac{u_{i-1}^j - 2u_i^j + u_i^j}{h^2} + \frac{u_{i-1}^{j+1} - 2u_i^{j+1} + u_{i+1}^{j+1}}{h^2} \right) + f_i^j .$$

The Crank–Nicholson scheme is six-point (connects six values of the desired function) in contrast to the four-point explicit and implicit schemes. To find a solution at the next time step, one can also use the sweep method here. Characteristically, this scheme is also absolutely stable.

[30] In a similar way, the second boundary value problem for the heat equation can be solved, as well as the boundary value problems for generalizations of this equation considered in the previous chapter. We also note that for the approximate solution of the multidimensional heat equation and other multidimensional partial differential equations, the **fractional step method** is used, in which the transition from one layer in time to another takes several steps, see [377]. Among other methods for approximate solving problems of mathematical physics, we also note the **finite element method**, see, for example, [54].

[31] Mathematical models of the migration of two or more species over the territory with other forms of interspecific interactions (see Chapter 7) are obtained in a similar way. The model includes terms described both migration (an analogue of heat conduction and diffusion) and interspecific interaction, as in models the predator is prey, niche, and symbiosis.

[32] If this territory is isolated from the outside world (for example, an island), then the second type boundary condition is used. The derivative of the density of the species at the border is zero here.

[33] For the phenomenon of drug resistance in the treatment of cancer, in particular, for hormone resistance, see [251]. Mathematical models of the tumor growth are considered, for example, in [186], and the models of the cancer treatment are given in [44], [217], [249].

[34] When describing the phenomenon of antibiotic resistance in Chapter 7, we considered a bactericidal antibiotic that kills bacteria. The considered type of hormone therapy is similar to bacteriostatic antibiotics, which do not kill bacteria, but suppress their birth rate.

[35] It is more accurate to consider the tumor not as a linear, but as a spatial object. In particular, in the simplest case, considering the shape of the tumor to be spherical, we can go to spherical coordinates. Using spherical symmetry, we come to one spatial coordinate characterizing the distance from a given point to the center of the sphere. In this case, Stefan condition will be set on the outer surface of the sphere, the position of which varies with time.

[36] Obviously, with $f(t) = 0$, there is no treatment. With positive values of this function, the denominator of the corresponding fraction turns out to be more than unity, which means that we observe a decrease in the growth of cancer cells.

[37] For a numerical analysis of the presented model, see [316].

Chapter 12

Wave processes

In Chapters 4 and 5 we analyzed mathematical models of oscillatory processes. The studied object there (pendulum, spring, electric circuit) was understood as a material point, i.e., had no spatial sizes. However, distributed objects can also be associated with oscillations. In particular, mechanical generalizations of the string and electrical oscillations of the wire are natural generalizations of the above processes. The periodically changing characteristics of these processes no depend only on time, but also on the spatial coordinate directed along the string or wire. Two-dimensional analogues of these processes are membrane oscillations or the propagation of waves on the surface of the water. Finally, sound and electromagnetic waves propagate in space. A brief review of mathematical models of the above processes and methods of their analysis is the subject of this chapter.

The main object of study here is the vibrating string equation[1], which is a distributed analogue of the spring oscillation equation. Like the heat equation, this is a second order partial differential equation with two independent variables. The equation is derived based on the law of momentum conservation.

The mathematical model of the considered process is a boundary value problem for the resulting equation. The solution to two such problems is found in this chapter. In particular, using the method of separation of variables described earlier in the analysis of transfer processes, we study the mathematical model of transverse vibrations of a string with fixed ends. In this case, the solution of the problem is obtained as a Fourier series. The analysis of an infinitely long string is carried out, using the D'Alembert method. This is associated with the reduction of the equation to canonical form with the subsequent finding of its analytical solution. As an application, the propagation of running waves is investigated.

The final section of the chapter is devoted to electrical vibrations in the wire. The mathematical model of this process is characterized by a system of telegraph equations[2], whence under certain conditions the equations for voltage in the circuit and electric current follow, which coincide in form with the vibrating string equation.

The Appendix establishes the law of energy conservation for an oscillating string. A general wave equation is also given, covering also the process of membrane oscillation and the propagation of sound waves. The oscillations of an elastic beam are described by a fourth order partial differential equation. The Maxwell equations, which constitute the most important mathematical model of electrodynamics, are also considered. Finally, the finite difference method for an approximate solution of the vibrating string equation is presented.

Lecture

1 Vibration of string

A natural generalization of spring oscillations (see Chapter 4) is the process of vibration of a ***string*** that is an object with sizes. Suppose the string is long and thin enough. Then it can be considered spatially one-dimensional and take into account the change in all characteristics only along the length of the string. In this case, the time t and the spatial variable x directed along the string are the independent variables. Thus, unlike a spring or a pendulum, a string can no longer be qualified as a material point, and its mathematical model cannot be reduced to a system of a finite number of ordinary differential equations. As in the case of a heated body or diffusing gas, we have a distributed object.

We restrict ourselves to considering only transverse vibrations of string [3], which are directed perpendicular to the x axis. The state function of the system here is the deviation u of the string from the x axis. The coordinate system is chosen so that when the string is located directly on the x axis (i.e., at $u = 0$) and is at rest, oscillations do not occur. Thus, the zero value of the state function of the system corresponds to the equilibrium position. Thus, the quantity $u(x, t)$ expresses the deviation of the string at point x at time t from the equilibrium position, see Figure 12.1.

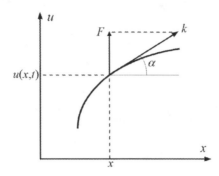

FIGURE 12.1: Transverse vibrations of string.

The equation of the string movement, like the equation of a material point movement, is derived on the basis of the law of momentum conservation[4]. The momentum of a body of mass m moving with velocity v is equal to the product $p = mv$. The mass of a homogeneous string of length Δx is determined by the formula $m = \rho\Delta x$, where ρ is the linear density, i.e., the quantity of substance per unit length[5]. Density is determined by the material of which the string consists. This is the parameter of the system. Therefore, the string momentum is $p = \rho v \Delta x$. Here, the value v represents the velocity of change of the state function u, i.e., its partial time derivative. This derivative here, as in the previous chapters, for brevity, we denote by u_t. Thus, we obtain the formula $p = \rho u_t \Delta x$. This is true if the characteristics of the string are constant in the considered interval. In reality, the function u, and hence its time derivative (and in the case of an non-homogeneous string, its density) vary from point to point. Such a change can be neglected only on a plot of arbitrarily small length dx. The momentum here is $dp = \rho u_t dx$.

Consider a segment of the string from point x to $x + \Delta x$. The corresponding momentum is found by integrating the last equality

$$p = \int_{x}^{x+\Delta x} \rho u_t dx.$$

However, to derive the state equation, we are not interested in the momentum itself, but in its change over a time interval from t to $t + \Delta t$. Then the desired change in momentum is

$$p_1 = \int_{x}^{x+\Delta x} \rho(x)\big[u_t(x, t + \Delta t) - u_t(x, t)\big]dx. \tag{12.1}$$

We found how the momentum changes in a given segment of a string over a specified time. Now it remains to be seen how this momentum has changed. If a constant force F acts on a body, then in a time Δt it gives it a momentum $p = F\Delta t$. In the case of a variable force, this equality only makes sense on an infinitesimal time interval dt. The momentum value corresponds to it $dp = Fdt$. Then for the time from t to $t + \Delta t$ we get the momentum

$$p = \int\limits_{t}^{t+\Delta t} Fdt.$$

To clarify this equality, it is necessary to determine the meaning of the force F. Assume that the only force acting on the string is due to its tension. Suppose the string is flexible enough, i.e., does not resist bending[6]. Then the force of its tension or simply, tension k is directed tangentially to the string profile, see Figure 12.1. The tension is determined by the material, is variable in the case of a non-homogeneous string and is a parameter of the system.

Since we take into account only the transverse vibrations of the string, the value for the momentum should include not the tension itself, but its projection onto the u axis. As a result, we obtain the equality $F = k\sin\alpha$, where α is the angle of inclination of the tangent to the string profile at a given point. We restrict ourselves to considering only small vibrations of the string[7]. The sine of a small angle is close enough to its tangent. However, the tangent of the inclination angle of the tangent to the curve characterizing the dependence of the function u on the argument x is equal to its derivative with respect to this argument $u_x = \partial u/\partial x$. As a result, the equality $F = k\tan\alpha = ku_x$ holds with rather high accuracy.

We are interested in the change in momentum on the segment $[x, x + \Delta x]$ under the action of tension forces. Then the required force F is equal to the difference of the forces at the ends of the considered segment, see Figure 12.2. Thus, we find the force

$$F = (ku_x)\big|_{x+\Delta x} - (ku_x)\big|_{x}.$$

Consequently, the change in the momentum of the considered string segment in this time interval is equal to

$$p_2 = \int\limits_{t}^{t+\Delta t} \left[(ku_x)\big|_{x+\Delta x} - (ku_x)\big|_{x}\right]dt. \tag{12.2}$$

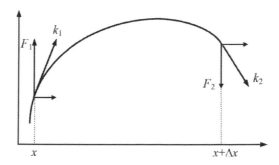

FIGURE 12.2: The action of tension forces on a segment.

We believe that the momentum of the system changes exclusively under the action of tension forces[8]. As a result, we obtain the **law of momentum of conservation** $p_1 = p_2$. Then, using conditions (12.1), (12.2), we have

$$\int\limits_{x}^{x+\Delta x} \rho(x)\big[u_t(x, t + \Delta t) - u_t(x, t)\big]dx = \int\limits_{t}^{t+\Delta t} \big[(ku_x)\big|_{x+\Delta x} - (ku_x)\big|_x\big]dt.$$

Transformed this equality by the main value theorem, similar to how it was done in previous chapters. We have

$$\rho(\xi)\big[u_t(\xi, t + \Delta t) - u_t(\xi, t)\big]\Delta x = \big[k(x + \Delta x)u_x(x + \Delta x, \tau) - k(x)u_x(x, \tau)\big]\Delta t,$$

where $\xi \in [x, x + \Delta x]$, $\tau \in [t, t + \Delta t]$. After dividing by $\Delta x \Delta t$, we get

$$\rho(\xi)\frac{u_t(\xi, t + \Delta t) - u_t(\xi, t)}{\Delta t} = \frac{k(x + \Delta x)u_x(x + \Delta x, \tau) - k(x)u_x(x, \tau)}{\Delta x}.$$

Passing to the limit as $\Delta x \to 0$, $\Delta t \to 0$, determine the equality

$$\rho\frac{\partial^2 u}{\partial x^2} = \frac{\partial}{\partial x}\Big(k\frac{\partial u}{\partial x}\Big), \tag{12.3}$$

called the **vibrating string equation**[9]. If the string is homogeneous, i.e., its characteristics are constant, then the equality (12.3) takes the form

$$u_{tt} = a^2 u_{xx}, \tag{12.4}$$

where $a = \sqrt{k/\rho}$.

The heat equation and the vibrating string equation not only describe different processes, but also relate to different types[10]. However, the procedure for their determining is largely the same, see Table 12.1. First, the change in a certain quantity (momentum, heat quantity) of a segment of a homogeneous object (string, body) is calculated over a time interval, see step 1. Then the same quantity is determined on an interval of arbitrarily small length, see step 2. Next an extensional segment of a non-homogeneous object is considered, step 3. At the fourth step, the considered value (momentum, heat quantity) is calculated over a time interval under the action of a constant specific action (force, heat flow). Then it is determined on a small time interval (step 5) and on an extensional time interval in the case of variable action (step 6). At the seventh step, one calculates a change in this value in an interval of object under considered influence (step 7). Next, the corresponding conservation law is applied (step 8), the mean theorem (step 9), and as a result of the passage to the limit, the state equation of the system is obtained (step 10).

The obtained relations, like the heat equation, are partial differential equations. To obtain a complete mathematical model, it is necessary to set the corresponding initial and boundary conditions. The vibrating string equation is of the second order in time and requires the specification of two initial conditions. We assume that at the initial time $t = 0$ the initial profile of the string, characterized by the function $\varphi = \varphi(x)$, and the velocity distribution $\psi = \psi(x)$ along its length are known[11]. Thus, the following equalities hold

$$u(x, 0) = \varphi(x), \quad u_t(x, 0) = \psi(x). \tag{12.5}$$

The studied equation has a second order in the spatial variable. In this regard, two additional conditions must be specified at the ends of the string, i.e., boundary conditions. Consider a string of length L. Suppose its left end corresponds to the point $x = 0$, and the

TABLE 12.1: Stages of deriving of state equations.

Step	Vibrations of string	Heat transfer
1	$p = \rho(u_t\vert_{t+\Delta t} - u_t\vert_t)\Delta x$	$Q = c\rho(u\vert_{t+\Delta t} - u\vert_t)\Delta x$
2	$dp = \rho(u_t\vert_{t+\Delta t} - u_t\vert_t)dx$	$dQ = c\rho(u\vert_{t+\Delta t} - u\vert_t)dx$
3	$p_1 = \int\limits_x^{x+\Delta x} \rho(u_t\vert_{t+\Delta t} - u_t\vert_t)dx$	$Q_1 = \int\limits_x^{x+\Delta x} c\rho(u\vert_{t+\Delta t} - u\vert_t)dx$
4	$p = F\Delta t$	$Q = q\Delta t$
5	$dp = Fdt$	$dQ = qdt$
6	$p = \int\limits_t^{t+\Delta t} ku_x dt$	$Q = \int\limits_t^{t+\Delta t} \lambda u_x dt$
7	$p_2 = \int\limits_t^{t+\Delta t} (ku_x\vert_{x+\Delta x} - ku_x\vert_x)dt$	$Q_2 = \int\limits_t^{t+\Delta t} (\lambda u_x\vert_{x+\Delta x} - \lambda u_x\vert_x)dt$
8	$p_1 = p_2$	$Q_1 = Q_2$
9	$\rho\dfrac{u_t\vert_{t+\Delta t} - u_t\vert_t}{\Delta t} = \dfrac{ku_x\vert_{x+\Delta x} - ku_x\vert_x}{\Delta x}$	$c\rho\dfrac{u\vert_{t+\Delta t} - u\vert_t}{\Delta t} = \dfrac{\lambda u_x\vert_{x+\Delta x} - \lambda u_x\vert_x}{\Delta x}$
10	$\rho\dfrac{\partial^2 u}{\partial t^2} = \dfrac{\partial}{\partial x}\left(k\dfrac{\partial u}{\partial x}\right)$	$c\rho\dfrac{\partial u}{\partial t} = \dfrac{\partial}{\partial x}\left(\lambda\dfrac{\partial u}{\partial x}\right)$

right end corresponds to the point $x = L$. At the left end, the first type condition can be specified

$$u(0,t) = \alpha_1(t),$$

i.e., the left end moves by the known law $\alpha_1 = \alpha_1(t)$. In particular, when $\alpha_1 = 0$, it is rigidly fixed in the equilibrium position. The second type condition is

$$ku_x(0,t) = \beta_1(t),$$

where the tension force β_1 is known at the left end of the string. If $\beta_1 = 0$, then the tension is absent, and the end is free.

There are other types of boundary conditions. However, the above are the most important. Similar conditions can be given at $x = L$. Depending on the nature of the investigated phenomenon, various combinations of conditions at the ends of the string are possible. In the simplest cases, both first type conditions of the are considered

$$u(0,t) = \alpha_1(t), \quad u(L,t) = \alpha_2(t) \qquad (12.6)$$

with known laws of ends movement, or both second type conditions

$$ku_x(0,t) = \beta_1(t), \quad ku_x(0,t) = \beta_2(t) \qquad (12.7)$$

with known forces at the ends.

The vibrating string equation with initial conditions (12.5) and boundary conditions (12.6) make up the ***first boundary value problem*** for this equation. When replacing equalities (12.6) with (12.7), we obtain the ***second boundary value problem***. Both of these problems are generally non-homogeneous. If the values in the right-hand sides of the boundary conditions are equal to zero, then we obtain homogeneous boundary value problems. In particular, the boundary conditions

$$u(0,t) = 0, \quad u(L,t) = 0 \qquad (12.8)$$

correspond to the fixed ends of the string, and the equalities

$$u_x(0,t) = 0, \quad u_x(L,t) = 0 \qquad (12.9)$$

correspond to the free ends.

A special problem arises when the string is so long that it can be considered infinite. Then the boundary of the string is absent and it is not entirely clear where additional conditions for the spatial variable should be specified. However, it turns out that in this case the boundary conditions are not needed at all. The equation of oscillation of an infinite string with initial conditions (12.5) makes up the **Cauchy problem** for this equation.

Vibrations of the string are described by a partial differential equation.

2   Vibrations of string with fixed ends

Consider the oscillations of a homogeneous string of length L with fixed ends with a known initial profile of the string and its velocity distribution. The mathematical model of the studied process is the homogeneous first boundary value problem (12.4), (12.5), (12.8). To solve it, we use the method of separation of variables[12], considered in the study of the heat equation, see Chapter 10.

Find the solution of the problem by the formula

$$u(x, t) = X(x)T(t). \tag{12.10}$$

Putting this value to the equality (12.4), we get

$$X(x)T''(t) = a^2 X''(x)T(t).$$

Now we have

$$\frac{T''(t)}{a^2 T(t)} = \frac{X''(x)}{X(x)}.$$

Equality of functions of various arguments is possible only in the case when the values included in its left and right sides are constants. Denoting the corresponding constant by λ, we get

$$T''(t) - \lambda a^2 T(t) = 0, \ t > 0, \tag{12.11}$$

$$X''(x) - \lambda X(x) = 0, \ 0 < x < L. \tag{12.12}$$

Thus, we obtain ordinary differential equations connected by a common constant λ. Substituting the function u from equality (12.10) into boundary conditions (12.8), we obtain the relations

$$X(0)T(t) = 0, \ \ X(L)T(t) = 0, \ \ t > 0.$$

Considering that the identity of the function T equal to zero by virtue of (12.10) will lead to the fact that the solution of the considered is identically equal to zero. This contradicts the initial conditions. As a result, we conclude

$$X(0) = 0, \ \ X(L) = 0. \tag{12.13}$$

Relations (12.12), (12.13) exactly coincide with the Sturm–Liouville problem considered in the study of the heat equation, see Chapter 10. It has a non-trivial solution for $\lambda = \lambda_k$ only, where

$$\lambda_k = -\left(\frac{k\pi}{L}\right)^2, \ k = 1, 2, \dots .$$

The corresponding solutions to problem (12.12), (12.13) are

$$X_k(x) = c_k \sin \frac{k\pi}{L} x,$$

where the constants c_k are arbitrary. The general solution of equation (12.11) for $\lambda = \lambda_k$ is

$$T_k(t) = a_k \sin \frac{ak\pi}{L} t + b_k \cos \frac{ak\pi}{L} t,$$

where a_k and b_k are arbitrary constants. Now we find

$$u_k(x,t) = X_k(x)T_k(t) = \left(\varphi_k \sin \frac{ak\pi}{L} t + \psi_k \cos \frac{ak\pi}{L} t \right) \sin \frac{k\pi}{L} x,$$

where $\varphi_k = a_k c_k$, $\psi_k = b_k c_k$. For any values of the number k and the constants φ_k and ψ_k, the function u_k satisfies the vibrating string equation (12.4) and boundary conditions (12.8). The the sum of all such solutions have similar properties[13]

$$u(x,t) = \sum_{k=1}^{\infty} \left(\varphi_k \sin \frac{ak\pi}{L} t + \psi_k \cos \frac{ak\pi}{L} t \right) \sin \frac{k\pi}{L} x. \tag{12.14}$$

It remains to choose the constants φ_k and ψ_k so as initial conditions (12.5) are satisfied.

Putting in equality (12.14) $t = 0$ and using the first equality (12.5), we obtain

$$u(x,0) = \sum_{k=1}^{\infty} \varphi_k \sin \frac{k\pi}{L} x = \varphi(x).$$

Multiplying this equality by $\sin n\pi x/L$ with arbitrary n and integrating the result by x, we get

$$\sum_{k=1}^{\infty} \varphi_k \int_0^L \sin \frac{k\pi}{L} x \sin \frac{n\pi}{L} x dx = \int_0^L \varphi(x) \sin \frac{n\pi}{L} x dx.$$

The integrals of the left-hand side of this equality were calculated during the analysis of the heat equation, see Chapter 10. As a result, we find the Fourier coefficients

$$\varphi_k = \frac{2}{L} \int_0^L \varphi(x) \sin \frac{k\pi}{L} x dx, \quad k = 1, 2, \dots . \tag{12.15}$$

Similarly, differentiating equality (12.10) with respect to time, setting $t = 0$ and using the second equality (12.5), we determine

$$u_t(x,0) = \sum_{k=1}^{\infty} \psi_k \frac{ak\pi}{L} \sin \frac{k\pi}{L} x = \psi(x).$$

Repeating the previous transformations, find the coefficients

$$\psi_k = \frac{2}{ak\pi} \int_0^L \psi(x) \sin \frac{k\pi}{L} x dx, \quad k = 1, 2, \dots . \tag{12.16}$$

Thus, the solution to the problem (12.4), (12.5), (12.8) is determined by formula (12.14), where the numbers φ_k and ψ_k are calculated by formulas (12.15) and (12.16).

Consider the partial case of the problem such that $L = \pi$, $\varphi(x) = \sin x$, $\psi(x) = 0$. Thus, we have a string of the specified length, fixed at the ends, which at the initial moment of time has a sinusoidal profile and zero velocity. According to formulas (12.15) and (12.16), all the Fourier coefficients of the function φ, starting from the second and all parameters

ψ_k are equal to zero, and the value of φ_1 is equal to one. Then the solution to the problem in question is

$$u(x, t) = \cos at \sin x.$$

Substituting this value into equalities (12.4), (12.5), (12.8), we see that the function u defined in this way is indeed a solution to the problem.

Figure 12.3 shows the profiles of the string, as well as its velocity

$$v(x, t) = u_t(x, t) = -a \sin at \sin x$$

at various points in time. Graphs of these functions are shown in time steps $t = \pi/4a$. Let us analyze the results.

We consider the free vibrations of a string. The ends of the string are rigidly fixed. At the initial moment of time, the string is removed from the equilibrium position and is at rest. As a result of this, a tension force acts on the string, directed toward the equilibrium position. The string begins to move in this direction, gaining velocity. At time $t = \pi/2a$, it reaches equilibrium, while gaining maximum velocity. Then it continues to move by inertia, i.e., deviates from balance in the opposite direction. As the string deviates from equilibrium, the tension force directed against the movement increases, which means that it slows down the string. At time $t = \pi/a$, the string stops. At this time, it is maximally deviated from the equilibrium state. In view of the tension of the string, it again begins to move toward equilibrium, gaining velocity. At $t = 3\pi/2a$, it comes to an equilibrium state and, continuing to move, at the moment of time $t = 2\pi/a$ comes into the same state in which it was originally. Next, the process resumes[14]. In this case, the parameter a determines the oscillation frequency of the string.

In a similar way, the process of vibration of a string with free ends can be investigated[15]. To study the forced vibrations of the string and the non-homogeneous boundary one can use the same methods as in the study of the corresponding problems for the heat equation[16]. Make the following actions.

1. Give a physical interpretation of the problem statement.

2. Using the method of separation of variables, find a solution to the boundary value problem.

3. Substituting the solution found in the equation and boundary conditions, make sure that this is really a solution to the problem.

4. Give a physical interpretation of the results.

5. Establish the effect of string density and tension on the results.

Task 12.1 *The first boundary value problem for the vibrating string equation*. Consider the equation $u_{tt} = k u_{xx}$ on the unit interval with homogeneous boundary conditions and initial conditions $u(x, 0) = 0$, $u_t(x, 0) = \sin 3\pi x$.

Solution of the first boundary value problem for the vibrating string equation is obtained as a Fourier series by the method of separation of variables.

3   Infinitely long string

The movement of a homogeneous string of infinite length is described by the equation

$$u_{tt} = a^2 u_{xx}, \tag{12.17}$$

with initial conditions

$$u(x, 0) = \varphi(x), \quad u_t(x, 0) = \psi(x), \tag{12.18}$$

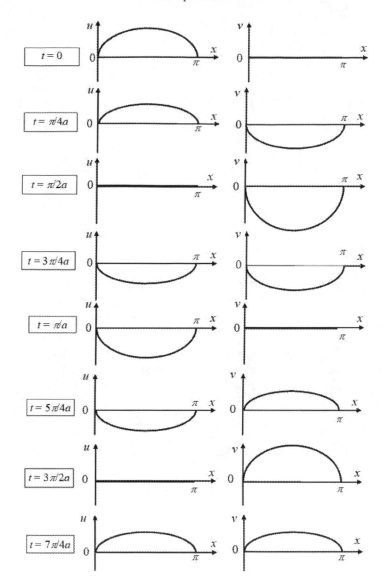

FIGURE 12.3: Position of the string and its velocity at different points in time.

where the parameter a, the initial form of string φ, and its initial velocity ψ are known.

To solve the studied **Cauchy problem**, we pass from the natural independent variables x, t to the new variables

$$\xi = \xi(x,t) = x + at, \quad \eta = \eta(x,t) = x - at,$$

which are called **characteristic**. We write the vibrating string equation in characteristic variables. Consider a new state function v so that

$$u(x,t) = v(\xi(x,t), \eta(x,t)).$$

Find the derivatives[17]

$$u_x = v_x = v_\xi \xi_x + v_\eta \eta_x = v_\xi + v_\eta,$$

$$u_t = v_t t = v_\xi \xi_t + v_\eta \eta_t = a(v_\xi - v_\eta).$$

Determine the second derivative

$$u_{xx} = (u_x)_x = (v_\xi + v_\eta)_x = v_{\xi\xi}\xi_x + v_{\xi\eta}\eta_x + v_{\eta\xi}\xi_x + v_{\eta\eta}\eta_x,$$

Taking into account the equality of the mixed derivatives $v_{\xi\eta}$ and $v_{\eta\xi}$, we have

$$u_{xx} = v_{\xi\xi} + 2v_{\xi\eta} + v_{\eta\eta}.$$

Now we find

$$u_{tt} = (u_t)_t = a^2(v_{\xi\xi}\xi_t + v_{\xi\eta}\eta_t - v_{\eta\xi}\xi_t - v_{\eta\eta}\eta_t).$$

Substituting the values of the second derivatives in the equality (12.18), we obtain

$$a^2(v_{\xi\xi} + 2v_{\xi\eta} + v_{\eta\eta}) = a^2(v_{\xi\xi} - 2v_{\xi\eta} + v_{\eta\eta}).$$

Thus, we have the equation

$$v_{\xi\eta} = 0, \tag{12.19}$$

which corresponds to the **canonical form** of the vibrating string equation.

From the equality

$$\frac{\partial}{\partial \xi}(v_\eta) = 0$$

it follows that the value in parentheses does not depend on the variable ξ. However, it may turn out to be a function of another variable η. Thus, by integrating relation (12.19) over the variable ξ, we establish the equality $v_\eta(\xi, \eta) = f(\eta)$ for any function f. Integrating this equality by η. We get

$$v(\xi, \eta) = \int f(\eta)d\eta + f_1.$$

The value f_1 here does not depend of η, but it can be the function of ξ. Denoting

$$f_2(\eta) = \int f(\eta)d\eta,$$

we find

$$v(\xi, \eta) = f_1(\xi) + f_2(\eta). \tag{12.20}$$

Thus, the solution of equation (12.9) is determined up to two arbitrary functions[18]. The function v defined by formula (12.20) is called the general solution of equation (12.19). Then the general solution of equation (12.17) is determined by the formula[19]

$$u(x, t) = f_1(x + at) + f_2(x - at). \tag{12.21}$$

We are interested, however, in a particular solution to the vibrating string equation that satisfies the initial conditions. We select the functions f_1 and f_2 in formula (12.21) in such a way as to ensure the validity of relations (12.18). We get the equalities

$$f_1(x) + f_2(x) = \varphi(x), \tag{12.22}$$

$$af_1'(x) - af_2'(x) = \psi(x), \tag{12.23}$$

where f_1', f_2' are the derivatives of the corresponding functions of one variable.

Integrating equality (12.23) from a value x_0 to an arbitrary x, we have

$$f_1(x) - f_2(x) = \frac{1}{a}\int_{x_0}^{x} \psi(y)dy + c, \tag{12.24}$$

where c is a constant.

Equalities (12.22), (12.24) are the system of linear algebraic equations for the unknown $f_1(x)$ and $f_2(x)$. Solving this system, we find the values

$$f_1(x) = \frac{1}{2}\varphi(x) + \frac{1}{2a}\int_{x_0}^{x}\psi(y)dy + \frac{c}{2}, \quad f_2(x) = \frac{1}{2}\varphi(x) - \frac{1}{2a}\int_{x_0}^{x}\psi(y)dy - \frac{c}{2}.$$

Note that the desired functions f_1 and f_2 are determined up to two constants x_0 and c.

Substitute the found values of the functions into equality (12.21). We get

$$u(x,t) = \left[\frac{1}{2}\varphi(x+at) + \frac{1}{2a}\int_{x_0}^{x+at}\psi(y)dy + \frac{c}{2}\right] + \left[\frac{1}{2}\varphi(x-at) - \frac{1}{2a}\int_{x_0}^{x-at}\psi(y)dy - \frac{c}{2}\right].$$

Thus, the unknown parameters x_0 and c disappeared during the transformations, and we find the unique solution to Cauchy problem (12.17), (12.18)

$$u(x,t) = \frac{\varphi(x+at) + \varphi(x-at)}{2} + \frac{1}{2a}\int_{x-at}^{x+at}\psi(y)dy.$$

This result is called the **D'Alembert formula**, and the method used to solve the problem is called the **D'Alembert method** or the **method of characteristics**.

We consider a partial case of problem (12.17), (12.18). Suppose that at the initial moment of time the string has the form $\varphi = \varphi(x)$, shown in Figure 12.4, and is at rest, i.e., has zero initial velocity. By the D'Alembert formula, the solution to the problem is

$$u(x,t) = \frac{1}{2}\varphi(x+at) + \frac{1}{2}\varphi(x-at).$$

Thus, the position of the string at time t is the sum of the quantities determined by the two terms on the right-hand side of the last equality. The first of them corresponds to the function φ, shifted to the left along the abscissa by the value of at and twice compressed along the ordinate. The second term differs from the first only in the direction of the shift.

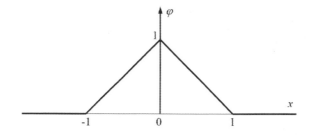

FIGURE 12.4: Initial form of the string.

Figure 12.5 shows the string profiles at different points in time with $a/4$ increments. As can be seen from the graphs, the position of the string is determined by the superposition of two half-waves, which gradually scatter in different directions. The observed phenomenon is called a **running wave**[20]. Note that the larger the value of the parameter a, the faster the wave front propagates. By equation (12.17), the coefficient a really has the dimension of velocity. This is the velocity of the running wave.

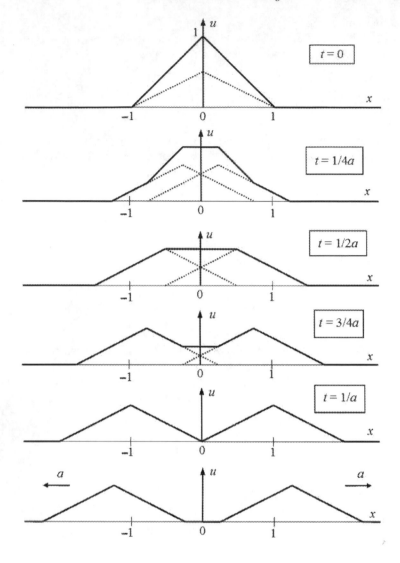

FIGURE 12.5: String profiles at different points in time.

Task 12.2 *Running wives.* Consider the equation $\rho u_{tt} = k u_{xx}$ for the infinite string with initial form shown in Figure 12.6 and zero initial velocity. Make the following actions.

1. Using the D'Alembert formula, find a solution of the problem.

2. Draw graphs of the form of the string at different points in time so as to evaluate the studied process.

3. Determine the influence of string density and tension on the results.

Solution of the Cauchy problem for the vibrating string equation
is determined by the D'Alembert method.
DAlembert's formula describes traveling waves.

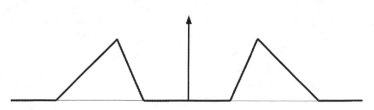

FIGURE 12.6: Initial form of the string for Task 12.2.

4   Electrical vibrations in wires

The vibrating string equation is a distributed analogue of the spring oscillation equation, and the vibrational circuit can be interpreted as an electrical analogue of the spring, see Chapter 4. There is an electrical string analogue, which is also a distributed analogue of a spring, see Figure 12.7. This is a thin long wire. As the current flows through the wire, the voltage V and the current strength I change, which depend on time t and on the spatial coordinate x directed along the wire. It is believed that the wire has inductance, capacitance, and resistance. In addition, a certain amount of energy is lost due to imperfections in the insulation of the wire. We establish the equation of state of the system on the basis of relations determining the balance of voltages and charges on the wire section.

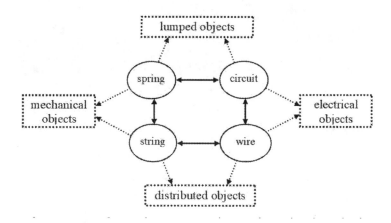

FIGURE 12.7: The analogy of a spring, string, circuit, and wire.

The voltage drop in a certain part of the wire is the sum of the voltages V_1 on the inductance and V_2 on the resistance. The inductance voltage is proportional to the velocity of change of current strength, i.e., its time derivative. We note, however, that since the object of study is distributed, the process parameter is not the wire inductance itself, but the inductance l of the unit length of the wire. If the current strength does not change along the length of the wire, then the inductance voltage for a wire of length Δx is $V_1 = lI_t\Delta x$. If the current strength depends on the variable x, then this equality remains valid only in the area of arbitrarily small length dx, i.e., $dV_1 = lI_t dx$. When considering the section of wire $x + \Delta x$ we get the voltage

$$V_1 = \int\limits_x^{x+\Delta x} lI_t dx.$$

The voltage at the resistance is determined by the Ohm's law, according to which the voltage is proportional to the current strength, $V_2 = RI$, where R is the electrical resistance. When considering an extended object (wire), the process parameter is not the value of R itself, but the resistance of a unit of length r. Then, in the case of constant current strength, the voltage across the resistance for a wire of length Δx is $V_2 = rI\Delta x$. In the case of a variable current strength, we obtain the relation $dV_2 = rIdx$. Then the voltage across the resistance for the section $x + \Delta x$ is determined by the formula

$$V_2 = \int_x^{x+\Delta x} rIdx.$$

The voltage drop in the considered section of the wire is the difference between the voltages at points x and $x + \Delta x$, see Figure 12.8. Therefore, the voltage balance on the wire $[x, x + \Delta x]$ is characterized by the formula

$$V(x,t) - V(x + \Delta x, t) = V_1 + V_2 = \int_x^{x+\Delta x} (lI_t + rI)dx.$$

Using the mean value theorem, we get

$$\frac{V(x,t) - V(x + \Delta x, t)}{\Delta x} = (lI_t + rI)\big|_{x=\xi},$$

where $\xi \in [x, x + \Delta x]$. Passing to the limit as $\Delta x \to 0$, we get

$$V_x(x,t) + lI_t(x,t) + rI(x,t) = 0. \tag{12.25}$$

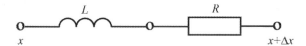

FIGURE 12.8: Part of wire with inductance and resistance.

Determine now the change in charge in the segment $x + \Delta x$ over time from t to $t + \Delta t$. The charge of the wire in time Δt under the condition of constant current strength is $q = -I\Delta t$. In the case of a variable current strength, this equality is satisfied only on an arbitrarily small time interval dt, i.e., $dq = -Idt$. Then for the time from t to $t + \Delta t$ we get a charge

$$q = -\int_t^{t+\Delta t} Idt.$$

Therefore, the change in charge in the segment $x + \Delta x$ in this time interval is

$$q_1 = -\int_t^{t+\Delta t} \big[I(x + \Delta x, t) - I(x,t)\big]dt.$$

The charge of the wire changes due to the fact that it has some electric capacity, as well as due to leakage due to imperfections in the insulation of the wire. As you know, the charge in the circuit is proportional to the voltage $q = CV$. If the voltage does not change, then the charge of a wire of length Δx is $q = cV\Delta x$, where c is the capacity of a unit of length of the wire. In the case of a variable voltage, this formula remains valid only when considering a portion of an arbitrarily small length $dq = cV\,dx$. Then the charge of the wire $[x, x + \Delta x]$ is equal to the integral

$$q_2 = -\int_x^{x+\Delta x} cV\,dx.$$

Thus, the following amount of electricity is consumed for recharging the considered part of the wire during the time from t to $t + \Delta t$

$$q_2 = -\int_x^{x+\Delta x} c\big[V(x, t + \Delta t) - V(x, t)\big]\,dx.$$

The loss of charge per unit time due to leakage through the insulation is proportional to the voltage $q = GV$, where G is the leakage coefficient. If the voltage is constant, then the charge $q = gV\Delta x\Delta t$, where g is the leakage coefficient per unit length of the wire, is lost in the time Δt in a section of length Δx. In the case of an variable voltage, we obtain the equality $dq = gV\,dx dt$. Thus, the loss of charge due to leakage in this section of the wire in the considered time interval is

$$q_3 = \int_t^{t+\Delta t}\int_x^{x+\Delta x} gV\,dxdt.$$

According to the law of conservation of charge, the charge changes due to recharging the wire and leakage, i.e., $q_1 = q_2 + q_3$. Then we get

$$\int_t^{t+\Delta t} \big[I(x + \Delta x, t) - I(x, t)\big]\,dt = \int_x^{x+\Delta x} c\big[V(x, t + \Delta t) - V(x, t)\big]\,dx + \int_t^{t+\Delta t}\int_x^{x+\Delta x} gV\,dxdt.$$

Using the mean value theorem, we obtain

$$\frac{I(x + \Delta x, \tau) - I(x, \tau)}{\Delta x} = c\frac{V(\xi, t + \Delta t) - V(\xi, t)}{\Delta t} + gV(\eta, \zeta),$$

where $\xi, \eta \in [x, x + \Delta x]$; $\tau, \zeta \in [t, t + \Delta t]$. Passing to the limit as $\Delta x \to 0$, $\Delta t \to 0$, we have

$$I_x(x, t) + cV_t(x, t) + gV(x, t) = 0. \tag{12.26}$$

Equations (12.25), (12.26) are the base of the mathematical model of the studied process and are called **telegraph equations**[21]

Determine separate equations for voltage and current. Differentiating equality (12.25) by x, and (12.26) by t, we get the equalities

$$V_{xx} + lI_{tx} + rI_x = 0, \quad I_{xt} + cV_{tt} + gV_t = 0.$$

Subtracting the second one multiplied by l from the first relation, we have

$$V_{xx} + rI_x - lcV_{tt} - lgVt = 0.$$

TABLE 12.2: Comparison of lumped and distributed models.

Characteristic	Circuit	Wire
inductance voltage	$V_1 = L\dot{I}$	$V_1 = \int\limits_{x}^{x+\Delta x} lI_t dx$
resistance voltage	$V_2 = RI$	$V_2 = \int\limits_{x}^{x+\Delta x} rI dx$
voltage balance	$V_1 + V_2 = V$	$V_1 + V_2 = V(x,t) - V(x+\Delta x, t)$
first equation	$L\dot{I} + RI = V$	$lI_t + rI = V_x$
charge	$\dot{q}_1 = -I$	$q_1 = \int\limits_{t}^{t+\Delta t} \big[I(x,t) - I(x+\Delta x, t)\big] dt$
charge on capacitor	$q_2 = CV$	$q_2 = \int\limits_{x}^{x+\Delta x} \big[V(x, t+\Delta t) - V(x,t) dt\big]$
charge balance	$q_1 = q_2$	$q_1 = q_2$
second equation	$C\dot{V} = -I$	$cV_t = -I_x$
state equation	$L\ddot{I} + R\dot{I} + C^{-1}I = 0$	$lI_{tt} + rI_t - c^{-1}I_{xx} = 0$

Substituting here the value of I_x from equality (12.26), we obtain

$$V_{tt} = a^2 V_{xx} - bV_t kV,$$

where

$$a = \frac{1}{\sqrt{lc}}, \quad b = \frac{r}{l} + \frac{g}{c}, \quad k = \frac{rg}{lc}.$$

The equations with respect to the current is obtained analogically

$$I_{tt} = a^2 I_{xx} - bI_t - kI,$$

If we neglect the loss of charge through the insulation and the resistance of the wire, then $r = 0$, $g = 0$. Then $b = 0$, $k = 0$, and the previous equations take the form

$$V_{tt} = a^2 V_{xx}, \tag{12.27}$$

$$I_{tt} = a^2 I_{xx}. \tag{12.28}$$

Equalities (12.27), (12.28), up to the notation, coincide with the vibrating string equation. The equations of voltage and current vibrations in wires are distributed analogues of the equation of the circuit, just as the vibrating string equation serves as a natural generalization of the equation of spring oscillation to the case of a distributed object. For the obtained equations, various boundary value problems can be posed, similar to those that were studied in the analysis of the string vibrations.

Comparative characteristics of lumped (circuit) and distributed (wire) electrical models are given in Table 12.2. The charge leakage through the insulation of the wire does not take into account here.

Electrical vibrations in a wire are described by telegraph equations.
From the telegraph equations it follows the equations for voltage and current, which coincide in form with the vibrating string equation.

Direction of further work. Having considered the transfer processes and wave processes, in the next chapter, we will analyze the third class of systems with distributed parameters. We are talking about stationary systems described by equations of elliptic type and having numerous applications.

Appendix

In the study of the pendulum oscillation, it was found that in the process its kinetic and potential energy change periodically, but their sum remains constant. Below we establish the law of conservation of energy for the process of the string vibration. In addition, a wave equation is presented, which is a multi-dimensional analogue of the string oscillation equation and describes, in the two-dimensional case, membrane vibrations, and in the three-dimensional case, the propagation of sound waves.

A generalization of the previously considered process of electrical oscillations in wires is the problem of the propagation of electromagnetic waves in space. This phenomenon is described by the Maxwell equations underlying the mathematical models of electrodynamics.

For an approximate solution of the heat equation in the previous chapter, we used the finite difference method. The final section of this chapter is devoted to the presentation of this method for the vibrating string equation.

1   Energy of vibrating string

The oscillations of the pendulum and the spring can be characterized using the law of conservation of energy, see Chapter 3. Determine the vibration energy of the string. It consists of kinetic and potential energy. The kinetic energy of a string depends on its velocity

$$K = \frac{m(u_t)^2}{2}.$$

Considering that the mass of a homogeneous string of length Δx is $\rho \Delta x$, we obtain

$$K = \frac{\rho(u_t)^2}{2}\Delta x.$$

This formula implies the constancy of the characteristics of the string along its length. In fact, such an equality can be satisfied only on an arbitrarily small segment of length dx, which leads us to the relation

$$dK = \frac{\rho(u_t)^2}{2}dx.$$

Then the kinetic energy of a string of length L is equal to the integral

$$K = \int\limits_0^L \frac{\rho(u_t)^2}{2}dx.$$

If at a fixed time t the string has a profile $u = u(x,t)$, then the corresponding potential energy coincides, up to the sign, with the work that needs to be done to bring the string from the equilibrium position at $t = 0$ to the state $u(x,t)$ in time t. The work is equal to the acting force multiplied by the distance traveled, which, in turn, is equal to the product of velocity and the time of movement. If, under the action of the force F, the string moved at a velocity ut during the time Δt, then the work $A = Fu_t\Delta t$ was performed. Given the well-known representation of the force F, we find the value of the work

$$A = \left(ku_x|_{x+\Delta x} - ku_x|_x\right)u_t\Delta t = \frac{ku_x|_{x+\Delta x} - ku_x|_x}{\Delta x}u_t\Delta x\Delta t.$$

This formula makes sense when it refers to arbitrarily small intervals dx and dt. At the same time we find work $dA - (ku_x)_x u_t dx dt$. Then, when moving a string of length L in time t, the work is

$$A = \int_0^t \int_0^L (ku_x)_x u_t dx dt.$$

Using the formula of integration by parts, we get

$$A = \int_0^t (ku_x) u_t \Big|_0^L - \int_0^t \int_0^L k u_x u_{xt} dx dt.$$

For simplicity, suppose that the ends of a string are rigidly fixed. Then their velocity is zero, and from the last formula we get

$$A = - \int_0^t \int_0^L k u_x u_{xt} dx dt.$$

Transform the value under the integral, using the equality $\frac{\partial}{\partial t}(u_x)^2 = 2u_x u_{xt}$. As a result, we conclude that the work spent on removing the system from the equilibrium position at the initial time to some current state is

$$A = \frac{1}{2} \int_0^t \int_0^L k \frac{\partial}{\partial t}(u_x)^2 dx dt = \frac{1}{2} \int_0^L k\big[u_x(x,0)^2 - u_x(x,t)^2\big] dx.$$

Given that at $t = 0$ the string was in equilibrium, we have $u_x(x,0) = 0$. Thus, the potential energy, which differs from work only by a sign, is equal to

$$U = \int_0^L \frac{ku_x(x,t)^2}{2} dx.$$

As a result, we find the vibrational energy of the string at an arbitrary time

$$E = \frac{1}{2} \int_0^L \big[\rho(u_t)^2 + k(u_x)^2\big] dx. \tag{12.29}$$

Determine now the connection between the vibrating string equation and the energy. Multiplying equation (12.3) by the derivative u_t and integrating along the string length, we get

$$\int_0^L \rho u_{tt} u_t dx = \int_0^L k u_{xx} u_t dx. \tag{12.30}$$

Transform the integrand on the left side of equality (12.30). We obtain

$$\int_0^L \rho u_{tt} u_t dx = \frac{1}{2} \frac{d}{dt} \int_0^L \rho(u_t)^2 dx = \dot{K}.$$

The value on the right-hand side of equality (12.30) is transformed, using the formula integration by parts

$$\int\limits_0^L ku_{xx}u_t dx = ku_x u_t\big|_0^L - \int\limits_0^L ku_x u_{xt}dx = -\frac{k}{2}\int\limits_0^L \frac{\partial}{\partial t}(u_x)^2 dx = -\dot{U}.$$

As a result, equality (12.30) takes the form

$$\dot{K} + \dot{U} = \dot{E} = 0.$$

Now we have

$$E(t) = const.$$

The result obtained corresponds to the ***law of conservation of energy*** for the vibrating string.

We now give an interpretation of the results presented in Figure 12.3 from the standpoint of the law of conservation of energy. Having a corresponding deviation from the equilibrium position, the string has a potential energy, the action of which is always directed toward equilibrium. In this case, kinetic energy is absent, since the string is at rest. Due to the potential energy, the string begins to move toward equilibrium. As it approaches the equilibrium state, the potential energy of the string decreases. However, the fact of movement itself indicates the appearance of a certain velocity, and hence kinetic energy. In accordance with the law of conservation of energy, the total mechanical energy of the system does not change. Thus, a decrease in potential energy is accompanied by a corresponding increase in its kinetic energy. Thus, approaching the equilibrium position, the string moves with positive acceleration.

At time $t = \pi/2a$, the string is in equilibrium state. This means that its potential energy is zero. Then the total energy of the string coincides with its kinetic energy, which in this case reaches its maximum. This is possible if the velocity also turns out to be maximum in absolute value. Note, however, that in this case the string velocity is negative. Having a sufficiently large kinetic energy, the string continues to move by inertia and begins to deviate from the equilibrium position in Figure 12.3, we observe the string moving downward. As the string deviates from the equilibrium state, potential energy appears and increases, the action of which is directed toward equilibrium, and therefore, against movement. As a result, the movement of the string slows down, its kinetic energy decreases, and its potential one increases.

At time $t = \pi/a$, the string velocity reaches zero. In this case, the kinetic energy is also equal to zero, and the potential reaches a maximum. As a result, the string begins to move in the opposite direction. At $t = 3\pi/2a$, it reaches a state of equilibrium, having at this moment in time the maximum velocity. Continuing its movement, by inertia, the string deviates from equilibrium and at time $t = 2\pi/a$ reaches its initial position. Further, the process of oscillation of the string resumes. In accordance with the results obtained, the kinetic and potential energies of the string are periodic functions of time with a period $t = 2\pi/a$. The parameter a in this case determines the frequency of the string.

Task 12.3 ***Energy of vibrating string.*** Consider a homogeneous string on a unit segment with fixed ends, which is initially in equilibrium and has an initial velocity $\varphi(x) = \sin 2\pi x$. Make the following actions.

1. Using the method of variable separation, find a solution of the problem.
2. Determine the change in time of the kinetic energy of the string and draw its graph.
3. Determine the change over time of the potential energy of the string and draw its graph.
4. To carry out calculations under conditions of stability violation of the explicit difference scheme.

TABLE 12.3: Wave processes.

Dimension	Object	Equation
0	spring	$u_{tt} = -\omega^2 u$
1	string	$u_{tt} = a^2 u_{xx}$
2	membrane	$u_{tt} = a^2(u_{xx} + u_{yy})$
3	acoustic wave	$u_{tt} = a^2(u_{xx} + u_{yy} + u_{zz})$

2   Mathematical models of wave processes

The vibrating string equation is a natural generalization of the equation of oscillation of the spring (material point, i.e., a zero-dimensional object) to the case of a distributed one-dimensional object. The two-dimensional case corresponds to the oscillation of the membrane that is a flat surface. In this case, the state function $u = u(x, y, t)$ describes the deviation of the membrane at the point with coordinates x, y from the equilibrium position at time t. This is characterized by the ***vibrating membrane equation***

$$u_{tt} = a^2(u_{xx} + u_{yy}),$$

which describes, in particular, the propagation of waves on the surface of a liquid. The process of wave propagation in space is associated with a function of four variables $u = u(x, y, z, t)$ and is described by the equation

$$u_{tt} = a^2(u_{xx} + u_{yy} + u_{zz}),$$

called the ***wave equation***. In particular, it underlies mathematical models of ***acoustics*** problems[22].

The sum of the second derivatives with respect to spatial variables is the Laplace operator and denoted by Δ. Then the wave equation is written as[23]

$$u_{tt} = a^2 \Delta u. \tag{12.31}$$

From here in the one-dimensional case the vibrating string equation, and in the two-dimensional case the vibrating membrane equation.

Comparative characteristics of wave processes are presented in Table 12.3.

Task 12.4 *Implicit difference scheme for the heat equation.* We consider the same for the first boundary value problem of heat equation (11.8) – (11.10), as in Task 11.3. Make the following analysis.

1. Choose the exact solution to the problem, as was done in Task 11.3, with the corresponding values of the functions included in the right-hand sides of the equation and boundary conditions.

2. Write down an algorithm for solving this problem based on an implicit difference scheme and a tridiagonal matrix algorithm, as well as an explicit difference scheme, see Task 11.3.

3. Create a program for solving the problem with the indicated algorithms.

4. To carry out calculations under conditions of stability violation of the explicit difference scheme.

5. Compare the calculation results by both methods with each other and with the exact solution to the problem.

3   Beam vibrations

An extremely interesting mathematical model is associated with the process of vibration of an elastic beam. The main difference between the beam oscillations and the previously considered process of vibrations of the string is due to the fact that the beam has resistance to bending. Based on the laws of the **elasticity theory**, it can be shown that small free vibrations of an elastic beam are described by the **vibratins beam equation**[24]

$$\rho S \frac{\partial^2 u}{\partial t^2} + \frac{\partial^2}{\partial x^2}\left(EJ \frac{\partial^2 u}{\partial x^2}\right) = 0,$$

where $u = u(x,t)$ is the deviation of the beam from the equilibrium state, ρ is its density, S is the cross-sectional area, E is **Young modulus**, and J is the moment of inertia of the beam section. For a homogeneous beam, the equation takes the form

$$u_{tt} + a^2 u_{xxxx} = 0,$$

where $a = \sqrt{EJ/\rho S}$.

The vibrating beam equation has a fourth order in spatial variable. Thus, the boundary value problem for this equation requires the specification of two initial and four boundary conditions. The situation with the initial conditions is quite clear. As in the case of a string, the initial beam profile φ and its initial velocity ψ are specified, which corresponds to the conditions

$$u(x,0) = \varphi(x), \quad u_t(x,0) = \psi(x).$$

Boundary conditions can be defined in various ways.

Typically, two conditions are set at each end of the beam. In particular, if the left end of the beam corresponding to the variable $x = 0$ is rigidly fixed, then the boundary conditions take the form

$$u(0,t) = 0, \quad u(0,t) = 0.$$

If the left end of the beam rests freely on some support, then we obtain the boundary condition

$$u(0,t) = 0, \quad u_x(0,t) = 0.$$

In the case of the free left end, the conditions are

$$u_x(0,t) = 0, \quad u_{xx}(0,t) = 0.$$

Similar conditions can be specified at the right end of the beam. In this case, various combinations of boundary conditions at the left and right ends of the beam are possible.

Task 12.5 *Vibrating beam equation.* Consider vibrations of a homogeneous beam of length L, the ends of which are freely supported. The initial position and velocity of the beam are known. Make the following analysis.

1. Apply the method of separation of variables for the vibration beam equation.
2. Analyze the corresponding Sturm–Liouville problem.
3. Find a solution of the problem as a Fourier series.
4. Find a solution to the problem for a beam of length π with an initial profile $\varphi(x) = \sin x$ and zero initial velocity.
5. Investigate the effect of parameter a on the results.
6. Give a physical interpretation of the results.

4   Maxwell equations

Previously, electrical vibrations in a wire were considered. In a general case, ***electro-magnetic waves*** are considered[25]. The electromagnetic field is characterized by four vector functions that are ***electric field strength*** $\mathbf{E}$, ***magnetic field strength*** $\mathbf{H}$, ***electric induction*** $\mathbf{D}$ and ***magnetic induction*** $\mathbf{B}$. Each of these values has three components.

The mathematical models of electrodynamics are based on ***Maxwell equations*** connecting the above values. These include the ***Gauss law***

$$\mathrm{div}\mathbf{D} = 4\pi\rho,$$

according to which an electric charge is a source of electric induction. Here ρ is the charge density, and the divergence div of the vector $\mathbf{D} = (D_1, D_2, D_3)$ is characterized by the equality[26]

$$\mathrm{div}\mathbf{D} = \frac{\partial D_1}{\partial x} + \frac{\partial D_2}{\partial y} + \frac{\partial D_3}{\partial z}.$$

Then we have the ***Gauss law of magnetic field***

$$\mathrm{div}\mathbf{B} = 0,$$

postulating the absence of magnetic charges. By the ***Faraday law of induction***

$$\mathrm{rot}\mathbf{E} = -\frac{1}{c}\frac{\partial \mathbf{B}}{\partial t}.$$

the change in magnetic induction generates a vortex electric field, where c is the ***velocity of light*** in vacuum, and the ***rotor*** rot of the vector $\mathbf{E} = (E_1, E_2, E_3)$ is characterized by the equality

$$\mathrm{rot}\mathbf{E} = \left(\frac{\partial E_3}{\partial y} - \frac{\partial E_2}{\partial z}, \frac{\partial E_1}{\partial z} - \frac{\partial E_3}{\partial x}, \frac{\partial E_2}{\partial x} - \frac{\partial E_1}{\partial y}\right).$$

Finally, the equality

$$\mathrm{rot}\mathbf{H} = \frac{1}{c}\frac{\partial \mathbf{D}}{\partial t} + \frac{4\pi}{c}\mathbf{j},$$

means that electric current and a change in electric induction generate a vortex magnetic field, where $\mathbf{j}$ is the ***density of the electric current***. The above relations are supplemented by ***material field equations***

$$\mathbf{D} = \varepsilon\mathbf{E}, \ \ \mathbf{B} = \mu\mathbf{H}, \ \ \mathbf{j} = \sigma\mathbf{E}, \tag{12.32}$$

where ε is the ***dielectric constant***, μ is the ***magnetic permeability***, and σ is the ***conductivity of the medium***.

Under certain conditions, the Maxwell equations can be simplified[27]. In particular, in a vacuum in the absence of charges and currents for $\varepsilon = 1$, $\mu = 1$, after excluding the induction of electric and magnetic fields using equalities (12.32), we obtain the following relations

$$\mathrm{div}\mathbf{E} = 0, \ \ \mathrm{div}\mathbf{H} = 0, \ \ \mathrm{rot}\mathbf{E} = -\frac{1}{c}\frac{\partial \mathbf{H}}{\partial t}, \ \ \mathrm{rot}\mathbf{H} = \frac{1}{c}\frac{\partial \mathbf{E}}{\partial t}. \tag{12.33}$$

Acting on the third equality (12.33) by the operator rot and taking into account the possibility of its permutation with the time derivative, we have

$$\mathrm{rot}\,\mathrm{rot}\mathbf{E} = -\frac{1}{c}\frac{\partial}{\partial t}\mathrm{rot}\mathbf{H} = -\frac{1}{c^2}\frac{\partial^2\mathbf{E}}{\partial t^2} \tag{12.34}$$

by fourth equality (12.33).

It can be shown that for any vector function $\mathbf{E}$, the following equality holds

$$\operatorname{rot}\operatorname{rot}\mathbf{E} = \nabla\operatorname{div}\mathbf{E} - \Delta\mathbf{E},$$

where ∇ is the **gradient**, i.e., the vector whose components are partial derivatives with respect to spatial variables, and Δ is the Laplace operator, i.e., sum of second order derivatives with respect to spatial variables. Then, taking into account the first equality (12.33), we transform the relation (12.34) to the following form

$$\mathbf{E}_{tt} = c^2\Delta\mathbf{E}. \tag{12.35}$$

Analogically, from the equalities (12.33) it follows

$$\mathbf{H}_{tt} = c^2\Delta\mathbf{H}. \tag{12.36}$$

Note that the strength of electric and magnetic fields are vector quantities. Thus, under the assumptions made, it turns out that all its components satisfy the wave equation (12.31), and the velocity of light acts as the parameter a here. In Section 3, it was noted that this coefficient is associated with the wave velocity. Thus, electromagnetic waves propagate at the light velocity.

5 Finite difference method for the vibrating string equation

As already noted, an analytical solution of differential equations can be found only in exceptional cases. In the previous chapter, the finite difference method for the approximate solving of the heat equation was described. It can also be used to solve boundary value problems for the vibrating string equation[28].

Considered the non-homogeneous vibrating string equation

$$u_{tt}(x,t) = a^2 u_{xx}(x,t) + f(x,t), \ \ 0 < x < L, \ 0 < t < T \tag{12.37}$$

with initial conditions

$$u(x,0) = \varphi(x), \ u_t(x,0) = \psi(x), \ 0 < x < L \tag{12.38}$$

and boundary conditions

$$u(0,t) = \alpha(t), \ u(L,t) = \beta(t), \ t > 0, \tag{12.39}$$

where the numbers a and L and the functions f, φ, ψ, α, and β are known.

The interval $[0,T]$ is divided into parts with a step $\tau = T/N$ by the points $t_j = \tau j$, $j = 0, ..., N$, and the interval $[0, L]$ is divided with a step $h = L/M$ by the points $x_i = hi$, $i = 0, ..., M$. A solution to the problem is sought at the grid nodes (x_i, t_j). The second derivatives in equation (12.37) are approximated as follows

$$\frac{\partial^2 u(x_i, t_j)}{\partial t^2} \approx \frac{u(x_i, t_{j-1}) - 2u(x_i, t_j) + u(x_i, t_{j+1})}{\tau^2},$$

$$\frac{\partial^2 u(x_i, t_j)}{\partial x^2} \approx \frac{u(x_{i-1}, t_j) - 2u(x_i, t_j) + u(x_{i+1}, t_j)}{h^2}.$$

Then we obtain the equality

$$\frac{u_i^{j-1} - 2u_i^j + u_i^{j+1}}{\tau^2} = a^2\frac{u_{i-1}^j - 2u_i^j + u_{i+1}^j}{h^2} + f_i^j,$$

where $u_i^j = u(x_i, t_j)$, $f_i^j = f(x_i, t_j)$. Now we find

$$u_i^{j+1} = 2u_i^j - u_i^{j-1} + a^2 \frac{u_{i-1}^j - 2u_i^j + u_{i+1}^j}{h^2} + f_i^j, \ i = 1, ..., M-1, \ j = 0, ..., N-1. \quad (12.40)$$

This formula makes it possible to determine the values of the desired function on the subsequent time layer with its known value on the two previous layers. All values on the zero layer in time are found from the first initial condition

$$u_i^0 = \varphi_i, \ i = 0, ..., M,$$

where $\varphi_i = \varphi(x_i)$. Replacing the derivative by the difference in the second condition (12.38), we get

$$u_i^1 = u_i^0 + \tau \psi_i, \ i = 0, ..., M,$$

where $\psi_i = \psi(x_i)$. Now one can find the desired function of the next (second) layer from the first to the penultimate value in accordance with formula (12.40). Its boundary values are found by the formulas

$$u_0^j = \alpha^j, \ u_M^j = \beta^j, \ j = 1, ..., N,$$

where, $\alpha^j = \alpha(t_j)$, $\beta^j = \beta(t_j)$. Thus, we find an approximate solution of problem (12.37) – (12.39) for all grid nodes[29].

Task 12.6 *Finite difference method for the vibrating string equation.* Consider boundary problem (12.37) – (12.39) on the unit interval for $a = 1$. Makes the following analysis.

1. Determine the function $v(x, t) = \sin \pi x \cos \pi t$ as the solution of the considered problem. Putting this value to equalities (12.37) – (12.39), find the corresponding functions f, φ, ψ, α, and β. Now for the found values of these functions, the considered boundary value problem for the vibrating string equation has exactly this solution.

2. Solve problem (12.37) – (12.39) with the tasks found on the previous paragraph in accordance with the explicit scheme of the difference scheme.

3. Compare the resulting approximate solution with the exact solution. Estimate the change of error from layer to layer by the time $\delta_j = \max\limits_{i=1,2,...,M-1} |u_i^j - v_i^j|$, where the approximate solution u_i^j is the approximate solution of the problem, and $v_i^j = v(x_i, t_j)$ with given function v.

4. Determine the effect of grid steps on the accuracy of the solution to the problem.

Notes

[1] The vibration string equation is one of the most important equations of mathematical physics and is considered within the framework of any course on partial differential equations, see, for example, [96], [147], [169], [192], [235], [269], [274], [349], [361].

[2] The telegraph equation is considered, for example, in [137], [285], [297].

[3] The longitudinal and torsional vibrations of the string also make sense. In the first case, deformation is carried out along the string (compression and tension), and in the second case, rotations around the x axis. In these cases, it is also possible to obtain an equation in the form of (12.5). The derivation of the equation of longitudinal vibrations of a string is given, for example, in [349].

[4] The equation of string oscillation can also be obtained by the principle of least action, see Chapter 16.

[5] We have already used the concept of linear density in determining the heat equation. The use of linear

rather than ordinary density (the amount of substance per unit volume) is explained by the fact that in this model the string has only length, but not volume.

[6] In the Appendix, the process of vibrations of elastic beam is considered. The indicated assumption will be removed there.

[7] We recall that in deriving the equation of pendulum oscillation, we also first determined a sine (nonlinear function), and only then, assuming small oscillations, we obtained the linear equation.

[8] When describing the oscillations of a pendulum or spring, we also took into account the friction force. As a result, an additional term proportional to velocity appeared in the corresponding equations, i.e., derivative of deviation. For string oscillation, the friction can also be taken into account. In this case, the term proportional to the time derivative of the string deviation of the equilibrium position will also enter the corresponding equation. In the presence of an external force acting on the string, a non-homogeneous equation for the vibrations of the string is obtained.

[9] In Chapter 16, the vibrating string equation will be derived from the principle of least action.

[10] In Chapter 10, we have already classified the equation

$$a_{11}u_{xx} + 2a_{12}u_{xy} + a_{22}u_{yy} = F(x, y, u, u_x, u_y),$$

where a_{11}, a_{12}, a_{22} are given values, and F is a known function. The sign of discriminant $D = a_{12}^2 - a_{11}a_{22}$ here is decisive. In particular, at $D = 0$ this equation is **parabolic**, which is characteristic of the heat equation. For $D > 0$, the equation is called **hyperbolic**, and for $D < 0$, it is **elliptic**. For the vibrating string equation $u_{tt} - a^2 u_{xx} = 0$ for $y = t$ we have $D = a^2$. Therefore, this equation is hyperbolic. It can be shown that, using a special transformation of independent variables $x = x(\xi, \eta)$, $y = y(\xi, \eta)$ a general hyperbolic equation can always be reduced to a canonical form

$$u_{\xi\eta} = \Phi(\xi, \eta, u, u_\xi, u_\eta).$$

This is how the Cauchy problem considered in Section 3 for the vibrating string equation is solved.

[11] For the heat equation, the time condition can be specified only at the initial time, and the boundary value problem with data at the final time is ill-posed. However, for the vibrating string equation, the position and velocity of the string can be specified both at the initial and at the final time. Thus, from the known values of the position of the string and its velocity at a given time, it is possible to restore the history of the system. As a result, a dynamic group of transformations can be defined to describe the process, see Chapter 5.

[12] The method of separation of variables for the vibrating string equation is given in [96], [235], [349], [361].

[13] As at a similar stage in the study of heat equation, here we need to justify the convergence of the considered Fourier series. In the same way, all further actions with infinite sums need strict substantiation. However, given that we are interested in the properties of mathematical models, and not in the theory of partial differential equations, we carry out all the corresponding calculations formally. Justification of the relevant procedures can be found in the previously cited literature.

[14] We know that the solution of the heat equation tends to its equilibrium position. Now we observe oscillations around the equilibrium. In particular, if one sets the position and velocity of the string to zero, then its position will not change over time, i.e., we are really a position of equilibrium. If we consider the forced vibrations of the string by the force that does not change, then the equilibrium position will be a profile of the string determined by the external action.

[15] If we use the variable separation method to solve the second boundary value problem for the vibrating string equation, then for the corresponding Sturm–Liouville problem we obtain the second type boundary conditions. As was shown in Chapter 10, this problem has the corresponding cosines as a nontrivial solution.

[16] Forced string vibrations are described by the non-homogeneous vibrating string equation. To solve this problem, the Fourier method described in Chapter 10 is used. The corresponding solution is sought in the form of the Fourier series in sines, i.e., by solutions of the corresponding Sturm–Liouville problem. In this case, the Fourier coefficients are determined from the solution of the Cauchy problems for second order non-homogeneous ordinary differential equations. In turn, the solution of the non-homogeneous boundary value problem for the vibrating string equation is sought as the sum of two functions, one of which is selected initially so that it satisfies the given boundary conditions. As a result of this, the second of the functions satisfies homogeneous boundary conditions.

[17] Here we use the theorem on the differentiation of **composite function**, see, for example, [13].

[18] This paradoxical, at first glance, result will not seem so surprising if we recall that the solution to an ordinary second order differential equation is found up to two arbitrary constants, i.e., quantities independent of this argument. In particular, the solution to the equation $u_t(x, t) = 0$ will be not only any constant, but also an arbitrary function of the unique variable x.

[19] When finding a general solution to the vibrating string equation, we used a fairly common idea. First,

such a change of variables is performed in which the studied problem takes on a fairly simple form. Then the problem is solved in new variables. Finally, the last step is to return to the original variables.

[20] Running waves can be observed on the surface of the water when any body (for example, stone) falls into it. The waves on the surface of the water are characterized by a two-dimensional analogue of the vibrating string equation, i.e., vibrating membrane equation. However, if we observe the water level in the vertical plane with the origin corresponding to the point of incidence of the body into the water, we will see running waves similar to those shown in Figure 12.5.

[21] When deriving telegraph equations, the magnetic field was not taken into account. The full mathematical model of electrodynamics includes Maxwell equations for electric and magnetic fields, see Appendix.

[22] General problems of wave physics are considered, for example, in [262], [268], [66], and the acoustics problems in [242]. The derivation of the equations of acoustics is given in Chapter 14. A peculiar wave process is associated with the Schrödinger equation, which will be considered in Chapter 15.

[23] Chapter 13 will also consider an non-homogeneous wave equation and steady vibrations associated with it.

[24] About elasticity theory, see [203], [296]. The derivation of the beam vibrating equation is given, for example, in [349]. A two-dimensional analogue of this equation related to the process of oscillation of an elastic plate is considered in Chapter 13. The problem of determining the form of bending of an elastic plate under the action of a constant force is also considered there.

[25] The problems of electrodynamics are considered, for example, in [204], [278], [343], [352], see also [116], [258].

[26] The concepts of gradient, rotor and divergence of vector analysis are discussed in more detail in Chapter 13.

[27] In a subsequent Chapter, in particular, the electrostatics problems will be considered, where the characteristics of the process do not change with time.

[28] About the finite difference method, see [243], [280], [301], [325], [340], [347].

[29] We used an *explicit scheme* for the vibrating string equation. It is stable under the condition $a\tau \leq h$. Absolutely stable is the corresponding *implicit scheme* in which, when approximating the second derivative with respect to the spatial variable, the value of the function is taken at the next time step. For practical use of the implicit scheme, the tridiagonal matrix algorithm described in Chapter 11 is used.

Chapter 13

Mathematical models of stationary systems

Two classes of systems with distributed parameters were previously considered. There are the transfer processes and wave processes. For the first-class models (for example, the heat equation), the system tends with time to the equilibrium position, which is a function of spatial coordinates. For models of the second class (for example, the equation of string vibration), oscillations are observed around the equilibrium position. This chapter explores various stationary systems whose states do not change over time and can be understood as equilibrium positions of the previously considered systems.

As already noted, the transfer processes are characterized by the tendency of the system to an equilibrium state if the conditions of the process do not change over time. In the one-dimensional case, it is described by an ordinary differential equation, and in the multidimensional case, by a partial differential equation. This applies, in particular, in stationary heat transfer, which is described in the simplest case by the Laplace or Poisson equations, see Section 1. Like the heat and vibration string equations, they are the most important equations of mathematical physics[1]. Writing down the Laplace operator in cylindrical and spherical coordinates, we can find solutions of this equation with cylindrical and spherical symmetry, see Section 2.

The phenomena described below are associated with various physical fields. Therefore, in Section 3 we give a description of the general concept of a vector field, as well as its main characteristics. There are rotor, flux, divergence, etc. Among vector fields, potential fields determined by the gradient of a function of spatial variables play a special role. The most important potential fields are the electrostatic and gravitational fields, considered in Sections 4 and 5, respectively. They are described by the same Laplace or Poisson equations with respect to the potential of the corresponding fields. Using the cylindrical and spherical coordinates, one determines the law of change in the potential of the electrostatic field of a point charge and charged wire, as well as the potential of the gravitational field determined by a material point, is determined.

The Appendix shows that the potential of the velocity field of a steady fluid flow is described by similar equations. Other equations related to stationary systems are also considered, and methods for solving such equations are also given.

Lecture

1   Stationary heat transfer

A rather long and thin body under the influence of a heat source that does not change with time and characterized by the density of heat sources $F = F(x)$, where x is the spatial coordinate directed along the body, is considered. It is required to establish the temperature distribution $u = u(x)$ provided that the body does not experience any other effects.

In accordance with the technique described in previous chapters, we select an interval $[x, x + \Delta x]$ along the length of the body. If the heat sources did not vary along the length of the body, then due to their action in this area, the quantity of heat $Q = F\Delta x$ would be released. With a variable heat source, this equality only makes sense on an arbitrarily small portion dx, i.e., the equality $dQ = Fdx$ holds. Then on the interval $[x, x + \Delta x]$, the

DOI: 10.1201/9781003035602-13

following heat quantity is released

$$Q_1 = \int\limits_{x}^{x+\Delta x} F(x)dx.$$

As noted in Chapter 10, heat fluxes occur in an unevenly heated body, directed from a hotter area to a colder one. For definiteness, we choose the direction of the heat flux coinciding with the direction of the x coordinate. Then the change in the heat quantity in the selected set is the difference between the heat flux entering the point x and the heat flux leaving the point $x + \Delta x$, which corresponds to the equality

$$Q_2 = q(x) - q(x + \Delta x),$$

where $q(x)$ is heat flux corresponding to the heat quantity[2] passing through point x.

In this case, the heat balance in the selected part of the body is characterized by the equality $Q_1 + Q_2 = 0$, i.e., as a result of the action of heat sources and heat fluxes, the heat quantity in the selected area does not change. Then after dividing by Δx we get

$$\frac{q(x) - q(x + \Delta x)}{\Delta x} + \frac{1}{\Delta x} \int\limits_{x}^{x+\Delta x} F(\xi)d\xi = 0.$$

We pass to the limit here[3] as $\Delta x \to 0$. The left-hand side results in the value of the derivative of the function q with a minus sign at the point x, and on the right-hand side as a result of passing to the limit, taking into account the main theorem, $F(x)$ is obtained. Thus, we have the equality

$$-\frac{dq(x)}{dx} + F(x) = 0. \tag{13.1}$$

As noted in Chapter 10, the flux q is directly proportional to the temperature difference and inversely proportional to the distance between the points, which can be written as the equality

$$q(x) = -\lambda(x)\frac{u(x) - u(x - \Delta x)}{\Delta x}.$$

In the limit, this gives Fourier law $q = -\lambda\frac{du}{dx}$. As a result, the equation (13.1) takes the form

$$\frac{d}{dx}\left(\lambda\frac{du}{dx}\right) = -F. \tag{13.2}$$

This is the stationary heat equation for an non-homogeneous body under the action of a heat source.

Note that equation (13.2) can be obtained in another way. In Chapter 10, the process of heat transfer in a one-dimensional body under the influence of heat sources was considered. It is described by the heat equation

$$c\rho\frac{\partial u}{\partial t} = \frac{\partial}{\partial x}\left(\lambda\frac{\partial u}{\partial x}\right) + F, \tag{13.3}$$

where c is the heat capacity, ρ is the density. If the characteristics of the process, i.e., the coefficients of the equation, the function F, as well as the behavior of the system at the boundary does not change with time, the system tends to the equilibrium. As the state reaches the equilibrium position, the derivative of the state function with respect to time tends to zero. Thus, the equilibrium position for equation (13.3) is the solution of equation (13.2).

Similar considerations can be made in the multidimensional case. Particularly, in the three-dimensional case, the non-homogeneous heat equation is

$$c\rho\frac{\partial u}{\partial t} = \frac{\partial}{\partial x}\left(\lambda\frac{\partial u}{\partial x}\right) + \frac{\partial}{\partial y}\left(\lambda\frac{\partial u}{\partial y}\right) + \frac{\partial}{\partial z}\left(\lambda\frac{\partial u}{\partial z}\right) + F.$$

If the characteristics of the system here do not change with time, then the system tends to a stationary position, which means that the derivative of the state function of the system tends to zero. As a result, we obtain the equality

$$\frac{\partial}{\partial x}\left(\lambda\frac{\partial u}{\partial x}\right) + \frac{\partial}{\partial y}\left(\lambda\frac{\partial u}{\partial y}\right) + \frac{\partial}{\partial z}\left(\lambda\frac{\partial u}{\partial z}\right) + F = 0.$$

If the body is homogeneous, we get

$$u_{xx} + u_{yy} + u_{zz} = f,$$

where $f = -F/\lambda$. Given the definition of the Laplace operator Δ, we obtain the formula

$$\Delta u = f \tag{13.4}$$

that is called the **Poisson equation**. This is the partial differential equation in which spatial coordinates turn out to be independent variables. In the absence of external heat sources, the function F vanishes. Then the equation (13.4) takes the form

$$\Delta u = 0 \tag{13.5}$$

that is called the **Laplace equation**. The obtained equations are the simplest equations of stationary heat transfer[4]. Note the Laplace equation has numerous physical applications (see the subsequent sections of this chapter), as well as an independent mathematical meaning[5].

As usual, the mathematical model of a system characterized by differential equations, both ordinary and partial derivatives, also includes additional boundary conditions. Let the Poisson or Laplace equation be considered in a set Ω bounded by a surface S. In the one-dimensional case, this surface consists of two points that are the ends of the segment Ω, in the two-dimensional case, this is a closed curve bounding the flat region Ω, and in the three-dimensional case, this the a surface bounding the body Ω.

At each point of region S, a state function value can be specified

$$u(x) = \varphi(x),\ x \in S, \tag{13.6}$$

where φ is known function defined on S and meaning the temperature at the boundary. The Laplace or Poisson equation with this condition makes up the **Dirichlet problem**[6]. The second most important boundary-value problem is the **Neumann problem**[7], where the heat flux, characterized by a function ψ, is specified at the boundary. This corresponds to the boundary condition

$$\lambda\frac{\partial u(x)}{\partial n} = \psi(x),\ x \in S, \tag{13.7}$$

on the left-hand side of which is the derivative with respect to the external normal to S, see Figure 13.1. Depending on the physical meaning of the problem, other types of boundary conditions are possible. In addition, mixed boundary conditions are admissible, for example, a condition of the first type is specified on a part of the boundary, and a second condition on the other[8].

The described boundary value problems are called **internal** since the processes occurring within the considered region are considered. **External boundary value problems** in which the behavior of the system outside this area is studied are also worthwhile[9], see Figure 13.1.

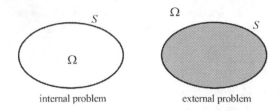

FIGURE 13.1: Internal and external problems.

Task 13.1 *Finite difference method for the stationary heat transfer*. Consider the equation $\lambda u_{xx} = \sin x$ on the interval $[0, \pi]$ with homogeneous boundary conditions. Make the following actions.

1. Find the analytical solution of the problem.
2. Write a finite-difference scheme for this problem.
3. Develop the tridiagonal matrix algorithm for solving the problem.
4. Carry out calculations for various values of the parameter λ.
5. Give a physical interpretation of the results.
6. Estimate the error in solving the problem depending on the number of partition points.

Stationary heat transfer is characterized by the Poisson equation.
In the absence of heat sources, this is described by the Laplace equation.
Dirichlet and Neumann problems are the most important for it.

2 Spherical and cylindrical coordinates

When solving concrete problems, sometimes it makes sense to switch from Cartesian coordinates to some others. Most commonly used are spherical and cylindrical coordinates.

To go to **spherical coordinates**, the following transformation of independent variables is used

$$x = r \sin \theta \cos \varphi, \; y = r \sin \theta \sin \varphi, \; z = r \cos \theta.$$

Spherical coordinates include the radial component r that is the distance from this point to the origin and two angles φ and θ, see Figure 13.2.

We write the Laplace operator in spherical coordinates. To do this, we find the derivative

$$\frac{\partial}{\partial x} = \frac{\partial}{\partial r} \frac{1}{\partial x / \partial r} + \frac{\partial}{\partial \theta} \frac{1}{\partial x / \partial \theta} + \frac{\partial}{\partial \varphi} \frac{1}{\partial x / \partial \varphi} =$$

$$\frac{1}{\sin \theta \cos \varphi} \frac{\partial}{\partial r} + \frac{1}{r \cos \theta \cos \varphi} \frac{\partial}{\partial \theta} + \frac{1}{r \sin \theta \sin \varphi} \frac{\partial}{\partial \varphi}.$$

Derivatives in other independent variables are found in a similar way. After finding the second derivatives, we establish the form of the Laplace operator in spherical coordinates

$$\Delta u = \frac{1}{r^2} \frac{\partial}{\partial r} \left(r^2 \frac{\partial u}{\partial r} \right) + \frac{1}{r^2 \sin \theta} \frac{\partial}{\partial \theta} \left(\sin \theta \frac{\partial u}{\partial \theta} \right) + \frac{1}{r^2 \sin^2 \theta} \frac{\partial^2 u}{\partial \varphi^2}.$$

Thus, the Poisson equation in spherical coordinates is

$$\frac{1}{r^2} \frac{\partial}{\partial r} \left(r^2 \frac{\partial u}{\partial r} \right) + \frac{1}{r^2 \sin \theta} \frac{\partial}{\partial \theta} \left(\sin \theta \frac{\partial u}{\partial \theta} \right) + \frac{1}{r^2 \sin^2 \theta} \frac{\partial^2 u}{\partial \varphi^2} = f. \qquad (13.8)$$

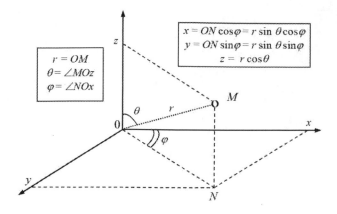

FIGURE 13.2: Spherical coordinates.

The **cylindrical coordinates** are defined by transformations

$$x = r \cos \varphi, \ y = r \sin \varphi.$$

They are characterized by the Cartesian coordinate z directed along the cylinder axis, the coordinate r corresponding to the distance from the given point to the cylinder axis, and the angular coordinate φ, see Figure 13.3. The Laplace operator in the cylindrical coordinates is

$$\Delta u = \frac{1}{r} \frac{\partial}{\partial r} \left(r \frac{\partial u}{\partial r} \right) + \frac{1}{r^2} \frac{\partial^2 u}{\partial \varphi^2} + \frac{\partial^2 u}{\partial z^2}.$$

Then the Poisson equation in the cylindrical coordinates is

$$\frac{1}{r} \frac{\partial}{\partial r} \left(r \frac{\partial u}{\partial r} \right) + \frac{1}{r^2} \frac{\partial^2 u}{\partial \varphi^2} + \frac{\partial^2 u}{\partial z^2} = f. \tag{13.9}$$

In particular, in the two-dimensional case, the z axis is absent and the cylindrical coordinates turn into **polar** ones. The corresponding Poisson equation has the form

$$\frac{1}{r} \frac{\partial}{\partial r} \left(r \frac{\partial u}{\partial r} \right) + \frac{1}{r^2} \frac{\partial^2 u}{\partial \varphi^2} = f.$$

A natural question arises, what is the sense of transformation from a fairly simple Poisson equation in Cartesian coordinates (13.4) to much more cumbersome equations with variable coefficients (13.8) and (13.9)? The answer to this question is be clarified below by the following tasks, the final sections of this chapter, as well as the Dirichlet problem for the Laplace equation in a circle considered in the Appendix.

Task 13.2 *Stationary temperature field in a ball.* A uniform ball is considered in the absence of external heat sources, at the boundary of which a constant temperature is given. Write the mathematical model of the system taking into account spherical symmetry, i.e., the dependence of temperature at an arbitrary point on the body solely on the distance from this point to the center of the ball.

If a system has some type of symmetry,
it makes sense to move from a Cartesian coordinate system
to another, consistent with this type of symmetry.

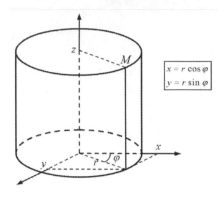

FIGURE 13.3: Cylindrical coordinates.

3   Vector fields

In the previous chapters of Part II, we considered non-stationary systems with distributed parameters described by partial differential equations. The corresponding state functions most often depended on two variables, that are time and spatial coordinates. These variables change independently of each other[10]. In stationary systems, two or even three independent variables are spatial and characterize the shape of the studied object. To describe such systems, some information is needed from **vector analysis**, a branch of mathematics related to the extension of methods and constructions of mathematical analysis to the multidimensional case[11].

Most of the phenomena described below are associated with the consideration of fields, for example, electric, gravitational, etc. All of them are the vector fields. The **vector field** is a mapping that transforms each point of the space to a vector with a begin at this point[12]. Let a three-dimensional Euclidean space be considered in which the point is a three-dimensional vector $\mathbf{r} = (x, y, z)$. Then the vector field is characterized by a vector function

$$\mathbf{A} = \mathbf{A}(\mathbf{r}) = \big(A_1(x, y, z), A_2(x, y, z), A_3(x, y, z)\big).$$

This equality can be determined in the following form

$$\mathbf{A} = A_1(x, y, z)\mathbf{i} + A_2(x, y, z)\mathbf{j} + A_3(x, y, z)\mathbf{k},$$

where $\mathbf{i} = (1, 0, 0)$, $\mathbf{j} = (0, 1, 0)$, $\mathbf{k} = (0, 0, 1)$. In physics, a significant number of vector objects are considered, for example, force, velocity, acceleration, etc. For example, the **force fields** characterized by a force vector $\mathbf{F}$. In particular, an electrostatic field is characterized by the electrostatic force, and the gravity field is determined by the gravitational one.

Note a simplest way to determine a vector field based on a scalar differentiable function of many variables defined in space. The **gradient** of a differentiable function of many variables is a vector whose components are the corresponding partial derivatives[13]. Consider a differentiable function $u = u(\mathbf{r}) = u(x, y, z)$. Then its gradient is a vector function[14]

$$\nabla u = \left(\frac{\partial u}{\partial x}, \frac{\partial u}{\partial y}, \frac{\partial u}{\partial z}\right).$$

This equality can also be written as

$$\nabla u = \frac{\partial u}{\partial x}\mathbf{i} + \frac{\partial u}{\partial y}\mathbf{j} + \frac{\partial u}{\partial z}\mathbf{k}.$$

In physics, the fields determined by the gradient of a scalar function are often considered. For example, according to the Fourier law, the heat flux is characterized by the following equality $\mathbf{q} = -\lambda \nabla u$, where λ is the thermal conductivity coefficient, and u is the temperature. Similarly, according to Fick's law, the diffusion flux is determined by the formula $\mathbf{q} = -D\nabla u$, where D is the diffusion coefficient and u is the concentration. A vector field A is called **potential** if it is determined by the gradient of a scalar value u, called the **potential**. Thus, the fields of heat and diffusion fluxes are potential.

Consider a vector field $\mathbf{A}$ determined by the potential u. Then, the formula $\mathbf{A} = \nabla u$ holds, which corresponds to the equalities

$$A_1 = \frac{\partial u}{\partial x}, \ A_2 = \frac{\partial u}{\partial y}, \ A_3 = \frac{\partial u}{\partial z}.$$

If these functions are continuously differentiable, then the mixed derivatives of u do not depend on the order of differentiation. For example, the mixed derivative u_{yz} can be interpreted to a varying degree both as a derivative of A_3 with respect to y, and as a derivative from A_2 by z. As a result, we obtain

$$\frac{\partial A_3}{\partial y} - \frac{\partial A_2}{\partial z} = 0, \ \frac{\partial A_1}{\partial z} - \frac{\partial A_3}{\partial x} = 0, \ \frac{\partial A_2}{\partial x} - \frac{\partial A_1}{\partial y} = 0.$$

As noted in the previous chapter, a vector function whose components are the left parts of the last equalities is called the **rotor** of the vector field. Thus, the rotor of a vector field $\mathbf{A}$ is the vector

$$\mathrm{rot}\mathbf{A} = \left(\frac{\partial A_3}{\partial y} - \frac{\partial A_2}{\partial z}\right)\mathbf{i} + \left(\frac{\partial A_1}{\partial z} - \frac{\partial A_3}{\partial x}\right)\mathbf{j} + \left(\frac{\partial A_2}{\partial x} - \frac{\partial A_1}{\partial y}\right)\mathbf{k}.$$

According to the previous equations for a potential vector field, the rotor is equal to zero at any point[15]. In particular, according to the Maxwell equations, in the stationary case, the rotor of the electric field is equal to zero. Thus, the electrostatic field turns out to be potential, see Section 4. In force fields, the condition of field potentiality means that the work with instantaneous movement of the particle affected by the field along a closed equality circuit to zero.

Define two important procedures over vectors. Let two vectors $\mathbf{a}=(a_1, a_2, a_3)$ and $\mathbf{b}=(b_1, b_2, b_3)$ be given. The **dot product** $\mathbf{a} \cdot \mathbf{b}$ of these vectors is the sum of the products of their components. Note the equality $\mathbf{a} \cdot \mathbf{b} = |\mathbf{a}||\mathbf{b}|\cos\theta$, where $|\mathbf{a}|$ is the **modulus of the vector** $\mathbf{a}$, see Figure 13.4a. The **cross product** $\mathbf{a} \times \mathbf{b}$ of these vectors is a vector perpendicular to both vectors whose length is equal to the area of the parallelogram formed by the original vectors, and the choice from two directions is determined so that the triple of vectors in the product in order and the resulting vector is right[16]. This means that from the end of the vector $\mathbf{a} \times \mathbf{b}$ the shortest rotation from the vector $\mathbf{a}$ to the vector $\mathbf{b}$ is visible to the observer counterclockwise, see Figure 13.4b.

Note that the rotor of the vector field $\mathbf{A}=(A_1, A_2, A_3)$ can be written as a cross product, as well as a determinant

$$\mathrm{rot}\mathbf{A} = \nabla \times \mathbf{A} = \begin{vmatrix} \mathbf{i} & \mathbf{j} & \mathbf{k} \\ \partial/\partial x & \partial/\partial y & \partial/\partial z \\ A_1 & A_2 & A_3 \end{vmatrix},$$

where on the right-hand side is the determinant of the corresponding matrix. We also note another equality connecting the rotor and gradient with the dot and cross product[17]

$$\frac{1}{2}\nabla|\mathbf{A}|^2 = \mathbf{A} \times \mathrm{rot}\mathbf{A} + (\mathbf{A} \cdot \nabla)\mathbf{A}.$$

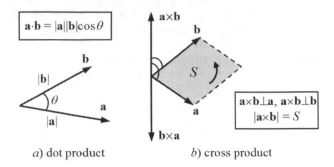

a) dot product   *b)* cross product

FIGURE 13.4: Dot and cross products of vectors.

To determine another important concept of vector analysis, we consider one hydrodynamic problem[18]. Consider the movement of a fluid. It can be described using the ***velocity field***, associating to each point of the considered volume, the velocity vector $\mathbf{v}$ of the particles of the fluid flow at this point. Let S be some surface bounding a part of the volume filled with fluid, and dS be an arbitrarily small element of this surface.

Let us try to estimate the amount of fluid flowing through a given surface. The movement of a fluid through this surface is characterized by the projection v_n of its velocity $\mathbf{v}$ onto the direction of the normal to this surface. In particular, during the time dt, a fluid passes through the platform dS, spaced at a distance of $|v_n|dt$ and filling the volume $|v_n|dtdS$. Thus, for the indicated time, a liquid with a volume $dV = |v_n|dtdS$ passes through this area. Thus, the volume of fluid that passes through the surface dS per unit time is $d\Phi = |v_n|dS$. As a result, the amount of fluid passing during the unit time through the surface S is equal to the surface integral

$$\Phi = \int_S |v_n|dS.$$

Obviously, the normal component of the velocity vector is $v_n = |v_n|\mathbf{n} = |\mathbf{v}|\cos\theta\mathbf{n}$, where $\mathbf{n}$ is the unit normal vector to the surface dS, and θ is the angle between the directions of the vectors $\mathbf{v}$ and $\mathbf{n}$, see Figure 13.5. Given that $\mathbf{n}$ is the unit vector, the last equality can be written as $|\mathbf{v}||\mathbf{n}|\cos\theta$ that is the dot product of the vectors $\mathbf{v}$ and $\mathbf{n}$. Thus, we find

$$\Phi = \int_S (\mathbf{v} \cdot \mathbf{n})dS.$$

The value Φ is called the flux of the velocity field through the surface S. In the general case, the ***flux of the vector field*** $\mathbf{A}$ through the surface S is the integral

$$\Phi = \int_S (\mathbf{A} \cdot \mathbf{n})dS.$$

Consider a point and describe around it a surface dS. Let us determine the ratio of the flux of the vector field $d\Phi$ through this surface to the volume ΔV bounded by this surface. The corresponding limit

$$\operatorname{div}\mathbf{A} = \lim_{\Delta V \to 0} \frac{\Delta \Phi}{\Delta V}$$

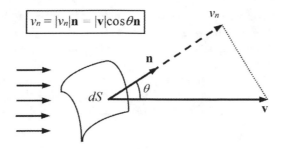

FIGURE 13.5: Fluid movement through the surface.

is called the **divergence** of the vector field $\mathbf{A}$. It can be shown that divergence is characterized by the equality[19]

$$\mathrm{div}\mathbf{A} = \frac{\partial A_1}{\partial x} + \frac{\partial A_2}{\partial y} + \frac{\partial A_3}{\partial z}.$$

This equality can also be written as the dot product $\mathrm{div}\mathbf{A} = \nabla\cdot\mathbf{A}$.

To clarify the meaning of divergence, we turn to the equation for the electric field strength $\mathrm{div}\mathbf{E} = 4\pi\rho/\varepsilon$, where ρ is the charge density and ε is the dielectric constant, see Maxwell equations, Chapter 12. Then the divergence of the electric field strength characterizes the charge density. In the general case, we can say that the divergence of the vector field is the distribution density of the field sources.

The vector field $\mathbf{A}$ is **solenoidal** if $\mathrm{div}\mathbf{A} = 0$. In particular, in the absence of electric charges, equality $\mathrm{div}\mathbf{E} = 0$ is true, and the electric field turns out to be solenoidal. According to the Maxwell equations, the magnetic field strength satisfies the equality $\mathrm{div}\mathbf{H} = 0$. Thus, the magnetic field is always solenoidal, and there are no magnetic charges.

These concepts are used to describe the various fields considered later.

> *A wide class of physical processes is characterized by vector fields.*
> *Rotor, flux, and divergence are characteristics of the vector fields.*

4 Electrostatic field

Consider the electrostatic field[20]. As noted in Chapter 12, the electric field strength $\mathbf{E}$ satisfies the equality

$$\mathrm{div}\mathbf{E} = 4\pi\varepsilon^{-1}\rho,$$

where ρ is the **charge density**, ε is the **dielectric constant**. To describe the field, a new scalar state function is introduced, called the **potential of the electrostatic field** u and related to the strength by the relation $\mathbf{E} = -\nabla u$. Substituting the value of the field strength in the previous equality, we obtain $\mathrm{div}\nabla u = -4\pi\varepsilon^{-1}\rho$. Obviously, the divergence of the gradient is the Laplace operator. Then we get

$$\Delta u = -4\pi\varepsilon^{-1}\rho.$$

Thus, the potential of the electrostatic field will satisfy the Poisson equation. In the absence of charges, we have the Laplace equation

$$\Delta u = 0.$$

As an example, we consider an electrostatic field created by a charge concentrated at a point. Select this point as the origin. Assume that the environment is homogeneous. Then the electrostatic field has a spherical symmetry, i.e., the field potential at a point depends solely on the distance from this point to the origin, where the charge is located. Therefore, any spherical surface centered at the origin will be an ***equipotential surface***, i.e., characterized by the same potential value. In this regard, we use the Laplace equation in a spherical coordinate system. According to equality (13.8), we obtain

$$\frac{1}{r^2}\frac{\partial}{\partial r}\left(r^2\frac{\partial u}{\partial r}\right) + \frac{1}{r^2\sin\theta}\frac{\partial}{\partial\theta}\left(\sin\theta\frac{\partial u}{\partial\theta}\right) + \frac{1}{r^2\sin^2\theta}\frac{\partial^2 u}{\partial\varphi^2} = 0.$$

By the spherical symmetry, the function u does not depend on the angular coordinates θ and φ, which means that its derivatives with respect to angular coordinates are equal to zero. As a result, we obtain the differential equation

$$\frac{\partial}{\partial r}\left(r^2\frac{du}{dr}\right) = 0. \tag{13.10}$$

After integration, we get

$$\frac{du}{dr} = \frac{c_1}{r^2},$$

where c_1 is a constant. Integrating again, we find the function

$$u(r) = -\frac{c_1}{r} + c_2, \tag{13.11}$$

where c_2 is a constant.

Formula (13.11) gives a general solution to equation (13.10). Now one must specify some additional information. Note that as one moves away from the charge, the electric field weakens and should disappear at a sufficiently large distance from it. This means that at r tends to infinity the electric field potential should tend to zero[21]. Ensuring this property requires setting $c_2 = 0$. Given the positivity of the potential, we conclude that the constant c_1 must be negative. In particular, for $c_1 = -1$ we obtain the equality

$$u(r) = \frac{1}{r}. \tag{13.12}$$

Formula (13.12) determines the potential of the field formed by a unit charge located at the origin[22]. In the general case, the value of the constant c_1 is determined here by the magnitude of the charge creating an electric field. According to the results obtained, the electric field potential of a point charge at an arbitrary point is inversely proportional to the distance from this point to the charge, see Figure 13.6. The above example shows the effectiveness of the transition to new independent variables. Using the spherical symmetry of the studied system, we transform the Laplace equation with three spatial coordinates to the ordinary differential equation (13.10). As a result, a solution to the problem in the three-dimensional case was found.

Consider now the electric field created by a thin infinite uniformly charged wire. In the case of a homogeneous conductive medium, the field potential is again described by the Laplace equation. We assume that the Cartesian coordinate z is directed along the wire. Obviously, the field potential of a uniformly charged wire changes equally in any direction from the wire and remains unchanged at any portion of its length. Thus, the field potential at an arbitrary point depends solely on the distance from this point to the wire. Therefore, the equipotential surfaces in this case are the lateral surfaces of all kinds of cylinders, the axis of which is the wire in question. In this regard, it makes sense to go to cylindrical coordinates.

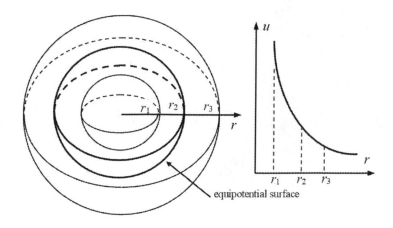

FIGURE 13.6: Potential of a point charge field.

By the equality (13.9), the Laplace equation in cylindrical coordinates is

$$\frac{1}{r}\frac{\partial}{\partial r}\left(r\frac{\partial u}{\partial r}\right) + \frac{1}{r^2}\frac{\partial^2 u}{\partial \varphi^2} + \frac{\partial^2 u}{\partial z^2} = 0.$$

By the cylindrical symmetry of the system, the derivatives of the function u with respect to the variables φ and z are equal to zero. As a result, we obtain the differential equation

$$\frac{\partial}{\partial r}\left(r\frac{\partial u}{\partial r}\right) = 0. \tag{13.13}$$

After integration, we get

$$\frac{du}{dr} = \frac{c_1}{r},$$

where c_1 is an arbitrary constant. Integrating this equality, we obtain

$$u(r) = c_1 \ln r + c_2, \tag{13.14}$$

where c_2 is a constant.

Formula (13.14) gives a general solution to equation (13.13). As in the case of the field of a point charge, we note that the field potential should decrease with distance from the wire and tends to zero at unbounded increasing of r. This leads to the equality $c_2 = 0$. The specific value of the parameter c_1 depends on the charge of the wire, or rather its charge density, i.e., charge unit of its length. In particular, assuming that $c_1 = -1$, we obtain the equality

$$u(r) = \ln \frac{1}{r}. \tag{13.15}$$

characterized the potential of an electric field formed by a wire of unit charge density[23]. According to the result, the potential decreases with increasing distance from the wire, see Figure 13.7.

As in the previous case, using the transition to new independent variables, it was possible to significantly simplify the form of the mathematical model by obtaining the ordinary differential equation (13.13). In practice, the replacement of variables is also carried out in the absence of pronounced symmetry in order to obtain the problem in the field of a simpler structure.

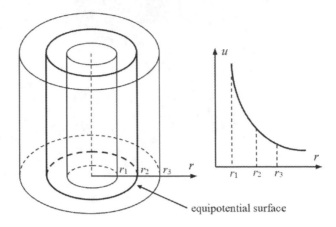

FIGURE 13.7: Potential of the field of an infinite wire.

Potential of the electrostatic field is described by the Laplace equation.
Potential of the field of a point charge is found by spherical coordinates.
Potential field of an infinite wire is found using cylindrical coordinates.

5 Gravity field

Consider the gravity field[24]. According to the ***law of universal gravitation***, the force of interaction of two bodies is directly proportional to the masses of these bodies and inversely proportional to the distance between them. We place the second body at the origin, and the position of the first body relative to the second is characterized by the vector $\mathbf{r}$. Given that force is a vector quantity, we have the equality

$$\mathbf{F} = -G\frac{mM}{|\mathbf{r}|^2}\mathbf{e},$$

where $\mathbf{F}$ is the force, M and m are the masses of the bodies, G is the gravitational constant, and e is the vector of unit length, directed from the first body to the second. Obviously, the equality $\mathbf{r} = -|\mathbf{r}|\mathbf{e}$ is true. Thus, the previous equality takes the form

$$\mathbf{F} = -G\frac{mM}{|\mathbf{r}|^3}\mathbf{r}. \tag{13.16}$$

By the Newtons second law, the acceleration that the first body acquires under the influence of gravitational force is characterized by the equality $\mathbf{F} = M\mathbf{a}$. As a result, taking into account formula (13.16), we find the acceleration of the first body relative to the second one

$$\mathbf{a} = -G\frac{m}{|\mathbf{r}|^3}\mathbf{r}. \tag{13.17}$$

Note that the result does not depend on the mass of the first body, i.e., all bodies under the influence of gravitational force acquire the same acceleration. Formula (13.17) sets the acceleration field of the gravitational force determined by the point body of mass m.

Surround the body generating the gravitational field with a spherical surface S of radius R, see Figure 13.8. As noted earlier, the field flux is determined by the formula

$$\Phi = \int_S \mathbf{a} \cdot \mathbf{n} dS.$$

Note that the acceleration of the body is directed toward the source of the gravitational field, while the external normal to the spherical surface is directed in the opposite direction, see Figure 13.8. Thus, the angle θ is equal to 2π, which means that $\cos\theta = -1$. Since the normal is a vector of unit length, we obtain

$$\mathbf{a} \cdot \mathbf{n} = |\mathbf{a}| = G\frac{m}{|\mathbf{r}|^2},$$

because of the equality (13.17). Then we get

$$\Phi = \int_S \frac{1}{|\mathbf{r}|^2} dS = \frac{Gm}{R^2} \int_S dS,$$

because the distance from any point on the spherical surface to its center is equal to the radius of the sphere. Under the integral, here we have the area of a spherical surface, which is equal to $4\pi R^2$. Now we have $\Phi = 4\pi Gm$ that is true for any spherical surface bounding a body of mass m, generating a gravitational field. Find the ratio

$$\frac{\Phi}{V} = 4\pi G\frac{m}{V},$$

where V is the volume bounded by the surface S. The ratio of body weight to its volume gives the average density value. Then, passing to the limit, when the volume is compressed to a point, taking into account the definition of divergence, we obtain the value of diva. The limit of the ratio of mass to volume gives the density ρ. As a result, we get the formula

$$\mathrm{diva} = 4\pi G\rho. \tag{13.18}$$

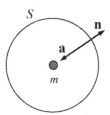

FIGURE 13.8: Sphere surrounding the particle.

To further transform the result, we return to the problem of the body falling under the influence of its weight, which was considered in Chapter 1. The potential energy of the body is determined by the formula $U = mgx$, where x is the vertical coordinate of the body and g is the gravity acceleration. The body weight is determined by the formula $F = -mg$, where the minus sign is due to the fact that the direction of the force and the vertical coordinate are directed in the opposite direction. From the last two equalities follows the equality

$$F = -\frac{dU}{dx}.$$

In the general case, the energy U is connected with the force $\mathbf{F}$ by the formula[25]

$$\mathbf{F} = -\nabla U. \tag{13.19}$$

TABLE 13.1:   Analogy between electrostatic and gravitational fields.

Characteristic	Electrostatics	Gravity
field	electrostatic field	gravity field
source	charge q	mass m
interaction force	electrostatic force $\mathbf{F}$	gravity force $\mathbf{F}$
interaction constant	dielectric constant ε	gravity constant G
interaction law	Coulomb's law	law of universal gravitation
law formula	$\mathbf{F}=\varepsilon^{-1}q_1q_2|\mathbf{r}|^{-3}\mathbf{r}$	$\mathbf{F}=Gm_1m_2|\mathbf{r}|^{-3}\mathbf{r}$
field strength	electrostatics field strength $\mathbf{E}$	acceleration $\mathbf{a}$
environment property	charge density ρ	density ρ
divergence formula	$\mathrm{div}\mathbf{E}=4\pi\varepsilon^{-1}\rho$	$\mathrm{div}\mathbf{a}=4\pi G\rho$
field potential	potential of electrostatic field u	potential of gravity field u
gradient formula	$\mathbf{E}=-\nabla u$	$\mathbf{a}=-\nabla u$
Poisson equation	$\Delta u = 4\pi\varepsilon^{-1}\rho$	$\Delta u = 4\pi G\rho$

It is characteristic of falling bodies that the ratio of potential energy to body mass is determined solely by the height of the body above the surface of the earth, i.e., characteristic of the gravitational field at point x. This quantity $u = gx$, and is called the **potential of the gravitational field**. In the general case, the potential of the gravitational field is related to the potential energy of a body having mass m, using the equality $U = mu$. Then the formula (13.19) takes the form $\mathbf{F}=-m\nabla u$.

By the Newtons second law, we have the equality $\mathbf{F}=m\mathbf{a}$. Then the previous equality implies $\mathbf{a}=-\nabla u$. Substituting this value into the formula (13.18), we find

$$\mathrm{div}\mathbf{a} = -\mathrm{div}\nabla u = 4\pi G\rho.$$

Since the divergence from the gradient is the Laplace operator, we obtain

$$\Delta u = 4\pi G\rho. \tag{13.20}$$

Thus, the potential of the gravitational field satisfies the Poisson equation[26].

Note the analogy between the characteristics of electrostatic and gravitational fields, presented in Table 13.1.

Task 13.3 *Gravitational field of a material point.* The gravitational field generated by the material point is considered. Follow these steps.

   1. Write the Poisson equation (13.20) in a spherical coordinate system.

   2. Using the spherical symmetry, reduce the resulting relation to an ordinary differential equation.

   3. Find a solution to the resulting equation, taking into account the fact that the field potential tends to zero with distance from its source.

   4. Give a physical interpretation of the results.

   5. Establish the effect of density on the results.

Gravitational field is potential.
Potential of the gravitational field is described by the Poisson equation.

Direction of further work. The Appendix considers the steady-state flow of an incompressible fluid. In the next chapter, we will analyze the non-stationary fluid flow, i.e., hydrodynamic problems.

Appendix

Below we consider another example of a physical system described by the Laplace equation. We are talking about the steady flow of an incompressible fluid. In addition, mathematical models of physical processes described by other stationary systems are considered. In particular, the problem of steady-state oscillations of various systems with distributed parameters is characterized by the Helmholtz equation, which is a generalization of the Poisson equation. An analysis of the bending of a thin elastic plate leads to a biharmonic equation, which is a fourth-order partial differential equation.

We also consider the simplest qualitative and quantitative methods for solving stationary systems. In particular, for the Dirichlet problem for the Laplace equation in a circle, the variable separation method used earlier for studying non-stationary systems with distributed parameters is applied. For an approximate solving of the considered problems, the establishment method can be used, which is based on the interpretation of the boundary value problem under study as the equilibrium position of a non-stationary system.

1 Stationary fluid flow

Consider problems associated with the movement of a fluid in space. It is characterized by the density ρ and the velocity vector $\mathbf{v}$, which are connected by the ***continuity equation***[27]

$$\frac{\partial \rho}{\partial t} + \mathrm{div}\rho\mathbf{v} = 0.$$

In the stationary case, we obtain the equality $\mathrm{div}\rho\mathbf{v}=0$. If the fluid is incompressible, then its density is constant, and the continuity equation takes the form $\mathrm{div}\mathbf{v}=0$. Therefore, the velocity field of the steady flow of an incompressible fluid is solenoidal.

To describe the stationary flow of an incompressible fluid, a new state function is introduced. This is denoted by u and called the ***potential of velocity field***, besides the following equality holds $\mathbf{v} = -\nabla u$. Substituting this value to the continuity equation, we have $\mathrm{div}\nabla u = 0$. Considering that the divergence from the gradient is the Laplace operator, we conclude that the potential of the velocity field for the stationary flow of an incompressible fluid satisfies the Laplace equation

$$\Delta u = 0.$$

Task 13.4 *Stationary fluid flow in a pipe*. The stationary flow of an incompressible fluid in an infinitely long pipe is considered. Follow these steps.

1. Write a mathematical model of the process, given that no liquid passes through the pipe walls.

2. Choose the coordinate system most suitable for the given case and give a statement of the problem in this coordinate system.

3. To simplify the formulation of the problem, based on symmetry.

4. Analyze the resulting model.

5. Give a physical interpretation of the results.

2 Steady oscillations

Define another equation for stationary systems. Consider a non-homogeneous wave equation that is

$$u_{tt} = a^2 \Delta u + F. \tag{13.21}$$

This describes forced mechanical, electromagnetic, or sound vibrations, see Chapter 12.

Suppose the function F is periodic in time, in particular, $F = f \sin \omega t$, where f is a function of spatial variables. The solution to equation (13.21) will be sought in the form $u = v \sin \omega t$, where v is the function of spatial variables. Thus, we consider **steady-state oscillations**, i.e., oscillations with the frequency of forced oscillations. Substituting this value into equality (13.21), we have

$$-\omega^2 v \sin \omega t = a^2 \Delta v \sin \omega t + f \sin \omega t.$$

Now the function v satisfies the equation

$$a^2 \Delta u + \omega^2 v + f = 0 \qquad\qquad (13.22)$$

called the **Helmholtz equation**. Thus, the analysis of steady-state oscillations leads to the need to analyze the Helmholtz equation[28].

3   Bending a thin elastic plate

Chapter 12 stated that the vibration of an elastic beam is described by a fourth-order equation

$$u_{tt} + a^2 u_{xx} = 0,$$

where the function u characterizes the deviation of the beam from the equilibrium position. In the case of rigid fastening of the ends of the beam $x = 0$ and $x = L$, this equation is supplemented by the boundary conditions

$$u(0,t) = 0, \ u_x(0,t) = 0, \ u(L,t) = 0, \ u_x(L,t) = 0$$

and the corresponding boundary conditions.

A two-dimensional analogue of this problem is the process of oscillation of a thin elastic plate, which is described by the following equations

$$\frac{\partial^2 u}{\partial t^2} + a^2 \left(\frac{\partial^4 u}{\partial x^4} + 2 \frac{\partial^4 u}{\partial x^2 \partial y^2} + \frac{\partial^4 u}{\partial y^4} \right) = 0.$$

The value in parentheses here corresponds to the twofold application of the Laplace operator for the two-dimensional case and is denoted by Δ^2. Thus, we get the equation

$$u_{tt} + a^2 \Delta^2 u = 0.$$

A two-dimensional analogue of the above boundary conditions is the vanishing of the considered function and its normal derivative on the boundary S of this spatial domain. As a result, we obtain the boundary conditions

$$u(x,y) = 0, \ \frac{\partial u(x,y)}{\partial n} = 0, \ x \in S,$$

corresponding to fixing the plate on the border. The mathematical model is also replenished with initial conditions characterizing the initial position of the plate and the distribution of its velocity at the initial moment of time. Obviously, if these values are identically equal to zero, then the plate will be in equilibrium, i.e., the oscillations are absent.

Suppose now that a constant external force acts on the plate, distributed according to some law $f = f(x,y)$. In this case, the equation of plate oscillation turns is non-homogeneous and takes the form

$$u_{tt} + a^2 \Delta^2 u = f.$$

Now, let us ask ourselves, what shape will the plate have under the influence of the indicated force, being in equilibrium? To answer this question, the second time derivative in the above relation should be equated to zero. As a result, we obtain a fourth-order equation

$$a^2 \Delta^2 u = f,$$

which is considered with the above boundary conditions[29].

4 Variable separation method for the Laplace equation in a circle

When analyzing the equations of heat conduction and string vibrations, we used the method of separation of variables. This method is also applicable to the analysis of stationary systems[30]. Consider the Laplace equation

$$u_{xx} + u_{yy} = 0 \tag{13.23}$$

in a circle of radius r, on the boundary S of which some function f is given. Thus, the Dirichlet problem is solved, consisting of the equation (13.23) with the boundary conditions

$$u(x, y) = f(x, y), \ (x, y) \in S. \tag{13.24}$$

The direct application of the method of separation of variables is difficult here, since in the boundary condition the variables x and y depend on each other. As a result, we pass to the polar coordinates by the equalities

$$x = \rho \cos \varphi, \ y = \rho \sin \varphi.$$

In the new variables, the Laplace equation (13.23) has the form

$$\frac{1}{\rho} \frac{\partial}{\partial \rho} \left(\rho \frac{\partial u}{\partial \rho} \right) + \frac{1}{\rho^2} \frac{\partial^2 u}{\partial \varphi^2} = 0, \ 0 < \rho < r, \ 0 < \varphi < 2\pi. \tag{13.25}$$

The boundary condition (13.24) is transformed the the equality

$$u(r, \varphi) = g(\varphi), \ 0 < \varphi < 2\pi, \tag{13.26}$$

where $g(\varphi) = f(r \cos \varphi, r \sin \varphi)$.

By the **method of separation of variables**, we find the solution of the problem (13.25), (13.26) as the product

$$u(\varphi, \rho) = R(\rho)\Phi(\varphi).$$

As a result, after dividing by $R\Phi/\rho^2$, we obtain

$$\frac{\rho \frac{d}{d\rho} \left(\rho \frac{dR(\rho)}{d\rho} \right)}{R(\rho)} = -\frac{\frac{d^2 \Phi(\varphi)}{d\varphi^2}}{\Phi(\varphi)}.$$

This equality is possible if its left and right sides are some constant, denoted by $-\lambda$. Thus, we obtain two ordinary differential equations

$$\Phi'' + \lambda\Phi = 0, \tag{13.27}$$

$$\rho \frac{d}{d\rho} \left(\rho \frac{dR}{d\rho} \right) - \lambda R = 0. \tag{13.28}$$

Taking into account that turning a circle through an angle of 2π does not change anything, we establish that relations (13.25), (13.26) are supplemented by the periodicity condition

$$u(\rho, \varphi) = u(\rho, \varphi + 2\pi).$$

Then we have the periodic boundary condition for the equation (13.27)

$$\Phi(\varphi) = \Phi(\varphi + 2\pi). \tag{13.29}$$

Obviously, the problem (13.27), (13.29) has a non-trivial solution only for $\lambda = k$, where k is an arbitrary positive integer. The corresponding solution is

$$\Phi_k(\varphi) = a_k \cos k\varphi + b_k \sin k\varphi,$$

where a_k and b_k are constants. The general solution of the equation (13.28) for $\lambda = k$ is

$$R_k(\rho) = c_k \rho^k + d_k \rho^{-k}, \tag{13.30}$$

where c_k and d_k are constants.

Note that, as we approach the center of the circle, the function R_k tends to infinity, which means that the product $R_k \Phi_k$ has a similar property. To avoid this, we put the parameter d_k equal to zero. As a result, we find

$$u_k = R_k \Phi_k = (\alpha_k \cos k\varphi + \beta_k \sin k\varphi)\rho^k,$$

where $\alpha_k = a_k c_k$, $\beta_k = b_k c_k$. The found function satisfies equation (13.25), is a periodic function of the angular coordinate, and has no singularities for $\rho = 0$ for any number k and any constants α_k and β_k. A similar property is possessed by their sum

$$u(\rho, \varphi) = \sum_{k=0}^{\infty} (\alpha_k \cos k\varphi + \beta_k \sin k\varphi)\rho^k. \tag{13.31}$$

It remains to choose here the parameters α_k and β_k in such a way as to ensure that the boundary condition (13.26) is satisfied. Assuming in the formula (13.31) $\rho = r$, we have

$$u(r, \varphi) = \sum_{k=0}^{\infty} (\alpha_k \cos k\varphi + \beta_k \sin k\varphi)r^k = g(\varphi).$$

From the theory of Fourier series, it is known that the function g can be represented as follows

$$g(\varphi) = \frac{\mu_0}{2} + \sum_{k=0}^{\infty} (\mu_k \cos k\varphi + \nu_k \sin k\varphi),$$

where

$$\mu_k = \frac{1}{\pi} \int_0^{2\pi} g(\varphi) \cos k\varphi, \ k = 0, 1, ...; \ \nu_k = \frac{1}{\pi} \int_0^{2\pi} g(\varphi) \sin k\varphi, \ k = 1, 2, ... $$

Comparing the available series, we find the coefficients

$$\alpha_0 = \frac{\mu_0}{2}, \ \alpha_k = \frac{\mu_k}{r^k}, \ \beta_k = \frac{\nu_k}{r^k}, \ k = 1, 2, ... $$

Thus, the solution to the problem (13.25), (13.26) is determined by the formula (13.31) with the Fourier coefficients determined by the above method.

Task 13.5 *External Dirichlet problem for the Laplace equation in a circle*. Consider the external Dirichlet problem in a circle. Make the following analysis.

1. Find the solution of the problem by the method of separation of variables. In this case, pay attention to the properties of the function R_k with an unlimited increase in the argument ρ.

2. Find the solution of the problem with the boundary condition $u(r, \varphi) = \cos 2\varphi$.

3. Substituting the found value in the equation and the boundary condition, make sure that this really is a solution to the problem.

4. Establish the behavior of the system with increasing variable ρ.

5. Establish the influence of the radius of the circle on the results.

5 Establishment method

It has already been noted many times that finding an analytical solution to differential equations is an exceptional case. We consider one general approximate method for solving boundary value problems for the studied equations. As is known, solutions of boundary value problems for the Poisson equation and other mathematical models of stationary systems can be obtained as equilibrium positions of the corresponding non-stationary systems. In particular, the equation (13.2), which is a one-dimensional analogue of the Poisson equation, with some boundary conditions, turns out to be the equilibrium position for the heat equation (13.3) with the corresponding boundary conditions, see Section 1.

Approximate methods for solving the last problem have been described before, in particular, the finite difference method. If we use the explicit difference scheme here (see Chapter 11), the result is a recurrence formula that allows us to calculate the value of the solution of the heat equation on the current layer in time from its known value in the previous time step. With respect to the initial (stationary) problem, the time steps of the non-stationary problem correspond to *iterations*. In particular, the initial distribution of the state function for the heat equation, which is known, corresponds to the initial iteration of the approximate solution algorithm for equation (13.2). Moving from layer to layer in time in the process of solving the heat equation, we approach the equilibrium position of the non-stationary system in the case of the effectiveness of the applied numerical algorithm[31]. We move from iteration to iteration, approaching the solution of the original problem. Thus, as an approximate solution to the original problem, the result of the corresponding iteration of the algorithm or, what is the same, an approximate solution of the heat equation at a sufficiently large moment in time can be chosen.

The described method for solving the boundary value problem for equation (13.2) is the *establishment method*[32]. Naturally, this problem is quite simple, can be solved in a simpler way (see, for example, Task 13.1), and even allows an analytical solution. However, this algorithm, in principle, is applicable for the approximate solving of a wide class of stationary systems, including multi-dimensional and even non-linear systems.

Task 13.6 *Establishment method for the non-homogeneous boundary value problem*. One consider the following boundary value problem for a second-order ordinary differential equation

$$u_{xx} = f(x), \ x \in (0, L), \ u(0) = a, \ u(L) = b,$$

where the function f and the numbers L, a, b are known. Make the following actions.

1. Write the boundary value problem for the heat equation, the equilibrium position for which is the solution of the given problem.

2. Write down an explicit difference scheme for solving the resulting heat equation with the corresponding initial and boundary conditions.

3. Create a program to solve the problem based on the specified algorithm.

4. Carry out the calculations in accordance with the developed program until the time difference between the values of the function on neighboring layers reaches 1 percent.

5. Find solutions of the given boundary value problem based on the tridiagonal matrix algorithm, as was done in Task 13.1 with the same partition of the interval $(0, L)$ and problem parameters.

6. Find an analytical solution of this problem and compare previously obtained approximate solutions.

Notes

[1] As noted earlier, the equation

$$a_{11}u_{xx} + 2a_{12}u_{xy} + a_{22}u_{yy} = F(x, y, u, u_x, u_y)$$

under the condition $a_{12}^2 - a_{11}a_{22} < 0$ is called **elliptic**. Elliptic equations, which include the Poisson and Laplace equations, make up the third type of equations of mathematical physics. The properties of such equations and methods for solving them are considered in any course of partial differential equations, see, for example, [96], [147], [169], [192], [235], [269], [274], [349], [361].

[2] If we consider a non-stationary process, then $q(x)$ is the quantity of heat passing per unit time, see Chapter 10. If the multidimensional case is also considered, then we are talking about the quantity of heat passing per unit time through a unit section.

[3] Chapter 19 discusses the problem of justifying this passage to the limit.

[4] A discrete model of stationary heat transfer is given in Chapter 17, see also Chapter 19.

[5] The importance of the Laplace equation is indicated, in particular, by the fact that its solutions have a special name. A function that is twice continuously differentiable in a certain region and satisfies the Laplace equation there is called a **harmonic function**. The properties of harmonic functions are studied in the theory of equations of mathematical physics, see, for example, [235], [274], [349], [361]. Of particular importance is the Laplace equation in the **theory of functions of a complex variable**, see, for example, [62]. In particular, the central concept of this theory is an **analytic function**. The function of the complex variable $f = f(z) = u(x, y) + iv(x, y)$ of the complex variable $z = x + iy$ is analytic in a set if it can be represented there as a converging Taylor series. The real and imaginary parts of the analytic function f, i.e., the functions u and v satisfy the Laplace equation. We also note that boundary value problems for the Laplace and Poisson equations appear in the calculus of variations as conditions for the extremum of some integral functionals, see Chapter 16.

[6] Chapter 19 will define the concepts of classical and generalized solutions to the Dirichlet problem for the Poisson equation.

[7] The Neumann problem for the Poisson equation has distinctive features from the Dirichlet problem, as well as from the boundary value problems considered earlier for the heat equation and string vibrations. Consider, for example, the Laplace equation in the one-dimensional case with homogeneous Neumann conditions. Thus, the boundary conditions assume that the derivative of the desired function is equal to zero on the left and right boundaries. Obviously, any constant is a solution to this problem. Thus, the solution of the homogeneous Neumann problem for the Laplace equation has an infinite number of solutions, although the number of boundary conditions here is exactly equal to the order of the equation. We now consider equation (13.2) with homogeneous conditions of the second type. Integrating this equality from zero (left boundary) to L (right boundary) and taking into account the boundary conditions, we obtain the equality

$$\int_0^L f(x)dx = 0.$$

Thus, if the function F from the right-hand side of the Poisson equation, i.e., does not satisfy the above equality, then the corresponding homogeneous Neumann problem has no solution at all. Thus, depending on the combination of parameters, the Neumann problem for the Poisson equation may turn out to be both unsolvable and non-unique.

[8] Chapter 19 will show that setting both boundary conditions on one part of the boundary in the absence of conditions on the other part of the boundary corresponds to an ill-posed boundary value problem.

[9] External boundary value problems arise, for example, when studying the flow of a fluid or gas around a body. This type of problem can be posed in the description of the stationary fluid flow considered in the Appendix.

[10] However, for the Stefan problem considered in Chapter 11, the spatial set in which the process is considered changes with time.

[11] The most important concepts of vector analysis are presented in any course of mathematical analysis, see, for example, [13], [61], [79], [148], [206], [344], [353]. The book [227] is devoted directly to the issues of vector analysis.

[12] The *scalar field* maps to each point of the considered set the value of a scalar function. Thus, determining a scalar field is equivalent to defining a function of many variables. Thus, the fields of temperatures, concentrations, etc. are scalar. We also note *non-stationary fields*, the characteristics of which vary with time.

[13] The concept of gradient is used in an essential way in the theory of extremum, indicating the direction of growth of the function of many variables, see Chapter 21.

[14] To denote the gradient of the function u, we also use the notation $\mathrm{grad}\,u$.

[15] Equality to zero at each point of the rotor of the vector field is a necessary but not sufficient condition for the potentiality of the field. Note that if the considered spatial domain is not *simply connected*, then the potential may turn out to be a multi-valued function. The concept of simply connectedness refers to *topology* and assumes that any closed curve on it can be continuously contracted to a point, see, for example, [165]. In particular, the spherical surface is simply connected, but the surface of the torus is not.

[16] To denote the cross product of vectors **a** and **b**, we also use the notation [**a**,**b**].

[17] The above formula will be used in a subsequent chapter to describe the movement of an ideal fluid.

[18] Chapter 14 is devoted to the problems of hydrodynamics.

[19] To justify the divergence formula, it is easiest to choose a parallelepiped with sides Δx, Δy, Δz as the volume ΔV. Then we should explicitly calculate the integrals of the dot product $\mathbf{A}\cdot\mathbf{n}$ along each of its faces. The sum of these integrals gives the value ΔV. Dividing it by the volume $\Delta V = \Delta x \Delta y \Delta z$ passing to the limit as ΔV tends to zero, we obtain the well-known representation of divergence.

[20] Problems of electrostatics are considered, for example, in [200], [278], [343], [352].

[21] The equality of the unknown function to zero at infinity actually represents the boundary condition for the equation considered in the infinite set.

[22] Formula (13.12) gives a *fundamental solution to the Laplace equation in space*. It is used, for example, to determine the *Green function* in the study of boundary value problems for the Poisson equation, see, for example, [235], [274], [349], [361].

[23] Formula (13.15) gives a *fundamental solution to the Laplace equation on the plan*. This is used, for example, to determine the *Green function* in the study of boundary value problems for the two-dimensional Poisson equation.

[24] The gravity field, see [200], [367].

[25] This relationship between force and potential energy is characteristic of any force field.

[26] Chapter 21 consider the inverse problems of gravimetry, which consist in reconstructing the density entering the Poisson equation from the results of measuring the gravitational field.

[27] The continuity equation and other aspects of hydrodynamics are discussed in Chapter 14.

[28] The Helmholtz equation is considered also for the problem of gas diffusion in a steady-state chain reaction, see [349]. The properties of the Helmholtz equation are considered, for example, in [285].

[29] In the absence of external force, we obtain the relation $\Delta^2 u = 0$, called the *biharmonic equation*. Its four times continuously differentiable solution is called the *biharmonic function*.

[30] The Fourier method and other methods for solving the considered equations, in particular, the variational method, the Galerkin, Ritz methods, and Green functions, are described in the previously mentioned literature on the theory of partial differential equations.

[31] Recall here that the explicit difference scheme for the heat equation is effective when the stability condition is satisfied, see Chapter 11.

[32] The idea of the establishment method is most easily described by the example of an approximate solution of nonlinear algebraic equations. Suppose you want to find a solution to the algebraic equation $f(x) = 0$, where the function f is quite complicated, so that finding an analytical solution to the equation is not possible. Obviously, if the number x is a solution to this algebraic equation, then it turns out to be the equilibrium position for the differential equation $\dot{x} = f(x)$. This equation can be solved approximately using the Euler method. Moreover, replacing the derivative of the function $x = x(t)$ at an arbitrary point

t_k with the corresponding difference, we obtain

$$\frac{x(t_k + \tau) - x(t_k)}{\tau} = f(x(t_k)),$$

where the parameter τ has the sense of the time step and is an algorithm parameter. The result is a recurrence formula

$$x_{k+1} = x_k + \tau f(x_k),$$

where $x_k = x(t_k)$. The last equality can be interpreted as an *iterative process* for solving the given algebraic equation. Moreover, the value x_0 corresponding to the initial value for the differential equation, in relation to the original algebraic equation, is interpreted as the initial approximation. Suppose now that the sequence $\{x_k\}$ defined in this way with an unlimited growth in the iteration number k converges to a value of x. Then, under the condition that the function f is continuous as a result of going over to the limit in the above recurrence formula, we obtain the equality $x = x + \tau f(x)$, and hence $f(x) = 0$, regardless of the parameter τ. Thus, the above algorithm in the case of its convergence guarantees the solution to the considered algebraic equation. Thus, the value of x_k can be chosen as its approximate solution for a sufficiently large iteration number k.

Chapter 14

Mathematical models of fluid and gas mechanics

In previous chapters, we analyzed the main classes of systems with distributed parameters described by partial differential equations. With the exception of the Maxwell equations mentioned in Lecture 12, the state of the systems was characterized by a unique state function or, in extreme cases, many functions of the same type. However, in practice, most often there are systems that are characterized by several state functions of different nature. This is true, in particular, for the problems of *fluid and gas mechanics* considered in this chapter.

In a previous chapter, the steady flow of an incompressible fluid was considered. The study of fluid movement refers to **hydrodynamics**[1]. The movement of a fluid or gas differs significantly from the movement of a solid, all points of which describe the same paths. This allows us to consider a solid body as a material point, and to describe the process using a lumped parameter system. In this case, we are dealing with distributed objects, i.e., process characteristics depend not only on time t, but also on the vector of spatial coordinates $\mathbf{r}=(x,y,z)$. The movement is accompanied by a change in the **velocity** vector $\mathbf{v}=(v_1,v_2,v_3)$ and two scalar functions that are the **density** ρ and the **pressure** p.

In Section 1, we establish the material balance in a moving fluid and obtain the continuity equation. This is a first order partial differential equation relating the density of a fluid and its velocity. Section 2 establishes the movement equations for an ideal fluid. These Euler equations are a system of first order partial differential equations with respect to the velocity of fluid, which also includes pressure. Supplemented by continuity and state equations, as well as boundary conditions, this system provides a mathematical model of the movement of an ideal fluid. By adding an additional term to the movement equations, one can describe the fluid movement in the field of gravity, see Section 3. In Section 4, we change an ideal fluid to a real one, also taking into account the viscosity properties. As a result, we obtain the Navier–Stokes equations for the fluid velocity, which have the second order and underlie the most important mathematical model of hydrodynamics.

The Appendix considers special problems of fluid and gas mechanics, in particular, the Burgers, Korteweg–de Vries equations, acoustics, boundary layer, and systems in which the fluid flow is accompanied by qualitatively different processes, in particular, with heat transfer and magnetic fields.

Lecture

1 Material balance in a moving fluid

We determine the change in mass in the process of fluid movement. If a fluid occupying a certain volume ΔV has a constant density ρ, then its mass is $\Delta m = \rho \Delta V$. If the density varies from point to point, then this equality makes sense only for an arbitrarily small volume dV. Thus, we obtain $dm = \rho dV$. Integrating this equality, we find the mass of fluid

---

DOI: 10.1201/9781003035602-14

contained in a given volume

$$m = \int_V \rho dV.$$

Then, given that the density is a variable, we find the change in the mass of the fluid over time from t to $t + \Delta t$

$$m_1 = \int_V \left[\rho(\mathbf{r}, t + \Delta t) - \rho(\mathbf{r}, t) \right] dV.$$

The change in mass in the volume occurs due to leakage through its surface. To estimate the quantity of substance passing through a given surface in the considered time interval, we assume that we have a sufficiently long thin cylinder. Then all the characteristics are considered averaged over the cylinder section and are functions of a unique spatial variable x directed along the cylinder axis. If the fluid moves with velocity v, then it passes the path $v\Delta t$ during the time Δt, and a fluid of volume $\Delta V = v\Delta S\Delta t$ flows out through the section ΔS, see Figure 14.1.

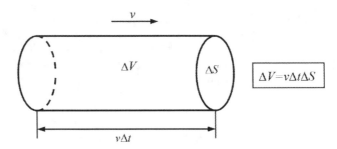

FIGURE 14.1: Volume of fluid flowing through the cross section.

If the density is constant, the mass of the fluid decreases by the value $\rho\Delta V = \rho v\Delta S\Delta t$. If the fluid flows into the considered volume, i.e., if the fluid velocity is negative, then the mass increases by the same amount. Consider the unit vector $\mathbf{n}$ of the outer normal to the surface. Obviously, it is equal to -1 when the fluid enters the considered volume and 1 when it exits. Then, combining the two directions of the velocity considered above, we can write $\Delta m = -\rho\mathbf{v}\cdot\mathbf{n}\Delta S\Delta t$, where the minus sign is explained by the fact that, at a positive velocity, the fluid flows from the given volume. A similar relationship holds true for any surface. If the process characteristics are variable, then the last relation makes sense on an arbitrarily small element of the surface dS and for a small time interval dt. As a result, we obtain the equality $dm = -\rho\mathbf{v}\cdot\mathbf{n}dSdt$.

To calculate the quantity of substance passing through the surface S over time from from t to $t + \Delta t$, it is necessary to integrate the last relation over this surface and over time. Thus, we find the value

$$m_2 = - \int_t^{t+\Delta t} \int_S \rho\mathbf{v} \cdot \mathbf{n}dSdt.$$

From the theory of integration, the connection between volume and surface integrals is known[2]. By the ***divergence theorem***, for any differentiable function $f = f(\mathbf{r})$, the following formula holds

$$\int_V \frac{\partial f}{\partial r_i} dV = \int_S f \cos\alpha_i dS, \; i = 1, 2, 3, \tag{14.1}$$

where α_i is the angle between the direction of the external normal and the axis r_i. Note that the i-th component of the vector $\mathbf{n}$ having unit length is $n_i = \cos\alpha_i$, $i = 1, 2, 3$. If, as f, we choose the components of the vector $\rho\mathbf{v}$ and add the corresponding equalities (14.1), then we find the value

$$m_2 = -\int\limits_{t}^{t+\Delta t}\int\limits_{S}\operatorname{div}\rho\mathbf{v}\,dV\,dt.$$

The material balance in the volume ΔV on the time interval $[t, t + \Delta t]$ is characterized by the equality $m_1 = m_2$, i.e., a change in the mass of fluid in the considered volume for a given time is carried out exclusively due to its inflow or outflow through the boundary of this volume. Using the formulas established above, we have

$$\int\limits_{V}\left\{[\rho(\mathbf{r}, t + \Delta t) - \rho(\mathbf{r}, t)] + \int\limits_{t}^{t+\Delta t}\operatorname{div}\rho\mathbf{v}\,dt\right\}dV = 0.$$

By the ***mean value theorem***, the value of the volume integral is equal to the integrand at a point of the given set, multiplied by the volume. After dividing by Δt we obtain the equality

$$\frac{\rho(\mathbf{y}, t + \Delta t) - \rho(\mathbf{y}, t)}{\Delta t} + \operatorname{div}\rho(\mathbf{y}, s)\mathbf{v}(\mathbf{y}, s) = 0,$$

where $\mathbf{y} \in V$, $s \in [t, t + \Delta t]$. Passing to the limit as $\Delta t \to 0$, we get

$$\rho_t + \operatorname{div}\rho\mathbf{v} = 0. \tag{14.2}$$

Due to the arbitrariness of the selected volume and time interval, it is performed at any point in the space-time region.

The last equality is called the ***continuity equation***. However, it also includes a velocity vector, which is also unknown. Thus, it must be used in conjunction with movement equations. In addition, as noted above, to describe the fluid movement, it is also necessary to determine the pressure. A fluid whose density is solely a function of pressure is called ***barotropic***, i.e., there is a functional relationship between pressure and density

$$p = f(\rho), \tag{14.3}$$

called the ***equation of state***, and the specific form of the function f depends on the features of the considered phenomenon.

For ***incompressible fluid***, the density does not change with pressure[3]. In this case, the continuity equation takes the form

$$\operatorname{div}\mathbf{v} = 0, \tag{14.4}$$

i.e., the velocity field is solenoidal.

> *Density of moving compressible fluid satisfies the continuity equation.*
> *Velocity field of incompressible fluid is solenoidal.*

2   Ideal fluid movement

Determine the movement equation using the ***law of conservation of momentum***. Consider again a volume ΔV bounded by the surface area ΔS. As already noted, the mass of fluid contained in this volume is $\Delta m = \rho\Delta V$. Using Newton's second law, we write the equality $\Delta m\mathbf{a} = \Delta\mathbf{F}$, where $\mathbf{a}$ is the acceleration vector, $\Delta\mathbf{F}$ is the vector of force acting on the given fluid volume. As a result, we obtain the equality $\mathbf{a}\rho\Delta V = \Delta\mathbf{F}$.

At this stage, we restrict ourselves to **ideal fluid** whose particles have an absolute mobility. Such a fluid does not resist shear forces. There are no tangential stresses in it, and the only force taken into account is determined by pressure[4]. **Pressure** p is a value numerically equal to the force acting on the surface unit of an allocated volume of fluid from the side of its environment. In the case when the pressure turns out to be constant, the pressure force $\Delta\mathbf{F}$ falling on the surface area ΔS is equal to the product of pressure and the area of this surface and is directed toward the considered volume under, i.e., opposite to the direction of the outer normal $\mathbf{n}$ to the given surface, see Figure 14.2. Thus, the force at constant pressure is determined by the formula $\Delta\mathbf{F} = -p\mathbf{n}\Delta S$, which implies the equality $\mathbf{a}\rho\Delta V = -p\mathbf{n}\Delta S$.

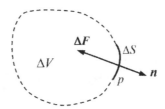

FIGURE 14.2: Action of the pressure force.

The resulting formula assumes that the characteristics of the system do not change in the allocated volume. In the case of variable characteristics, the above equality makes sense if it refers to an arbitrarily small volume of fluid dV bounded by the surface dS

$$\mathbf{a}\rho dV = -p\mathbf{n}dS, \tag{14.5}$$

For further transformations, it is necessary to clarify the value of the acceleration $\mathbf{a}$. It does not express a change in the velocity $\mathbf{v}$ of the fluid at a fixed point, but a change in the velocity of a particle moving in space. If the trajectory of movement is characterized by the vector function $\mathbf{r} = \mathbf{r}(t) = (r_1, r_2, r_3)$, then the acceleration is equal to

$$\mathbf{a} = \frac{d}{dt}\mathbf{v}(\mathbf{r}, t) = \frac{\partial\mathbf{v}}{\partial t} + \sum_{i=1}^{3}\frac{\partial\mathbf{v}}{\partial r_i}\frac{dr_i}{dt} = \frac{\partial\mathbf{v}}{\partial t} + \sum_{i=1}^{3}v_i\frac{\partial\mathbf{v}}{\partial r_i}.$$

The second summand at the right-hand side here is the sum of the products of two vectors components, i.e., their dot product. The first of the vectors here is the velocity, and the second is the vector of partial derivatives, i.e., the gradient. Thus, the previous equality takes the form $\mathbf{a} = \mathbf{v}_t + (\mathbf{v}\cdot\nabla)\mathbf{v}$. Substituting this value in the formula (14.5), we obtain

$$\rho\big[\mathbf{v} + (\mathbf{v}\cdot\nabla)\mathbf{v}\big]dV = -p\mathbf{n}dS.$$

To go to a volume V bounded by a surface S, we integrate

$$\int_V \rho\big[\mathbf{v} + (\mathbf{v}\cdot\nabla)\mathbf{v}\big]dV = -\int_S p\mathbf{n}dS.$$

Using the formula (14.1) in vector form, we get

$$\int_S p\mathbf{n}dS = \int_V \nabla p dV.$$

Then one transforms the previous equality to the form

$$\int_V \left\{\rho[\mathbf{v}_t + (\mathbf{v} \cdot \triangledown)\mathbf{v}] + \nabla p\right\}dV = 0.$$

Using the mean value theorem, we obtain the equality

$$\mathbf{v}_t + (\mathbf{v} \cdot \triangledown)\mathbf{v} = -\rho^{-1}\nabla p. \tag{14.6}$$

The resulting relation is called the ***Euler equation*** or ***movement equation of ideal fluid***. This is a system of first order partial differential equations.

The Euler equation can also be written in another form using the vector analysis formula

$$\frac{1}{2}\nabla|\mathbf{v}|^2 = (\mathbf{v} \cdot \triangledown)\mathbf{v} + \mathbf{v} \times \operatorname{rot}\mathbf{v}$$

from the previous chapter. Then from the equality (14.6) it follows[5]

$$\frac{\partial \mathbf{v}}{\partial t} - \mathbf{v} \times \operatorname{rot}\mathbf{v} = -\nabla\left(\frac{p}{\rho} + \frac{|\mathbf{v}|^2}{2}\right). \tag{14.7}$$

Relations (14.2), (14.3), as well as (14.6) or (14.7) constitute the ***hydrodynamic equations for ideal fluid***. Supplemented with appropriate boundary conditions, they form a mathematical model of the studied process.

Task 14.1 ***Euler equation in a cylindrical coordinates***. Write the Euler equation in a cylindrical coordinate system. Consider the case of axisymmetric flow, when the fluid characteristics are independent of angular coordinate.

Movement of ideal fluid is described by the Euler equation.

3 Ideal fluid under the gravity field

Equation (14.6) was obtained under the assumption that the unique force acting on moving fluid is due to pressure. However, in practice, we often cannot neglect the influence of gravity. In this case, when describing the processes occurring in the volume ΔV, an additional term $\Delta\mathbf{F}_g$, appears, which characterizes the weight of the considered fluid volume. Obviously, we have $\Delta\mathbf{F}_g = \Delta m\mathbf{g} = \rho\mathbf{g}\Delta V$, where Δm is the mass of fluid in the volume ΔV, and $\mathbf{g}$ is the vector of gravitational acceleration. Then we get

$$\mathbf{a}\rho\Delta V = -p\mathbf{n}\Delta S + \rho\mathbf{g}\Delta V.$$

Considering an arbitrarily small volume dV, we obtain

$$\mathbf{a}\rho dV = -p\mathbf{n}dS + \rho\mathbf{g}dV.$$

After integration and using of divergence theorem we have

$$\int_V \left\{\rho[\mathbf{v}_t + (\mathbf{v} \cdot \nabla)\mathbf{v} - \mathbf{g}] + \nabla p\right\}dV = 0.$$

Using the mean value theorem, we obtain[6]

$$\mathbf{v}_t + (\mathbf{v} \cdot \nabla)\mathbf{v} = \mathbf{g} - \rho^{-1}\nabla p. \tag{14.8}$$

This relation describes the movement of fluid in a gravitational field and is the ***movement equation of ideal fluid under gravitational field***. We also note another form of this equality

$$\frac{\partial \mathbf{v}}{\partial t} - \mathbf{v} \times \mathrm{rot}\mathbf{v} = \mathbf{g} - \nabla\left(\frac{p}{\rho} + \frac{|\mathbf{v}|^2}{2}\right).$$

that is an analogue of the equation (14.7). Consider the stationary case. Then the fluid velocity does not change with time, and we get

$$\mathbf{v} \times \mathrm{rot}\mathbf{v} = \nabla(\rho^{-1}p + |\mathbf{v}|^2/2) - \mathbf{g}. \tag{14.9}$$

In hydrodynamics, the concept of ***streamlines*** is widely used. These are such curves whose tangents indicate the direction of the velocity vector at the point of tangency at a given moment in time[7]. In stationary flow, the streamlines do not change in time and correspond to the trajectories of the movement of fluid particles[8]. Multiply the last equality scalarly by the unit vector **l** of the tangent to the streamline at each point. We get

$$(\mathbf{v} \times \mathrm{rot}\mathbf{v}) \cdot \mathbf{l} = \nabla(\rho^{-1}p + |\mathbf{v}|^2/2) \cdot \mathbf{l} - \mathbf{g} \cdot \mathbf{l}. \tag{14.10}$$

Transform this equality, see Figure 14.3. As you know (see the previous chapter), the dot product of two vectors is equal to the product of their moduli and the cosine of the angle between them. As already noted, the direction of the vectors **v** and **l** coincide. The vector product of two vectors is perpendicular to each of them, see the previous chapter. Then the cosine of the angle between the vectors **v**×rot**v** and **l** is zero, which means (**v**×rot**v**)·**l**=0. For an arbitrary function a, the equality $\nabla a \cdot \mathbf{l} = \nabla a \cos\theta$ is true, because **l** is the unit vector. The resulting value is the projection of the vector modulus of ∇a on the vector direction **l**, which corresponds to the derivative of the function a in the direction l. Finally, the gravitational acceleration vector **g** is directed downward, i.e., in the direction opposite to the vertical z coordinate. Then the cosine of the angle between the vectors **g** and **l** is $-\partial z/\partial l$. As a result, we find $\mathbf{g}\cdot\mathbf{l} = -g\partial z/\partial l$.

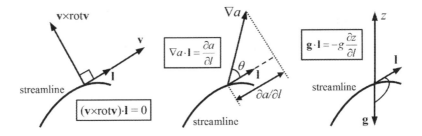

FIGURE 14.3: Transformation of the equality (14.10).

Thus, one transforms the equality (14.10) to the formula

$$\frac{\partial}{\partial l}\left(\frac{v^2}{2} + \frac{p}{\rho} + gz\right) = 0.$$

After integration, we get the formula

$$\frac{v^2}{2} + \frac{p}{\rho} + gz = \mathrm{const}, \tag{14.11}$$

called the ***Bernoulli equation***. Thus, along the streamline, the value on the left-hand side of equality (14.11) does not change. After multiplying by the density, this equality takes the form[9]

$$\frac{\rho v^2}{2} + p + \rho g z = \text{const.}$$

In the resulting equality, each term has the meaning of energy per unit volume. In this case, the first term is related to kinetic energy, the third to potential, and the second to the work of pressure forces.

Task 14.2 *Torricelli formula.* The ideal fluid flowing from a small hole in the wall of an open vessel is considered. It is required to determine the dependence of the fluid flow velocity on the excess of the height of the fluid in the vessel over the height of the hole. Follow these steps.

1. Select the location of the hole as the origin of the vertical coordinate z.

2. Consider the left-hand side of equality (14.11) at the point z corresponding to the fluid level in the vessel, considering the pressure there to be atmospheric and the fluid velocity to be zero.

3. Consider the left-hand side of equality (14.11) at the point $z = 0$, again considering p equal to atmospheric pressure.

4. Equate the obtained expressions in accordance with the Bernoulli equation.

5. From the obtained equality, find the dependence of the velocity of fluid outflow from the hole depending on the fluid level in the vessel, which gives the *Torricelli formula*.

Consider the **hydrostatic problem of incompressible fluid**, characterizing a stationary fluid of constant density under the gravity field. Then the vector equation (14.11) takes the form $\rho^{-1}p = \mathbf{g}$. Given that the acceleration vector of gravity is directed strictly down vertically, we arrive at a system of differential equations

$$\frac{\partial p}{\partial x} = 0, \quad \frac{\partial p}{\partial y} = 0, \quad \frac{\partial p}{\partial z} = -\rho g. \tag{14.12}$$

From the first two equalities it follows that the pressure does not depend on the x and y coordinates, i.e., determined only by the height z. Integrating the last equality, we determine the law of pressure $p(z) = p_0 - \rho g z$, where p_0 is fluid pressure at a height of $z = 0$, which corresponds to atmospheric pressure, if the free surface of the fluid is chosen as the reference point.

Task 14.3 *Barometric formulas.* Equalities (14.12) can also be used for a compressible fluid. However, in this case, the density is not a constant value. The aim of the study is to determine the law of change in the atmospheric pressure of a gas and its density from height. Make these steps.

1. Using the **Clapeyron equation** $p = \mu^{-1}RT\rho$, where μ is the molecular mass of the gas, R is the universal gas constant, and T is the temperature, exclude the density from the third equality (14.12).

2. Assuming that the atmosphere is in an *isothermal state*, i.e., the gas temperature is constant, find a solution to the third equation (14.12), i.e., dependence of atmospheric pressure on altitude at a given temperature. In this case, p_0 is the pressure on the earth's surface $z = 0$ is considered known.

3. Using the Clapeyron equation, establish the law of change in gas density depending on height.

4. Analyze the obtained relations, called the **barometric formulas**.

Movement of ideal fluid under the gravity field
is described by a system of first order partial differential equations.

TABLE 14.1: Analogy between transfer processes.

Branch of physics	Thermophysics	Molecular physics	Hydrodynamics
phenomenon	heat conduction	diffusion viscous	fluid movement
state function	temperature	concentration	velocity
measure heat	quantity	mass	momentum
measure flux	heat flux	diffusion flux	momentum flux
transfer coefficient	thermal conductivity	diffusion coefficient	viscosity
state equation	heat equation	diffusion equation	Navier–Stokes equations

4 Viscous fluid movement

So far, we have been considering the ideal fluid. It was assumed that the only force acting on the fluid is due to pressure, and the transfer of momentum was carried out due to the mechanical movement of various parts of the fluid from one place to another and the action of pressure, as well as gravity, if one is taken into account. Real fluid is **viscous**. The viscosity effect consists in the presence of additional momentum transfer from the region with a higher velocity to the region with a lower velocity, which leads to energy dissipation. Here there is a movement of the various parts of the fluid relative to each other. In a viscous fluid, due to friction between the various fluid layers, tangential stresses arise. In this regard, the forces acting on the surface, limiting a certain volume of fluid, can be directed not only normal to the surface, but also in any direction. Allowance for these effects makes it necessary to add additional terms to the Euler equations.

The movement of incompressible viscous fluid is described by the **Navier–Stokes equations**[10]

$$\mathbf{v}_t + (\mathbf{v} \cdot \nabla)\mathbf{v} = \nu\Delta\mathbf{v} - \rho^{-1}\nabla p, \qquad (14.13)$$

where ν is the **kinematic viscosity** coefficient or just **viscosity**. Obviously, the main part of the Navier–Stokes equations, characterized by the higher derivatives entering into these relations, is similar to the equations of heat conduction and diffusion, see Chapter 10. The heat equation describes the redistribution of temperatures (state function of system) in various parts of the body due to the transfer of energy (measure of system), due to the phenomenon of heat conduction. The diffusion equation describes the redistribution of concentration (state function of system) in various parts of the region due to the transfer of matter (measure of system) due to the phenomenon of diffusion. The Navier–Stokes equations describe the redistribution of velocities (state functions of system) in different parts of the fluid due to momentum transfer (measure of system), due to the viscosity of the fluid. A comparative description of these processes is given in Table 14.1[11].

The density of an incompressible fluid is constant. Therefore, the system of equations (14.13) includes four unknown functions, particularly, the velocity vector and pressure. As a result, the three Navier–Stokes equations are supplemented by the continuity equation in the form (14.4), i.e.,

$$\mathrm{div}\mathbf{v} = 0. \qquad (14.14)$$

Equations (14.13), (14.14) with the corresponding boundary conditions constitute a mathematical model of the movement of incompressible viscous fluid[12].

Obviously, for $\nu = 0$, i.e., when there is no viscosity, the Navier–Stokes equations are reduced to the Euler equation. Thus, the Euler equations can be understood as a degenerate case of the Navier–Stokes equations with zero viscosity.

If the fluid is compressible, then the continuity equation is performed in the form (14.2), and the movement equations are

$$\rho\left[\mathbf{v}_t + (\mathbf{v} \cdot \nabla)\mathbf{v}\right] = \eta\Delta\mathbf{v} - \nabla p + \left(\zeta + \frac{\eta}{3}\right)\nabla\mathrm{div}\mathbf{v},$$

where η is the **dynamic viscosity**, and ζ is the **bulk viscosity**. In the case of viscous fluid moving in the field of gravity or any other force[13], the equations under consideration are supplemented by a term that takes this effect into account, as was done for an ideal fluid, see Section 3.

Consider a particular case. Let there be a stationary fluid flow between two fixed parallel planes in the presence of a pressure gradient. We select the x coordinate in the direction of fluid movement. We assume that the y coordinate is directed from one plane to another so that the corresponding coordinates of the planes correspond to $y = 0$ and $y = h$. In this case, there is only the fluid velocity v in the direction of the x axis that depends only on the y coordinate. Then the first of the Navier–Stokes equations takes the form

$$v_{yy} = \eta^{-1}p_x, \tag{14.15}$$

where $\eta = \rho\nu$ is the **dynamic viscosity**, second equation is transformed to $p_y = 0$, and third one is trivial.

By the last equality, the pressure does not depend on y, which means that it changes only in the direction of x axis. Thus, on the left-hand side of the equality (14.15), there is a function of y, and the value on the right one depends from x. Therefore, both sides of this equality are constants. Then the pressure is a linear function. Integrating the equation (14.15) twice, we find

$$v(y) = \frac{1}{2\eta}\frac{dp}{dx}y^2 + ay + b,$$

where a are b arbitrary constants.

To find these constants, one should set the boundary conditions for equation (14.15). It is quite natural that the fluid vanishes at the boundary of the plates, i.e., $v(0) = 0$, $v(h) = 0$. Using the first equality, we find $b = 0$. From the equality

$$v(h) = \frac{1}{2\eta}\frac{dp}{dx}h^2 + ah + b,$$

it follows that

$$a = -\frac{1}{2\eta}\frac{dp}{dx}h.$$

Thus, we find

$$v(y) = -\frac{1}{2\eta}\frac{dp}{dx}(h - y)y.$$

By this formula, the fluid velocity is distributed according to a parabolic law, which corresponds to the **Poiseuille law**, see Figure 14.4. The maximum value of the velocity is observed in the middle of the fluid layer[14]

$$v = -\frac{h^2}{8\eta}\frac{dp}{dx}.$$

Note that if the pressure gradient is negative (the pressure on the left is greater than the pressure on the right), then the fluid flows in the direction of x.

Task 14.4 *Steady flow of fluid in a pipe*. Find the law of change in the velocity of viscous fluid during its steady flow in a cylindrical pipe. Make the following actions.

1. Write the Navier–Stokes equations in a cylindrical coordinate system.

2. Simplify the equations, assuming that the fluid moves only in the direction of the z coordinate corresponding to the axis of the cylinder and depends only on the radial coordinate.

3. On the pipe surface, set the fluid velocity to zero.

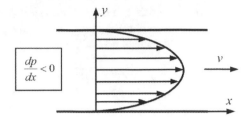

FIGURE 14.4: Steady flow of viscous fluid between the plates.

4. Find the law of change of velocity from the radial coordinate. Explain the results.

5. Find the maximum fluid velocity.

6. Establish the influence of the pipe radius, the viscosity of the fluid and the pressure gradient on the law of movement of fluid and explain the results.

It should be borne in mind that the Navier–Stokes equations are nonlinear and differ in a high degree of complexity. Their analytical solution to the system can only be found in exceptional cases[15].

Movement of viscous fluid is described by the Navier–Stokes equations.

Direction of further work. We continue to consider mathematical models characterized by partial differential equations. The direction of the last chapter covers the most important classes of systems with distributed parameters. In particular, the potential of the velocity field of incompressible fluid is described by the Laplace equation characterizing a wide class of stationary systems. Sound propagation is described by the wave equation and is a prime example of wave processes. The movement of incompressible viscous fluid is described by the Navier–Stokes equations and is associated with transfer processes. The final chapter of Part II deals with the problems of quantum mechanics. The corresponding mathematical models are based on the amazing Schrödinger equation, which is a partial differential equation containing an imaginary unit as a coefficient.

Appendix

We consider here special problems of fluid and gas mechanics. In particular, the Burgers equation describes the movement of viscous fluid in the one-dimensional case and can be analyzed using the method of characteristics. The problem of the propagation of surface waves leads to the Korteweg–de Vries equation, which is of the third order and describes the solution solution. The problem of flowing around a solid body by a fluid or gas stream is described by the Prandtl equations, which are a special case of the Navier–Stokes equations. A mathematical description of the propagation of sound waves leads to the already known wave equation. The phenomenon of thermal convection is described by the system of Navier–Stokes equations and heat equation, and the movement of conducting fluid in a magnetic field is described by the system of Navier–Stokes and Maxwell equations.

1   Burgers equation

The movement of viscous fluid in the one-dimensional case is described using the **Burgers equation**

$$v_t + vv_x = \nu \Delta v.$$

If the fluid is weakly viscous, then the term on the right-hand side of the equation can be neglected. Thus, the movement equation takes the form

$$v_t + vv_x = 0. \tag{14.16}$$

The equation (14.16) is studied using the **method of characteristics**. According to this method, in the x, t plane, such curves $x = \xi(t)$ are determined, called **characteristics**, along which the solution of this equation does not change. Thus, the equality

$$\frac{d}{dt} v\big(\xi(t), t\big) = 0.$$

Calculating the derivative on the left-hand side of the equality, we obtain

$$\frac{\partial v}{\partial t} + \frac{\partial v}{\partial x} \frac{d\xi}{dt} = 0.$$

Comparing the obtained formula with equality (14.16), we conclude that $d\xi/dt = v$. Thus, $\xi(t) = vt + c$, where c is an arbitrary constant. Therefore, the characteristics are described by the equality $x = vt + c$.

Based on the obtained results, we conclude that the value v of the solution of equation (14.16) at the point with coordinates $(vt + c, t)$ does not depend on t. To find the concrete value of a constant, the use of boundary conditions is required. Let the fluid velocity at the initial time be distributed according to the law $\varphi = \varphi(x)$, i.e., we have the initial condition

$$v(x, 0) = \varphi(x). \tag{14.17}$$

It follows that $v(c, 0) = \varphi(c)$. Thus, the following equality holds

$$v(vt + c, t) = v(c, 0) = \varphi(c).$$

Now we get $v(x, t) = \varphi(x - vt)$. The resulting relationship gives an implicit definition of the fluid velocity function[16]. However, it only makes sense when the characteristics do not intersect. If the characteristics intersect, then the corresponding Burgers equation does not have a solution understood in the classical sense[17], which is associated with the occurrence of a **shock wave**.

Task 14.5 *Method of characteristics*. For the equation $v_t + av_x = 0$ in the unit square with some constant a, perform the following steps.

1. Find an equation of arbitrary characteristic.

2. To establish possible options for setting boundary conditions, with the corresponding boundary value problem has a unique solution. Pay attention to the influence of the sign of number a.

3. Find a solution to the boundary value problem in the square.

4. Choose boundary conditions so that the problem has no solution.

2 Surface wave movement

The process of propagation of long waves on the surface of fluid is considered. In this case, we are interested in processes occurring only in the vertical plane, characterized by the horizontal coordinate x and vertical coordinate z. We confine ourselves to the consideration of the only state function of the system, for which the vertical component v of the velocity vector is chosen. In the simplest case, it is described by the movement equation

$$\rho(v_t + vv_x) = F,$$

where ρ is the constant fluid density, and F is the acting force.

The velocity v is the time derivative of the vertical coordinate characterizing the deviation of the fluid surface from the equilibrium state, i.e., the value z_t. As a result, the movement equation takes the form

$$\rho(z_{tt} + z_t z_{tx}) = F,$$

The tension force proportional to the second derivative of the deflection z_{xx} and the elastic force proportional to the fourth derivative z_{xxxx} (see Chapter 12) are selected as F. Thus, we get the equation

$$\rho(z_{tt} + z_t z_{tx}) = c_1 z_{xx} + c_2 z_{xxxx},$$

where c_1 and c_2 are constants.

In the case of long waves, the values $z_t z_{tx}$ and z_{xxxx} turn out to be small in comparison with the other terms of the last equation. Then, with a certain choice of units of measure, this equality can be written in the form

$$z_{tt} + 12\varepsilon z_t z_{tx} = z_{xx} + 2\varepsilon z_{xxxx}, \tag{14.18}$$

where ε is a small positive parameter.

For $\varepsilon = 0$, we have the classical string vibration equation $z_{tt} = z_{xx}$, which, in accordance with the D'Alembert method, has a general solution

$$z(x,t) = f_1(x + t) + f_2(x - t),$$

where f_1 and f_2 are arbitrary functions, see Chapter 12. Thus, we are dealing with traveling waves, which move with time in different directions along the x axis. A similar behavior of the system is observed if in equation (14.18) only the zero terms with respect to ε are taken into account. However, if the indicated two waves do not touch, then the influence of the first-order terms with respect to ε becomes significant. Moreover, each of these waves can be considered independently.

Consider a solution to equation (14.18), which is sufficiently close to the function of the argument[18] $x - t$. As such, the function $z(\xi, \eta) = z(x - t, \varepsilon t)$ can be chosen. Find the derivatives

$$z_t = -z_\xi + \varepsilon z_\eta, \quad z_{xt} = -z_{\xi\xi} + \varepsilon z_{\xi\eta},$$

$$z_{tt} = z_{\xi\xi} - 2\varepsilon z_{\xi\eta} + \varepsilon^2 z_{\eta\eta}, \quad z_{xx} = z_{\xi\xi}, \quad z_{xxxx} = z_{\xi\xi\xi\xi}.$$

Note that terms of the zeroth order with respect to ε cancel out. Then, neglecting all of order ε^2 and higher, after dividing by 2ε we get

$$z_{\xi\eta} + 6z_\xi z_{\xi\xi} = z_{\xi\xi\xi\xi}.$$

Defining a new variable $u = z_\xi$, we have the formula

$$u_\eta + 6uu_\xi = u_{\xi\xi\xi}.$$

called the ***Korteweg–de Vries equation***. It underlies the study of the movement of long surface waves. Various exact solutions are obtained for this equation, in particular, of the soliton type[19]. A ***soliton*** is a stable single wave propagating in a nonlinear medium. Solitons, being waves, in their behavior resemble particles. In particular, they retain their structure when interacting with each other and some external disturbances.

3 Boundary layer model

In problems of a fluid flowing around a body (for example, a ship or an aircraft), the effect of viscosity is manifested only in a rather thin layer near the surface of the body, called the ***boundary layer***[20]. Under these conditions, it is possible to simplify the mathematical model of the process, dividing the region under consideration into two parts. Outside the boundary layer, i.e., in most of the flow, one can proceed from the hypothesis of an ideal fluid and describe the process with substantially simpler Euler equations. In the boundary layer itself, viscosity cannot be neglected. However, due to the small thickness of the layer, the corresponding Navier–Stokes equations can be simplified.

For simplicity, we consider the problem of flow around a flat body. Let this plane be characterized by x and z coordinates, and the y axis be perpendicular to this plane. In this case, we assume that the fluid moves in the direction of the x axis. In this case, the characteristics of the process do not change along the z axis. Denoting by u and v the components of the velocity vector, we obtain the following system of equations

$$u_t + uu_x + vu_y = \nu(u_{xx} + u_{yy}) - \rho^{-1}p_x, \tag{14.19}$$

$$v_t + uv_x + vv_y = \nu(v_{xx} + v_{yy}) - \rho^{-1}p_y, \tag{14.20}$$

$$u_x + v_y = 0. \tag{14.21}$$

Since the main movement is in the direction of the x axis, the velocity v is small compared to u. The main changes in velocity, on the contrary, occur in the direction of the y axis, since the velocity at the streamlined surface is zero, and at a small distance from it, it is equal to the velocity of the main fluid flow. Then, in equation (14.19), the second derivative u_{xx} can be neglected, and in equation (14.20), all terms including the velocity v turn out to be quite small[21]. Therefore, the only term $\rho^{-1}p_y$ that does not contain this velocity is also quite small, which means that we can assume that the pressure depends only on the variable x. As a result, the equation (14.19) reduces to the form

$$u_t + uu_x + vu_y = \nu u_{yy} - \rho^{-1}p_x, \tag{14.22}$$

and the equality (14.20) is trivial.

The formulas (14.21), (14.22) are called the ***Prandtl equations*** and describe the fluid movement in the boundary layer. The pressure included in the last equality is independent of the variable y. Therefore, it depends on the x coordinate in the same way as outside the boundary layer, where the laws of movement of an ideal fluid apply.

4 Acoustic problem

In Chapter 12, acoustic problems were mentioned. Now we determine the mathematical model of these processes[22]. Consider the propagation of sound in a gas. This is described by the equations of hydrodynamics (14.2), (14.3), (14.6) under additional conditions that allow us to significantly simplify the mathematical model. In particular, the equation of state (14.3) has the form

$$p = p_0 \left(\frac{\rho}{\rho_0} \right)^\gamma,$$

where p_0, ρ_0 are initial pressure and density, $\gamma = c_p/c_v$, c_p and c_v are heat capacity of gas at constant pressure and volume. The indicated values are considered process parameters.

The second assumption is that all the characteristics of the process are quite small, and in this regard, squares and higher degrees of state functions can be neglected. Then in the Euler equations (14.6) we can neglect the nonlinear term and obtain the equality

$$\mathbf{v}_t = -\rho^{-1}\nabla p. \tag{14.23}$$

Determine a new unknown variable $s = (\rho - \rho_0)/\rho_0$, called the **gas condensation** and is a relative change in its density. Then the following equality holds $\rho = \rho_0(1 + s)$ and the formula for determining pressure is $p = p_0(1 + s)^\gamma$.

Using the formula of the Taylor series[23], we get

$$(1 + s)^\gamma = 1 + \gamma s + \gamma(\gamma - 1)s^2 + \dots.$$

Neglecting all terms, starting from the square s^2, with a sufficiently high degree of accuracy, we establish the validity of the relation, $p = p_0(1 + \gamma s)$, which implies that $\nabla p = p_0 \gamma \nabla s$. Then we can transform the value on the right-hand side of equality (14.23)

$$\frac{1}{\rho}\nabla p = \frac{p_0 \gamma}{\rho_0(1 + s)}\nabla s = \frac{p_0 \gamma}{\rho_0}\left(1 - s + \frac{1}{2}s^s + \dots\right)\nabla s,$$

where the Taylor series is applied again. Neglecting the higher order terms with respect to unknown functions, we reduce equality (14.23) to the form

$$\mathbf{v}_t = -a^2 \nabla s, \tag{14.24}$$

where $a^2 = p_0\gamma/\rho_0$.

Turn to the continuity equation (14.2). Given the existing representation of density, we obtain

$$\rho_0 s_t = \operatorname{div}\left[\rho_0(1 + s)\mathbf{v}\right].$$

Dividing by ρ_0 and neglecting the product of unknown functions, we obtain

$$s_t + \operatorname{div}\mathbf{v} = 0. \tag{14.25}$$

Exclude the velocity from the system (14.24), (14.25). Acting as the divergence operator on equality (14.24), we have

$$\operatorname{div}\mathbf{v}_t = -a^2 \operatorname{div}\nabla s.$$

Given that the divergence from the gradient is the Laplace operator, we arrive at the equality

$$\operatorname{div}\mathbf{v}_t = -a^2 \Delta s.$$

Differentiating the equality (14.25) with respect to time, we obtain

$$s_{tt} = -\operatorname{div}\mathbf{v}_t.$$

Thus, we get the formula

$$s_{tt} = a^2 \Delta s$$

that is the **wave equation**. Using the existing equalities, we conclude that the density and pressure also satisfy the wave equations, i.e.,

$$\rho_{tt} = a^2 \Delta \rho, \quad p_{tt} = a^2 \Delta p.$$

Establish the equation for velocity. Differentiating equality (14.24) with respect to time, we have

$$\mathbf{v}_{tt} = -a^2 \nabla s_t.$$

Acting by the operator ∇ on the formula (14.25), we obtain

$$\nabla s_t = -\nabla \mathrm{div} \mathbf{v}.$$

Thus, we get the equation

$$\mathbf{v}_{tt} = a^2 \nabla \mathrm{div} \mathbf{v}.$$

The resulting formulas are called the **acoustic equations** and describe the propagation of sound waves. The coefficient a has the dimension of velocity and is called the **sound velocity** of a gas.

5   Thermal convection

We considered earlier the problems of thermophysics and hydrodynamics. However, in applications, it is possible that both of these branches of physics must be applied simultaneously. The phenomenon of thermal convection is considered, in which heat transfer occurs due to the movement of the fluid flow. This process can be described using the **Boussinesq approximation**. The corresponding mathematical model includes the Navier–Stokes equations and heat equation, supplemented by terms describing the mutual influence of thermal and hydrodynamic effects. The result is a system of equations

$$\mathbf{v}_t + (\mathbf{v} \cdot \nabla)\mathbf{v} = \nu \Delta \mathbf{v} - \rho^{-1}\nabla p - \beta u \mathbf{g},$$

$$\mathrm{div} \mathbf{v} = 0,$$

$$u_t + \mathbf{v} \cdot \nabla u = a^2 \Delta u,$$

with corresponding boundary conditions, where $\mathbf{v}$ is the fluid velocity vector, ν is the kinematic viscosity, ρ is the density, p is the pressure, β is the coefficient of volume expansion, u is the temperature deviation from its equilibrium state, $\mathbf{g}$ is the gravity acceleration vector, and a thermal diffusivity coefficient.

Task 14.6 *Temperature distribution in fluid for the Poiseuille flow.* The steady flow of viscous fluid in a long cylindrical pipe is considered, on the walls of which a constant temperature is given. Make the following steps.

1. Determine the fluid flow velocity in accordance with the Poiseuille law, as was done in Task 14.4.

2. Write down the heat equation (see above) in steady state in cylindrical coordinates under the condition that the temperature at a given point of the pipe depends exclusively on the distance from this point to the axis of the cylinder, substituting the velocity value from the previous step into it.

3. The resulting ordinary second order differential equation expressing the dependence of temperature on the radial coordinate can be solved at a known temperature on the pipe surface and the temperature on the axis of the cylinder being finite.

4. Give a physical interpretation of the results.

6 Problems of magnetohydrodynamics

In the previous section, we investigated the process at the junction of hydrodynamics with thermophysics. This time we have to get acquainted with the mathematical models of ***magnetohydrodynamics***, which lies at the junctions of hydrodynamics with electrodynamics[24]. The subject of its study is the dynamics of conducting fluid or gas in a magnetic field. Examples of such media are various kinds of plasma, liquid metals, and salt water.

When a conductive fluid is in a magnetic field, an electric field is induced during its movement. Due to the electric current, forces arise that affect the fluid movement, and also a change in the magnetic field itself. Thus, for the mathematical description of the considered process, it is necessary not only to involve the equations of hydrodynamics and electrodynamics, but also to include additional terms in these equations that take into account the mutual influence of electric and magnetic fields and fluid movement.

The standard mathematical model of magnetic hydrodynamics includes the movement equations of compressible viscous fluid, supplemented by a term that takes into account the influence of electromagnetic forces

$$\rho\left[\frac{\partial \mathbf{v}}{\partial t} + (\mathbf{v} \cdot \nabla)\mathbf{v}\right] = \eta \Delta \mathbf{v} - \nabla p - \left(\zeta + \frac{\eta}{3}\right)\nabla \mathrm{div}\mathbf{v} + \frac{1}{4\pi}\mathbf{H} \times \mathrm{rot}\mathbf{H},$$

compressible fluid continuity equation

$$\rho_t + \mathrm{div}\rho\mathbf{v} = 0,$$

equation of state

$$p = f(\rho),$$

and Maxwell equations for magnetic field, supplemented by a term that takes into account the movement of the fluid

$$\frac{\partial \mathbf{H}}{\partial t} = -\frac{1}{\sigma}\frac{c^2}{4\pi}\mathrm{rot}(\nabla \times \mathbf{H}) + \mathrm{rot}(\mathbf{v} \times \mathbf{H}),$$

with the corresponding initial and boundary conditions, where ρ is the density of the medium, $\mathbf{v}$ is the velocity vector, p is the pressure, ζ is the bulk viscosity, η is the dynamic viscosity, $\mathbf{H}$ is the magnetic field strength, and σ is the conductivity of the medium.

Notes

[1] The problems of fluid and gas mechanics are considered, for example, in [33], [95], [201], [218], [292], [308], [368], see also [115], [258], [374].

[2] Corresponding formulas are given in any course in mathematical analysis, see, for example, [13], [61], [79], [148], [206], [344], [353].

[3] For an incompressible fluid, the ***sound velocity***, i.e., the velocity of propagation of elastic waves in a medium is infinite. Thus, any disturbance is instantly transmitted throughout the stream. Since the sound velocity in real fluids and gases is not infinite, the model of an incompressible fluid is applicable only in cases where the speed of the particles of the medium is small in comparison with the velocity of sound, i.e., with a small ***Mach number***.

[4] In the following sections of this chapter, we will also take into account the force of gravity and the friction force associated with the viscosity of fluid.

[5] We act with the rot operator on equality (14.7). Since derivatives with respect to spatial variables and time are permutable, the operators rot and differentiation with respect to time are also permutable. Since

the rotor of the potential field is equal to zero (see Chapter 13), the action of the rotor on the gradient on the right-hand side of equality (14.7) gives zero. As a result, we arrive at the equation

$$\frac{\partial}{\partial t}\text{rot}\mathbf{v} = \text{rot}(\mathbf{v} \times \text{rot}\mathbf{v}),$$

including exclusively fluid velocity. One obvious corollary follows from this. Assume that the fluid flow at the initial time is potential, i.e., $\text{rot}\mathbf{v} = 0$. Then from the previous equality it follows that in the future the fluid flow will remain potential.

[6] If the fluid is in a force field, then the corresponding Euler equation is again written in the form of (14.8), but $\mathbf{g}$ is understood to be the strength vector of this field. In particular, the Appendix addresses the problems of magnetic hydrodynamics, in which electromagnetic forces act on fluid.

[7] The streamlines are described by a system of differential equations

$$\frac{dx}{dv_1} = \frac{dy}{dv_2} = \frac{dz}{dv_3}.$$

[8] In the non-stationary case, this is no longer the case.

[9] Naturally, this is not the constant from the previous equality.

[10] The derivation of the Navier–Stokes equations is given, for example, in [201], [218], [308], [368].

[11] Table 14.1 is an extension of Table 10.1 of Chapter 10.

[12] The study of the Navier–Stokes equations is often carried out using the transition to the dimensionless variables. Let consider the problem of flowing around a body, characterized by size L, by an incompressible viscous fluid flow having a velocity U at a sufficient distance from the body. We pass from the independent variables $\mathbf{r}=(r_1, r_2, r_3)$ and t and state functions v and p to the new independent variables ξ and τ and state functions u and q using the equalities $\mathbf{r}=L\xi$, $t = T\tau$, $\mathbf{v}=U\mathbf{u}$, $p = Pq$, where the constants T and P will be selected so that the result is as simple as possible. We get the equalities

$$\frac{\partial \mathbf{v}}{\partial t} = UT^{-1}\frac{\partial \mathbf{u}}{\partial \tau}, \quad \frac{\partial \mathbf{v}}{\partial r_i} = UL^{-1}\frac{\partial \mathbf{u}}{\partial \xi_i}, \quad \frac{\partial p}{\partial r_i} = PL^{-1}\frac{\partial q}{\partial \xi_i}, \quad \frac{\partial^2 \mathbf{v}}{\partial r_i} = UL^{-2}\frac{\partial^2 \mathbf{u}}{\partial \xi_i^2}.$$

As a result, the Navier–Stokes equations take the form

$$\mathbf{u}_t + UL^{-1}T(\mathbf{u} \cdot \nabla)\mathbf{u} = \nu L^{-2}T\Delta\mathbf{u} - PT(\rho LU)^{-1}\nabla q.$$

Determining $T = U^{-1}L$, $P = \rho U^2$, $\text{Re} = UL\nu^{-1}$, we get

$$\mathbf{u}_t + (\mathbf{u} \cdot \nabla)\mathbf{u} = \text{Re}^{-1}\Delta\mathbf{u} - \nabla q.$$

The obtained Navier–Stokes equations in dimensionless form include a single parameter Re, called the *Reynolds number*.

[13] In Section 6, the influence of an electromagnetic field will be an external force.

[14] Here we use the well-known property according to which the derivative of the function at the point of extremum vanishes. Problems of extremum theory are considered in Chapters 15 and 21.

[15] Among the special cases of the Navier–Stokes equations that can be solved analytically, the *Couette–Taylor flow* of viscous fluid arising under the action of viscous friction forces between two coaxial cylinders rotating at different velocities, as well as the *Couette flow* of viscous fluid between two parallel walls, one of which moves relative to another. It should be borne in mind that the limitations of our capabilities for effective Navier Stokes analysis are due to objective reasons. For sufficiently small Reynolds numbers, the system describes the *laminar flow* of fluid when the fluid moves in layers without mixing. Known analytical solutions of the Navier–Stokes equations are established only under these conditions. However, for large Reynolds numbers, the flow becomes *turbulent*. Various fluctuations of a chaotic nature arise in the system. The phenomenon of turbulence is widely encountered in practice, as well as in the process of numerical solution of the system, however, a significant number of problems remain unexplored in this direction. A qualitative analysis of the Navier–Stokes equations is carried out, for example, in [192], [215], [345], and an approximate solving to these problems are carried out, for example, in [9], [292], [345]. We also note that the problem of the existence of a global solution of the Navier–Stokes equations in space remains open and is among the seven Mathematical Millennium Problems.

[16] If the initial velocity distribution is linear, i.e., $\varphi(x) = ax + b$, where a and b are constants, then the boundary value problem (14.16), (14.17) has a solution $v(x, t) = (ax + b)/(ax + 1)$.

[17] The concept of a classical solution to the problem of mathematical physics, as well as its generalized solution, is defined in Chapter 19.

[18] Similar calculations can be done for the function of argument $x + t$.

[19] For the Korteweg–de Vries equation and solitons see [86], [255], [381].

[20] For boundary layer theory see [201], [218], [308], [303].

[21] According to equality (14.21), the derivative v_y in equation (14.20) is not small, being equal to $-u_x$. However, the product vvy small, because of the smallness of the first factor.

[22] Acoustic problems are considered, for example, in [201], [242].

[23] The Taylor series at a point a of a function f infinitely differentiable in a neighborhood of this point is the sum

$$f(x) = \sum_{k=0}^{\infty} \frac{f^{(n)}(a)}{n!} (x - a)^n,$$

where $f^{(n)}(a)$ is the derivative of order n of a given function at this point, and the symbol ! denotes factorial, see the above mathematical analysis manuals. If the number x is close enough to a, then to find the approximate value of $f(x)$, one can use a finite number of members of this series.

[24] Mathematical models of magnetic hydrodynamics are considered, in [69], [204].

Chapter 15

Mathematical models of quantum mechanical systems

The laws of quantum mechanics differ significantly from the laws of classical mechanics. The mathematical apparatus of quantum mechanics is based on the amazing Schrödinger equation. This is a specific partial differential equation with respect to a complex-valued function called the wave function. This characteristic allows, in particular, estimating the probability of finding a quantum mechanical particle in a given volume.

Below we establish the Schrödinger equation for a free particle and for a particle moving in an external field. Then the problem of passing a potential barrier is investigated. It is shown that, in contrast to the classical case, even a relatively slow quantum mechanical particle is capable of overcoming a sufficiently high potential barrier.

The Appendix considers the concept of a normalized wave function and analyzes the movement of a particle in a potential well with infinite walls.

Lecture

1 Quantum mechanics problems

The problems of **quantum mechanics** introduce us to a wonderful world. The usual laws of nature are largely unsuitable there. Everywhere we are trapped by surprises, entering into obvious contradictions with established life experience and common sense[1]. In particular, it turns out that, due to some incomprehensible reasons, we cannot determine accurately both the coordinate and momentum of the studied object. For some reason, any particle invariably possesses wave properties. Oddly enough, an electron can be located only in some pre-fixed orbits around the nucleus. A particle with a relatively low energy can overcome a seemingly insurmountable potential barrier.

All this, at first glance, seems completely ridiculous and causes strong internal protest. One gets the impression that in reality we are faced only with temporary troubles caused by the imperfection of our knowledge, that as information accumulates, these paradoxical statements will find a rational explanation, and the long-awaited order will reign in the microcosm.

However, we have to come to terms with the idea that our knowledge and experience need a radical correction when addressing the processes of the microcosm. There are other laws that, despite their seeming paradox, are not less strict than the usual postulates of our world. This means that there must certainly be a special mathematical apparatus that would allow describing the most amazing phenomena of quantum physics.

In classical physics, it is generally accepted that any characteristic of a system can be measured with any degree of accuracy. Of course, in each specific experiment a certain measurement error is always allowed. But as the measuring equipment improves with the support of the high purity of the experiment, this error, in principle, can be infinitely reduced. At

DOI: 10.1201/9781003035602-15

the same time, as a result of the analysis of the processes of the microcosm, it was found that there is a fundamental limit of the accuracy of measuring physical quantities. Quantitatively, it is characterized by the ***Heisenberg uncertainty principle*** $\Delta x \Delta p \geq \hbar/2$, where Δx and Δp are errors in determining the coordinate and momentum of a particle, and $\hbar$ is a universal physical constant called ***Planck constant***[2].

Note that the uncertainty principle does not impose fundamental restrictions on the accuracy limit for measuring the coordinate or momentum separately. However, the simultaneous determination of these quantities with an arbitrary degree of accuracy is impossible in principle. Due to the significant smallness of the Planck constant in classical physics, the uncertainty relation does not appear. However, when describing the processes of the microcosm, there is no longer any possibility to neglect this limitation[3].

The uncertainty relation actually sets the limit of applicability of such concepts of classical physics as the position and momentum of a particle. Thus, complications arise not in the microcosm itself, but when we try to perceive it, who live according to completely different laws. By imposing on the microcosm the concepts of coordinates, impulse, trajectory, etc., so familiar to us and clearly not peculiar to it, we have to pay in full for these obviously dubious actions. The laws of quantum physics are no less strict than the classical laws. One should be surprised here by our attempt to impose the rules of the game known to us on a qualitatively different world.

The laws of quantum physics are objective. Therefore, there must be some kind of mathematical relationships that describe these laws.

> ***The laws of quantum mechanics differ significantly***
> ***from the laws of classical mechanics.***

2 Wave function

The manifestation of the uncertainty principle is one of the fundamental law of quantum mechanics that is the ***wave-particle duality***. As soon as we were unable to localize a quantum-mechanical particle in space, we will no longer be able to perceive it as a material point. The investigated particle will now be some kind of blurred object that does not have clearly defined boundaries. In a sense, it resembles a wave, which is also an object with not very precisely designated dimensions.

In quantum physics, it is generally accepted that any particle has certain wave properties. Mathematically, these properties can be described using some mysterious complex-valued function ψ, called the ***wave function***, which depends on the time t and the coordinate vector $\mathbf{x} = (x_1, x_2, x_3)$ of the particle. The physical meaning of the wave function is as follows. The probability of finding a particle at time t in an arbitrarily small volume $d\mathbf{x} = dx_1 dx_2 dx_3$ is proportional to the value $|\psi(\mathbf{x}, t)|^2 d\mathbf{x}$, which includes the square of the modulus of the wave function[4]. Thus, the probability of finding a particle characterized by a wave function ψ at time t in a certain spatial region V turns out to be proportional to the integral[5]

$$\int_V |\psi(\mathbf{x}, t)|^2 d\mathbf{x}.$$

From the above sense of the wave function, it follows that it is impossible to localize a quantum mechanical particle in space. For simplicity, we restrict ourselves to explaining this property for the spatially one-dimensional case. Suppose that a particle on a straight line with a wave function $\psi = \psi(x, t)$ at some time instant t is located directly at a specific point

x_0. Then, proceeding from the physical meaning of the wave function, for any arbitrarily small number $\varepsilon > 0$ we get the equalities

$$\int\limits_{-\infty}^{x_0-\varepsilon} |\psi(x,t)|^2 dx + \int\limits_{x_0+\varepsilon}^{\infty} |\psi(x,t)|^2 dx = 0, \quad \int\limits_{x_0-\varepsilon}^{x_0+\varepsilon} |\psi(x,t)|^2 dx = c,$$

where c is a positive constant. By the first of them, the function ψ is equal to zero outside the interval $(x_0 - \varepsilon, x_0 + \varepsilon)$. From the second equality and the mean value theorem it follows that it takes the value $c/2\varepsilon$ at a point of this interval. Noting the arbitrary smallness of ε, we conclude that that no function can have the specified properties[6].

Note another interesting property of the wave function. Let a wave function ψ be given. We define a new function φ by the equality

$$\varphi(\mathbf{x}, t) = \exp(i\theta)\psi(\mathbf{x}, t),$$

where i is the imaginary unit, θ is a real number. Using the **Euler formula**

$$\exp(i\theta) = \cos\theta + i\sin\theta,$$

we get the equality

$$|\varphi(\mathbf{x}, t)|^2 = |\cos^2 q + \sin^2 q| \, |y(x,t)|^2 = |\psi(\mathbf{x}, t)|^2.$$

As noted earlier, not the wave function itself has physical meaning, but only the square of its modulus. Consequently, the probability densities corresponding to the wave functions φ and ψ characterize the same state of the quantum mechanical system. Thus, two wave functions differing by a complex factor equal to unity in absolute value correspond to the same state of the particle. Thus, it turns out that the state of the particle determines the wave function up to the mentioned factor, which is one of the natural manifestations of the exotic properties of the mathematical apparatus of quantum mechanics[7].

A wave characterized by a wave function is called a **de Broglie wave** or **probability wave**. According to the ideology of wave-particle duality, a particle and its de Broglie wave are actually one and the same object.

Below we will establish an equation whose solution is the wave function[8].

> *State of a quantum mechanical system is described by a wave function.*
> *Wave function at any point at any time takes on complex values.*
> *Integral of the square of the modulus of the wave function*
> *characterizes the probability of finding a particle in a given volume.*
> *Wave function is determined up to a multiplier with a modulus equal to 1.*

3 Schrödinger equation

By its nature, the wave function seems to be periodic in each argument. For simplicity, we will look for it in the following form

$$\psi(\mathbf{x}, t) = \psi_0(t)\psi_1(x_1)\psi_2(x_2)\psi_3(x_3), \tag{15.1}$$

where ψ_j is a periodic function. The simplest examples of periodic functions are sine and cosine. However, it turns out to be more convenient here the exponent with an imaginary exponent associated with trigonometric functions. This is explained by the fact that we decided to look for the function ψ in the form (15.1). The product of the exponent again gives the exponent, which makes it possible to simplify further calculations. At the same time, sine and cosine clearly do not possess such a property. Thus, we can try to search for the function ψ_j in the form $\psi_j(y) = \exp(ik_j y)$, where k_j is a real number, $j = 0, 1, 2, 3$. The fact that the resulting function ψ turns out to be complex-valued should not bother us too much. Recall that the physical meaning is not the wave function itself, but only the square of its modulus, i.e., real value. Under the assumptions made, formula (15.1) takes the form

$$\psi(\mathbf{x}, t) = \exp\left[i\left(k_0 t + \sum_{j=1}^{3} k_j x_j\right)\right].$$

Denote $\omega = -k_0$, $k = (k_1, k_2, k_3)$. Now the wave function is determined by the formula

$$\psi(\mathbf{x}, t) = \exp\left(i\mathbf{x} \cdot \mathbf{k} - i\omega t\right), \tag{15.2}$$

where we have the dot product of vectors under the exponent. The vector $\mathbf{k}$ and the number ω included in equality (15.2) are called the ***wave vector*** and the ***frequency*** of the function ψ.

In quantum mechanics, it is asserted that the frequency ω is proportional to the total energy E of a moving particle, and the wave vector $\mathbf{k}$ is proportional to its momentum $\mathbf{p}=(p_1, p_2, p_3)$, and the corresponding coefficients in both cases are the Planck constant $\hbar$, i.e., $E = \hbar\omega$, $\mathbf{p}=\hbar\mathbf{k}$. Under these conditions, formula (15.2) takes the form

$$\psi(\mathbf{x}, t) = \exp\left(i\frac{\mathbf{x} \cdot \mathbf{k}}{\hbar} - i\frac{Et}{\hbar}\right). \tag{15.3}$$

The energy of a particle moving in the absence of any external forces is the sum of its rest energy E_0 and kinetic energy K. According to the ***Einstein formula***, the rest energy is $E_0 = mc^2$, where m is the particle's rest mass, and c is the velocity of light in vacuum. The kinetic energy is determined by the formula

$$K = \frac{m|\mathbf{v}|^2}{2} = \frac{|\mathbf{p}|^2}{2m},$$

where $\mathbf{v}$ is the particle velocity. Thus, its total energy is

$$E = mc^2 + \frac{|\mathbf{p}|^2}{2m}.$$

From the equality (15.3) it follows

$$\psi(\mathbf{x}, t) = \exp\left(i\frac{\mathbf{x} \cdot \mathbf{p}}{\hbar} - i\frac{|\mathbf{p}|^2 t}{2m\hbar} - i\frac{mc^2 t}{\hbar}\right) = \exp\left(-i\frac{mc^2 t}{\hbar}\right)\exp\left(i\frac{\mathbf{x} \cdot \mathbf{p}}{\hbar} - i\frac{|\mathbf{p}|^2}{2m\hbar}t\right).$$

On the right-hand side of this equality, the second factor depends on the state of the particle, i.e., from its position and impulse. The first factor does not depend on the state and is a complex-valued function equal in absolute value to one. As noted earlier, this value has no effect on the state of the system. Therefore, we can choose only the second factor as this wave function[9]. Thus, we arrive at the equality[10]

$$\psi(\mathbf{x}, t) = \exp\left(i\frac{\mathbf{x} \cdot \mathbf{p}}{\hbar} - i\frac{|\mathbf{p}|^2 t}{2m\hbar}\right). \tag{15.4}$$

Let us now try to find a fairly simple equation, the solution of which will be the function ψ defined above. Find the time derivative

$$\psi_t = -i\frac{|\mathbf{p}|^2}{2m\hbar}. \tag{15.5}$$

Derivatives with respect to spatial variables are determined in a similar way

$$\psi_{x_j} = i\frac{p_j}{\hbar}\psi, \; j = 1, 2, 3.$$

Calculating the second derivative and summing over j, taking into account the definition of the Laplace operator, we have

$$\Delta\psi = -\frac{|\mathbf{p}|^2}{\hbar^2}\psi. \tag{15.6}$$

Comparing formulas (15.5), (15.6), we obtain the equality

$$i\hbar\psi_t = -\frac{\hbar^2}{2m}\Delta\psi. \tag{15.7}$$

Thus, the wave function determined by formula (15.4) satisfies partial differential equation (15.7), called the **Schrödinger equation** for a free particle, i.e., particle moving in the absence of external forces[11]. Equality (15.7) could be simplified by reducing it by Planck constant. However, for further research, it is more convenient to save it in the above form[12]. Note that, in contrast to all previously considered mathematical models, we are dealing with a complex-valued state function ψ. This circumstance, however, will not be a serious hindrance to further research.

The Schrödinger equation was obtained on the basis of completely formal mathematical reasoning. However, surprisingly, it really has a wonderful physical meaning and numerous applications[13].

Task 15.1 *Schrödinger equation.* Equation (15.7) for the function ψ includes an imaginary unit. Therefore, its solution, i.e., the wave function at an arbitrary point at any time is a complex number. Thus, the representation $\psi = u + iv$ takes place, where u and v are functions of coordinates and time, taking only real values. The following steps are required.

1. Substitute the indicated representation into equality (15.7) and, separating the real and imaginary parts, obtain a system of two equations connecting the functions u and v.

2. From the obtained system of equations, derive the relations that each of these functions satisfies separately.

3. Indicate which of the equations considered in the previous chapters is similar to the established equations for the functions u and v in the spatially one-dimensional case.

> *The wave function satisfies the Schrödinger equation.*
> *The Schrödinger equation for a free particle is*
> *a linear homogeneous second order partial differential equation,*
> *including the imaginary unit.*

4 Particle movement under an external field

Formula (15.7) describes the movement of a free quantum mechanical particle, i.e., a particle that is not affected by any external forces. Let us now turn to analyzing of the behavior of a particle under an external field[14] characterized by the potential energy U. In this case, the energy of the system will already have the form[15]

$$E = K + U = \frac{|\mathbf{p}|^2}{2m} + U.$$

Substituting the energy value into equality (15.3), we find the value of the wave function

$$\psi(\mathbf{x}, t) = \exp\left(i\frac{mc^2 t}{\hbar}\right) \exp\left(i\frac{\mathbf{x} \cdot \mathbf{p}}{\hbar} - i\frac{|\mathbf{p}|^2}{2m\hbar}t - i\frac{U}{\hbar}t\right).$$

After differentiation we obtain

$$\psi_t = -i\frac{|\mathbf{p}|^2}{2m\hbar} - i\frac{U}{\hbar}\psi, \quad \Delta\psi = -\frac{|\mathbf{p}|^2}{\hbar}\psi.$$

Thus, we get the equality

$$i\hbar\psi_t = -\frac{\hbar^2}{2m}\Delta\psi + U\psi. \tag{15.8}$$

This is called the **Schrödinger equation** for a particle under an external field. In the absence of an external field, i.e., at $U = 0$ it coincides with the corresponding equation for a free particle. Relation (15.8) was obtained under the assumption that the potential energy is constant. However, this equation remains valid even in the case when the value of U depends on the coordinate of the particle.

The derivation of the Schrödinger equation was based on purely formal mathematical reasoning. However, it can be used to establish some physical effects that actually exist in nature and are fundamentally unpredictable using classical mechanics. One of these is the passage of a potential barrier by a particle.

Movement of a particle by an external field is described by the Schrödinger equation, which includes the potential energy of the field.

5 Potential barrier

Consider the problem of a particle passing through a potential barrier, see Figure 15.1. In classical mechanics, a particle either overcomes a barrier if its energy is large enough, or is reflected from it if the particle's energy is less than the size of the potential barrier. Let us investigate this process from the point of view of quantum mechanics.

Suppose that the particle moves in the direction of the x axis, and at the point $x = 0$ there is a potential barrier, i.e., when passing through this point, the potential energy changes abruptly from zero to some constant positive value V. In the one-dimensional case, the Schrödinger equation is written as follows

$$i\hbar\psi_t = -\frac{\hbar^2}{2m}\psi_{xx} + U\psi. \tag{15.9}$$

Here the function U is assumed to be zero at $x < 0$ and takes a value V at $x > 0$, see Figure 15.2. If the value of the barrier V exceeds the energy E of the particle moving toward the barrier, then, according to the laws of classical mechanics, the particle can be in the region $x < 0$, but does not fall into the region $x > 0$. In particular, moving inside the allowed region and pushing against the potential barrier, the particle is reflected from it and continues to move in the opposite direction.

Let us investigate the Schrödinger equation (15.9) with the indicated relation between the process parameters. The solution to this equation will be sought in the form

$$\psi(x, t) = \varphi(x) \exp\left(-i\frac{E}{\hbar}t\right),$$

where φ an unknown function. The choice of just such a representation of the wave function

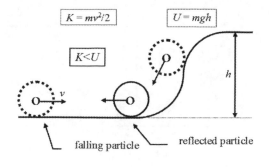

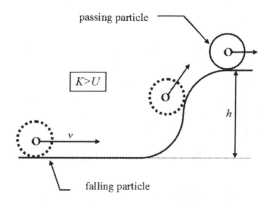

FIGURE 15.1: Classical particle with small energy does not pass the potential barrier.

is explained by the previously established character of its dependence on time. As a result of substituting the function ψ into equality (15.9), we obtain the ordinary differential equation

$$-\frac{\hbar^2}{2m}\frac{d^2\varphi}{dx^2} = \big[E - U(x)\big]\varphi. \tag{15.10}$$

The functions φ and ψ differ only by an exponential factor, which, as is already known, does not affect the value of the square of the modulus, and hence the state of the system. Thus, the function φ can also be considered the wave function of the considered particle.

For $x < 0$, equation (15.10) takes the form

$$\frac{\hbar^2}{2m}\frac{d^2\varphi}{dx^2} + E\varphi = 0.$$

Its general solution is determined by the formula[16]

$$\varphi(x) = a\exp(ikx) + b\exp(-ikx), \tag{15.11}$$

where a, b are arbitrary constants,

For $x > 0$, equation (15.10) is transformed to

$$\frac{\hbar^2}{2m}\frac{d^2\varphi}{dx^2} - (V - E)\varphi.$$

It has the general solution

$$\varphi(x) = c_1\exp(qx) + c_2\exp(-qx), \tag{15.12}$$

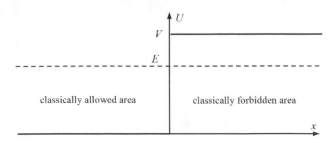

FIGURE 15.2: Potential barrier.

where c_1, c_2 are arbitrary constants, $q = \sqrt{2m(V - E)}$.

Let us establish the values of the constants included in relations (15.11), (15.12). Note that the second term on the right-hand side of the last equality, and hence the square of the modulus of the function φ, exponentially increases with increasing coordinate x. Then the probability of detecting a particle far from the potential barrier increases indefinitely, which means that it will inevitably exceed unity, which is impossible for a quantity that has the meaning of probability. To avoid this obviously absurd situation, we assume $c_2 = 0$.

The values of other constants in equalities (15.11), (15.12) are found from the condition of continuity of the function φ and its derivative at the point $x = 0$. We set

$$\varphi_-(0) = \varphi_+(0), \quad \varphi'_-(0) = \varphi'_+(0),$$

where φ_- and φ_+ denote the left and right limits of the function φ at a given point. Then the following equalities hold $a + b = c_1$, $ik(a - b) = c_1 q$. Now we find the values

$$a = \frac{k + iq}{2k}c_1, \quad b = \frac{k - iq}{2k}c_1.$$

It remains to determine the constant c_1. However, equation (15.10) is linear and homogeneous, and therefore, multiplying its solution by a constant, we again obtain a solution to the same equation. Having determined the value $c_1 = a^{-1}$, we establish the equality

$$\varphi(x) = \exp(ikx) + A\exp(-ikx), \ x < 0, \tag{15.13}$$

where

$$A = \frac{1 - i\sqrt{V/E - 1}}{1 + i\sqrt{V/E - 1}}.$$

Similarly, we find the function

$$\varphi(x) = \exp(qx), \ x > 0, \tag{15.14}$$

where

$$B = \frac{2}{1 + i\sqrt{V/E - 1}}.$$

The function φ, defined by equality (15.13), is a superposition of two waves. The first term $\exp(ikx)$ corresponds to a wave propagating to the right (toward the barrier), and the second term proportional to $\exp(-ikx)$ corresponds to a wave propagating to the left (away from the barrier). Taking into account the equality $|A| = 1$, we will establish that the amplitudes of these waves coincide. It is this circumstance that justifies the choice of the parameter c_1.

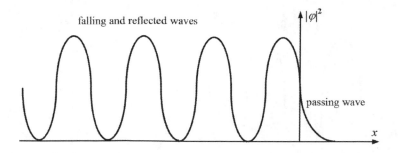

FIGURE 15.3: Squared modulus of the solution of equation (15.10).

Thus, the wave function at $x < 0$ is a superposition of the incident and reflected waves, see Figure 15.3. This result is in good agreement with the prediction of classical mechanics, according to which a particle hits a potential barrier and is reflected from it.

The decisive difference between a quantum mechanical particle and a classical one is observed in the region $x > 0$, which is forbidden by the laws of classical mechanics. As can be seen from formula (15.14), the particle passes the potential barrier with a certain probability, thus, it finds itself in a certain vicinity of it in the zone forbidden from the classical point of view. Characteristically, the function φ rapidly (exponentially) decreases to zero, see Figure 15.3. Thus, the probability of finding a quantum mechanical particle in the depth of the considered region (far enough from the barrier) is practically zero, i.e., the particle cannot be there in any way.

Thus, the de Broglie wave, meeting a potential barrier, is partially reflected, partially penetrates through it. Thus, with a certain degree of probability, a particle with a relatively low energy can overcome a sufficiently high potential barrier[17]. A similar phenomenon, called the **tunneling effect**, is actually observed in practice. It does not fit into the laws of classical mechanics and indicates a fairly high efficiency of the Schrödinger equation.

Suppose now that the potential barrier is not bounded, i.e., the value of V is arbitrarily large. Passing to the limit in equalities (15.13), (15.14) for $V \to \infty$, we find the values $\varphi(x) = \exp(ikx) + \exp(-ikx)$ for $x < 0$ and $\varphi(x) = 0$ for $x > 0$. The results obtained indicate that even a quantum mechanical particle is not able to overcome an infinite potential barrier, see Figure 15.4. Its properties do not differ in any way from the properties of an ordinary classical particle, i.e., the de Broglie wave in this case is completely reflected from the barrier. Obviously, the greater the height of the potential barrier, the less the probability of a particle passing through it[18]. A similar result is obtained for a finite barrier in the case when Planck constant tends to zero[19].

Estimate the probability of a particle passing through a potential barrier. This is proportional to the integral

$$P = \int_0^\infty |\varphi|^2 dx = |B|^2 \int_0^\infty \exp(-2qx)dx = \frac{1}{2q}|B|^2 = \frac{2E}{Vq}.$$

Taking into account the value of the parameter q, we find the number

$$P = \frac{2E\hbar}{V\sqrt{2m(V - E)}}.$$

Hence, it follows that the probability of a particle passing through a potential barrier increases with an increase in the particle energy E, as well as with a decrease in the barrier

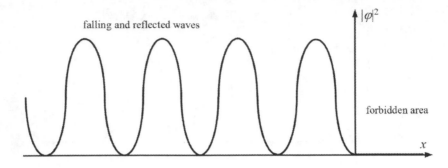

FIGURE 15.4: Square of the modulus of the equation solution for infinite barrier or for $\hbar \to 0$.

depth V, the particle mass m and the energy difference $V - E$. This value is rather small due to the essential smallness of the Planck constant. The results obtained convincingly show that in the macrocosm one can safely use the laws of classical mechanics[20].

Task 15.2 *Potential barrier for high energy particle.*. The case is considered when the energy of a moving particle exceeds the value of the potential barrier. In this case, the classical particle will surely overcome the potential barrier. Using the Schrödinger equation, show that a quantum mechanical particle with a certain probability can be reflected from the barrier.

> *In accordance with the tunneling effect,*
> *a quantum mechanical particle can overcome a potential barrier*
> *when the energy is less than the barrier height.*

Direction of further work. In the previous chapters, we have looked at various mathematical models described by ordinary differential equations and partial differential equations. However, other forms of mathematical models are also possible. In Part III, we consider some special forms of mathematical models that require the use of an additional mathematical apparatus that goes beyond the theory of ordinary differential equations and partial differential equations.

Appendix

We consider here the concept of a normalized wave function. It is remarkable in that the integral of the square of its modulus is exactly equal, and not only proportional to the probability of finding a particle in the region of integration. We also get acquainted with the quantization of the energy of a particle in a potential well with infinitely high walls, which is to a certain extent similar to the application of the apparatus of Fourier series for problems of mathematical physics considered in the previous chapters.

1   Wave function normalization

It was noted earlier that the integral over a certain volume of the square of the modulus of the wave function of a certain particle is proportional to the probability of finding this particle in a given volume. Consider the one-dimensional case for simplicity. Then the probability of finding a particle on a certain interval $[a, b]$ at time t is equal to the integral

$$P(a, b) = c \int_a^b |\psi(x, t)|^2 dx, \tag{15.15}$$

where c is a constant. Obviously, the particle must be located somewhere, i.e., the probability of its finding on the whole number line is equal to one. Thus, the following equality holds[21]

$$c \int_a^b |\psi(x, t)|^2 dx = 1.$$

Now we can find the considered constant[22].

Let us now define the function $\eta(x, t) = \sqrt{c}\psi(x, t)$ with defined before constant c. Multiplying equalities (15.7) or (15.8) by the root of c, we conclude that the function η satisfies the same Schrödinger equation as ψ, and therefore is the wave function of the same particle. Besides, it satisfies the equalities

$$\int_{-\infty}^{\infty} |\eta(x, t)|^2 dx = 1, \quad P(a, b) = \int_a^b |\eta(x, t)|^2 dx.$$

Thus, the integral over an interval of the square of the modulus η is exactly equal (not only proportional) to the probability of finding a particle in this interval, as a result of which it is more convenient to work with such a wave function in practice. The function η is called the ***normalized wave function***[23].

2 Particle movement in a well with infinitely high walls

Consider the Schrödinger equation (15.8) for the spatially one-dimensional case in a well of width L with infinitely high walls. The corresponding potential energy is zero on the interval $(0, L)$ and infinity outside this interval.

Determine again the solution to the equation in the form

$$\psi(x, t) = \varphi(x) \exp\left(-i\frac{E}{\hbar}t\right),$$

where E is a particle energy, and φ is an unknown function. As a result of substitution of this expression in equality (15.8), taking into account the distribution of potential energy, we obtain the ordinary differential equation

$$\frac{\hbar^2}{2m}\frac{d^2\varphi}{dx^2} + E\varphi = 0, \ 0 < x < L. \tag{15.16}$$

Find its general solution

$$\varphi(x) = c_1 \sin \lambda x + c_2 \cos \lambda x, \tag{15.17}$$

where $\lambda = \sqrt{2mE}/\hbar$, c_1 and c_2 are arbitrary constants.

It was noted earlier that the particle is in a given area. As a result, the wave function is zero outside the interval $(0, L)$. In this regard, equation (15.6) is considered with the boundary conditions $\varphi(0) = 0$, $\varphi(L) = 0$. Setting $x = 0$ in equality (15.7), we conclude that $c_2 - 0$. As a result of the definition $x = L$, in equality (15.7), we obtain $c_1 \sin \lambda L = 0$. The equality $c_1 = 0$ leads to the equality of the wave function to zero everywhere. This corresponds to the absence of a particle in the considered region, which contradicts the problem statement. Thus, we arrive at the equality $\sin \lambda L = 0$. It is realized when $L = k\pi$, where the number k takes any of the values 1, 2, 3,... . Taking into account the definition of the constant λ, we conclude that the energy of a particle located in the considered region can only take on the values

$$E_k = \frac{\hbar^2}{2m} \left(\frac{k\pi}{L}\right)^2, \ k = 1, 2, \dots .$$

The corresponding solution of the equation (15.6) is

$$\varphi_k(x) = c_k \sin \frac{k\pi}{L} x,$$

where c_k is a constant. The we find the wave function

$$\psi_k(x) = c_k \sin \frac{k\pi}{L} x \exp\left(-i\frac{E}{\hbar}t\right).$$

To determine the normalized wave function, we find the integral

$$\int_0^L |\psi_k(x,t)|^2 dx = \left|c_k \exp\left(-i\frac{E}{\hbar}t\right)\right| \int_0^L \sin^2 \frac{k\pi}{L} x dx = (c_k)^2 \frac{L}{2}.$$

Equating this value to one, we define the constant c_k. Thus, the corresponding normalized wave function is

$$\psi_k(x) = \sqrt{\frac{2}{L}} \sin \frac{k\pi}{L} x \exp\left(-i\frac{E}{\hbar}t\right).$$

The obtained results show that the energy of the considered particle cannot be arbitrary. It can only take on the values E_k. The discreteness of the possible values of the energy of a quantum mechanical particle essentially distinguishes it from a particle in classical mechanics[24].

Naturally, due to the linearity and homogeneity of equation (15.16), any linear combination of its solutions will also be a solution to the problem. Thus, the general solution to the equation can be sought in the form of a series

$$\psi(x,t) = \sum_{k=1}^{\infty} a_k \psi_k(x,t),$$

where a_k are constants. Thus, we have come to represent the solution of the problem in the form of a Fourier series, well known from the previous chapters.

Notes

[1] A systematic presentation of quantum mechanics is given, for example, in [55], [199], [214], [299], [370], see also [258], [285]. The mathematical apparatus of quantum mechanics is described in more depth in [221].

[2] The uncertainty principle connects not only the coordinate and momentum of a particle. A similar condition is also satisfied, for example, by the energy–time pair

[3] The uncertainty principle is to a certain extent related to the peculiarity of setting up the experiment in the microcosm. Indeed, any measurement is the interaction of the research object with the measuring equipment. Naturally, any measurement introduces certain disturbances in the system. For problems of classical physics, they are usually quite small in comparison with the absolute values of the characteristics of the process. However, in the microcosm the situation is changing significantly. The more accuracy we want to achieve in the measurement process, the greater perturbations we are forced to introduce into the studied process. In particular, with the highest accuracy in measuring the position of a particle, we perturb the system so much that its momentum changes significantly. If we try to accurately measure the momentum, then in the process of its direct measurement, the particle will change its position in space. We emphasize once again that the point here is not in the resolution of the measuring equipment or in our inability to maintain the high purity of the experiment. The difficulties that arise are not subjective, but objective, reflecting the real properties of the surrounding world. Note, however, that the above reasoning is not entirely correct, since any impacts on the system introduce perturbations in its states, although not every pair of characteristics is related by the uncertainty relation.

[4] The most important concepts of probability theory are described in Chapter 18, where stochastic models will be discussed. The concept of **probability** meets all the requirements for measures that we encountered in Chapter 11, see also [129]. In particular, probability is a functional, since its argument will be a random event, and its values will be numbers. In addition, the probability is always non-negative and additive in the sense that the probability of occurrence of at least one of two independent events is equal to the sum of the probabilities of these events. Note that, like any measure, the probability is associated with the integration.

[5] In probability theory, the characteristic that determines the probability that some random variable takes values from an arbitrarily small volume element is called the **probability density**. So, the probability density of the location of a particle in a particular area is proportional to the square of the modulus of the wave function.

[6] However, **distributions**, in particular the δ-**function** (see Chapter 21), possess these properties. It is characteristic that this concept was introduced by Paul Dirac, one of the creators of quantum mechanics.

[7] Each wave function characterizes some state of particle movement. However, the inverse statement is not true, since a particular state of particle motion determines the wave function up to a constant complex factor with a modulus equal to unity. In particular, we use this property to obtain the wave function in the form (15.4).

[8] In all the preceding cases, the derivation of the state equations of the system was carried out on the basis of the causal relationship acting in each specific case. In this case, we first select the form of the wave function so that it possesses the noted set of properties, and then we simply select a fairly simple equation, the solution of which will be the selected wave function.

[9] The fact that the wave function determined by formula (15.4) does not depend on the velocity of light, although this quantity was present in previous calculations, is of fundamental importance. This circumstance emphasizes that quantum mechanics is a nonrelativistic theory. Simultaneous accounting of quantum mechanical and relativistic effects is carried out within the framework of **quantum field theory** and is associated with significant difficulties, see, for example, [229].

[10] Suppose now that a particle on a straight line has a specific momentum p. Then, using equality (15.4) in the one-dimensional case, we find

$$\int\limits_{-\infty}^{\infty} |\psi(x,t)|^2 dx = \int\limits_{-\infty}^{\infty} \left| \exp\left(i\frac{2xp}{\hbar}\right)\right|^2 dx.$$

The integral on the left must be equal to one, since the particle is anywhere. However, the integral on the right is not limited. This contradiction indicates the fundamental impossibility of accurately determining the momentum of the particle.

[11] Schrödinger equation (15.7) coincides up to coefficients with the heat equation. However, the presence of an imaginary unit in it fundamentally changes its properties. It turns out that the Schrödinger equation is much closer to mathematical models of wave processes, rather than transfer processes. In particular, it characterizes wave effects rather than the tendency of the system to equilibrium. In contrast to the heat equation, the Schrödinger equation is reversible in time, i.e., from a given value of the wave function, the history of the system can be reconstructed. In a sense, Task 15.1 clarifies the reason for these properties.

[12] See, in particular, the derivation of formula (15.8).

[13] As noted in Chapter 1, the pursuit of simplicity that we have used in our previous calculations is in excellent agreement with the general principles of modeling natural phenomena.

[14] Naturally, external forces acting on a particle are due to the presence of other particles. Thus, for a more accurate description of the process, the interaction between the de Broglie waves of these particles should be described. This description is typical of quantum field theory. However, in this case we are interested

in the movement of a single particle, while the influence of other particles is characterized by the potential energy.

[15] The absence of internal energy in the expression under consideration is explained by the fact that, as was established above, it does not affect the square of the modulus of the wave function, which, in fact, has a physical meaning.

[16] Indeed, the corresponding characteristic equation has purely imaginary roots.

[17] In Chapters 18 and 20, we will also encounter the lack of determinism of processes. This is true, for example, for the Euler and Benard problems. They are characterized by non-linear differential equations that admit a non-unique solution. In practice, in each specific case, naturally, one of them is realized, and the output to a certain solution is due to a combination of some random factors affecting the process. Thus, when setting up a series of experiments under seemingly identical conditions, we can observe different outcomes corresponding to possible solutions. However, in this case, the nature of the phenomenon is completely different. We investigate the linear Schrödinger equation, which itself admits a well-defined solution. Uncertainty arises not in the process under study itself, but when trying to interpret it using classical concepts such as the coordinate or momentum of a particle. Thus, probabilistic effects do not appear in the microcosm itself, but in the process of its interaction with the objects of the macrocosm.

[18] One can see some analogy between the movement through the barrier of a quantum-mechanical particle and the stream of classical particles. In particular, when a larger number of molecules with a relatively low average energy move, it may turn out that some part of the molecules with a high velocity can overcome the barrier, although most of it will be reflected from it. In the problem under consideration, the barrier with a nonzero probability can be overcome by a single particle with low energy.

[19] This fact is a manifestation of the fact that quantum mechanics goes over into classical mechanics when passing to the limit $\hbar \to 0$.

[20] In quantum mechanics, only average values or **expected value** of various characteristics make sense. For simplicity, we restrict ourselves to considering the spatially one-dimensional case. Let F be some characteristic of an object with a wave function $\psi = \psi(x, t)$. Then its mathematical expectation is

$$E[F] = \int_{-\infty}^{\infty} \psi^*(x) F \psi(x) dx,$$

where ψ^* is the quantity complex conjugate to ψ. Note that the mean values satisfy relations similar to the equation of movement

$$\frac{d}{dt} E[x] = \frac{1}{m} E[p], \quad \frac{d}{dt} E[p] = -E\left[\frac{dU}{dx}\right],$$

where x, p, U are coordinate, momentum, and potential energy. At, these equalities are transformed into the classical equations if $\hbar \to 0$

$$\dot{x} = \frac{1}{m} p, \quad \dot{p} = -\frac{dU}{dx}.$$

In particular, if $\hbar \to 0$, the constant q in equality (15.14) tends to zero, and hence $\varphi(x) = 0$. Thus, the particle certainly does not penetrate into the forbidden band, see Figure 15.4. These properties indicate that classical mechanics is the limiting case of quantum mechanics as the Planck constant tends to zero.

[21] In this case, the set of integration is infinite, as a result of which this integral is **improper**.

[22] This is true if the integral included in this equality makes sense. However, if this property is violated, then it follows from the indicated equality that $c = 0$. Then, in accordance with formula (15.15), the probability of finding a particle on an arbitrary interval is zero, which cannot be. Thus, the indicated integral is probably finite, and the last equality can be used to find the constant under consideration. We also note that the set of all functions integrable with the modulus square is a **functional space** L_2, widely used not only in quantum mechanics, but also in functional analysis and its many applications, see, for example, [153], [185], [283].

[23] From the above reasoning it follows that the value of c is a function of time. However, it can be deduced from the Schrödinger equation that it will still be a constant. All the conclusions made remain valid not only in the one-dimensional, but also in the general case.

[24] In functional analysis, the eigenvalues of a linear operator A are numbers λ for which the equation $Au = \lambda u$ has a nonzero solution u. Equation (15.16) can be written in the form $A\varphi = E\varphi$, where the operator A is characterized by the equality

$$A\varphi = -\frac{\hbar^2}{2m} \frac{d^2\varphi}{dx^2}.$$

Thus, the admissible values of the particle energy are the eigenvalues of the operator defined by the Schrödinger equation.

Part III

Other mathematical models

In all the previous cases, we considered continuous deterministic systems described by differential equations. However, in many cases, mathematical models can be described by other means. In particular, in Chapter 16, various extremum problems act as mathematical models. We consider discrete systems in Chapter 17 and stochastic systems in Chapter 18.

Chapter 16

Variational principles

Until now, differential equations only have been the mathematical models of the studied processes. In this lecture, the laws of evolution are derived from extremum problems, which themselves can be considered as mathematical models of systems. This applies, for example, to the brachistochrone problem, in which it is required to determine a curve in the vertical plane, moving along which the body will get from one given point to another in a minimum time, see Section 1. It is a special case of the Lagrange problem related to the **calculus of variations**[1]. The Lagrange problem can be reduced to a differential equation, called the Euler equation, see Section 2. The problem of finding the curve of the smallest length is an application of this technique, see Section 3. The equation of movement for the process of falling body considered in Chapter 1 turns out to be the Euler equation for the problem of minimizing an integral called the action, see Section 4. In the general case, minimizing the action of a system is the principle of least action, which is one of the deepest laws of physics[2], see Section 5. As an application, Section 6 considers the strings vibration process described in Chapter 12.

The Appendix establishes a connection between variational principles and conservation laws, and also considers physical processes, for the description of which other problems of the calculus of variations are used. In addition, some methods are given for the approximate solving of problems of finding an extremum.

Lecture

1 Brachistochrone problem

Let the points A and B be given in the vertical plane that do not lie on the straight line perpendicular to the earth's surface. It is necessary to find a curve connecting these points, such that, moving along which, under the action of its own weight, the body get from one point to another in a minimum time. This problem is called the **brachistochrone problem**[3].

Place point A at the origin. The y coordinate is directed vertically downward, and the x coordinate is horizontal, see Figure 16.1. Let (X, Y) denote the coordinates of point B. Then the problem is to find a function $y = y(x)$ that satisfies the conditions

$$y(0) = 0, \ y(X) = Y. \tag{16.1}$$

Determine the velocity of the falling body at a point with coordinate y. Using the equalities $\dot{y} = v$, $\dot{v} = g$, find the derivative

$$\frac{dy}{dv} = \frac{dv/dt}{dy/dt} = \frac{g}{v}.$$

Thus, the equality $vdv = gdy$ is true. Integrating the resulting formula, we establish the equality $v^2/2 = gy + c$, where c is an arbitrary constant. Considering that at $y = 0$, i.e., the

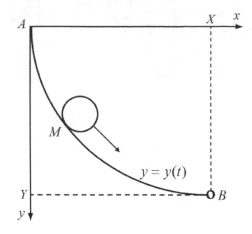

FIGURE 16.1: Movement of the body for the brachistochrone problem.

body is at rest at point A, we conclude that the constant c is equal to zero. Then the velocity of a falling body with a vertical coordinate y is determined by the formula $v = \sqrt{2gy}$.

If we exclude the influence of the resistance force, then both the free falling of the body and its rolling along a certain arc $y = y(x)$ are determined exclusively by the action of the gravitational force. Then the velocity of the body at a fixed height in both cases is the same and equal to v. Let s denote the length of the arc of the given curve from the point M, where the body is at time t to the origin of coordinates, i.e., point A. Then the velocity of the body is

$$\frac{ds}{dt} = v = \sqrt{2gy}. \tag{16.2}$$

The path traversed by the body, and hence the time spent to overcome it, essentially depends on the desired function $y = y(x)$. Suppose that the body, located at time t at point M with coordinates (x, y), moving along the considered curve for a sufficiently small time interval Δt moves to a point M' with coordinates $(x + \Delta x, y + \Delta y)$, see Figure 16.2.

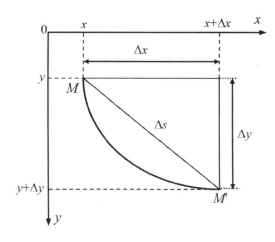

FIGURE 16.2: Calculating of arc length.

If Δt is small enough, the length of the arc connecting the points M and M' is arbitrarily close to the length of the segment MM', equal to

$$\Delta s = \sqrt{\Delta x^2 + \Delta y^2} = \sqrt{1 + \left(\frac{\Delta y}{\Delta x}\right)^2}\,\Delta x.$$

As the length of the time interval Δt tends to zero, the point M' tends to M, and we obtain the following value for the arc element

$$ds = \sqrt{1 + y'(x)^2}dx, \tag{16.3}$$

where $y' = dy/dx$. As a result, relation (16.2) can be written in the form

$$\sqrt{\frac{1 + y'(x)^2}{2gy(x)}}dx = dt. \tag{16.4}$$

The resulting value connects the horizontal coordinate of the moving body x with time t. Note that at $t = 0$, i.e., at the initial moment of time, the body, being at the point A, has the coordinate $x = 0$. At the final time T it gets to the point B with the horizontal coordinate X, see Figure 16.1. Then, as a result of integrating equality (16.4), we establish the equality[4]

$$\int_0^X \sqrt{\frac{1 + y'(x)^2}{2gy(x)}}dx = \int_0^T dt = T.$$

The brachistochrone problem is reduced to minimizing the integral T on the set of functions y satisfying condition (16.1). In the calculus of variations, a problem of this type is called the **Lagrange problem**. The dependence $T = T(y)$ is called the **functional** because it associates a function y with the value of the integral $T(y)$, i.e., number. Note that, as a mathematical model of the considered physical phenomenon, we have not an equation of one form or another, but the problem of finding the extremum of the functional[5].

Mathematical model of the system is a minimization problem
for an integral functional on the set of functions
that take known values at the ends of a given interval.
The brachistochrone problem is the special case of the Lagrange problem.

2 Lagrange problem

We established that the problem of minimizing some integral functional, which is one of the Lagrange problems, acts as a mathematical model of the considered process. To analyze such problems, some information from the **extremum theory** is required. The simplest extremum problem is to minimize a function $f = f(x)$. As is known, if the minimum of the function f is attained at a point x, then the derivative of this function vanishes at this point[6], i.e., the equality $f'(x) = 0$ is true. This is an algebraic equation with respect to the desired minimum point called the **stationary condition**[7].

Consider the problem of finding a function $x = x(t)$ on a segment $[t_1, t_2]$ that minimizes the functional

$$S = S(x) = \int_{t_1}^{t_2} L[t, x(t), \dot{x}(t)]dt$$

and satisfies the boundary conditions

$$x(t_1) = x_1, \ x(t_2) = x_2, \tag{16.5}$$

where L is a given function, and x_1 and x_2 are known numbers. It is called the **Lagrange problem**.

Suppose a function x is its solution. Then the inequality $S(x) \leq S(y)$ holds for all functions y satisfying conditions (16.5). In particular, as y one can choose the function $x + \sigma h$, where σ is a number, and h is an arbitrary function that vanishes at the boundaries of the given segment[8]. Thus, the inequality $S(x) \leq S(x + \sigma h)$ is true.

Fixing a function h with the above properties, we define a function[9] $f = f(\sigma)$ by the formula $f(\sigma) = S(x + \sigma h)$. As a result, the previous inequality takes the form $f(0) \leq f(\sigma)$. Taking into account the arbitrariness of the number σ, we conclude that the function f has a minimum at the point zero. Therefore, the stationary condition $f'(0) = 0$ is true. Find the derivative

$$f'(0) = \lim_{\sigma \to 0} \frac{f(\sigma) - f(0)}{\sigma} = \lim_{\sigma \to 0} \int_{t_1}^{t_2} \frac{L(t, x + \sigma h, \dot{x} + \sigma \dot{h}) - L(t, x, \dot{x})}{\sigma} dt.$$

Using the **Taylor's formula**, determine the value

$$L(t, x + \sigma h, \dot{x} + \sigma \dot{h}) = L(t, x, \dot{x}) + L_x(t, x, \dot{x})\sigma h + L_{\dot{x}}(t, x, \dot{x})\sigma \dot{h} + \eta(\sigma),$$

where $\eta(\sigma)/\sigma \to 0$ as $\sigma \to 0$, L_x and $L_{\dot{x}}$ are partial derivatives of L with respect to the second and third arguments. Now we have

$$f(\sigma) - f(0) = \sigma \int_{t_1}^{t_2} \left[L_x(t, x, \dot{x})h + L_{\dot{x}}(t, x, \dot{x})\dot{h} + \eta(\sigma)/\sigma \right] dt.$$

After dividing by σ and passing to the limit as $\sigma \to 0$ we get[10]

$$f'(0) = \int_{t_1}^{t_2} L_x(t, x, \dot{x})h \, dt + \int_{t_1}^{t_2} L_{\dot{x}}(t, x, \dot{x})\dot{h} \, dt.$$

Transform the second integral using the integration by parts, taking into account the equality of the function h to zero at the boundaries, we obtain

$$\int_{t_1}^{t_2} \left[L_x(t, x, \dot{x}) - \frac{d}{dt} L_{\dot{x}}(t, x, \dot{x}) \right] h \, dt. \tag{16.6}$$

By the **fundamental lemma of calculus of variations**[11], if for a continuous function $g = g(t)$ the following equality holds

$$\int_{t_1}^{t_2} g(t)h(t) \, dt = 0$$

for any sufficiently smooth function h equal to zero on the boundaries of the integration interval, then $g(t) = 0$ everywhere on the considered interval. Applying this result to equality (16.6), we conclude that the solution of the Lagrange problem satisfies the equality

$$L_x(t, x, \dot{x}) - \frac{d}{dt}L_{\dot{x}}(t, x, \dot{x}) = 0 \qquad (16.7)$$

called the **Euler equation** or **Euler–Lagrange equation**[12]. Differential equation (16.7) has the second order and is considered with boundary conditions (16.5), i.e., we have the boundary value problem[13]. Therefore, the solution of the Lagrange problem satisfies the Euler equation with the corresponding boundary conditions[14].

Lagrange problem is to minimize some integral functional.
Lagrange problem is reduced to the Euler differential equation.

3   Shortest curve

As an application, consider the geometric problem of finding a curve of the smallest length passing through two given points. Let points on the plane with coordinates (t_1, x_1) and (t_2, x_2) be given. It is required to find a curve $x = x(t)$ such that the length of the arc connecting these points was minimal. Thus, the curve x satisfies equalities (16.5). In Section 1, it was established that the length ds of an arbitrarily small segment of the arc of a curve is characterized by equality (16.3). As applied to this notation, this equality takes the form $ds = \sqrt{1 + \dot{x}^2}$. As a result of its integration, we obtain the following formula for calculating the length of the curve x on the interval $[t_1, t_2]$

$$S(x) = \int_{t_1}^{t_2} \sqrt{1 + \dot{x}^2}dt.$$

Thus, finding the curve of minimum length reduces to minimizing the functional S on the set of functions $x = x(t)$ satisfying conditions (16.5).

The of minimum length is found from the solution of the Lagrange problem with the function $L = \sqrt{1 + \dot{x}^2}$. The corresponding Euler equation is

$$\frac{d}{dt}\frac{\dot{x}}{\sqrt{1 + \dot{x}}} = 0.$$

Now we find $\frac{\dot{x}}{\sqrt{1+\dot{x}}} = c$, where c is a constant. The resulting equality is an algebraic equation for the derivative $\dot{x}$. Then this derivative will be equal to a constant c_1. As a result, we conclude that the desired function is determined by the formula $x(t) = c_1 t + c_2$, where c_2 is a constant. The concrete values of the constants are determined from the boundary conditions. Thus, the curve of the shortest length connecting two given points is a straight line.

Task 16.1 *Brachistochrone problem.* Determine the Euler equation for the brachistochrone problem.

Curve of the shortest length connecting two given points is a straight line.

4   Body falling problem and the concept of action

Let us return to the study of the process of falling of a body under its own weight, discussed in Chapter 1. As already noted, this process is described by the equation of movement

$$\ddot{x} = -g, \qquad (16.8)$$

where $x = x(t)$ is the body height above the ground that changes over time, g is the acceleration of gravity. Note that it is a corollary of Newton's second law[15] $m\ddot{x} = -P$, where m is the mass, P is the weight of the body. We rewrite this equality as

$$-P - m\ddot{x} = 0.$$

Let us try to interpret it as the Euler equation

$$L_x(t, x, \dot{x}) - \frac{d}{dt} L_{\dot{x}}(t, x, \dot{x}) = 0$$

for some function L. Comparing the last two equalities, it is natural to assume that the following conditions are true

$$L_x(t, x, \dot{x}) = -P, \quad L_{\dot{x}}(t, x, \dot{x}) = m\dot{x}. \qquad (16.9)$$

These equalities can be interpreted as differential equations for the function L. Integrating the first equality over x, we conclude

$$L(t, x, \dot{x}) = -Px + c_1, \qquad (16.10)$$

where c_1 is a value that does not depend of x. But it can depend of other arguments of the function L[16]. Analogically, after integration equality (16.9) by $\dot{x}$ we get

$$L(t, x, \dot{x}) = \frac{m\dot{x}^2}{2} + c_2, \qquad (16.11)$$

where c_2 does not depend of $\dot{x}$, but it can be depending of other arguments of the considered function[17].

From equality (16.10), it follows that the dependence of L from x is determined by the first summand of its right-hand side. The dependence of L from $\dot{x}$ is determined by the first summand of its right-hand side of equality (16.11). Thus, the body falling equation is the Euler equation for the function

$$L(t, x, \dot{x}) = \frac{m\dot{x}^2}{2} - Px + f(t)$$

with an arbitrary function f. Note that the third term does not depend on x. Therefore, it does not affect the minimization of the corresponding integral. Thus, the body falling equation can be interpreted as the Euler equation for the problem of minimizing the functional

$$S(x) = \int_{t_1}^{t_2} \left[\frac{m\dot{x}(t)^2}{2} - Px(t) \right] dt.$$

The first term under the integral is the kinetic energy $K(t)$ of the falling body, and the second one is the potential energy $U(t)$. Thus, we have the equalities

$$K = \frac{m\dot{x}}{2}, \quad U = Px.$$

In mechanics, the difference $L = K - U$ between the kinetic and potential energies is called the **Lagrangian** of the system[18], and the integral of the Lagrangian is called the **action** of the system, denoted by S. Thus, the equation of movement (16.7) can be obtained as a consequence of the problem of minimizing the action, i.e., among all possible variants of evolution of the system is realized that corresponds to the minimum of action. This statement, called the **principle of least action**, can be understood as a mathematical model of the considered process of the body falling.

Body falling equation is the Euler equation for a Lagrange problem.

5 Principle of least action

Consider a body of mass m, which moves in space in a certain field. The movement of a point in space is described by its coordinates, i.e., components of the vector function $\mathbf{x}=(x_1, x_2, x_3)$. The potential energy of the field U is characterized by the position of the body in space, i.e., there is a dependence $U = U(\mathbf{x})$. Body weight generally changes with time, i.e., the dependence $m = m(t)$ is given[19]. In particular, in the above problem of a body falling, the body mass was constant, the motion took place in a gravitational field with a change in only the vertical coordinate x, and the potential energy was determined by the equality $U = mgx$. Let us define a mathematical model of the process, from which we can find the law of movement of the body, i.e., the dependence

Try again to determine the **action** of the system on the time interval $[t_1, t_2]$ by the formula

$$S(\mathbf{x}) = \int_{t_1}^{t_2} L dt,$$

where L is the **Lagrangian** of the system, defined as the difference between the kinetic energy K and the potential one U.

The **kinetic energy** in the previous case was determined by the formula $K = m\dot{x}^2/2$. However, for the considered system, the coordinate, and hence its velocity, i.e., the derivative of the coordinate are vectors. The vector analogue of the square of a scalar value is the square of the modulus of the corresponding vector. Thus, the kinetic energy is determined by the formula

$$K = \frac{m|\dot{\mathbf{x}}|^2}{2}.$$

Based on the results obtained, we find the action

$$S(\mathbf{x}) = \int_{t_1}^{t_2} \left(\frac{m|\dot{\mathbf{x}}|^2}{2} - U(\mathbf{x}) \right) dt.$$

The **principle of least action**, also called the **Hamilton's principle**, says that the law of the system evolution, i.e., the vector function $\mathbf{x}=\mathbf{x}(t)$ is a solution to the problem of minimizing the functional S, i.e., corresponds to the minimum of system action[20].

Try to establish, using the previously described technique, what relations the solution of the posed minimization problem satisfies. In this case, the integrand of the functional to be minimized, i.e., Lagrangian, has the form

$$L(t, \mathbf{x}, \dot{\mathbf{x}}) = \frac{m|\dot{\mathbf{x}}|^2}{2} - U(\mathbf{x}).$$

Note that the direct use of the Euler equation (16.7) here is not possible, since the Lagrange problem considered in Section 2 contained a single unknown function. However, it can be shown that relation (16.7) is also valid in this case if we give it a vector interpretation

$$L_{\mathbf{x}}(t, \mathbf{x}, \dot{\mathbf{x}}) - \frac{d}{dt} L_{\dot{\mathbf{x}}}(t, \mathbf{x}, \dot{\mathbf{x}}) = 0. \tag{16.12}$$

Here the Lagrangian depends on the vector quantities $\mathbf{x}$ and $\dot{\mathbf{x}}$. In this regard, the partial derivatives of L included in the last equality also turn out to be vector. Calculating the partial derivatives of the Lagrangian, we arrive at the system of differential equations

$$\frac{d}{dt}\left[m(t)\frac{d\dot{\mathbf{x}}}{dt}\right] + \nabla U(\mathbf{x}) = 0.$$

This equality can be transformed by introducing the force vector using the equality $\mathbf{F} = -\nabla U$. In particular, in the problem of falling body, the weight $P = -mg$ was considered, equal to the sign of the derivative with respect to x of the potential energy $U = mgx$. Taking into account the representation of the potential energy gradient, we bring the above equality to the form

$$\frac{d}{dt}\left[m(t)\frac{d\dot{\mathbf{x}}}{dt}\right] = \mathbf{F}(t).$$

As a result, we get the vector form of **Newton second law**, as a consequence of the principle of least action.

Task 16.2 *Euler equations for the vector case.* Consider the minimization problem for the functional

$$S(\mathbf{x}) = \int_{t_1}^{t_2} L(t, \mathbf{x}, \dot{\mathbf{x}})dt$$

on the set of n-th order vector functions taking given values at the ends of the interval under consideration, where L is a known function of its general arguments. Following the procedure described in Section 2, establish the system of Euler equations (16.12). For this, the function $f = f(\sigma)$ is again defined and its derivative at zero is equated to zero. As a result, a vector analogue of equality (16.6) is obtained, from which, by appropriately choosing the vector function h, the desired result is obtained.

> *Movement of a body in space under the action of an arbitrary force*
> *can be described using the principle of least action.*
> *As a consequence, the vector form of Newton's second law is obtained.*

6 Vibrations of string

Let us establish the principle of least action in relation to the process of string oscillation, discussed in Chapter 12. It is described by a function $u = u(x, t)$, which characterizes the deviation of the string from the equilibrium position at the point x at the time t.

The kinetic energy of the string depends from its velocity. We have

$$K = \frac{m(u_t)^2}{2},$$

where m is the mass of string, and u_t is the derivative of the function u with respect to t, i.e., the velocity. As noted earlier, the mass of a homogeneous string of length Δx is determined by the formula $m = \rho \Delta x$, where ρ is its linear density. Thus, the kinetic energy of the selected section of the string is $K = \rho(u_t)^2/2\Delta x$. The resulting formula assumes the invariability of the characteristics of the string along its length. In our case, this formula is correct if we refer it to an arbitrarily small section of length dx. As a result, we bring to the equality $dK = \rho(u_t)^2/2dx$. Then the kinetic energy of the string of length X is equal to the integral

$$K = \int_0^X \frac{\rho(u_t)^2}{2} dx.$$

The potential energy of a section of the flexible string is proportional to its elongation. According to the previously established formula (16.3), a section with a length dx of the string having a profile $u = u(x,t)$ in a deformed state is stretched to the value $ds = \sqrt{1 + (u_x)^2}$, where u_x is the derivative of the function u with respect to the variable x. As in Chapter 12, we restrict ourselves to small vibrations of the string when the derivative u_x is small enough. Then, using the Taylor's formula, we have

$$\sqrt{1 + (u_x)^2} = 1 + \frac{(u_x)^2}{2} + \ldots \approx 1 + \frac{(u_x)^2}{2},$$

i.e., we neglect terms of higher degrees relative to u_x. Thus, the string elongation is $ds - dx = (u_x)^2/2$. As a result, we find the potential energy of the string section $dU = k(u_x)^2/2dx$, where the coefficient k is the string tension. Now we determine the potential energy of the string of length X

$$U = \int_0^X k \frac{(u_x)^2}{2} dx.$$

Following the previously described method, we find the Lagrangian of the system

$$L = K - U = \frac{1}{2} \int_0^X \left[\rho(u_t)^2 + k(u_x)^2 \right] dx.$$

As a result, the action of the system on the time interval from 0 to T is equal to

$$S = S(u) = \frac{1}{2} \int_0^T \int_0^X \left[\rho(u_t)^2 + k(u_x)^2 \right] dx dt.$$

According to the principle of least action, the string moves in such a way that the action of the system during the movement is minimal. Thus, the mathematical model of the string movement turns out to be the problem of minimizing the functional S.

Let us define an analogue of the Euler equation for this problem. We introduce again the function $f(\sigma) = S(u + \sigma h)$, where u is a solution to the problem, h is an arbitrary function, and σ is a number. Then the function f has its minimum at zero, which means that its derivative is zero at this point. Find

$$f(\sigma) = f(0) + \frac{1}{2} \int_0^T \int_0^X \left\{ 2\sigma \left(\rho u_t h_t - k u_x h_x \right) + \sigma^2 \left[\rho(h_t)^2 - k(u_x)^2 \right] \right\} dx dt.$$

Now we get

$$f'(0) = \int_0^T \int_0^X \left(\rho u_t h_t - k u_x h_x \right) dx dt = 0.$$

This equality is true for all functions h. We choose as h an arbitrary function that vanishes at the boundaries of the considered space and time interval. Then after integration by parts in the last equality we have

$$\int_0^T \int_0^X (\rho u_{tt} - k u_{xx}) h \, dx \, dt = 0.$$

Applying the fundamental lemma of the calculus of variations[21], taking into account the arbitrariness of the function h, we get the equality

$$\rho u_{tt} - k u_{xx} = 0.$$

Determining $a^2 = k/\rho$, we obtain the **vibrating string equation**

$$u_{tt} = a^2 u_{xx},$$

see Chapter 12. Thus, the principle of least action allows you to determine the equation that describes the law of movement of the string.

> *String vibrations can be described using the principle of least action.*
> *As a consequence, the vibrating string equation is obtained.*

Direction of further work. In all the previously considered mathematical models, the state of the system was characterized by functions, the arguments of which changed in a continuous manner. However, in practice, discrete systems are often encountered, which are characterized not by functions, but by vectors. Such models will be discussed in the next chapter.

Appendix

In Section 2, when studying the Lagrange problem, the Euler equation was obtained. In a particular case, its first integral can be determined for it. For the movement of a body in space considered in Section 5, this allows us to establish the law of conservation of energy. A similar idea is used to derive the laws of light refraction based on Fermat's principle. The methods of the calculus of variations are also applicable in the case when the values of the desired function at the boundaries of a given region are not fixed in advance, as well as in the presence of some additional restrictions. In particular, the problem of the fastest crossing of a boat from one river bank to another is investigated, as well as the process of pendulum oscillations. The final section is devoted to approximate methods for solving extremum problems[22]. Along the way, the concept of the derivative of a functional is introduced, with the help of which many results of the calculus of variations can be unified.

1 Law of conservation of energy

Note an important special case of the Lagrange problem considered in Section 2. Suppose that the function L under the integral does not depend on the variable t, i.e., $L = L(x, \dot{x})$. Find the derivative

$$\frac{d}{dt}\left(\dot{x}\frac{\partial L}{\partial \dot{x}} - L\right) = \ddot{x}\frac{\partial L}{\partial x} + \dot{x}\frac{d}{dt}\left(\frac{\partial L}{\partial \dot{x}}\right) - \frac{\partial L}{\partial x}\dot{x} - \frac{\partial L}{\partial \dot{x}}\ddot{x} = -\dot{x}\left[\frac{\partial L}{\partial x} - \frac{d}{dt}\left(\frac{\partial L}{\partial \dot{x}}\right)\right].$$

If the function x satisfies the Euler equation (16.7), then the final value is zero. Thus, we get the first order equation

$$\dot{x}\frac{\partial L}{\partial \dot{x}} - L = c, \tag{16.13}$$

where c is an arbitrary constant.

In the theory of differential equations, some function that does not change on any solution of the considered equation or system of equations is called the **first integral** of the system. Thus, the first integral for the Euler equation here is the value on the left-hand side of equality (16.13).

In particular, in the problem of a body falling, the function L is determined by the formula

$$L(t, x, \dot{x}) = \frac{m\dot{x}^2}{2} - Px$$

and does not depend on time. Definite the value

$$\dot{x}\frac{\partial L}{\partial \dot{x}} - L = m\dot{x}^2 - \frac{m\dot{x}^2}{2} + Px = \frac{m\dot{x}^2}{2} + Px.$$

Thus, the first integral for the equation of falling body is the sum of kinetic and potential energy, and equality (16.13) describes the **law of conservation of energy**.

Let us now consider the Lagrange problem in which the state x is a vector, and the integrand L again does not explicitly depend on the variable t. Then equality (16.13) also remains valid if the first term in its left-hand side is interpreted as the dot product of the corresponding vectors[23]. As an example, consider the problem described in Section 5 of the movement of a material point in space under the action of an arbitrary force.

The function L here is determined by the formula

$$L(t, \mathbf{x}, \dot{\mathbf{x}}) = \frac{m|\dot{\mathbf{x}}|^2}{2} - \mathbf{F} \cdot \mathbf{x}.$$

For the applicability of the described method, we assume that the mass of the body and the force acting on it do not change over time. Then we find the first integral of the system

$$\dot{\mathbf{x}} \cdot L_{\dot{\mathbf{x}}} - L = \sum_{i=1}^{n} m\dot{x}_i^2 - \frac{m|\dot{\mathbf{x}}|^2}{2} + \mathbf{F} \cdot \mathbf{x} = \frac{m|\dot{\mathbf{x}}|^2}{2} + \mathbf{F} \cdot \mathbf{x}.$$

Now we obtain

$$\frac{m|\dot{\mathbf{x}}|^2}{2} + \mathbf{F} \cdot \mathbf{x} = c.$$

The first term on the left-hand side represents the kinetic energy of the system, and the second one is the potential energy determined by the given force. Thus, the first integral of the system is again the energy integral, and the last relation gives the **law of conservation of energy**[24].

2 Fermat's principle and light refraction

Consider the process of propagation of light in an inhomogeneous medium related to **geometric optics**[25]. The basic physical law here is **Fermat's principle**, according to which light propagates from one point to another along a path that corresponds to the minimum time to overcome it. Hence, in particular, it follows that in a homogeneous medium, light propagates in a straight line, since the velocity of light in a homogeneous medium is constant.

Let us establish a mathematical model of the process under consideration. Let the light propagate on the plane along some curve $y = y(x)$, and the start and end points are known. The velocity of light is determined by the formula $v = ds/dt$, where the function $s = s(t)$ characterizes the path traveled. As noted earlier, the length of an arc element of a given curve corresponding to a segment dx is $ds = \sqrt{1 + y'(x)^2}\,dx$. Then from the previous equality it follows

$$dt = \frac{\sqrt{1 + y'(x)^2}}{v}\,dx.$$

Suppose that the initial time $t = 0$ corresponds to a point with coordinates (x_1, y_1), and the final time $t = T$ corresponds to a point (x_2, y_2). Then, as a result of integration, we find the time of movement of light along the curve $y = y(x)$ from the initial point to the final one

$$T = T(y) = \int_{x_1}^{x_2} \frac{\sqrt{1 + y'(x)^2}}{v}\,dx. \tag{16.14}$$

According to Fermat's principle, the light moves from one point to another in such a way that the functional (16.14), which describes the time of movement, is the smallest. Thus, the mathematical model of the considered process is the Lagrange problem, which consists in minimizing a given functional with known boundary conditions.

Let us establish the law of light refraction using Fermat's principle. Suppose the point $y = 0$ corresponds to the border of two media, see Figure 16.3. In each of them, the velocity of light is a constant. Thus, we have the following representation of the velocity of light included in formula (16.14): $v = v_1$ if $y > 0$ and $v = v_2$ if $y < 0$.

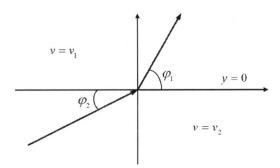

FIGURE 16.3: Light refraction.

The integrand for the considered variational problem does not depend directly on the variable x. Then the Euler equation can be reduced to the first order equation (16.13) $y'L_{y'} - L = c$. In this case, it takes the form

$$\frac{y'^2}{v\sqrt{1 + y'^2}} - \frac{\sqrt{1 + y'^2}}{v} = c.$$

Then we have

$$v\sqrt{1 + y'^2} = c_1, \tag{16.15}$$

where $c_1 = -1/c$. Solving the differential equation (16.15) with the appropriate boundary conditions, one can find the desired law of movement $y = y(x)$. Note that the derivative of the function is equal to the tangent of the trajectory inclination angle, i.e., $y' = \tan\varphi$. Considering the well-known formula

$$\frac{1}{\sqrt{1+\tan^2\varphi}} = \cos\varphi,$$

transform the equality (16.15) to $v/\cos\varphi = c_1$. Now we get the formula

$$\frac{v_1}{\cos\varphi_1} = \frac{v_2}{\cos\varphi_2},$$

connecting the angles of incidence and refraction with the velocity of light. This is the law of *refraction of light*, called *Snell's law*.

3   River crossing problem

Consider the problem of crossing from one river bank to another by a boat in the minimum time. The river is assumed to be X wide and have straight banks. Select the coordinate system such that the x coordinate is directed across the river, and the y coordinate is along the river along one of its banks, see Figure 16.4. The velocity of the river $v = v(x)$ is known. The velocity of the boat exceeds the maximum velocity of the river and is considered constant. It is required to choose such a boat course, i.e., function $y = y(x)$ so that, moving along it, it swam from one bank to another in the minimum time.

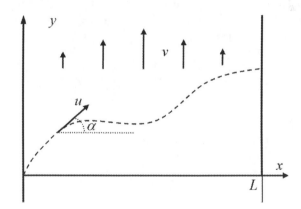

FIGURE 16.4: Movement of the boat.

Choosing the starting point of the boat for the origin of coordinates, we have the boundary condition

$$y(0) = 0. \tag{16.16}$$

The boat velocity in the transverse and longitudinal directions is characterized by the differential equations

$$\frac{dx}{dt} = u\cos\alpha, \quad \frac{dy}{dt} = u\sin\alpha + v, \tag{16.17}$$

where α is the angle determined by the course of the boat. These relations imply the equality

$$y'(x) = \frac{dy}{dx} = \frac{u\sin\alpha + v}{u\cos\alpha}.$$

Hence, after squaring, it follows

$$y' - \frac{v}{u \cos \alpha} = \frac{\sin \alpha}{\cos \alpha}$$

Hence, after squaring, it follows

$$y'^2 - \frac{2vy'}{u \cos \alpha} + \frac{v^2}{u^2 \cos^2 \alpha} = \frac{\sin^2 \alpha}{\cos^2 \alpha} = \frac{1}{\cos^2 \alpha} - 1.$$

Denoting $z = (u \cos \varphi)^{-1}$, we obtain the following square equation with respect to z

$$(v^2 - u^2)z^2 - 2vy'z + (y'^2 + 1) = 0.$$

Now we have

$$z = \frac{1}{u \cos \alpha} = \frac{-vy' \pm \sqrt{v^2 y'^2 - (v^2 - u^2)(y'^2 + 1)}}{v^2 - u^2}.$$

Note that the minus sign here leads to a negative value of the angle α and does not give direction toward the other side of the river.

Using the first of equalities (16.17), we define

$$dt = \frac{1}{u \cos \alpha} dx = \frac{-vy' + \sqrt{v^2 y'^2 - (v^2 - u^2)(y'^2 + 1)}}{v^2 - u^2} dx.$$

During the time from zero to T, choosing a certain route of movement, the boat travels along the x coordinate from zero to the value X. Then, integrating the previous relation, we find the time of the boat's movement

$$T = T(y) = \int_0^X \frac{-vy' + \sqrt{v^2 y'^2 - (v^2 - u^2)(y'^2 + 1)}}{v^2 - u^2} dx.$$

Thus, to find the function y, it is required to minimize the functional T on the set of all functions satisfying the boundary condition (16.16).

Note that in this case the mathematical model of the system does not correspond to the Lagrange problem, since there is only one boundary condition. However, to study the problem, one can nevertheless use the methodology described in Section 2.

Let us minimize the functional

$$T(y) = \int_0^X L[x, y(x), y'(x)] dx$$

on the set of functions $y = y(x)$ satisfying the boundary condition $y(0) = y_1$. Suppose that a function y is a solution to this problem. Then the inequality $T(y) \leq T(y + \sigma h)$ holds for any number σ and any function $h = h(x)$, which vanishes at $x = 0$. We define the function $f(\sigma) = T(y + \sigma h)$. Obviously, it has a minimum at zero, which implies the equality $f'(0) = 0$. Repeating the transformations from Section 2, we find

$$f'(0) = \int_0^X \left[L_y(x, y, y')h + L_{y'}(x, y, y')h' \right] dx.$$

After integration by parts, taking into account the boundary condition for the function h, we obtain

$$\int_0^X \left[L_y(x, y, y') - \frac{d}{dx} L_{y'}(x, y, y') \right] h dx + L_{y'}(x, y, y') h \bigg|_{x=X} = 0. \qquad (16.18)$$

which differs, up to notation, from equality (16.6) only in the presence of the second term on the left-hand side.

The function h is arbitrary here. Choosing it equal to zero for $x = X$, we obtain

$$\int_0^X \left[L_y(x, y, y') - \frac{d}{dx} L_{y'}(x, y, y') \right] h dx = 0.$$

Hence, by the fundamental lemma of the calculus of variations, Euler equation (16.7) follows. However, equality (16.19) remains valid for functions h that are nonzero at the right end of this interval. At the same time, the first term vanishes due to the Euler equation. As a result, we obtain the equality $L_{y'}(x, y, y')h = 0$ for $x = X$. Hence, since h is arbitrary, the formula follows

$$L_{y'}\big((X, y(X), y'(X))\big) = 0$$

called the ***transversality condition***. Thus, the solution of the problem satisfies the second order differential Euler equation with a given boundary condition at the left end of this interval and the transversality condition at its right end.

Return to the river crossing problem. The function L here does not directly depend on y. Then we obtain the Euler equation

$$\frac{d}{dx} \frac{\partial}{\partial y'} \left[\frac{-v(x)y' + \sqrt{v(x)^2 y'^2 - (v(x)^2 - u^2)(y'^2 + 1)}}{v(x)^2 - u^2} \right] = 0.$$

Now we get

$$\frac{1}{u^2 - v^2} \left[\frac{u^2 y'}{\sqrt{u^2(y'^2 + 1) - v^2}} - v \right] = c_1,$$

where c_1 is an arbitrary constant. From the transversality condition, it follows

$$\frac{\partial}{\partial y'} \left[\frac{-vy' + \sqrt{v^2 y'^2 - (v^2 - u^2)(y'^2 + 1)}}{v^2 - u^2} \right] \bigg|_{x=X} = 0.$$

Then the constant c_1 of the previous equality is zero. Thus, we obtain the first order differential equation

$$\frac{u^2 y'}{\sqrt{u^2(y'^2 + 1) - v^2}} - v = 0.$$

After squaring, we find the value $y' = v/u$. Integrating this equation taking into account boundary condition (16.16), we obtain

$$y(x) = \frac{1}{u} \int_0^x v(\xi) d\xi.$$

This equality gives the desired trajectory of the boat.

Task 16.3 *Bolza problem.* Consider the problem of minimizing the above functional of a general form without any boundary conditions, called the ***Bolza problem***. Show that its solution satisfies the Euler equation and two transversality conditions.

4 Pendulum oscillations

In Chapter 3, we considered the process of pendulum oscillation. Try to analyze it based on the principle of least action. Therefore, we consider the movement of the pendulum in the vertical plane, characterized by horizontal and vertical coordinates x and y, which vary with time. In this case, the equilibrium of the pendulum is selected as the origin of coordinates. The kinetic energy of the pendulum is determined by the velocity vector $\mathbf{v}=(\dot{x},\dot{y})$

$$E = \frac{m\mathbf{v}^2}{2} = \frac{m}{2}(\dot{x}^2 + \dot{y}^2),$$

where m is the mass of pendulum. The potential energy of the pendulum is determined by its weight and is equal to $U = -mgy$, where g is the acceleration of gravity, and the minus sign is due to the direction of the vertical coordinate and weight. Thus, the Lagrangian of the system is

$$L = \frac{m}{2}(\dot{x}^2 + \dot{y}^2) + mgy.$$

The action of the system on an interval $[t_1, t_2]$ is

$$S(x,y) = \int\limits_{t_1}^{t_2} L\, dt = \int\limits_{t_1}^{t_2} \left[\frac{m}{2}(\dot{x}^2 + \dot{y}^2) + mgy\right] dt.$$

In this problem, the coordinates of the pendulum are not independent of each other. The pendulum is invariably located at some distance l from the suspension point, as a result of which its coordinates at any time are related by the equality

$$x^2 + y^2 = l^2. \tag{16.19}$$

According to the principle of least action, the pendulum moves from one given point at a time instant t_1 to another given point at a time instant t_2 in such a way that the action of the system is minimized and condition (16.19) is satisfied. So, the problem for a conditional extremum turns out to be a mathematical model of the considered process.

Let us establish a connection between this model and the one that was obtained in Chapter 3. Direct use of the previously described technique is not possible due to the existing limitation (16.19). However, we again define the function $f(\sigma) = S(\mathbf{u}+\sigma\mathbf{h})$, where the pair $\mathbf{u}=(x,y)$ means the solution of the considered problem, σ is an arbitrary number, and the components of the vector $\mathbf{h}=(h_1,h_2)$ vanish on the boundaries of this interval and are such that the corresponding quantity $\mathbf{u}+\sigma\mathbf{h}$ satisfies condition (16.19)[26]. Under these conditions, the function f will again have a minimum at zero. Then we can equate to zero its derivative at zero. As a result, we obtain the equality

$$\int\limits_{t_1}^{t_2} \left[(\dot{x}h_1 + \dot{y}h_2) + gh_2\right] dt = 0.$$

After integration by parts, taking into account the equality to zero of the vector $\mathbf{h}$ at the boundary, we have

$$\int\limits_{t_1}^{t_2} \ddot{x}h_1\, dt + \int\limits_{t_1}^{t_2} (\ddot{y} - g)h_2\, dt = 0. \tag{16.20}$$

In the resulting relation, the components of the vector $\mathbf{h}$ are not arbitrary, since for equality (16.19) to be valid, they must satisfy the condition

$$(x + \sigma h_1)^2 + (y + \sigma h_2)^2 = l^2.$$

Subtracting from this equality (16.19), after dividing σ and passing to the limit as σ tends to zero, we have

$$xh_1 + yh_2 = 0. \tag{16.21}$$

Thus, the components of the vector $\mathbf{h}$ turn out to be related by equality (16.21). We assume that the function h_2 is arbitrary here, and h_1 is determined by this condition. Multiplying the last equality by a function $\lambda = \lambda(t)$, integrating it with respect to t and adding the result with equality (16.20), we obtain

$$\int_{t_1}^{t_2} (\ddot{x} + \lambda x)h_1 dt + \int_{t_1}^{t_2} (\ddot{y} + \lambda y - g)h_2 dt = 0. \tag{16.22}$$

We select the function λ in such a way that the first integral in equality (16.22) vanishes, i.e., the following equality holds

$$\ddot{x} + \lambda x = 0. \tag{16.23}$$

As a result, equality (16.22) takes the form

$$\int_{t_1}^{t_2} (\ddot{y} + \lambda y - g)h_2 dt = 0.$$

Here the function h_2 is arbitrary. Then, using the fundamental lemma of the calculus of variations, we obtain

$$\ddot{y} + \lambda y - g = 0. \tag{16.24}$$

Thus, three unknown functions x, y, and λ are related by differential equations (16.23), (16.24) with the corresponding boundary conditions and algebraic equation (16.19)[27]. We guarantee the validity of condition (16.19) by passing to the polar coordinate system using the equalities $x = l\sin\theta$, $y = l\cos\theta$. Then equalities (16.23) and (16.24) take the form

$$-l\sin\theta\dot{\theta}^2 + l\cos\theta\ddot{\theta} + \lambda l\sin\theta = 0, \quad -l\cos\theta\dot{\theta}^2 - l\sin\theta\ddot{\theta} + \lambda l\cos\theta + g = 0.$$

Multiplying the first of these equalities by $\cos\theta$, and the second by $\sin\theta$ and and differing it, we have

$$\ddot{\theta} + \omega^2 \sin\theta = 0,$$

where $\omega^2 = g/l$. The obtained result is the **pendulum oscillation equation** discussed in Chapter 3. In the case of small oscillations, we can assume that the sine of the angle is close enough to the value of the angle itself, and the equation for the oscillation of the pendulum takes the usual form

$$\ddot{\theta} + \omega^2\theta = 0.$$

Thus, the described mathematical models can be obtained as a consequence of the considered variational principle.

5 Approximate solution of minimization problems

As already noted, the necessary condition for the extremum of the function $f = f(x)$ at a point x is the equality to zero of the derivative of this function at this point $f'(x) = 0$. This is an algebraic equation with respect to the desired point. If the function f is simple enough, then the solution can be found directly. However, for rather complex functions, the solution can be determined only approximately, for example, using the following algorithm

$$x_{k+1} = x_k - \alpha_k f'(x_k), \ k = 0, 1, ..., \tag{16.25}$$

where k is the number of iteration, α_k is a positive parameter. This is called the **gradient method**[28].

The practical implementation of the algorithm (16.25) does not cause much difficulty. However, above we considered mathematical models related to the minimization of functionals. To extend the considered algorithm to general problems, it is sufficient to define the concept of the derivative of the functional. We restrict ourselves to considering the case when the functional is defined on some set (space) V with dot product[29]. In particular, if we are dealing with a function of one variable, then the dot product is an ordinary product. If a function of many variables is considered, then its argument is a vector, and the dot product is characterized by the usual equality

$$\mathbf{x} \cdot \mathbf{y} = \sum_{i=1}^{n} x_i y_i.$$

If we consider a functional that depends on functions defined on an interval $[t_1, t_2]$, then the dot product can be defined as follows

$$x \cdot y = \int_{t_1}^{t_2} x(t)y(t)dt.$$

The **Gâteaux derivative** of a functional S at a point x is an element $S'(x)$ of a space V such that satisfies the equality

$$\lim_{\sigma \to 0} \frac{S(x + \sigma h) - S(x)}{\sigma} = S'(x) \cdot h$$

for all h. The derivative of the functional is that as a result of the indicated passage to the limit it turns out to be scalarly multiplied by an arbitrary h. It is easy to verify that for a function of one variable the Gâteaux derivative coincides with the ordinary derivative, for a function of several variables $f = f(x)$ the Gâteaux derivative is the gradient $\nabla f(x)$, and for the previously considered functional from the Lagrange problem the derivative is

$$S'(x) = L_x(t, x, \dot{x}) - \frac{d}{dt} L_{\dot{x}}(t, x, \dot{x}). \tag{16.26}$$

A natural application of the Gâteaux derivative in extremum theory is the possibility of extending the known results of minimizing functions of one variable to extremal problems of a general form. In particular, it is easy to show that the necessary condition for the minimum of the functional S at some point x (depending on the situation, a point is understood as a number, vector, function) is the **stationary condition**[30] $S'(x) = 0$. For a function of one variable, the last relation is reduced to the usual stationary condition, which is an algebraic equation. For a function of many variables $f = f(x)$, we obtain the equality $\nabla f(x) = 0$, which is a system of algebraic equations. In the case of the functional from the Lagrange problem, to obtain the stationary condition, it is sufficient to equate to zero the expression on the right-hand side of equality (16.26), which results in the Euler equation (16.7), which is an ordinary differential equation. Thus, with the help of differentiation of functionals, it is possible from a unified standpoint to describe a significant number of results of the extremum theory, and also to establish a deep connection between it and the theory of equations[31].

For an approximate solution of the problem of minimizing the functional, one can use the gradient method, which is a natural generalization of the algorithm (16.25)

$$x_{k+1} = x_k - \alpha_k S'(x_k), \ k = 0, 1, \dots.$$

In particular, for the problem of minimizing a function of many variables, we obtain the equality[32]

$$\mathbf{x}_{k+1} = \mathbf{x}_k - \alpha_k \nabla f(\mathbf{x}_k), \ k = 0, 1, \dots .$$

Task 16.4 *Dirichlet integral*. Consider the Dirichlet integral

$$S(u) = \int_\Omega \Big[\sum_{i=1}^n \Big(\frac{\partial u}{\partial x_i} \Big)^2 + 2 f u dx \Big],$$

where n is the dimension of the area Ω, f is a known function. Determine the derivative of this functional and write down the corresponding stationary condition. In the process of finding the derivative, the Green's formula is used, which is a multidimensional analogue of the integration by parts formula.

Notes

[1] About the *calculus of variations*, see [39], [89], [109], [169], [285], [298], [378].

[2] About the principle of least action, see [197], [198].

[3] The desired curve is called **brachistochrone**. Note that if the points A and B lie on the same vertical line, then, obviously, the body will simply fall down. In fact, we get the problem considered in Chapter 1 of the fall of a body under its own weight. The development of the calculus of variations as an independent branch of mathematics is connected with the brachistochrone problem.

[4] Note that we are using the same procedure as for deriving equations of mathematical physics in Part 2. Indeed, first we select some interval of length Δx and establish a certain relation there. Then we note that, since the considered characteristics are in fact variables, the obtained relation is valid only in a section of arbitrarily small length dx. Then, because of integrating this equality, we establish a relation that is valid on the our interval.

[5] We will consider the problems of finding the extremum of functions and functionals in Chapters 17 and 21.

[6] Indeed, if x is a minimum point of the function f, then $f(x) \le f(y)$ for all y. Choosing $y = x + h$, we establish the inequality $f(x + h) - f(x) \ge 0$ for all h. If the number h is positive, then after dividing the last inequality by h and passing to the limit as h tends to zero, we obtain $f'(x) \ge 0$. If the number h is negative, then after dividing the same inequality by h and passing to the limit, we establish the opposite inequality $f'(x) \le 0$. Since h is arbitrary, we conclude that both obtained inequalities hold. This is possible only if $f'(x) = 0$, which is the stationary condition. Naturally, in obtaining it, it is assumed that the given function is differentiable at the considered point.

[7] The resulting equation is not equivalent to the original extremum problem, being a **necessary** but not a **sufficient minimum condition**. In particular, the stationarity condition is satisfied not only by the minimum points, but also by the maximum points of the function (for example, for the function $f(x) = -x^2$), not only by the points of the **absolute**, but also by the **local minimum** (the minimum points are not in the entire domain of definition of the function, but only in the neighborhood of this point (see, for example, for the function $f(x) = 3x^4 - 8x^3 - 6x^2 + 24x$), and even an inflection point (for example, for the function $f(x) = x^3$).

[8] The value y determined in this way is called the **variation** of the function x.

[9] Indeed, for a still unknown but potentially existing solution x and a fixed function h, one can calculate the value of the corresponding integral, which will depend on the number σ.

[10] In the calculus of variations, the resulting limit is called the **variation of the functional**, and in functional analysis, the **derivative of the functional** I at the point x **in the direction** h.

[11] For the fundamental lemma of the calculus of variations, see [109], [169].

[12] It was established earlier that the problem of minimizing a function is reduced to the stationary condition, which is an algebraic equation. Now the problem of minimizing the integral functional has been transformed to the ordinary differential equation. In the process of solving the problem of minimizing

an integral functional that depends on a function of several variables, a partial differential equation is obtained, see Appendix. Thus, we are convinced of a serious connection between the theories of equations and extremum problems.

[13] We have already determined boundary value problems for ordinary differential equations in the study of problems of mathematical physics, see Chapters 10 and 12. Additional properties of such problems, including those for nonlinear equations, are considered in Chapter 19.

[14] Naturally, the Euler equation, like the stationary condition, is only a necessary minimum condition. In the calculus of variations, other extremum conditions are also known, including sufficient see [89], [109], [169], [298].

[15] The minus sign is explained by the fact that the direction of the x coordinate and the action of the gravitational force are opposite.

[16] Indeed, differentiating equality (16.10) with respect to x, we obtain the first equality (16.9).

[17] Indeed, differentiating equality (16.11) with respect to $\dot{x}$ we obtain the second equality (16.9).

[18] The Lagrangian as the difference between kinetic and potential energy may seem like a very strange characteristic. Indeed, what is the point of subtracting different types of energy? However, let us imagine the movement of a body raised to a height. Initially, the body is at rest, which means it has zero kinetic energy. At the same time, being raised to a certain height, the body has a certain potential energy. Falling, the body acquires a certain velocity, which invariably increases. In the process of falling, the height of the body above the earth's surface decreases, which means that its potential energy also decreases. Thus, a decrease in potential energy is accompanied by an increase in kinetic energy. A similar situation was observed during the movement of the pendulum, see Chapter 4. If the pendulum is deflected from equilibrium and released, then the initial moment of movement it has some potential energy, and the kinetic energy is equal to zero. As the pendulum approaches equilibrium, its potential energy decreases, and the kinetic energy increases. At the moment of reaching the equilibrium position, the potential energy of the pendulum is zero, and the kinetic energy is maximum. Continuing its movement by inertia, the pendulum deviates from equilibrium, acquiring some potential energy. In turn, its kinetic energy decreases. The maximum potential energy is reached at the moment of reaching the minimum kinetic energy, i.e., when the pendulum stops deviating from equilibrium. Thus, the behavior of kinetic and potential energies turns out to be opposite, which means that kinetic energy and potential energy, taken with the opposite sign, behave in the same way. Consequently, the Lagrangian of the system, i.e., the difference between kinetic and potential energies, as well as its integral, i.e., the action of the system do have some physical meaning. The action of the system is taken as a measure of the movement of the considered body, and the principle of least action is a fundamental law of nature and a mathematical model of a wide class of physical phenomena.

[19] For example, Chapter 1 considered at the movement of a missile, the mass of which changed due to the combustion of fuel.

[20] Along with the principle of least action, mechanics also uses the principle of possible work, the ***principle of least coercion***, the ***principle of least curvature***, etc. The variational principles of mechanics establish properties that make it possible to distinguish the true movement of an object from its other possible states. In addition to classical mechanics, the variational principle is also used in electrodynamics, optics, quantum mechanics, the theory of relativity, etc. Variational principles of mechanics can be found, for example, in [47], [110], [198], [197].

[21] Naturally, we use here the two-dimensional form of the fundamental lemma of the calculus of variations.

[22] Approximate solution of extremum problems, see [102], [112], [328].

[23] To justify this property, it is sufficient to perform the same actions as in the scalar case.

[24] In physics, the laws of conservation of momentum, angular momentum, mass, charge, etc. are also considered. In accordance with ***Noether's theorem***, they are associated with various symmetries of physical systems, see [109], [254].

[25] About ***geometric optics***, see [163].

[26] At this stage, it does not matter how the specified condition can be met.

[27] The method used for solving problems for conditional extremum is called the Lagrange multiplier method, see [39], [89], [109], [169], [298], [378].

[28] Indeed, the sequence of parameters $\{\alpha_k\}$ is chosen to converge with a limit α. Suppose the sequence $\{x_k\}$ has a limit x. Then, passing to the limit in equality (16.25), we obtain $x = x - \alpha f'(x)$. Therefore, x is a solution of the stationary condition.

[29] The sets with dot product are called ***unitary spaces***. With some additional restriction, they are ***Hilbert spaces***, which are one of the central concepts of functional analysis and have numerous applications, see [153], [185], [283]. The above definition of the derivative of a functional can be extended to a wider class of topological vector spaces. Derivatives of general operators are defined in a similar way, i.e., transformations

connecting two sets of arbitrary nature, [153], [185]. The book [313] is devoted to the application of various forms of operator derivatives in extremum theory.

[30] It is easy to verify that if the functional is ***convex***, i.e., for any arguments x and y and any number σ from the interval [0,1], the following inequality holds $S(\sigma x + (1 - \sigma)y) \leq \sigma S(x) + (1 - \sigma)S(y)$, then the stationary condition is necessary and a sufficient minimum condition, i.e., any solution to it minimizes the functional.

[31] ***Variational inequalities*** are a natural generalization of the stationary condition to the case of functional minimization problems on a convex subset of the space under consideration, see [215]. Variational inequalities by themselves can serve as mathematical models of physical processes, [81], [113], [177]. An example is the problem of the distribution of fluid pressure in a volume bounded by a thin semi-permeable membrane.

[32] This explains the use of the term "gradient method".

Chapter 17

Discrete models

In all the preceding cases, we considered continuous mathematical models. The state functions depended there on one or several arguments varying continuously in a certain region. However, in practice, situations often arise with state of the system that changes discretely. In this case, the independent variables take on a finite or countable set of values. Maybe the independent variable n can take, for example, the values 1,2, ..., N, and the state function is a **vector** $\mathbf{x} = (x_1, x_2, ..., x_N)$. Maybe the value of the independent variable n can be, for example, any natural number, and the state function turns out to be a **sequence** $\{x_n\}$. Sometimes, we can have many independent variables. In particular, in the case of two discrete variables, the state of the system is described by a **matrix** X with elements x_{ij}. There can also be many state functions. This chapter discuss discrete models related to various subject areas[1].

Section 1 discusses the simplest discrete models of population dynamics. We consider analogues of the Malthus and the Verhulst models discussed in Chapter 7 and represent some kind of recurrence relations. Within certain limits, they retain the properties of their continuous analogues, but for some values of the system parameters, they have significantly different properties. In Section 2, we consider a discrete model of heat transfer in a one-dimensional body in a steady state.

In contrast to these problems, the following sections are associated with the mathematical description of systems that do not have natural continuous analogues. They relate to **operations research**[2]. Section 3 considers the transport problem, which consists in the rational distribution of products from points of production to points of consumption. Section 4 describes the traveling salesman problem, which consists in choosing the optimal route for a certain number of cities. Finally, Section 5 presents the prisoner's dilemma related to game theory. The theoretical basis for such problems is the methods of **discrete mathematics**[3].

The Appendix discusses a discrete model of epidemiology and provides additional information on the above objectives.

Lecture

1 Discrete population dynamics models

Chapter 7 discussed population dynamics problems in which the independent variable, i.e., time changed in a continuous manner. However, in some cases, models make sense in which the time changes discretely[4]. This is to some extent due to the fact that the population itself is a discrete object, i.e., the number of species is an integer value. In addition, generation-to-generation transitions are discrete. Finally, the development of the system is monitored at fixed points in time, i.e., discrete.

Consider the simplest problem of population dynamics when there is a single biological species. Its state is characterized by a number x_n at time n. The unit of time can be chosen so that n takes the values 0,1,2, etc. The main parameter of the system is the growth of the species k, which characterizes the change between the number of born and died individuals in one-time step per one individual. Then kx_n is the change in the size of the population

DOI: 10.1201/9781003035602-17

TABLE 17.1:   Discrete and continuous Malthus models.

Characteristic model	**Discrete model**	**Continuous model**
model	$x_{n+1} - x_n = kx_n,\ x_0 = a$	$\dot{x}(t) = kx(t),\ x(0) = a$
independent variable	discrete time n	continuous time t
state function	sequence $\{x_n\}$	function $x = x(t)$
$k > 0$	unlimited population growth	unlimited population growth
$k = 0$	stationary sequence	equilibrium state
$-1 < k < 0$	convergence of sequence population extinction	tending to equilibrium population extinction
$k \leq -1$	model is not applicable	population extinction

consisting of x_n individuals in one-time step. Thus, we get the equality

$$x_{n+1} - x_n = kx_n, \quad n = 0, 1, \dots . \tag{17.1}$$

Add the initial condition

$$x_0 = a, \tag{17.2}$$

where the initial number a is known.

The formula (17.1) is the ***difference equation*** and is a discrete analogue of a differential equation $\dot{x} = kx$. Supplemented by the initial condition (17.2), it turns out to be a discrete analogue of the corresponding Cauchy problem. The continuous system is described by the function $x = x(t)$. Now we have a set of all possible values $\{x_n\}$, that is the ***sequence***. Since the indicated differential equation corresponds to the Malthus model considered in Chapter 7, problem (17.1), (17.2) can be called the ***discrete Malthus model***.

Like its continuous analogue, this problem has an explicit solution

$$x_n = (k+1)x_{n-1} = (k+1)^2 x_{n-2} = \dots = (k+1)^n x_0 = (k+1)^n a.$$

If in the continuous case we are interested in the behavior of the function $x = x(t)$ with an unbounded increase in the argument t, then in the discrete case we study the behavior of the sequence $\{x_n\}$ with an unbounded increase in the number n, which corresponds to the problem of ***sequence convergence***[5].

Obviously, for $k > 0$, the values of x_n increase indefinitely, which corresponds to an unlimited growth of the population, regardless of the initial state of the system. In the case $k = 0$, the values of x_n do not change, i.e., the system is in equilibrium. For $-1 < k < 0$, the sequence $\{x_n\}$ has the number 0 as its limit; the population is dying out. All these results correspond to similar properties of the continuous Malthus model. However, for $k = -1$, the solution to problem (17.1), (17.2) vanishes in one-time step, and for $k < -1$ it can already take negative values, which makes no sense at all. Thus, the discrete Malthus model makes sense only for $k > -1$, and the properties of the continuous and discrete models do not fully coincide. Comparative analysis of discrete and continuous Malthus models is given in Table 17.1.

Like the continuous case, it is natural, after analyzing the model with constant growth, which allows for unlimited increasing of population, to turn to a model in which the value of k is a decreasing function of the system state. Thus, it is assumed that with increasing of population, its growth decreases due to the limited availability of food and other vital resources. We get

$$x_{n+1} - x_n = (\varepsilon - \gamma x_n)x_n, \quad n = 0, 1, \dots$$

Here, the parameter ε characterizes the natural growth of the species, and γ is a decrease in growth due to the limited life resources. Setting $y_n = \gamma r^{-1} x_n$, we have the difference equation

$$y_{n+1} = r(1 - y_n)y_n, \quad n = 0, 1, \dots, \tag{17.3}$$

where $r = 1 + \varepsilon$. Formula (17.3) is called the **discrete Verhulst equation** or the **logistic equation**. It is supplemented by an initial condition (17.2).

Determine properties of this model. Convergence of the sequence $\{y_n\}$ to a limit y, that is, when the system tends to an equilibrium state, it means that the point y is a solution to the algebraic equation $y = f(y)$, where $f(y) = r(1 - y)y$. This equation has two solutions $y = 0$ and $y = (r - 1)/r$. By the nonnegativity of the species number, we conclude that for $r < 1$, only a trivial equilibrium position is possible, corresponding to the extinction of the population. This result is quite natural, since the inequality $r < 1$ corresponds to negative natural growth, i.e., the predominance of mortality over births.

For $r > 1$, a nontrivial equilibrium position is possible $y = (r - 1)/r$. However, the question arises whether it is sustainable, i.e., is the sequence $\{y_n\}$ converge to it? Obviously, convergence is realized if the value $|f'(y)|$ is less than one, i.e., the function f changes rather slowly near the equilibrium position, see Figure 17.1.

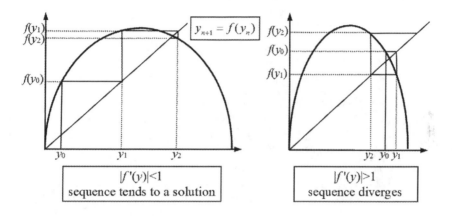

FIGURE 17.1: Sequence $\{y_n\}$.

Find the derivative $f'(y) = r(1 - 2y)$. Taking into account the equality $y = (r - 1)/r$, we conclude that the condition $|f'(y)| < 1$ is fulfilled for $1 < r < 3$. Thus, in the given range of variation of the parameter r, the sequence determined from equalities (17.3) converges. This means the tendency of the number of a species to a position of equilibrium at which the available vital resources are exactly enough for the viability of the given population. This behavior of the system is fully consistent with the properties of the continuous Verhulst equation, see Chapter 7.

However, for values $r > 3$, this equilibrium position is not stable, as a result of which the behavior of the system turns out to be not obvious. Let us return to considering the equality $y = f(y)$. Substituting $f(y)$, in its right-hand side instead of y, we obtain $y = f(f(y))$. Thus, the equality $y = r^2 y(1 - y)[1 - r(1 - y)y]$ is true. The obtained algebraic equation of the fourth order has four solutions. It can be verified that for values $3 < r < 1 + \sqrt{6}$ two such solutions turn out to be stable, and the values of the sequence $\{y_n\}$, that is, the population size will fluctuate endlessly between these values. A further increase in the parameter r leads to the need to consider the equations $y = f(f(f(y)))$, $y = f(f(f(f(y))))$, etc. In this case, the population size will fluctuate around four, eight, etc. values. However, starting from a value[6] of r, oscillations no longer occur, and the system is **chaotic**. A small change in the initial conditions leads here to significant changes in the behavior of the system[7].

Task 17.1 *Discrete Verhulst model*. Perform calculations by the formula (17.3) for r values equal to 0.5, 2, 3.3, 3.5, 3.8.

Population dynamics processes are described by difference equations.
Properties of the discrete and continuous Malthus models are similar,
except for the case when the discrete model is not applicable.
Properties of the discrete and continuous Verhulst models
are similar for small values of the growth.
With its increasing, the number of equilibrium positions
doubles and the system behaves chaotic.

2 Discrete heat transfer model

In Chapter 13, we considered stationary heat transfer in a sufficiently long and thin body in the presence of heat sources. Their action is characterized by the density of heat sources $F = F(x)$, where x is a spatial coordinate directed along the body. Then, in an interval $[x, x + \Delta x]$, the quantity of heat is

$$Q_1 = \int\limits_{x}^{x+\Delta x} F(x)dx.$$

Like Chapter 13, we choose the direction of the heat flux coinciding with the direction of the x coordinate. Then the change in the quantity of heat in the selected area is the difference between the heat flux entering the point x and the heat flux leaving the point Δx, which corresponds to the equality $Q_2 = q(x) - q(x + \Delta x)$, where $q(x)$ is the heat flux, corresponding in this case to the amount of heat passing through point x. We again assume that all thermophysical processes are exhausted by the presence of heat fluxes and the action of heat sources, which corresponds to the equality $Q_1 + Q_2 = 0$. Then, after dividing by Δx, we obtain

$$\frac{q(x) - q(x + \Delta x)}{\Delta x} + \frac{1}{\Delta x}\int\limits_{x}^{x+\Delta x} F(x)dx = 0. \tag{17.4}$$

Obviously, the heat flux from one point to another is directly proportional to the temperature difference and inversely proportional to the distance between the points. Taking into account the chosen direction of the heat flow, we conclude that to calculate the heat flux at the point x, it is necessary to consider the temperature difference at this point and some previous point, for example, $x - \Delta x$. As a result, we get the formula

$$q(x) = -\lambda(x)\frac{u(x) - u(x - \Delta x)}{\Delta x}, \tag{17.5}$$

where u is a body temperature, and λ is a coefficient of thermal conductivity.

In Chapter 13, to derive a mathematical model in relations (17.4), (17.5), the passage to the limit was carried out. As a result, a differential equation was obtained for the unknown temperature u. However, it is possible to do without the passage to the limit. Let us divide the interval $[0, L]$, where L is the body length, into M equal parts[8] with a step $h = L/M$. Then in relations (17.4), (17.5), as x, we can choose the values $x_i = ih$, $i = 1, ..., M - 1$.

Denote

$$u_i = u(x_i),\ q_i = q(x_i),\ \lambda_i = \lambda(x_i),\ f_i = -\frac{1}{h}\int\limits_{x_i}^{x_{i+1}} F(x)dx.$$

Then formulas (17.4), (17.5) take the form

$$\frac{q_{i+1} - q_i}{h} = -f_i, \quad q_i = -\lambda_i \frac{u_i - u_{i-1}}{h}.$$

Denoting[9]

$$\delta_x q_i = \frac{q_{i+1} - q_i}{h} = -f_i, \quad \delta_{\bar{x}} u_i = \frac{u_i - u_{i-1}}{h},$$

we get

$$\delta_x(\lambda_i \delta_{\bar{x}} u_i) = f_i, \quad i = 1, ..., M - 1. \tag{17.6}$$

Equalities (17.6) can be interpreted as a system of equations for the unknown values of u_i. Obviously, we have here $M - 1$ equation for $M + 1$ unknowns $u_0, u_1,...,u_M$. The two missing conditions are obtained from the boundary conditions. The system of difference equations (17.6) with boundary conditions constitutes a **discrete model of stationary heat transfer**[10]. In contrast to the discrete biological models considered above, in this case the independent variable (argument) is selected from a finite set, and the state of the system is characterized by a vector with components u_i.

We have already considered such a problem in Chapter 12 when using an implicit finite difference scheme to solve the heat equation[11]. However, in this case, the difference equations (17.6) have a direct physical meaning, being the basis of the mathematical model of the considered heat transfer process.

Task 17.2 *Discrete heat transfer model*. Calculate difference equations (17.6) by the tridiagonal matrix algorithm described in Chapter 12. Suppose the presence of a known heat flow on the left boundary of the body and the condition of heat insulation on its right boundary.

> *Stationary heat transfer is described by a discrete model,*
> *which is a system of difference equations with boundary conditions.*

3 Transportation problem

A qualitatively different class of mathematical models is associated with the transportation problem[12]. Consider points of production and points of consumption of any product. Each point of production is characterized by its own volume of production, and each point of consumption is characterized by its need for a given product. In this case, it is assumed that the total volumes of production and consumption coincide. The cost of transporting a unit of production from each point of production to each point of consumption is known. It is required to distribute the output of all points of production between points of consumption in such a way that the total transport costs are minimal.

Determine a mathematical model of the considered system. Let us denote by m the number of points of production, and by n the number of points of consumption. The available volume of production of the i-th producer is denoted by a_i, $i = 1, ..., m$, and the required volume of consumption by the j-th consumer is denoted by b_j, $j = 1, ..., n$. The equality of the total volumes of production and consumption corresponds to the equality

$$\sum_{i=1}^{m} a_i = \sum_{j=1}^{n} b_j.$$

Suppose x_{ij} units of production are sent from the i-th producer to the j-th consumer.

Then, in order to dilute all the products from each point of production, the fulfillment of the consumption condition is required corresponds to

$$\sum_{j=1}^{n} x_{ij} = a_i, \ i = 1, ..., m. \tag{17.7}$$

In order for each consumer to receive the required number of products, the equality

$$\sum_{i=1}^{m} x_{ij} = b_i, \ j = 1, ..., n. \tag{17.8}$$

Obviously, adding up all equalities (17.7) and all equalities (17.8), we get the same result, which corresponds to the above equality of the total volumes of production and consumption.

The cost of transporting a unit of production from the i-th point of production to the j-th point of consumption is denoted by c_{ij}. Then the total cost of transporting all products will be equal to

$$S = \sum_{i=1}^{m} \sum_{j=1}^{n} c_{ij} x_{ij}. \tag{17.9}$$

Thus, the problem is to find such values of x_{ij}, satisfying conditions (17.7), (17.8), which provides a minimum for S. This is the **transportation problem**. In this case, the independent variables are the numbers of production and consumption points with a finite number of values $i = 1, ..., m$, $j = 1, ..., n$, and the state of the system is characterized by all possible values x_{ij}, that form the corresponding **matrix** X.

Similar to the models considered in the previous sections, the state of the system here is characterized by a finite number of values, as a result of which we refer this formulation of the problem to discrete models. At the same time, it is related to the problems discussed in Chapter 16, the need to find the extremum of the corresponding value. An example of solving a transportation problem is given in the Appendix.

Transportation problem is the distribution
of product points between consumption points.
This is a minimization of linear function, related by linear equalities.

4   Traveling salesman problem

Let us now turn to the well-known traveling salesman problem[13]. There are n cities. The traveling salesman must go around all these cities at least once and return to the original city. It is required to choose the optimal route of movement. The distance traveled, the cost of movement, the time it takes to complete the route, etc. can be used as an optimality criterion.

The mathematical model is based on the representation of the system in the form of a **graph**[14], which is a set of **vertices**, i.e., cities $i = 1, 2, ..., n$, which the traveling salesman must visit, and the set of **paths** (i, j) from the i-th vertex to the j-th. Each path (i, j), called an **edge** of the graph, corresponds to some criterion of profitability, i.e., weight c_{ij}, which is some non-negative number. This can be the distance between cities, route time, trip cost, etc. An example of a graph for the traveling salesman problem is shown in Figure 17.2. Thus, the solution to the traveling salesman problem is to choose the optimal route for all the vertices of this graph. The optimality criterion here is the total weight of all paths of the selected route, which must be minimized.

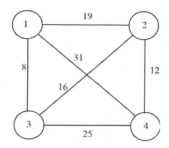

FIGURE 17.2: Graph of a symmetric traveling salesman problem.

If the distance between cities is chosen as the criterion of profitability, then the condition $c_{ij} = c_{ji}$ for all values of $i, j = 1, 2, ..., n$ is satisfied. In this case, the corresponding graph is called **undirected**, and the traveling salesman problem is called **symmetric** (see, for example, Figure 17.2). However, in a number of cases, the **asymmetric traveling salesman problem** corresponding to a **directed graph** also makes sense. This is possible when some cities are connected by one-way roads or when there are differences in the cost of traveling the route in the forward and backward directions.

The existing criteria of profitability can be characterized by a matrix with elements c_{ij}, which turns out to be symmetric or asymmetric, depending on the corresponding properties of the considered problem. In particular, for the problem characterized by Figure 17.2, the corresponding matrix is

$$C = \begin{pmatrix} 0 & 19 & 8 & 31 \\ 19 & 0 & 16 & 12 \\ 8 & 16 & 0 & 25 \\ 31 & 12 & 25 & 0 \end{pmatrix}.$$

Most often, the case is considered when the route must pass through each city exactly once. The corresponding route on the graph is called a **Hamiltonian cycle**. Without loss of generality, we can assume that each of the vertices is connected with all other vertices, which corresponds to the concept of a **complete graph**. Indeed, if a pair of vertices turns out to be disconnected, then a sufficiently large number can be artificially chosen as the corresponding criterion of profitability. As a result, such a route will certainly not be included in the optimal travel route for a traveling salesman. Thus, the **traveling salesman problem** is reduced to finding the Hamiltonian cycle of the complete graph of the least weight.

The traveling salesman problem can also be formulated analytically. In this case, the state of the system, i.e., the chosen route is characterized by a matrix X with components x_{ij} equal to 1 if the route includes a transition from the i-th city to the j-th city (the edge of the graph), and 0 otherwise. In this case, the criterion for the optimality of the route is the sum

$$\sum_{i=1}^{m} \sum_{j=1}^{n} c_{ij} x_{ij}.$$

However, it is necessary to impose constraints on the matrix components under which the selected system of edges composes a route corresponding to the statement of the traveling salesman problem.

First of all, we note that each city on the route is visited only once, i.e., the corresponding vertex of the graph must have a single incoming edge and a single outgoing edge. This corresponds to the fulfillment of the equalities

$$\sum_{i=1}^{n} x_{ij} = 1, \; j = 1, ..., n; \quad \sum_{j=1}^{n} x_{ij} = 1, \; i = 1, ..., n.$$

Note that if these relations are satisfied, the matrix X may not correspond to the route of the cities according to the conditions of the traveling salesman problem. In particular, the entire system of cities can disintegrate into separate cycles, see Figure 17.3. To avoid the occurrence of cycles, additional conditions are imposed on the components of the matrix X in the form of inequalities[15].

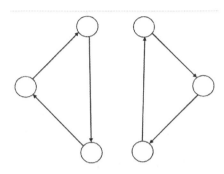

FIGURE 17.3: Graph with cycles not forming a route.

The traveling salesman problem has other interpretations. Let, in particular, there is a machine for processing n different parts. To reconfigure the machine from the i-th part to the j-th part, it takes time c_{ij}. The challenge is to select such a sequence of machining of all parts in order to minimize the total operating time of the machine. Note that the immediate processing time of parts can be ignored here, since it remains unchanged for any processing sequence, since all parts must be processed. The condition for the return of the traveling salesman to the original city in this case corresponds to the requirement to return the machine at the end of work to its original state for processing the next batch of parts.

The traveling salesman problem, in principle, can be solved by a simple enumeration of all possible routes. At the same time, being in each city, the traveling salesman can choose any of the cities that he has not yet visited. Thus, for the passage of n cities in the asymmetric traveling salesman problem, there exist $(n-1)!$ routes, i.e., the number of possible options depends factorially on the number of cities. This circumstance makes it practically impossible to solve the problem by brute force for a sufficiently large number of cities[16]. As a result, special algorithms are usually used to solve the problem[17].

> *Traveling salesman problem is to choose the optimal route*
> *for passing a certain number of points once.*
> *Traveling salesman problem is solved by means of graph theory.*

5   Prisoner's dilemma

In the previous two sections, the mathematical models of the system assumed the choice of one of the possible scenarios for the development of events, proceeding from some criterion of the optimality of this choice. However, in practice, situations often arise when the choice is made by two or more independent parties, each of which has its own preference. In this case, a specific scenario is the result of the choice of all parties involved. The problems arising in this case are solved on the basis of **game theory**[18]. Consider a fairly simple game problem called the **prisoner's dilemma**.

Two criminals were arrested for similar petty crimes. The police suspect them of more serious crimes, which they may have committed together. However, it has no direct evidence. In this regard, the police isolated the criminals from each other and offered each of them a deal with the investigation. If both criminals do not go through the deal, then they will be sentenced to 1 year in prison for a minor crime. If one of them testifies against the other, and the second is silent, then the first will be released in accordance with the deal, and the second will be sentenced to 10 years. Finally, if both agree to a deal, then their punishment will be mitigated and sentenced to 3 years instead of the prescribed 10.

Each of the criminals does not know which choice the other has made. Any of them can reason like this. Suppose that the second one was silent. If I also keep silent, I will be sentenced to a year. However, if I go to the deal, then I will be released. This means that in the case when the second criminal refused the deal, it is more profitable for me to testify. What will happen if he testified against me? If I keep silent, I will be sentenced to 10 years, and if I go to the deal, then only 3 years. Therefore, in both cases it is more profitable to make a deal. Moreover, since the second criminal thinks in the same way, the criminals testify against each other and receive three years.

However, other behavior of criminals is also possible. Each of them knows that if they both keep silent; they will receive only one year. Therefore, it may make sense to take a chance in the hope that the second criminal will do the same.

Thus, different logic of action is possible in a game problem. In the first case, the best possible option is guaranteed, regardless of the action of the other player. In the second, the absolutely best option is chosen, but it can only be achieved if the second player is selected accordingly.

Let us now give a mathematical model of the considered system. There are two players, each of whom has a certain set of **strategies**, which in our case consists of two options. Let us denote by H_1 and H_2 the sets of strategies of the first and second players. The **utility functions** of both players S_1 and S_2 are set, which assign to any pair of strategies x_1 and x_2 of the players, called a **situation**, some numbers $S_1(x_1, x_2)$ and $S_2(x_1, x_2)$. In particular, in the considered example, strategy 1 of each of the players (criminals) consists in refusing a deal with the consequence, and strategy 2 in their consent to the deal. We have the equalities $S_1(1, 1) = S_2(1, 1) = 1$, $S_1(2, 2) = S_2(2, 2) = 3$, $S_1(1, 2) = S_2(2, 1) = 10$, $S_2(1, 2) = S_1(2, 1) = 0$.

The question arises, which situation should be considered optimal? As you can see from the considered example, everything depends on what should be understood as optimality. In the first case, such was the situation (2.2), which has the following properties

$$S_1(2, i) \leq S_1(1, i), \ S_2(i, 2) \leq S_2(i, 2), \ i = 1, 2.$$

Here it is unprofitable for each of the players to change his strategy if the other player does not change his strategy. This situation is called the **Nash equilibrium**.

The second behavior of the criminals corresponds to the case when they both refuse the deal. Moreover, there is no other situation that would be preferable for both players at the same time, i.e., such a situation (j, k) so that the inequalities $S_i(j, k) < S_i(1, 1)$, $i = 1, 2$ are satisfied. This situation is called **Pareto optimal**. Here, an improvement in the position of one of the players is inevitably accompanied by a deterioration in the position of the other player.

The prisoner's dilemma shows that the quite natural concepts of Nash equilibrium and Pareto optimality can contradict each other. Like other classes of models, the considered system can have different practical interpretations. For example, suppose there are two warring countries that are currently in a state of truce. Each country is developing its own long-term development strategy. It can start an arms race, or it can use the funds it has for peaceful development. What choice another country will make remains unknown. In this regard, each country thinks in this way. If my enemy is arming, and we do not, he will attack us and destroy us. Therefore, we need to arm ourselves. If the enemy does not arm himself, and we do the same, then there will be no war. At the same time, we will raise our economy, which is certainly good. But if in this case we manage to arm ourselves, then over time we will be able to attack the enemy and finally defeat him, which is clearly preferable. Therefore, in both cases it is more profitable to arm. Thus, the Nash equilibrium in this case corresponds to the arms race. Of course, it would be better if both countries develop peacefully, corresponding to Pareto optimality. However, this situation is possible only with the right choice of strategies of both countries at the same time.

A close example. There are two warring factions that have agreed to exchange prisoners. According to the agreement, they bring the prisoners to some secluded place, where the exchange is carried out. Each of the groups can either act in accordance with the accepted agreement, or bring armed soldiers under the guise of prisoners. If both groups bring prisoners, then an exchange is made, which is good for both sides. If both groups bring soldiers, then no one wins, but neither loses anything. If one of the groups brings prisoners, and the second brings soldiers, then the soldiers fight off their prisoners. As a result, the first group does not free its prisoners and is left without foreign prisoners, i.e., obviously loses. At the same time, the second wins, since it frees its prisoners, leaving behind the prisoners of the foreign side. If both groups, not trusting each other, strive not to lose in any case, then they choose the Nash equilibrium. Each of them brings soldiers to the meeting so that the exchange of prisoners is not implemented. Pareto optimality in which prisoners are exchanged is beneficial to both parties. However, this situation is not sustainable, as the violator gains a unilateral advantage.

The next example is economical. Two people entered into a barter deal. According to the agreement, they exchange bags containing the packaged goods that are the subject of the transaction. Each of them can stick to the agreement or break the deal by putting worthless cargo in the bag. Everyone's reasoning is quite natural. If that second person gave me the goods, and I do the same, then we both get what we want, which is good. But if at the same time I deceive him, then I will get an even greater gain. If he deceived me, then I am left with nothing, and it is more profitable for me to deceive him as well. At least that way I won't lose anything. Thus, the Nash equilibrium corresponds to the case when they both deceive each other. The Pareto optimality corresponding to the implementation of the transaction would be more profitable for them. But at the same time, there is still a risk of giving up the goods without receiving anything in return.

Another economic example. The two firms manufacture the same product and can charge either high prices or low prices. If both firms set low prices, then each of the firms earns a certain profit. If they both set high prices, then the profit will be significantly higher. However, if one firm sets high prices, and the second proposed low prices, then the first firm does not buy goods at all, and the second gets the maximum profit by selling all its goods. Here, the Nash equilibrium corresponds to low prices assigned by both firms, at which each of them receives a guaranteed profit regardless of the actions of the competitor. However, firms can enter into cartels by charging high prices, which corresponds to the Nash optimality. However, the risk remains that the other firm, by lowering prices, will ruin a competitor.

A comparative description of the different interpretations of the prisoner's dilemma is given in Table 17.2. General definitions of Nash equilibrium and Pareto optimality are given in the Appendix.

TABLE 17.2: Interpretations of the prisoner's dilemma.

Players	Game	Nash equilibrium	Pareto optimality
prisoners	deal with police	consent to a deal	refusal to testify
warring countries	development strategy	arms race	peaceful development
warring factions	prisoner exchange	violation of agreement	exchange execution
merchants	barter transaction	breach of agreement	fulfillment of an agreement
competing firms	price assignment	low prices	cartel agreement

Task 17.3 *Competition between two firms with three pricing options.* There are two firms that produce the same product. Each of the firms independently of each other can assign low, medium and high prices, which correspond to the values 1, 2, and 3 of the unit cost. The lower the price level, the higher the demand for the product. In particular, if both firms set low prices, then each of them will sell 12 units of production, if average, then 8, and if high, then 6. If one firm sets low prices, and the other proposed average, then the first will sell 17 units of production, and the second sells 5. If one firm sets average prices, and the other sets high, then the first sells 9 units of production, and the second sells four. If one firm charges low price and the other charges high, then the former sells 20 units, and the latter nothing. Give a mathematical description of this system, as well as find the Nash equilibrium and the Pareto optimal situation.

Prisoner's dilemma relates to game theory.
Criteria for choosing a strategy by players in game theory are
Nash equilibrium and Pareto optimality.

Direction of further work. In the previous chapters, we considered various deterministic mathematical models. However, systems are possible, the state of which depends on some random factors. Such tasks are the subject of the next chapter.

Appendix

The ideas described above in the lecture are developed below. As noted in Chapter 7, mathematical models of epidemiology can be seen as developing models of population dynamics. In this regard, the discrete model of the epidemic development considered below serves as a continuation of the discrete models of population dynamics discussed in Section 1. Then, using a particular example, the solution of the transport problem posed in Section 3 is described. The production planning model, also related to linear programming, is adjacent to the transport problem. Finally, the final section provides some of the game theory considerations associated with the prisoner's dilemma described in Section 5.

1 Discrete model of epidemic propagation

Chapter 7 examined the mathematical model for the propagation of the epidemic. At the time, the entire population was divided into compartments depending on the state of people in relation to a given disease. We consider another model of this phenomenon here.

* This result was obtained with the support of the project AP09260317 *"Development of an intelligent system for assessing the development of* COVID-19 *epidemics and other infections in Kazakhstan"* of al-Farabi Kazakh National University.

Its distinctive feature, which predetermined the choice of a discrete rather than continuous model, is the following circumstance. Any sick person after a certain time, much less than the time at which the process is being investigated, will necessarily either recover or die. In any case, after some time, he leaves for sure the compartment of patients, moving to another group. Similarly, any person who has been in contact with the sick, after some (obviously not very large) number of days, will either get sick or no longer get sick. Thus, he is guaranteed to leave this compartment. In this regard, it seems natural to measure time discretely, directly putting into the model the time spent by a person in a particular compartment.

In this case, the entire population is divided into the following compartments:

S: susceptible (healthy, but potentially ill);

E: exposed (healthy, in contact with ill);

A: asymptomatic (infected, asymptomatic);

I: mildly ill (mild patients undergoing treatment at home);

H: hospitalized (seriously ill, hospitalized);

R: recovered (recovered, who have no signs of illness);

D: died.

When describing the mathematical model, natural assumptions are made[19]. In particular, it is assumed that the population is homogeneous, i.e., any differences in age, sex, health status, etc. are not taken into account. The population is considered isolated, i.e., the arrival and departure of people outside the territory belonging to this population is not taken into account. Natural fertility and mortality are also not taken into account, because of which the total population N, which also includes the deaths as a result of the epidemic, remains unchanged. The population is considered to be large enough that various possible but unlikely situations are not taken into account. It is assumed that at the start of the study, the population is already covered by the epidemic. In addition, mutation of the virus and vaccination of the population are not taken into account.

It is assumed that the susceptible person becomes infected by passing through the exposed group. At the same time, only asymptomatic (to a greater extent) and mildly ill (to a lesser extent) are infected, since hospitalized patients are under control and to a lesser extent are ill. It is also assumed that all who have recovered are immunized, i.e., are not susceptible to disease.

When determining the model, the following intergroup transitions are taken into account, see Figure 17.4:

se: the susceptible can become exposed.

es: the exposed may not get sick and become susceptible again,

ea: the exposed may become asymptomatic,

ei: the exposed can become mildly ill,

eh: the exposed may become hospitalized,

ar: the asymptomatic can recover,

ae: the asymptomatic can become mildly ill,

ih: the mildly ill patient can be hospitalized,

ir: the mildly ill patient can recover,

hr: the hospitalized patient can recover,

hr: the hospitalized patient can die.

By p_{es}, p_{ea}, etc., we denote the proportions of exposed, passing over time into compartments of susceptible, asymptomatic patients, etc. The proportions of patients of different compartments who move to other groups have a similar meaning. All these quantities lie between zero and one, and the obvious equalities hold

$$p_{es} + p_{ea} + p_{ei} + p_{eh} = 1, \quad p_{ar} + p_{ae} = 1, \quad p_{ir} + p_{ih} = 1, \quad p_{hr} + p_{hd} = 1.$$

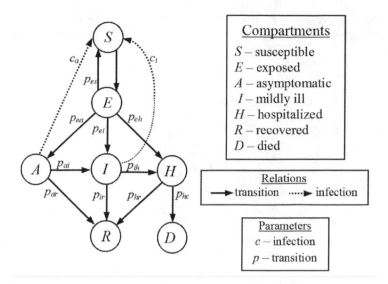

FIGURE 17.4: Relationship between population compartments of the epidemic model.

Thus, any exposed person will either not get sick at all, or get sick in one form or another. Other equalities have a similar meaning.

Each compartement is assigned a function of a discrete argument (time in days), which characterizes the number of people in a given compartment. Thus, under S_k, E_k, etc. means the number of susceptible, exposed, etc. on the k-th day from the start of the analysis[20].

It is assumed that exposed, as well as all compartments of patients are in this group for a limited time, different for each compartment[21]. In particular, under n_e, we mean the number of days during which the exposed either gets sick in one form or another, or probably is not get sick anymore; and n_a is the number of days during which asymptomatic patients will develop the disease, or they will finally recover. The values n_i and n_h have a similar meaning.

Further denote by e_k^j, a_k^j, etc., the number of exposed people on the j-th day since the time of contact, the number of asymptomatic patients on the j-th day since the onset of the disease, etc. at time k. In accordance with the above assumption, the number of contact and various patients on the current day is made up of the number of people in a given compartment who are there on different days of being in this group. Thus, the following equalities hold

$$E_k = \sum_{j=1}^{n_e} e_k^j, \ A_k = \sum_{j=1}^{n_a} a_k^j, \ H_k = \sum_{j=1}^{n_h} h_k^j, \ I_k = \sum_{j=1}^{n_i} i_k^j. \tag{17.10}$$

Naturally, every exposed on the j-th day of being in the exposed compartment in a day goes to the category of exposed of the $(j+1)$-th day of being in this group if it was not the last day of being in the compartment. In the latter case, he leaves this compartment. The same applies to all compartment of patients. Thus, we get the equalities

$$e_{k+1}^{j+1} = e_k^j, \ j = 1, ..., n_e - 1. \tag{17.11}$$

The following equalities have the analogical sense

$$a_{k+1}^{j+1} = a_k^j, \ j = 1, ..., n_a-1; \ h_{k+1}^{j+1} = h_k^j, \ j = 1, ..., n_h-1; \ i_{k+1}^{j+1} = i_k^j, \ j = 1, ..., n_i-1. \tag{17.12}$$

The number of susceptible at the next moment in time is equal to the number of susceptible at the previous moment of time minus new exposed people at this moment of time plus the number of uninfected exposed of the last day at the previous moment in time:

$$S_{k+1} = S_k - e_{k+1}^1 + p_{es}e_k^{n_e}. \tag{17.13}$$

The number of exposed people at the next time is equal to their number at the previous time plus the number of new exposed people at this moment in time minus the number of exposed on the last day at the previous time, i.e., who have left the exposed group at the moment:

$$E_{k+1} = E_k + e_{k+1}^1 - e_k^{n_e}. \tag{17.14}$$

The following equalities have the analogical sense

$$A_{k+1} = A_k + a_{k+1}^1 - a_k^{n_a}, \quad H_{k+1} = H_k + h_{k+1}^1 - h_k^{n_h}, \quad I_{k+1} = I_k + i_{k+1}^1 - i_k^{n_i}. \tag{17.15}$$

Further, the number of recovered people at the next time point is equal to their number at the previous point in time plus all recovered patients of the last day of illness at this time:

$$R_{k+1} = R_k + p_{ar}a_k^{n_a} + p_{ir}i_k^{n_i} + p_{hr}h_k^{n_h}. \tag{17.16}$$

Finally, the number of deaths at the next time point is the sum of the number of deaths at the previous time point and the number of new deaths, i.e., hospitalized ill patients on the last day of illness at this point in time who died:

$$D_{k+1} = D_k + p_{hd}h_k^{n_h}. \tag{17.17}$$

It remains to indicate the formulas for calculating the number of exposed people and various forms of patients on the first day at the next time. In particular, the number of exposed on the first day, i.e., newly contacted at a later point in time is determined by the formula

$$e_{k+1}^1 = (c_a A_k + c_i I_k)S_k/N, \tag{17.18}$$

where c_a, c_i are positive constants, characterizing the contagiousness of asymptomatic and mildly ill. Number of asymptomatic first day, i.e., newly ill at the next time point is equal to the number of those who left the exposed group at the previous time point who fell ill in an asymptomatic form, i.e.,

$$a_{k+1}^1 = p_{ea}e_k^{n_e}. \tag{17.19}$$

The number of mildly sick on the first day, i.e., newly ill at the next time point is equal to the number of those who left the exposed and asymptomatic groups at the previous time point who fell ill in a mild form, i.e.,

$$i_{k+1}^1 = p_{ei}e_k^{n_e} + p_{ai}a_k^{n_a}. \tag{17.20}$$

The following equalities have the analogical sense

$$h_{k+1}^1 = p_{eh}e_k^{n_e} + p_{ih}i_k^{n_i}. \tag{17.21}$$

Formulas (17.10)–(17.21) with the corresponding initial conditions constitute a mathematical model of the considered process and allow one to determine eight functions of the state of the discrete argument k, characterizing the number of people changing over time in all groups of the population[22].

Task 17.4 *Discrete epidemiology model*. Carry out calculations in accordance with formulas (17.10)–(17.21) with some initial conditions. Obtain experimentally the equilibrium position of the system. Interpret the results.

2 Potential method for solving a transportation problem

Let us return to the consideration of the transport problem posed earlier. So, there is m the number of points of production and n is the number of points of consumption. It is required to choose the distribution of x_{ij} units of production sent from the i-th producer to the j-th consumer, assuming that all production volumes and all consumption volumes are known, respectively

$$a_i = \sum_{j=1}^{n} x_{ij}, \ i = 1, ..., m, \ \ b_j = \sum_{i=1}^{m} x_{ij}, \ j = 1, ..., n$$

such that general transport costs

$$S = \sum_{i=1}^{m} \sum_{j=1}^{n} c_{ij} x_{ij}$$

will be minimal, where c_{ij} is the cost of transporting a unit of production from the i-th producer to the j-th consumer.

Let us describe one algorithm for solving a transport problem using one specific example. Suppose there are three producers and five consumers, i.e., $m = 3$, $n = 5$. The production volumes are characterized by vector $\mathbf{a} = (200,150,250)$, the consumption volumes are determined by the vector $\mathbf{b} = (100,80,90,150,180)$, and the cost of transportation are the matrix

$$C = \begin{pmatrix} 10 & 12 & 5 & 6 & 4 \\ 8 & 6 & 12 & 11 & 7 \\ 7 & 8 & 9 & 7 & 10 \end{pmatrix}.$$

Solving the problem begins with drawing up a basic transportation plan, which is determined in accordance with the minimum element method. For this, the minimum of the cost values is selected, which corresponds to the element $c_{15} = 4$. Then the corresponding (first) manufacturer sends to the corresponding (fifth) manufacturer the products he needs, i.e., at this stage of the study, $x_{15} = 180$ is assumed, see Table 17.3. Thus, the needs of the fifth manufacturer are fully satisfied, and the first manufacturer still has 20 units of products.

Further, from the remaining elements of the matrix C, the smallest element is selected, i.e., $c_{13} = 5$. The respective (first) manufacturer sends to the respective (third) manufacturer all the products he has, i.e., 20 units, i.e., we put $x_{13} = 20$. Thus, the first manufacturer has completely distributed its products, and the third consumer has yet to receive 70 product units.

We again choose the smallest of the remaining elements of the matrix C. These are the values of c_{14} and c_{22}, equal to 6. However, the first producer has already distributed his products, so we choose the second producer and the second consumer. Thus, we assume $x_{22} = 80$, which fully satisfies the needs of the second consumer, and the second manufacturer has another 70 units of products.

The smallest of the remaining elements of the matrix C are c_{25}, c_{31}, and c_{34}, which are equal to 7. The first of them does not fit, since the fifth consumer already receives the required volume of production. Any of the two remaining elements can be selected. We define $x_{31} = 100$, thereby completely satisfying the first consumer, and the third manufacturer has 150 product units. Now the minimum of the remaining elements of the matrix C is c_{34}. As a result, we choose $x_{34} = 150$, which satisfies the needs of the fourth consumer and is exactly equal to the remaining volume of the third producer. As a result, 70 units of products remain with the second manufacturer, which meets the needs of the third consumer. Thus, we set $x_{23} = 70$. All other elements of the matrix X are set equal to zero. The corresponding

TABLE 17.3: Baseline plan.

	1	2	3	4	5	a
1	10	12	5	6	4	
			20		180	200
2	8	6	12	11	7	
		80	70			150
3	7	8	9	7	10	
	100			150		250
b	100	80	90	150	180	

TABLE 17.4: Baseline plan.
First iteration

	1	2	3	4	5	a
1	10	12	5	6	4	
			90		110	200
2	8	6	12	11	7	
		80			70	150
3	7	8	9	7	10	
	100			150		250
b	100	80	90	150	180	

baseline plan is shown in Table 17.3, where in each cell corresponding to indices i and j, the top row contains the value of c_{ij}, and the bottom row contains x_{ij} if it is not zero. The costs corresponding to the baseline plan are $S = 5 \cdot 20 + 4 \cdot 180 + 6 \cdot 80 + 12 \cdot 70 + 7 \cdot 100 + 7 \cdot 150 = 3890$.

The rectification of the baseline plan is carried out in accordance with the potential method. Each producer and consumer are assigned a potential value, respectively, u_i and v_j. For each of the six occupied cells, it is assumed that the equalities $u_i + v_j = c_{ij}$ are satisfied. We have six equations for eight unknowns. Therefore, we choose two of them arbitrarily. We set $u_1 = 0$. Then $v_3 = c_{13} - u_1 = 5$, $v_5 = c_{15} - u_1 = 4$, $u_2 = c_{23} - v_3 = 7$, $v_2 = c_{22} - u_2 = -1$. There remain the equalities $u_3 + v_1 = c_{31} = 7$, $u_3 + v_4 = c_{33} = 7$. Choosing $u_3 = 0$, we find $v_1 = v_4 = 7$.

For free cells, we calculate the estimates by the formulas $s_{ij} = c_{ij} - (u_i + v_j)$. We have $s_{11} = 10 - 7 = 3$, $s_{12} = 12 + 1 = 13$, $s_{14} = 6 - 7 = -1$, $s_{21} = 8 - 7 + 1 = 2$, $s_{24} = 11 - 7 - 7 = -3$, $s_{25} = 7 - 7 - 4 = -4$, $s_{32} = 8 + 1 = 9$, $s_{33} = 9 - 5 = 4$, $s_{35} = 10 - 4 = 6$. The smallest value here is $s_{25} = -4$. In this regard, the change in the base plan is carried out by transferring goods to this particular cell. We draw your attention to the cost 7 corresponds to it, while a cell with a significantly higher cost of 12 turns out to be occupied. We move the 70 units of production available there to the cell under consideration, i.e., we set $x_{23} = 0$, $x_{25} = 70$. Then, to satisfy the fifth consumer, the first consumer sends him 70 less units of production, i.e., $x_{15} = 110$. The first producer sends these 70 units to the third consumer, satisfying his needs. As a result, $x_{13} = 90$. is found. The results of the first iteration are presented in Table 17.4. Now we calculate the cost of transportation $S = 5 \cdot 90 + 4 \cdot 110 + 6 \cdot 80 + 7 \cdot 70 + 7 \cdot 100 + 7 \cdot 150 = 3610$, while for the base case the value was 3890.

Using the described technique, we perform the second iteration. For eight potentials of producers and consumers, we again have six equations $u_i + v_j = c_{ij}$ for occupied cells. We choose two values of consumer potentials $v_1 = 0$ and $v_5 = 0$. Then from the available equations we find the values $u_1 = 4$, $u_2 = 7$, $u_3 = 7$, $v_2 = -1$, $v_3 = 1$, $v_4 = 0$. For free cells, according to the formula $s_{ij} = c_{ij} - (u_i + v_j)$ calculate the values $s_{11} = 6$, $s_{12} = 9$, $s_{14} = 2$, $s_{21} = 1$, $s_{23} = 4$, $s_{24} = 4$, $s_{32} = 2$, $s_{33} = 1$, $s_{35} = 3$. Since all these values are positive, we conclude that the results obtained according to the first iteration are a solution to the problem.

3 Production planning

The transport problem refers to ***linear programming***, which involves solving the problem of minimizing some linear function of many variables while fulfilling a certain number of linear constraints in the form of equalities and inequalities. Let us give an example of another problem of this type.

There is some production that produces m types of products. For their production, n types of resources are used. According to the production technology, a_{ij} units of the j-th resource are required to produce one unit of the i-th product. The reserves of the j-th resource are limited by a value b_j. The cost of one unit of the i-th product is equal to c_i. The problem is to determine the most profitable production plan, i.e., such quantity x_1, x_2,..., x_m units of each type of product, which delivers the maximum total cost of the output.

The cost of manufactured products is calculated by the formula

$$S = c_1 x_1 + c_2 x_2 + ... + c_m x_m. \tag{17.22}$$

The existing constraints on resource reserves are characterized by inequalities

$$a_{1j} x_1 + a_{2j} x_2 + ... + a_{mj} x_m \leq b_j, \; j = 1, 2, ..., n. \tag{17.23}$$

Finally, all x_i values must be non-negative, i.e.,

$$x_i \geq 0, \; i = 1, 2, ..., m. \tag{17.24}$$

Thus, the ***production planning problem***[23] is to find such integer values x_1, x_2,..., x_m, which satisfy inequalities (17.23), (17.24) and minimize the function S defined by formula (17.22).

4 Concepts of game theory

In ***game theory***, a process is considered in which two or more parties are involved[24]. Each of them pursues its own interests and implements a certain strategy leading to a win or lose, depending on the joint actions of all parties involved. A ***game in normal form*** is a set of players, each of which has a set of pure strategies H_i and a payment function S_i, $i = 1, ..., m$, where m is the number of players. The outcome of the game $s = (s_1, ..., s_m)$, where $s_i \in H_i$, $i = 1, ..., m$ is a specific set of pure strategies of all players. The utility function S_i associates each outcome s with some number $S_i(s)$.

In particular, for the prisoner's dilemma $m = 2$, the sets H_1 and H_2 coincide and consist of two elements that is 1, consisting in the refusal of a transaction with a consequence and 2, consisting in consent to the transaction. Any pair $s = (i, j)$ corresponds to a specific outcome, $i, j = 1, 2$.

When choosing strategies, each player is guided by his own interests. In this case, the key concept is the ***dominance of strategies***. In particular, the strategy s_i of the i-th player dominates his strategy s_i' if, for an arbitrary choice of the strategies of the other players, the value of the utility function when choosing a strategy s_i turns out to be no less than when choosing. We denote by **s** the outcome that includes the strategy of the i-th player s_i, and by **s**′ the outcome obtained from **s** by replacing strategy s_i with s_i'. Thus, the dominance of the strategy presupposes the fulfillment of the inequality $S_i(\mathbf{s}) \leq S_i(\mathbf{s}')$ for all possible outcomes **s**′. Dominance is strict if a strict inequality holds for all possible outcomes in the last relation. Otherwise, dominance is weak. If all players have strictly dominant strategies, then the appropriate outcome is a ***Nash equilibrium***. In particular, for the prisoner's dilemma, we are talking about minimizing utility functions, and strategies

2 of both players will be strictly dominant. The Nash equilibrium[25] corresponds to outcome (2.2), for which both utility functions take the value 3.

The Nash equilibrium may not correspond to the best value of all utility functions. In particular, in the prisoner's dilemma, outcome (1,1) corresponds to the values 1 of the utility functions. But the strategies that define it are not dominant for both players. This outcome is realized only with an agreed choice of both of them. The outcome **s** is *Pareto optimal* if there is no outcome **s′** such that the inequality $S_i(\mathbf{s}') < S_i(\mathbf{s})$ holds for all i.

Task 17.5 Consider a generalized version of the game discussed in Section 5. There are three firms that make the same product. Each of the firms can assign two options for prices that are low or high, which correspond to values 1 and 2. The lower the price level, the greater the demand for the product. In particular, if all firms set low prices, then each of them will sell 10 units of the product. If two firms set low prices, and the third sets high, then the first two firms sell 13 units of the product, and the third firm does not sell anything. If one firm sets low prices, and the other two set high, then the first will sell 16 units of production, and the other two sets 2. Finally, if all firms charge high prices, then each of them sells 6 units of product. Find outcomes corresponding to Nash equilibrium and Pareto optimality.

Notes

[1] Numerous discrete models are considered, for example, in [25], [110], [237], [240]. Stochastic discrete systems are discussed in Chapter 18.

[2] *Operations research* is a branch of applied mathematics that studies mathematical methods for making optimal decisions in various situations, see [144], [121].

[3] For *discrete mathematics* see [36], [167], [375].

[4] Discrete biological models are considered, for example, in [45], [85], [244], [287], [310].

[5] The sequence convergence problem is one of the central problems in mathematical analysis, see, for example, [13], [344], [353].

[6] This happens approximately at $r=3.56995$.

[7] Mathematical chaos theory is discussed, for example, in [6], [239].

[8] If necessary, you can divide the area into unequal parts.

[9] Values δ_x, $\delta_{\bar{x}}$ are difference analogues of the derivative. They are widely used in computational mathematics (in particular, in the finite difference method, see [243], [280], [301], [325], [340], [347]) and are called, respectively, "forward difference" and "backward difference".

[10] We will use the discrete stationary heat transfer model in Chapter 19 to justify the derivation of the corresponding continuous model described earlier in Chapter 13.

[11] Both continuous and discrete mathematical models of heat transfer are based on relations (17.4), (17.5), which are, in fact, difference rather than differential. Note that earlier we first went to the limit in them to obtain a model in the form of differential equations, and then we discretized these equations to obtain a practical result. In this sense, it is perhaps more logical to use a discrete model, which does not require either the passage to the limit or the approximation of derivatives.

[12] The transport problem is discussed in [263], [305]. It refers to linear programming, since both the minimized value and all existing constraints are linear. On the other hand, the transport problem belongs to integer programming, since its solution is a vector whose components take only integer values. In principle, solutions to such problems can be found by a simple enumeration of possible options, but, as a rule, this is ineffective, since it requires significant computational costs. General questions of combinatorics are discussed, for example, in [359], [286].

[13] For the traveling salesman problem, see [63].

[14] For graph theory see [41], [71], [174]

[15] The corresponding inequalities are given, for example, in [63].

[16] However, it is quite simple to estimate the limits of the minimum value of the route optimality criterion. In particular, the most advantageous transition from the i-th city corresponds to the minimum of the value c_{ij} for all values $j = 1, ..., n$. Thus, the most profitable traveling salesman route can never cost less than the sum of all these values. Similarly, you can choose the most advantageous of the options for moving to the j-th city, i.e., the minimum of the value c_{ij} over all values $i = 1, ..., n$, and then sum all the obtained values. Naturally, the solution to the problem cannot give the value of the optimality criterion below this result. To obtain an upper estimate, it is sufficient to choose a route in an arbitrary way and find the corresponding value of the optimality criterion.

[17] For algorithms for solving the traveling salesman problem, see [63], [263], [305].

[18] For game theory, see Fernandez, Isaacs, Osborne.

[19] If necessary, many of these assumptions can be removed by introducing additional terms into the mathematical model that take into account the influence of the relevant factors.

[20] If the study is carried out over a sufficiently long time interval, the time step increases.

[21] The time spent in a particular population group can be considered as a random variable, see Chapter 18.

[22] This model is considered in [318]

[23] Problem solving techniques for production planning, see [144], [121].

[24] Numerous examples of the application of game theory to economics are given in [21], [87], [111], [184], [363]. It is also applied in [70] sociology, [48], [57] biology, [11] warfare, and others.

[25] The solution to a game problem often comes down to the consistent elimination of dominated strategies. If we can remove all strictly dominated strategies, the result is a single Nash equilibrium. Removing all weakly dominated strategies also leads to a Nash equilibrium, but it may not be the only one.

Chapter 18

Stochastic models

Lecture

All the preceding lectures dealt with deterministic systems. In them, in the case of repeated experiments under the same conditions, the same result is observed. The state of the deterministic systems can be characterized using one or more state functions (in special cases, vectors, sequences, matrices, etc.) that take quite definite specific values at any point in its domain for any set of system parameters. Mathematical models of such systems, with all their diversity, were tasks that, in principle, allow to restore a specific law of change of the corresponding state functions.

However, there is a wide class of processes in which, for a reason, setting up an experiment under the same conditions leads to different results, and it is not possible to predict the occurrence of an event as a result of a single experiment. These systems are called **stochastic**. It is important to note that although an accurate prediction of the outcome of a random event is impossible here, there is much to be said about its properties. In this regard, stochastic systems can also be an object of mathematical analysis. They are described by qualitatively different models, the analysis of which requires the use of special methods of **probability theory**[1] and **mathematical statistics**[2].

In this lecture, fairly simple stochastic systems are described. We will consider mainly stochastic analogues of the simplest model of population dynamics that is the Malthus model[3]. In particular, Section 1 examines the evolution of a biological species when the birth of an individual is a random event. As a result, the size of the population turns out to be a random value, and its change over time is a random process. The mathematical model of this process is a system of differential-difference equations with respect to the probability that the number of a species at a particular moment in time takes on a certain value. The quantitative characteristics of the considered random process are determined. For a numerical analysis of the system, the Monte Carlo method described in Section 2 can be used. Section 3 considers a population in which the death of an individual is a random event, and in Section 4 both birth and death are random. The Appendix considers the known mathematical models of various processes, in which some parameters are random. In addition, discrete stochastic systems are presented[4].

1 Stochastic model of pure birth

Consider a biological species in favorable conditions. As noted in Chapter 7, its number $x = x(t)$ in the simplest case is described by the Malthus model determined by the differential equation $\dot{x} = kx$ with a positive value k. According to this equality, for an arbitrarily small-time interval dt, the population size increases by the amount $kx dt$, i.e., the change in the population size for a specified time is proportional to the values of this number and the time interval. Suppose that at the initial time $t = 0$, the population has an initial number of m. Then, over time, there is an exponential growth of the population by the equality $x(t) = me^{kt}$. The results obtained also apply to the model of pure birth, when the mortality of a species is not taken into account at all, and the coefficient k characterizes its fertility.

Let us now turn to a stochastic analogue of this process. The birth of an individual at one time or another is a **random event**[5]. As a result, the size of a given population $x(t)$ at an arbitrary moment of time t is no longer be a number, as in the deterministic model, but

DOI: 10.1201/9781003035602-18

turns out to be a ***random value***[6]. Then the change with time in the population size is not a specific function $x = x(t)$, but is a ***random process***[7]. The mathematical model in the deterministic case is a problem of determining the law of change in the state function of the system with time. The stochastic model determines the change over time in the ***probability*** that the specified random variable takes one or another value[8].

Obviously, the probability that for a population size n within a sufficiently small-time interval Δt one individual is born, the larger the population and the given time interval, the higher. In the simplest case, it can be considered equal to the value of $kn\Delta t$, where k is a positive constant characterizing the fertility[9]. Let $p_n(t)$ be the probability that at time t there are exactly n individuals[10] that is denoted by the equality $p_n(t) = \mathbf{P}\{x(t) = n\}$. Since the interval Δt is chosen sufficiently small, we assume that only one event can occur during this time. In particular, it is possible, although not necessary, the birth of one new individual. Then the presence of n individuals at the subsequent time $t + \Delta t$ is possible either in the case when at the previous time t the number of individuals was also n, and during the time Δt the birth of a new individual did not occur, or when the number of individuals was $n - 1$, and during this time one individual was born.

It is known that the probability of two ***inconsistent events***[11] is equal to the sum of the probabilities of these events. In particular, the presence of different population sizes at the same time is incompatible. Then the probability $p_n(t + \Delta t)$ that at the moment of time $t + \Delta t$ there are exactly n individuals in the population is the sum of the probability of event A, consisting in the fact that earlier the population size was equal to $n - 1$, and during this time one individual was born, and the probability of event B, according to which the population consisted of n individuals, and no one was born during the time Δt. Thus, the equality holds $p_n(t + \Delta t) = p(A) + p(B)$, where $p(A)$ and $p(B)$ denote the probabilities of the corresponding events.

Each of these two events is the result of the simultaneous execution of two simpler events. In particular, for event A, we are talking about the birth of one individual for a given time and about the size of the population $n - 1$. But the probability of the first of these events depends on the realization of the second event, i.e., we are dealing with ***conditional probability***[12]. In this case, the probability of event A is equal to the product of the specified conditional probability, i.e., the quantity $k(n - 1)\Delta t$, and the probability of the presence of the number $n - 1$ at the moment of time t, i.e., the quantities $p_{n-1}(t)$. Thus, we find the probability $p(A) = p_{n-1}(t)k(n - 1)\Delta t$.

Similarly, event B assumes the absence of the birth of an individual for a given time and the presence at time t of the number n. The probability of no offspring for one individual during the time Δt is $(1 - k)\Delta t$. Then, for the number n, the probability of no birth for the considered time is $(1 - k)n\Delta t$. As a result, we find $p(B) = p_n(t)(1 - k)n\Delta t$. Now we define

$$p_n(t + \Delta t) = p_{n-1}(t)k(n - 1)\Delta t + p_n(t)(1 - k)n\Delta t.$$

Dividing the resulting equality by Δt, we have

$$\frac{p_n(t + \Delta t) - p_n(t)}{\Delta t} = k\big[p_{n-1}(t)(n - 1) - p_n(t)n\big].$$

Passing to the limit as Δt tending to zero, we obtain the differential equation

$$\dot{p}_n(t) = k\big[p_{n-1}(t)(n - 1) - p_n(t)n\big]. \tag{18.1}$$

It allows us to determine the change over time in the probability that the population size is n, provided that the probability $p_{n-1}(t)$ is known.

Equality (18.1) makes sense for $n > 1$, since for $n = 0$ (in the absence of individuals) birth is impossible. Note that at the moment $t + \Delta t$ the number equal to 1 can be only in the case when at the previous moment there was the same number, and during this time the birth of a new individual did not occur. As a result, we obtain $p_1(t + \Delta t) = p_1(t)(1 - k)\Delta t$. Hence, after dividing by Δt and passing to the limit, we obtain the equation

$$\dot{p}_1(t) = -kp_1(t). \tag{18.2}$$

Let us assume for simplicity that at the initial time $t = 0$ there is only one individual. Thus, we have the initial conditions

$$p_1(0) = 1; \quad p_n(0) = 0, \; n > 1, \tag{18.3}$$

because the presence of one individual at the initial moment of time is a reliable event, and a different initial number is impossible.

Thus, the considered mathematical model is an infinite system of differential equations (18.1), (18.2) with initial conditions (18.3) with respect to the probabilities of the presence of one or another population size, i.e., the probability distribution of the corresponding random variable at an arbitrary time[13]. In the considered system, the independent variables are the time t and the number of individuals n. Based on their properties, this system can be considered as a ***differential-difference system***[14]. From it, we can find the corresponding probabilities and establish the properties of the investigated random process. Note that the considered random variable is discrete, since it can only take individual values $1, 2, \dots$.

The solution to equation (18.2) with the first initial condition (18.3) is determined by the formula $p_1(t) = e^{-kt}$. This implies, in particular, that the probability that the number of the species remains equal to 1 tends to zero over time, i.e., the population is likely to increase. Substituting this value into formula (18.1) for $n = 2$, we have

$$\dot{p}_2(t) + 2kp_2(t) = ke^{-kt}.$$

Solving this equation with initial condition, we have[15] $p_2(t) = (1 - e^{-kt})e^{-kt}$.

Obviously, the function p_2 is equal to zero at the initial time, is positive for all t, and tends to zero as $t \to \infty$. Therefore, somewhere it reaches its maximum value. To find it, determine $\dot{p}_2(t) = -ke^{-kt} + 2ke^{-2kt} = 0$. We have $e^{-kt} = 1/2$. Thus, the maximum of the function p_2 is attained at the time $\ln 2/k$, and is equal to $1/4$. So, the probability that the population size is equal to two is initially equal to zero, then increases up to the time $\ln 2/k$, at the same time reaching the value of $1/4$, after which it decreases and tends to zero with an unlimited increase in time.

Solving equation (18.1) with initial condition for all n, we get

$$p_n(t) = (1 - e^{-kt})^{n-1}e^{-kt}, \; n = 1, 2, \dots. \tag{18.4}$$

It is easy to verify[16] that the function p_n, being equal to zero initially, it increases until the time $\ln n/k$, reaching the value $(n - 1)^{n-1}/n^n$, after which it decreases and tends to zero with an unlimited increase in time, see Figure 18.1. Thus, the random variable $x(t)$ is characterized by the probability distribution[17] $p_n(t)$.

When describing a random variable, one can often restrict oneself to analyzing some of its numerical characteristics, called ***moments***. The most important of these characteristics is the ***expected value*** of the random variable, which corresponds to its mean value. The expected value of a discrete random variable $x(t)$, taking values n with probability $p_n(t)$, $n = 1, 2, \dots$, is determined by the formula

$$\mathbf{E}[x(t)] = \sum_{n=1}^{\infty} np_n(t) = e^{-kt} \sum_{n=1}^{\infty} n(1 - e^{-kt})^{n-1}.$$

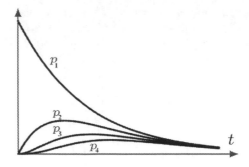

FIGURE 18.1: Probability distribution in the pure birth model.

Now we find[18] $\mathbf{E}[x(t)] = e^{kt}$. It is characteristic that the deterministic model of Malthus with a positive increase in k and a single initial number gives exactly the same law of change in the number of the species. Thus, the considered random process behaves on average in the same way as the deterministic model. Naturally, the larger the population, the closer the values of the corresponding random variable to its expected value, which means that the better the deterministic model describes the process under study.

The **variance** is a measure of the deviation of the values of a random variable x from its expected value. It is determined by the formula

$$\mathbf{Var}(x) = \mathbf{E}\big[(x - \mathbf{E}x)^2\big] = \mathbf{E}(x^2) - \big[\mathbf{E}(x)\big]^2.$$

For the considered random variable $x(t)$, after simple transformations, we find[19] $\mathbf{E}[x(t)] = e^{kt}(e^{kt} - 1)$. Thus, at the initial moment of time, the variance is zero, since the initial state of the system is known. However, over time, the value of a random variable can deviate arbitrarily far from its expected value.

The obtained results can easily be extended to the general case when the initial number of a species takes on a certain value m. Since mortality is not taken into account, the probability that at some time the number takes the value $m - 1$ is be equal to zero. Form equality (18.1) for $n = m$ it follows the equation

$$\dot{p}_m(t) = -mkp_m(t)$$

that is the extension of equality (18.2). Solving it with initial condition $p_m(0) = 1$, we find $p_m(t) = e^{-mkt}$. Thus, the probability that the population size will be equal to m decreases from unity at the initial moment of time to zero with an unlimited increase in time.

Solving equation (18.1) with zero initial condition sequentially for the values $m + 1$, $m + 2$, etc., we obtain the equality

$$p_n(t) = \binom{n - 1}{n - m}(1 - e^{-kt})^{n-m}e^{-mkt} \tag{18.5}$$

that is true for all $n = m, m + 1, ...$, where the numbers $\binom{n-1}{n-m}$ are binomial coefficients[20]. This result generalizes formula (18.4). Repeating the previous reasoning, we can make sure that the probability that the population size will be equal to n initially equals zero, increases up to the time $\ln(m/n)/k$, and then tends to zero with an unlimited increase in time.

For the random variable $x(t)$, we can find the expected value and variance

$$\mathbf{E}[x(t)] = me^{kt}, \quad \mathbf{Var}[x(t)] = me^{kt}(e^{kt} - 1).$$

TABLE 18.1: Deterministic and stochastic pure birth models.

Characteristic	Deterministic model	Stochastic model
independent variables	time	time and number of individuals
number value $x(t)$	number	random value
state function $x = x(t)$	function	random process
general law $kxdt$	population growth over time dt at number x	probability of the birth of an individual in time dt for the number x
initial number $x(0)$	number	number
state equation	differential equation	differential-difference system
result $m\exp(-kt)$	law of population change	expected value

TABLE 18.2: Stochastic models of pure birth.

Characteristic	Biology	Epidemiology	Physics
process	population growth	epidemic development	chain reaction
random event	birth of an individual	human infection	nuclear decay
random value	population size	number of infected	number of free neutrons
probable outcome	exponential population growth	exponential increased incidence	nuclear explosion explosion

Thus, in the general case, the expected value for the population size in the case of a stochastic analogue of the Malthus model of pure birth also coincides with the change in the population size in the deterministic model. Moreover, over time, the values of $x(t)$ can deviate more and more from its expected value. A comparative analysis of the deterministic and stochastic models is given in Table 18.1.

Earlier, we have repeatedly encountered a situation when the same mathematical model described various processes. Another interpretation of the pure birth model is the ***epidemic propagarion***. Here, the infection of a new person at one point or another is considered a random event. Then the number of infected people $x(t)$ at time t turns out to be a random variable, and the function $x = x(t)$ is a random process. Obviously, the probability that with the number of infected n within a sufficiently short time interval Δt one more person become infected, the more infected and the time interval. As a result, the probability $p_n(t)$ that there is exactly n infected at time t is described by relations (18.1)–(18.3). A more complex stochastic model of the propagation of the epidemic is discussed in the Appendix.

The physical analogue of this process is a ***nuclear chain reaction***, for example, the fission of nuclei of heavy elements[21]. It is a sequence of single nuclear reactions, each of which is caused by a particle, in particular, a neutron, which appeared as a reaction product at the previous step of the sequence. A single fission of a nucleus with the formation of free neutrons at one time or another can be considered a random event. As a result, the number $x(t)$ of free neutrons at time t turns out to be a random variable. The probability of decay of one nucleus within a sufficiently short time interval will be the higher, the freer neutrons are available and the given time interval. The exponential growth of the population and the development of the epidemic here corresponds to a nuclear explosion. One can describe also **chemical chain reactions** in which free atoms or radicals play the role of neutrons.[22]

A comparative analysis of stochastic models of pure birth is carried out in Table 18.2.

Stochastic model of pure birth describes the process
in which the birth of an individual is a random event.
Model is characterized by differential-difference equations.
Expected value of this process corresponds to the Malthus model.
Over time, large enough deviation of population size from its expected value is possible.

2 Monte Carlo method

In the previous section, the random process of pure birth was considered. We determined the probability distribution of the population size for it, and after that, the corresponding expected value and variance. This made it possible to establish some regularities of the studied system. However, for a practical analysis of the considered process, it is desirable to determine how the population size, which is a random variable at each moment in time, changes over time. To solve such problem, the *Monte Carlo method* is used[23]. It is based on obtaining a sufficiently large number of realizations of a certain random process so that its probabilistic characteristics coincide with the corresponding probabilistic characteristics of the studied process.

Let, in particular, there is a random variable x with a given value m of the expected value. Let us try to find such a random variable that has m as its expected value with a sufficiently high degree of accuracy. The theoretical basis for solving this problem is the concept of the normal probability distribution and the central limit theorem.

In the previous section, we considered random variables with a discrete set of values. However, a certain extended area often acts as a set of values. Thus, we are talking about a *continuous random variable*. The probability that a random variable x takes a value from some segment $[a, b]$ is

$$\mathbf{P}\big(a \leq x \leq b\big) = \int\limits_a^b f(\xi)d\xi.$$

Here, the function f is the main characteristic of the considered value. This is called the *probability density* function and has the following properties

$$f(\xi) \geq 0, \quad \int\limits_{-\infty}^{\infty} f(\xi)d\xi = 1.$$

The corresponding expected value is calculated by the formula

$$\mathbf{E}(x) = \int\limits_{-\infty}^{\infty} \xi f(\xi)d\xi,$$

and the variance is determined in the same way as before, i.e., $\mathbf{Var}(x) = \mathbf{E}(x^2) - [\mathbf{E}(x)]^2$.

Of particular interest is the *normal distribution*, for which the probability density is determined by the formula (see Figure 18.2)

$$f(\xi) = \frac{1}{\sigma\sqrt{2\pi}} \exp\left(-\frac{(\xi - \mu)^2}{2\sigma^2}\right),$$

where the constants μ and σ are parameters of the random variable corresponding to its expected value and square root of the variance. Note the *three-sigma rule*, according to which the probability that a random variable x differs from its expected value μ by no more than 3σ is approximately 0.9973, i.e., $\mathbf{P}\big(|x - \mu| \leq 3\sigma\big) \approx 0.9973$. Thus, practically all values of the random variable lie on the interval $(\mu - 3\sigma, \mu + 3\sigma)$. The extreme importance of the normal distribution is clarified with the help of the *central limit theorem*.

Let random variables $x_1, ..., x_N$ be given, which are assumed to be independent[24] and have the same distribution. Then they all have the same expected value μ and variance σ^2. It follows directly from the definition of these characteristics that their sum $\rho_N = x_1 + ... + x_N$ has an expected value $N\mu$ and a variance $N\sigma^2$. Consider now a random variable ζ_N, which

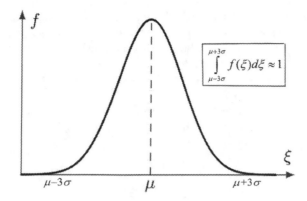

FIGURE 18.2: Normal distribution probability density.

has a normal distribution with the same mean and variance values. Then, according to the central limit theorem, for sufficiently large values of N for any interval $[a, b]$, the probability that ρ_N lies on this interval is approximately equal to

$$\mathbf{P}\big(a \le \rho_N \le b\big) \approx \int_a^b f(\xi)d\xi,$$

where f is the distribution density of ζ_N. Thus, the sum of a sufficiently large number of independent random variables has a distribution that is fairly close to normal.

Now let us return to the problem of determining a random variable x with an expected value m. Let $x_1, ..., x_N$ be independent random variables with distribution coinciding with the distribution of x. According to the central limit theorem, for a sufficiently large N, the distribution of the sum $\rho_N = x_1 + ... + x_N$ of random variables turns out to be approximately normal with the expected value $\mu = Ny$ and variance $\sigma^2 = N\delta^2$, where δ^2 is the variance of x. Then, in accordance with the three-sigma rule, determine the relation

$$\boldsymbol{P}\big(|\rho_N - Nm| \le 3\delta/\sqrt{N}\big) \approx 1.$$

Now we get

$$\boldsymbol{P}\Big(\Big|\frac{1}{N}\sum_{i=1}^N x_i - m\Big| \le 3\delta/\sqrt{N}\Big) \approx 1.$$

Let N values of a random variable with expected value m be found. According to this relation, the arithmetic mean of these values is approximately equal to m with a degree of accuracy determined by the variance of a given random variable and the number of selected values. The error of this approximation tends to zero with an increase in the number N. Thus, for stochastic modeling of a random variable with a given expected value, it is required to select a sufficiently large number of values of the random variable with the same distribution, take their arithmetic mean and obtain the corresponding accuracy estimate. So, we can, in principle, determine with the desired accuracy the desired random variable, if we have an effective way of choosing its values for a given probability distribution.

Consider a discrete random variable x taking values $x_1, ..., x_n$ with probabilities $p_1, ..., p_n$, respectively. Let us define the points

$$y_0 = 0, \ y_1 = p_1, \ y_2 = p_1 + p_2, \ ..., \ y_{n-1} = p_1 + p_2 + ... + p_{n-1}, \ y_n = 1$$

on the segment [0,1]. Obviously, the length of the interval (y_{i-1}, y_i) is equal to p_i, $i = 1, ..., n$. We now arbitrarily choose a point from the unit interval[25]. If this is a point of the interval (y_{i-1}, y_i), then its value x_i is chosen as the value of the random variable x in this experiment. Naturally, the probability of realizing precisely this value in this experiment is determined by the length of the corresponding interval, which is equal to pi in accordance with the existing probability distribution.

Now let the random variable x be distributed continuously on some interval (a, b) with the probability density $f = f(\xi)$. Obviously, the probability p that the considered random variable is lie on the interval (a, c) is equal to

$$p = \int_a^c f(\xi)d\xi. \tag{18.6}$$

Given an arbitrary number p from the unit segment, we consider the formula (18.6) as an equation with respect to the number c. Its solution is chosen as the value of the considered random variable[26].

Thus, there are ways to determine the values of a random variable with a given probability distribution, which makes it possible to use the Monte Carlo method to model the considered random variable. This method is used to simulate various random processes and in the following sections.

Task 18.1 *Pure birth model*. In accordance with the Monte Carlo method, calculate a random process of pure production with the initial condition $x(0) = m$. The values of the population size are considered at times $t_j = j\tau$, $j = 1, 2, ...$, where τ is a small enough time step. The number of time moments is chosen so large that it is possible to judge the development of a random process from them. In this case, the following actions are performed.

1. In accordance with formula (18.5), the probabilities $p_n(t_j)$ are determined that the random variable $x(t_j)$ takes values n. In this case, the value of the random variable $x(t_{i-1})$ at the previous moment of time is chosen as the minimum value of n, and the maximum value of n at which the inequality $p_n(t_j) > 0.01$.

2. For the obtained probability distribution, in accordance with the described technique, a sufficiently large number N of values of a given random variable are selected[27]. Their arithmetic mean is chosen as the value of the random variable $x(t_j)$.

3. A graph of the dependence of the obtained values of the population size on time is plotted in comparison with the corresponding expected value.

4. The standard deviation of the obtained result from the expected value at different points in time is determined. Results are compared with known variance.

5. It is checked to what extent the estimated time of reaching a population of a particular size is consistent with the maximum probability of reaching this value.

Monte Carlo method is used to simulate random variables.
Method involves calculating a large number of realizations
of a random variable with the same probabilistic characteristics
as the given random variable.

3   Stochastic model of population death

Let us now consider the process of population death. According to Malthus model (see Chapter 7), this process is described by a differential equation $\dot{x} = -kx$ with a positive coefficient k. By this equality, for an arbitrarily small-time interval dt, the population size decreases by the value $kxdt$, i.e., the change in the population size for a specified time is directly proportional to the value of this number and the value of the time interval. Then, over time, the extinction of the population is observed in accordance with the equality $x(t) = me^{-kt}$, where m is the initial population size.

Consider a stochastic analogue of this process. The probability that at a population size n within a sufficiently small-time interval Δt one individual will die the higher, the larger the population and the given time interval. Thus, it is equal to the value of $kn\Delta t$, where k is some positive constant. Let us again denote by $p_n(t)$ the probability that at time t there are exactly n individuals. Since the interval Δt is small enough, we assume that during this time, although not necessarily, the death of only one individual can occur. Under these conditions, the presence of n individuals at the next time instant $t + \Delta t$ is possible either in the case when at the previous time t the number of individuals was equal to n, i.e., no one died, or when the number of individuals was $n + 1$ earlier, and during the time Δt one individual died.

Repeating the reasoning from the previous section, we obtain the equality

$$p_n(t + \Delta t) = p_n(t)(1 - k)n\Delta t + p_{n+1}(t)k(n + 1)\Delta t.$$

Dividing by Δt and passing to the limit as Δt tends to zero, we obtain the differential equation

$$\dot{p}_n(t) = k\big[p_{n+1}(n + 1) - p_n(t)n\big]. \tag{18.7}$$

It is assumed that at the initial moment of time the population size is equal to m. As a result, we get the initial conditions

$$p_m(0) = 1; \quad p_n(0) = 0, \, n < m. \tag{18.8}$$

Obviously, the population size does not increase over time, which means that the probability that at some point in time the number is exceed m is zero. In this case, the population size $x(t)$ at time t is again a discrete random variable with a finite set of values $0, 1, ..., m$. Setting $n = m$ in equality (18.7), we obtain

$$\dot{p}_m(t) = -kmp_m(t).$$

Solving this equation with the initial condition (18.8), we find $p_m(t) = e^{-mkt}$, which coincides with the analogous value in the pure birth model. Substituting the result into equation (18.7) with $n = m - 1$ and solving this equation with a zero initial condition, find the value p_{m-1}. Repeating this procedure many times, we find all values

$$p_n(t) = \binom{m}{n}(e^{kt} - 1)^{m-n}e^{-mkt}, \, n = m, m - 1, ..., 0,$$

where $\binom{m}{n}$ are binomial coefficients. These values form the corresponding probability distribution[28].

Note that the probability p_m that the population size remains the same as at the initial moment of time decreases and tends to zero. Obviously, for $0 < n < m$, the function p_n vanishes at the initial time, takes positive values for $t > 0$, and tends to zero with time. Therefore, it reaches its maximum somewhere. To find this value, calculate the derivative

$$\dot{p}_n(t) = \binom{m}{n}(m - n)(e^{kt} - 1)^{m-n-1}ke^{-mkt} - \binom{m}{n}(e^{kt} - 1)^{m-n}mke^{-mkt}.$$

Equating this value to zero, find the point $\ln(m/n)/k$, at which the probability of equality n of the population size is maximum. Finally, the probability of a zero-population size is

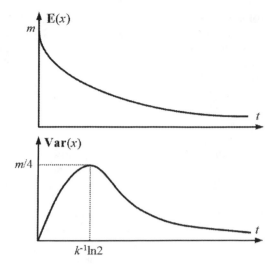

FIGURE 18.4: Expected value and variance in the model of population death.

of unstable nuclei will satisfy the system of differential-difference equations (18.7), and the death of the entire population corresponds to the decay of all unstable nuclei. There are other interpretations of the considered model[30].

Task 18.2 *Model of population death.* Using the Monte Carlo method, calculate the random process of population death by analogy with Task 18.1. In this case, when calculating the value of $x(t_i)$, the minimum value of n is chosen as the minimum n, at which the inequality $p_n(t_i) > 0.01$ is satisfied, and the value of the random variable $x(t_{i-1})$ is taken as its maximum value.

> *Population death model describes a random process*
> *in which the death of an individual is a random event.*
> *Model is characterized by differential-difference equations.*
> *Expected value of the process corresponds to the Malthus model.*

4   Stochastic Malthus model

Let us now consider the general case when there is both birth and death of individuals of the considered biological. In the simplest case, it corresponds to the Malthus model $\dot{x} = (a - b)x$, where the coefficient a characterizes the fertility, and b is the mortality of the species. At the initial size of the species m, the solution to this equation is determined by the formula $x(t) = me^{(a-b)t}$, see Chapter 7. For $a > b$, an exponential growth of the population is observed here, as in Section 1, for $a < b$, the population dies out, as in Section 2, and the case $a = b$ corresponds to a constant population size.

Let us turn to the stochastic analogue of this system. Assume that the probability of both birth and death of one individual within a sufficiently small-time interval Δt is the higher, the larger the population and the given time interval. In particular, we assume that the probability of the birth of one individual in time Δt for a population size n is equal to $an\Delta t$, and the corresponding probability of death is $bn\Delta t$. As in the previous cases, we believe that the time interval Δt is so small that during this time, the birth or death of only one individual can occur, although not necessarily. Then the presence of exactly n individuals at the next time instant $t + \Delta t$ is possible either in the case when at the previous time t the number of individuals was equal to n, i.e., no one was born or died, either when the number of individuals was earlier equal to $n - 1$, and during the time Δt one individual was born, or when the number of individuals was earlier equal to $n + 1$, and during the time Δt one individual died. Keeping the previous notation, we arrive at the equality

$$p_n(t + \Delta t) = p_n(t)(1 - a - b)n\Delta t + p_{n-1}(t)a(n - 1)\Delta t + p_{n+1}(t)b(n + 1)\Delta t.$$

After dividing by Δt and passing to the limit, determine the equation

$$\dot{p}_n(t) = a\big[p_{n-1}(n - 1) - p_n(t)n\big] + b\big[p_{n+1}(n + 1) - p_n(t)n\big] \qquad (18.9)$$

that is the extension of the equations (18.1) and (18.7).

Consider the case $n = 0$. The absence of individuals at a subsequent moment in time is possible either when they were absent at the previous moment in time, or there was one individual that died during the time t. As a result, we obtain the equality $p_0(t + \Delta t) = p_0(t) + p_1(t)b\Delta t$. Hence follows the equation

$$\dot{p}_0(t) = bp_1. \qquad (18.10)$$

It is assumed that the population size at the initial moment is equal to m. Then equations (18.9), (18.10) are supplemented by the initial conditions

$$p_m(0) = 1; \quad p_n(0) = 0, \; n \neq m. \qquad (18.11)$$

Problem (18.9) – (18.11) can no longer be solved explicitly, as the corresponding equations for the processes of pure birth and death, since, according to equality (18.9), the population size can turn out to be either more or less than the initial number. Nevertheless, in this case, for a random value of the abundance $x(t)$ at time t, one can find the expected value and the variance[31]

$$\mathbf{E}[x(t)] = me^{(a-b)t}, \qquad (18.12)$$

$$\mathbf{Var}[x(t)] = m\frac{a + b}{a - b}e^{(a-b)t}\big[e^{(a-b)t} - 1\big]. \qquad (18.13)$$

There are extensions of the corresponding results of the previous sections. Besides, determine the probability of the population absence at the time t

$$p_0(t) = \left[b\frac{e^{(a-b)t} - 1}{ae^{(a-b)t} - b}\right]^m, \; a \neq b; \quad p_0(t) = \left(\frac{at}{at + 1}\right)^m, \; a = b. \qquad (18.14)$$

According to formula (18.12), for $a > b$, i.e., in conditions of excess fertility over mortality, the expected result is unlimited population growth. For $a < b$, i.e., in conditions of excess of mortality over fertility, the expected value of the population size tends to zero, which corresponds to the extinction of the population. Finally, for $a = b$, the equality $\mathbf{E}[x(t)] = m$ is true that is the expected value of the size of the population at any time is its initial size. The results obtained are consistent with the deterministic Malthus model.

By the formula (18.13), for $a > b$, the variance increases indefinitely with time, i.e., possible arbitrarily large deviations of the population size from its expected value. For $a < b$, the variance tends to zero, which means that over time, the population size deviates less and less from its expected value. Thus, the death of the population is probably observed. Particularly difficult is the case $a = b$, in which both the expression in the denominator of the fraction and the expression in square brackets vanish. It can be shown that in this

TABLE 18.3: Deterministic and stochastic Malthus models.

Situation	Deterministic model	Stochastic model
mortality exceeds fertility	population extinction	guaranteed population extinction
mortality coincides with fertility	population immutability	guaranteed population extinction
fertility exceeds mortality	unlimited population growth	expected population growth in case of its possible death

case the variance is[32] $\mathbf{Var}[x(t)] = 2amt$. Thus, with $a = b$, over time, the population size can deviate more and more from its expected value, and the more, the larger its initial size and the parameter a, which in this case characterizes both fertility and mortality. This result indicates the difference between the behavior of the system in the stochastic and deterministic models.

Let us now turn to formula (18.14), which characterizes the probability of zero population size at time t. For $a < b$ we have

$$\lim_{t \to \infty} p_0(t) = \lim_{t \to \infty} \left[b \frac{e^{(a-b)t} - 1}{ae^{(a-b)t} - b} \right]^m = 1.$$

Thus, if the fertility exceeds mortality, the population will certainly die out. Therefore, in this case, the predictions of the deterministic and stochastic models coincide.

For $a = b$ we find

$$\lim_{t \to \infty} p_0(t) = \lim_{t \to \infty} \left(\frac{at}{at + 1} \right)^m = 1.$$

Thus, in the case of coincidence of fertility and mortality in the stochastic model, the population also dies out over time, while the deterministic model predicts the immutability of the population. This surprising conclusion is consistent with the fact that variance, when fertility and mortality coincide, grows indefinitely over time. So, in this case, in the stochastic model, the population is not viable.

Finally, if the fertility exceeds the mortality, the Malthus model shows the unlimited population growth. However, for the stochastic model, the following result is observed here

$$\lim_{t \to \infty} p_0(t) = \lim_{t \to \infty} \left[b \frac{e^{(a-b)t} - 1}{ae^{(a-b)t} - b} \right]^m = \left(\frac{b}{a} \right)^m.$$

This means that, in spite of the expected population growth, in this case, there is also a probability of population death. This probability is the greater, the closer the mortality is to the fertility and the smaller the initial population size. Naturally, for sufficiently large values of the initial number, the probability of extinction of the population turns out to be negligible even if the mortality of the species is sufficiently close to its fertility. For example, if the mortality differs from the fertility by only 1 percent with the initial population size of $2^{10} = 1024$ (not such a large number), the probability of death of the population is only $3.39 \cdot 10^{-5}$.

Comparative analysis of deterministic and stochastic models is presented in Table 18.3.

Task 18.3 *Stochastic Malthus model*. In accordance with the Monte Carlo method, calculate the random process described by the Malthus stochastic model by analogy with Task 18.1. When calculating the value of $x(t_i)$, here as the minimum (respectively, maximum) n, the minimum (respectively, maximum) value of n is chosen, at which the inequality $p_n(t_i) > 0.01$ is satisfied. Calculations are carried out in the case of the predominance of fertility, mortality, and the equality of these characteristics.

*Stochastic Malthus model describes a random process,
in which the birth and death of individuals are random events.
Expected value of the process corresponds to the Malthus model.
Extinction of the population here is possible
even if excess of fertility over mortality.*

Direction of further work. We have covered the various special classes of mathematical models, completing Part III of the course. In the final part, some additional issues will be considered, in particular, the properties of mathematical models, optimal control problems and the identification of systems.

Appendix

In this lecture, we investigated systems in which random events occur. However, stochastic models also appear in the case when some parameters in the considered before systems are given randomly. In particular, in Section 1 we again turn to the Malthus model. In this case, the increase in the population size is a random value. As a result, the number of the species also turns out to be a random variable, but with a continuous rather than discrete distribution. Various mathematical models with random parameters related to physics, chemistry and epidemiology are discussed in Section 2. The final two sections describe discrete stochastic systems related to economics and physics.

1   Malthus model with random population growth

We turn again to the Malthus model, which is characterized by the equation

$$\dot{x} = kx, \ t > 0. \tag{18.15}$$

However, an increase in the number of the form $k = k(t)$ at an arbitrary moment t is a random variable. Suppose also that all random variables $k(t)$ are independent of each other and have the same normal distribution with expected value μ and variance σ^2.

If the initial population number is m, then the solution of the equation (18.15) is

$$x(t) = m \exp \int\limits_0^t k(\tau)d\tau.$$

It can be shown that the integral on the right-hand side of this equality, equal to $\ln x(t)/m$, also has a normal distribution[33], but with the expected value μt and variance $\sigma^2 t$. The probability distribution of a random variable, the natural logarithm of which has a normal distribution, is called **lognormal**[34]. Taking into account the definition of the probability density of a lognormal distribution, we conclude that the ratio $x(t)/m$ has a probability density

$$f(\xi) = \frac{1}{\sigma\xi\sqrt{2\pi t}} \exp\left(-\frac{(\ln\xi - \mu t)^2}{2\sigma^2 t}\right).$$

Let us establish the properties of the function $f = f(\xi)$. It is defined on the set of non-negative numbers, is a smooth function, takes non-negative values, vanishes at $\xi = 0$ and has an integral over the entire domain of definition equal to one, see Figure 18.5. Then

$f(\xi) \to 0$ as $\xi \to \infty$. Thus, it must have at least one maximum. Zeroing its derivative, we find its only maximum point $\xi^* = \xi^*(t) = \exp[(\mu - \sigma^2)t]$. Obviously, with an unlimited increase in time, this value tends to zero for $\mu < \sigma^2$ and to infinity for $\mu > \sigma^2$.

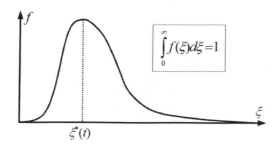

FIGURE 18.5: Probability density of a lognormal distribution.

It follows from the definition of the probability density that the probability that the $x(t)/m$ does not exceed a sufficiently small number ε is

$$p_\varepsilon(t) = \int_0^\varepsilon f(\xi)d\xi.$$

Then, for sufficiently large t, in the case of $\mu < \sigma^2$, the value of $p_\varepsilon(t)$ turns out to be arbitrarily close to unity (see Figure 18.6a), and in the case $\mu > \sigma^2$ this is arbitrarily close to zero (see Figure 18.6b). Hence it follows that if the inequality $\mu < \sigma^2$ is satisfied, the probability of extinction of the population tends to zero with an unlimited increase in time. Taking into account that the parameter μ is the expected value of population growth, we conclude that in the considered model, the extinction of the population is possible even with positive values of this increase. It should be noted that in a real situation, various kinds of fluctuations are inevitable. However, for sufficiently large populations, their influence is insignificant, and the observed values of the population size are quite close to the corresponding expected values. In the case of small populations, we can no longer neglect these phenomena.

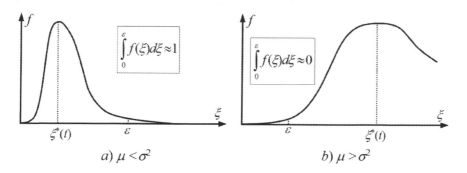

a) $\mu < \sigma^2$ b) $\mu > \sigma^2$

FIGURE 18.6: Estimation of the probability of population extinction.

Task 18.4 *Malthus model with random population growth*. Calculate the considered random process according to the Monte Carlo method. The analysis is carried out when the conditions $\mu < 0$, $\mu = 0$, $0 < \mu < \sigma^2$, and $\mu > \sigma^2$.

2   Models with random parameters

In the previous section, the standard Malthus model was considered with a random parameter. This type of problem arises for almost any previously described systems. Here are some typical examples related to different subject areas.

Chapter 1 discussed the problem of ***missile flight***. The missile was launched at an angle to the horizon and flew in a vertical plane under the action of thrust. At some point in time, the fuel ran out, the missile continued its flight by inertia and after a while landed due to the action of gravitational force. Air resistance was not taken into account here. Suppose that we decided to take into account the influence of the wind acting in the horizontal plane. In this case, the strength of the wind and its direction are considered to be random values.

Unlike the problem discussed in Chapter 1, the missile movement is considered in three-dimensional space. In this case, as the origin of coordinates, as before, the launch point of the missile on the ground is selected, and as the x axis is the direction in the horizontal plane corresponding to the direction of launch of the missile. The vertical coordinate z is directed upward (in the previously considered model it was denoted by y), and the y-axis is perpendicular to the other two axes. The equations of movement of the system are derived on the basis of Newton's second law, with the projections of the traction force and wind force acting in the x direction, the projection of the wind force in the y direction, and the projection of the traction force and weight in the z direction. As a result, we obtain the following system of differential equations

$$m\ddot{x} = F_t \cos\varphi + F_w \cos\alpha, \quad m\ddot{y} = F_w \sin\alpha, \quad m\ddot{z} = F_t \sin\varphi - mg,$$

where m is the missile mass, F_t is the module of thrust force, F_w is the modulus of wind force, φ is the angle in the vertical plane between the x-axis and the direction of missile launch, α is the angle in the horizontal plane between the x-axis and the direction of the wind, g is the acceleration of gravity. The above equations are considered with zero initial conditions up to a certain moment in time, after which the thrust force is assumed to be zero. The process ends at the moment of landing, i.e., when the vertical coordinate of the missile reaches zero. A feature of this model is the fact that the quantities F_w and α are random. For simplicity, we assume that their distribution functions remain constant.

Task 18.5 *Missile flight with random wind.* For the system considered above, specific values of the system parameters are determined, including the distributions of the magnitude and direction of the wind. The resulting system is solved numerically, and the values F_w and α are played at each time step. Calculations are carried out up to the moment the missile lands, determining the maximum height of the missile's rise, the coordinates of its landing point, and the time of movement. For comparison, calculations are carried out with the same parameters in the absence of wind with an explanation of the results obtained.

The second example relates to chemistry. Chapter 6 investigated the ***synthesis reaction*** A+B→C described by the differential equations

$$\dot{a} = -kab, \quad \dot{b} = -kab, \quad \dot{c} = kab.$$

with the corresponding initial conditions, where a, b, and c are the concentrations of the corresponding substances, and k is a process parameter called the reaction rate constant. The constant k can be determined in accordance with the ***Arrhenius equation***

$k = \alpha \exp(-R/ET)$, where α is the preexponential factor, R is the universal gas constant, E is the activation energy, and T is the temperature. Here R is the known constant, the temperature can be measured quite accurately, and the quantities α and E are assumed to be random. Other chemical reactions with random reaction characteristics can be investigated in a similar manner.

Task 18.6 *Synthesis reaction with a random reaction rate constant.* For the system considered above, specific values of the parameters of the system are specified, including the distributions α and E. The resulting system is solved numerically, and at each time step the values of the indicated quantities are played out. For comparison, calculations are carried out with the same parameters when their expected values are chosen as α and E.

The third example considered relates to epidemiology. Chapter 7 introduced the **SIR model** of epidemic propagation. It was described by the equations[35]

$$\dot{S} = -\beta N^{-1} IS, \ \dot{I} = \beta N^{-1} IS - \gamma I, \ \dot{R} = \gamma I.$$

with the corresponding initial conditions, where S, I and R are the numbers of susceptible, infected and recovered, N is the total population, i.e., the sum of individuals of all three above mentioned groups, β is a parameter characterizing the intensity of contacts between individuals, and γ is a parameter characterizing the intensity of recovery of infected individuals. The initial states of the system, as well as the population size, are known. However, the parameters β and γ can be random variables[36].

Task 18.7 *SIR model with random parameters.* For the system considered above, specific values of the system parameters are specified, including the distributions of β and γ. The resulting system is solved numerically, and at each time step, the values of the indicated quantities are played out. Calculations are carried out until the time when all patients will recover with a sufficiently high degree of accuracy. For comparison, calculations are carried out with the same parameters when their expected values are chosen as β and γ.

3 Discrete model of selling goods

Consider the simplest problem of selling goods[37]. We assume that the probability of selling a unit of goods per unit of time is k. Let us denote by $p_j(n)$ the probability that j units of goods are sold at time n. Suppose that at the initial time $n = 0$ the goods were still on sale. Then we have the equalities

$$p_0(0) = 1, \ p_j(0) = 0, \ j > 0.$$

At the moment of time 1, either nothing has yet been sold with probability $1 - k$, or one unit of goods has been sold with probability k. Then

$$p_0(1) = 1 - k, \ p_1(1) = k, \ p_j(1) = 0, j > 1.$$

At time 2, either nothing is still sold with probability $(1 - k)^2$ (there were no sales in both time intervals), or 1 unit of goods was sold with probability $2(1 - k)k$ (perhaps, the sale took place on the first interval and did not take place in the second, and, possibly, vice versa), or 2 units with probability k^2 (during both time intervals the sale took place). Then

$$p_0(2) = (1 - k)^2, \ p_1(2) = 2k(1 - k), \ p_2(2) = k^2, \ p_j(0) = 0, \ j > 2.$$

At the moment of time 3, it may turn out that nothing has been sold with probability $(1-k)^3$ (for all three time intervals, the sale did not take place), or 1 unit of product was sold with probability $3(1-k)^2k$ (in one of three intervals time, the sale took place, and the rest did not), or 2 units with a probability of $3(1-k)k^2$ (for any two of the three time intervals, the sale took place, but not in one), or 3 units with a probability k^3 (in each of intervals, the sale took place).

Similarly, at an arbitrary time n, goods may be unsold with probability $(1-k)^n$, or 1 unit of a product may be sold with a probability $\binom{n}{1}(1-k)^{n-1}k$ or either 2 units with probability $\binom{n}{2}(1-k)^{n-2}k^2$,..., or n units with probability k^n. As a result, for each moment of time n, we find the probabilities[38]

$$p_j(n) = \binom{n}{j}(1-k)^{n-j}k^j,$$

where $j = 0,...,n$. At later times j, this probability is zero. Note that if in Sections 1, 3 and 4 systems with continuous time and discrete distribution of random variables were considered, and in Sections 1 and 2 time and random variables were distributed continuously, then in this case, both time and random variables are distributed discretely.

For a random variable x_j describing the amount of goods sold at time j, one can find the expected value $\mathbf{E}[x_j] = kj$ and variance $\mathbf{Var}[x_j] = k(1-k)j$. Thus, the number of goods sold grows linearly on average. However, over time, more and more deviations of the number of goods sold from its expected value are possible.

4 Passage of a neutron through a plate

In nuclear physics, problems arise in the interaction of neutrons with matter[39]. Let us consider the simplest problem of passing a neutron flux through an infinite plate of a given thickness h. It is assumed that neutrons move in a direction perpendicular to the plate surface. As the neutron moves inside the plate, it interacts with the atoms of the substance that makes up the plate. A characteristic of this interaction is the cross section Σ, which is the ratio of the number of neutrons that have experienced interaction to the total number of neutrons passing through a unit area. This value is a parameter of the process under consideration.

When a neutron interacts with an atom of the plate substance, either it is absorbed by the atom or scattered, in which the neutron continues its movement, but in a new direction. In this case, the interaction cross section Σ is the sum of the absorption cross section Σ_c and the scattering cross section Σ_c. Thus, the ratio Σ_c/Σ describes the probability of absorption of a neutron by an atom, and Σ_s/Σ describes the probability of its scattering. It is also assumed that in the case of scattering, the energy of a neutron does not change after collision with an atom, and any direction of its further movement is equally probable. The length of the neutron path between two collisions λ is called the **mean free path**. We will assume that it is a random variable with a probability density $f(\xi) = \Sigma e^{-\Sigma \xi}$.

The movement of an individual neutron through the plate can have three different outcomes: either it is absorbed by the substance of the plate, or it passes through it, or it is reflected from it and continue moving in the opposite direction. The aim of the study is to assess the likelihood of each of these outcomes depending on the above process parameters. To solve this problem, the movement of a neutron is repeatedly played out, starting with its hitting the plate and ending with either its absorption or going beyond the plate in one direction or another. Then the frequency of each of the considered outcomes is calculated. Note that the considered process at each step of the neutron movement is associated with three independent random variables: the mean free path, the direction of movement, and the realization of absorption or scattering.

Choose the x coordinate directed across the plate, and the point $x = 0$ corresponds to the neutron entry into the plate. The entire process of a neutron moving inside a plate is made up of its individual movements between scattering by atoms of the plate. Let us denote by x_j the coordinate of the neutron in the direction of the x axis at the point of its j-th scattering, assuming $x_0 = 0$. Suppose x_j is known. Determining the subsequent value of x_{j+1} is reduced to playing out the free path and direction of travel.

Using the Monte Carlo method, choose the value of a continuous random variable λ with the above probability density f. By the formula (18.6), the solution l of the equation is chosen as its value

$$\int\limits_0^l f(\xi)d\xi = \int\limits_0^l \Sigma e^{-\Sigma\xi}d\xi = p,$$

where p is an arbitrary number from a unit segment. The last equality is reduced to the form $1 - e^{-\Sigma l} = p$.

Find $l = -\ln(1-p)/\Sigma$. Obviously, the value $q = 1 - p$ also turns out to be an arbitrary number from the unit segment. Thus, as the value of the mean free path λ, one can choose the number $l = -\ln q/\Sigma$ with an arbitrary q from the unit segment.

Thus, we know how far the neutron will travel before the next collision with the atom of the plate. However, it is still necessary to determine the direction of its movement. Equality of all directions of movement corresponds to the choice of an arbitrary angle φ between this direction and the x-axis, which is equivalent to an arbitrary choice of the cosine of this angle from the interval $[-1, 1]$. The density of the uniform probability distribution over some segment of length l is constant and equal to $1/l$. Then, according to formula (18.6), the solution s of the equation

$$\int\limits_{-1}^s f(\xi)d\xi = \int\limits_{-1}^s \frac{1}{2}d\xi = p,$$

where p is an arbitrary number from the unit segment. Find $\cos\varphi = 2p - 1$.

Knowing the direction of movement of the neutron and the distance traveled, find the coordinate of its next interaction in the direction of the x-axis (see Figure 18.7) $x_{k+1} = x_k + c\cos\varphi$. After that, a check is carried out for the possibility of completing the movement of the neutron in the plate. First of all, an arbitrary number is selected from the interval $[0,1]$. If it turns out to be less than the ratio Σ_c/Σ, then it is assumed that the neutron is absorbed by the atom. If this did not happen, but the inequality $x_{k+1} < 0$, is satisfied, then the neutron is reflected from the plate. If this did not happen, but the inequality $x_{k+1} > h$, is true, then the neutron flew through the plate. Otherwise, the neutron continues its motion inside the plate, and a search for a new point of collision of the neutron and the atom of the plate is carried out[40]. After multiple calculations of the neutron movement, the frequencies (the ratio of this outcome to the total number of experiments) of absorption of the neutron, its reflection and transmission through the plate are determined, which were the purpose of the study.

Task 18.8 *Passage of a neutron through a plate*. Perform system calculations based on the described algorithm. Set influence on the result of parameters h, Σ, Σ_c.

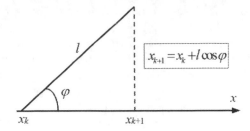

FIGURE 18.7: Movement of a neutron between two interactions.

Notes

[1] **Probability theory** is a branch of mathematics that studies random events, random variables, their properties and operations on them, see [37], [98], [120], [172].

[2] **Mathematical statistics** is a branch of mathematics that develops methods for recording, describing and analyzing observational and experimental data in order to build probabilistic models of mass random phenomena, see [49], [146], [205].

[3] Various stochastic models of biological processes are discussed in [7], [57], [244], [287], [342]. A wide class of stochastic models is also described in [11], [110], [230], [237].

[4] Chapter 19 discusses mathematical models that are problems with non-unique solutions. Each solution corresponds to a possible implementation of the physical processes described by them. The implementation of this or that variant here is carried out randomly.

[5] An event is **random** if it is realized as a result of some experiment, the result of which is not possible to predict accurately.

[6] The **random value** is a central concept in the theory of probability. It characterizes a random event using the concept of probability, which is a measure of the possibility of this event occurring.

[7] The **random process** is a family of random variables with some parameter, usually with the meaning of time, see [35], [76], [161].

[8] **Probability** is a quantification of the likelihood of a random event occurring. The probability of a **certain event** (it most likely occurs) is assumed to be equal to one, and the probability of an **impossible event** (it certainly does not occur) is assumed to be zero. An event that can happen, but does not necessarily happen, has a probability that lies between zero and one. Empirically, the probability is related to the frequency of occurrence of the event under consideration in a sufficiently large series of similar experiments. Probability is axiomatically defined using measure theory, see [37], [129].

[9] Chapter 7 discussed the Verhulst model, in which the derivative of the species number was quadratically dependent on the abundance itself. A stochastic model of population development with a quadratic dependence of the probability of birth and death of an individual on the number of a species is considered, for example, in [288].

[10] A transformation that associates each value of some random variable with the probability that this value takes on a given value is called the **probability function**.

[11] Events are **inconsistent** if the probability that they occur simultaneously as a result of a single experiment is zero.

[12] The **conditional probability** is the probability of one event occurring if the second event occurs. In particular, here we consider the probability of birth of an individual in time dt, provided that the population size takes the given values.

[13] A **probability distribution** is a law that describes the range of values of a random variable and the corresponding probabilities of these values.

[14] **Differential-difference equations** see [29].

[15] Indeed, the equality $\dot{p}_2(t) + 2kp_2(t) = e^{-kt}$ can be transformed to $\frac{d}{dt}\left(p_2 e^{2kt}\right) = ke^{kt}$. After integration with initial condition, we get $p_2(t)e^{2kt} = e^{kt} - 1$. Now we obtain $p_2(t) = (1 - e^{-kt})e^{-kt}$. The equation (18.1) with arbitrary n can be analyzed analogically.

[16] Indeed, find the derivative

$$\dot{p}_n(t) = (n-1)(1 - e^{-kt})^{n-2} + ke^{-2kt} - k(1 - e^{-kt})^{n-1}e^{-kt} = 0.$$

So, we have $e^{-kt} = 1/n$. Thus, the function p_n has the maximum $(n-1)^{n-1}/n^n$ at the time $\ln n/k$.

[17] In this case, we are dealing with a discrete distribution of the population size at an arbitrary time moment t. This random variable can take values n equal to $0,1,2, ...$, moreover, the probability that it takes a specific value x lies on the interval $[0,1]$, and the sum of all such probabilities is equal to one. In particular, using the property of a geometric progression, we establish that the sum over all n probabilities that the population size is equal to n is equal to one at any time

$$\sum_{n=1}^{\infty} p_n(t) = e^{-kt} \sum_{n=1}^{\infty} (1 - e^{-kt})^{n-1} = 1.$$

[18] Indeed, determine $z = 1 - e^{-kt}$. By the inequality $0 < z < 1$, we have

$$\mathbf{E}[x(t)] = e^{-kt} \sum_{n=1}^{\infty} nz^{n-1} = e^{-kt} \frac{d}{dz} \sum_{n=1}^{\infty} z^n = e^{-kt} \frac{d}{dz}\left(\frac{1}{1-z}\right) = e^{-kt}\frac{1}{(-z)^2} = e^{-kt}.$$

[19] Indeed, we have the equality

$$\mathbf{Var}[x(t)] = \mathbf{E}[x(t)]^2 - \left\{\mathbf{E}[x(t)]\right\}^2 = \mathbf{E}[x(t)]^2 - e^{-2kt}.$$

Using the definition of the expected value, we find

$$\mathbf{E}[x(t)] = \sum_{n=1}^{\infty} n^2(1 - e^{-kt})^{n-1}e^{-kt} = e^{-kt} \sum_{n=1}^{\infty} n^2 z^{n-1},$$

where $z = 1 - e^{-kt}$. By the obvious equality

$$n^2 z^{n-1} = \frac{d^2}{dz^2} z^{n+1} - \frac{d}{dz} z^n,$$

determine

$$\sum_{n=1}^{\infty} n^2 z^{n-1} = \frac{d^2}{dz^2} \sum_{n=1}^{\infty} z^{n+1} - \frac{d}{dz} \sum_{n=1}^{\infty} z^n = \frac{d^2}{dz^2}\left(\frac{1}{1-z} - z\right) - \frac{d}{dz}\left(\frac{1}{1-z}\right) = \frac{1+z}{(1-z)^3}.$$

Now we obtain

$$\mathbf{Var}(x) = e^{2kt}(2 - e^{-kt}) - e^{2kt} = e^{kt}(e^{kt} - 1).$$

[20] The **binomial coefficients** $\binom{n}{k}$ appear in the process of expanding the Newton binomial $(1 + x)^n$ in powers of x

$$(1 + x)^n = \sum_{k=0}^{n} \binom{n}{k} x^k.$$

They are determined by the formula $\binom{n}{k} = n!/k!(n-k)!$, where $k!$ denotes the factorial of a number, i.e., the product of all natural numbers from 1 to k, and $0! = 1$.

[21] Nuclear chain reactions see [34], [334].

[22] About chemical chain reactions see [194].

[23] Monte Carlo method see [101], [329]. Various applied problems solved using the Monte Carlo method are given, for example, in [31], [116]. In fact, it makes sense to talk not about one method, but about a group of Monte Carlo methods.

[24] Random variables are **independent** if the occurrence of one of them does not change the probability of the occurrence of the other.

[25] Arbitrary selection of a point from a unit interval on a computer is carried out using a random number generator.

[26] If there are difficulties with finding the value of c from equation (18.6), then other search methods are used, see [101], [329].

[27] Monte Carlo methods have an order of precision $1/\sqrt{N}$.

[28] Taking into account the binomial formula, we obtain

$$\sum_{n=0}^{m} p_n(t) = e^{-mkt} \sum_{n=0}^{m} \binom{m}{n} (e^{kt} - 1)^{m-n} = e^{-mkt}\left[1 + (e^{kt} - 1)\right]^m = 1.$$

Thus, the probability that the random variable $x(t)$ takes one of the values $0, 1, ..., m$ is equal to one.

^{29} About radioactive decay see [34], [334].

^{30} In particular, in [35], the problem of ruining a player is considered, and in [319], the problem of ruining a company.

^{31} These results are obtained by the method of generating functions, see [49].

^{32} It is necessary to find the limit

$$\lim_{b \to a} \mathbf{Var}[x(t)] = 2am \lim_{b \to a} [e^{(a-b)t} - 1])/(a - b).$$

Here the numerator and denominator of the fraction tend to zero, i.e., we are dealing with $0/0$ uncertainty. In accordance with ***L'Hôpital's rule***, if there are functions $f = f(b)$ and $g = g(b)$ such that $f(a) = 0$ and $g(a) = 0$, but $g'(a) \neq 0$, then we get

$$\lim_{b \to a} f(b)/g(b) = \lim_{b \to a} f'(b)/g'(b),$$

if the last limit exists. For our case $f(b) = e^{(a-b)t} - 1$, $g(b) = a - b$. Find $f'(b) = -te(a-b)t$, $g'(b) = -1$. Now we obtain $\lim_{b \to a} \mathbf{Var}[x(t)] = 2amt$.

^{33} This result is a consequence of the central limit theorem.

^{34} For ***lognormal distribution*** see [67].

^{35} One can write a stochastic analogue of the SIR model, like other models of epidemiology, considering infection, recovery, death, etc., a person on a certain interval of time by a random event.

^{36} Chapter 17 considered the model of the development of an epidemic with a finite time spent in a group of contact and various forms of patients. In this case, the time spent in a particular category was considered known. In reality, these values are random. As a result, it seems expedient to consider this model with the corresponding random parameters. Stochastic models of epidemiology are discussed in [320].

^{37} Stochastic models of economic systems are discussed, for example, in [78], [92], [335]. In [173], [70] are described stochastic models of sociology.

^{38} The considered probability distribution is called ***binomial***. Using the binomial formula, for any time n we have

$$\sum_{j=0}^{n} p_j(n) = \sum_{j=0}^{n} \binom{n}{j} (1 - k)^{n-j} k^j = 1,$$

i.e., we are really dealing with a probability distribution.

^{39} The Monte Carlo method in the problem of neutron transmission through a plate is considered, for example, in [329].

^{40} The collection of all values $\{x_k\}$ from the entry of a given neutron into the plate to a specific outcome is a discrete random process called a ***Markov chain***. This means that the probability of each subsequent event is determined solely from the state achieved in the previous step.

Part IV

Additions

Earlier it was noted that the use of mathematical methods for the analysis of various phenomena of nature and society is carried out in three stages. First, a mathematical model of the process is determined by translating the laws established by various sciences into the language of mathematics. Then a qualitative and quantitative analysis of the resulting mathematical problem is used. After that, the interpretation of the established results is given. In the previous chapters, we were engaged in the construction of various mathematical models of different systems, and made conclusions about the features of the corresponding processes, based on the properties of the models that describe them. In principle, one could limit ourselves to this, which explains the title of this part of the book. However, it would also be desirable to answer additional questions.

First, one would like to find out what mathematical properties the model under consideration possesses and what mathematical requirements are imposed on models in general? Further, it is necessary to find out how it is possible to adapt the existing mathematical model describing a phenomenon, in principle, to a specific situation? Finally, how can one go from solving forecasting problems, i.e. elucidation of the development of events under certain conditions of the process, to the control problems, i.e., the choice of these conditions in order to achieve this or that result? The final part of the book is devoted to a discussion of these issues.

Chapter 19

Mathematical problems of mathematical models

As we know, because of modelling the studied phenomenon, a certain mathematical problem is obtained. Let us ask ourselves the question, what general mathematical requirements should be presented to such problems? Based on common sense, one can have the following conclusion. Naturally, the solution of the problem must exist; otherwise, we simply have nothing to work with. Apparently, the solution of the problem should be unique, since, having set up, in principle, some kind of experiment, we get a concrete result. Finally, the parameters of the system, as a rule, are determined with some error. As a consequence, one should be sure that this error will not cause a large distortion of the results.

Unfortunately, these natural requirements, which underlie the concept of the problem well-posedness, are not always realized. Sometimes a reasonably posed problem turns out to be insoluble. The solution of the problem is sometimes not unique. The existing solution does not necessarily depend continuously on the process parameters. Finally, the question of what is the solution of the problem is often nontrivial. The decisive factor here is not the fact that ill-posed problems exist, and not even that the overwhelming majority of mathematical problems turn out to be ill-posed, but the fact that such problems can have a direct physical meaning. In this case, the ill-posedness of the problem turns out to be not a disadvantage, but an advantage of the considered mathematical model.

This chapter provides different types of ill-posed problems. Examples of unsolvable problems, problems with local but not global solvability, non-uniqueness of the problem solutions and the absence of a continuous dependence of the solution on the system parameters are given. The final section discusses classical and generalized solutions to the problem. In Appendix, boundary value problems for nonlinear differential equations are given, which have an essentially not unique solution. Problems are described for which the lack of uniqueness of the solution is consistent with the natural physical meaning of the described phenomenon. In addition, the physical meaning of the generalized solution and the method for substantiating the procedure for deriving mathematical models are discussed.

Lecture

1 Cauchy problem properties for differential equations

Part I considered a large class of processes described by ordinary differential equations with initial conditions, i.e., Cauchy problems. However, these problems sometimes have very nontrivial properties[1]. The general **Cauchy problem** for first-order differential equations has the form

$$\dot{u}(t) = f(t, u(t)), \ t > 0; \ u(0) = u_0, \tag{19.1}$$

where f is a known function, u_0 is a number. Let us show that, depending on the choice of f and u_0, this problem can have very unexpected properties.

Let us define, in particular, $u_0 = 0$; and suppose $f(t, u)$ is equal to -1 for $u \geq 0$ and

DOI: 10.1201/9781003035602-19

1 for $u < 0$. Let the function $u = u(t)$ be a solution of problem (19.1). Consider a time interval $(0, T)$, where this function is continuous and does not change sign. If the function u is not negative, then, proceeding from the form of the equation and the function f, we establish that its derivative is equal to -1 for all $t \in (0, T)$. Thus, in an arbitrarily small neighborhood of zero, the function u must decrease. Then, taking into account the equality $x(0) = 0$, we conclude that this function will certainly become negative, which contradicts the assumption about its behavior, see Figure 19.1a. If the function u is negative on $(0, T)$, then its derivative is equal to one, i.e., the function itself must increase. However, an increasing function satisfying the condition $u(0) = 0$, cannot be negative, see Figure 19.1b. Thus, we again come to a contradiction. As a result, we conclude that this Cauchy problem does not have a solution that is a continuous function even on a small-time interval.

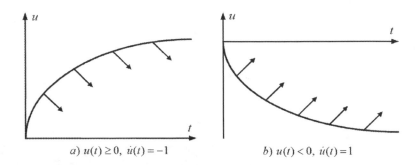

a) $u(t) \geq 0$, $\dot{u}(t) = -1$ b) $u(t) < 0$, $\dot{u}(t) = 1$

FIGURE 19.1: Phase velocity and change in the function are inverse directions.

The established result is explained by the discontinuity of the function f. However, even in the absence of obvious jumps, a similar situation can be encountered. Let the equalities $u_0 = 1$, $f(t, u) = 1/t$ be true. If now there is a solution of the Cauchy problem (19.1), then it is equal to unity at zero and increases arbitrarily rapidly in an arbitrarily small neighborhood of zero. Naturally, no normal function can have such properties. Thus, there is no solution of the problem in this case.

The problems that have arisen are related to the unboundedness of the terms of the equation in the vicinity of zero. However, the absence of both unboundedness and jumps does not insure us from unexpected situations. Let us define the values $u_0 = 1$, $f(t, u) = u^2$. It is easy to verify that the solution of problem (19.1) is $u(t) = 1/(1 - t)$. Obviously, as the argument t approaches unity, this solution increases indefinitely. Thus, the solution of the problem exists only for $t < 1$. In this case, the Cauchy problem is said to have only a **local solution**. A solution that exists at any time interval is called **global**. In this case, it is absent.

A completely unexpected result is observed for the values of the parameters $u_0 = 0$, $f(t, u) = \sqrt{|u|}$. A direct check can make sure that the function $u_\tau(t)$ is equal to zero for $0 \leq t \leq \tau$ and $(t - \tau)2/4$ for $t > \tau$ for any non-negative τ is a solution of the corresponding problem, see Figure 19.2. Thus, this problem has an infinite set of solutions[2].

Thus, Cauchy problem (19.1) may or may not have a solution. An existing solution can be both global and local. If a solution exists, then it is not necessarily the only one[3]. It should be borne in mind that the existence of a unique global solution, like any positive property, is an exception, not a rule[4]. Note that the negative mathematical properties of the model may be a consequence of the corresponding properties of the process under study. In particular, the existence of only a local solution can be associated with the explosion phenomenon, when the characteristics of the process change extremely rapidly over a

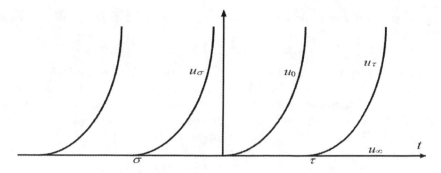

FIGURE 19.2: Set of solutions of Cauchy problem is infinite.

small-time interval, after which the system changes so much that the existing model is no longer able to describe it[5]. If the problem does not have a unique solution, then, perhaps, in fact, there are various forms for the system evolution, see Appendix.

> *Solution of the Cauchy problem for differential equations may not exist, be local and not unique.*

2 Properties of boundary value problems

In differential equations theory, along with the Cauchy problem, ***boundary value problems*** are also considered, in which additional data to the equation are given not at initial point of a given interval, but at both its ends. Such problems are encountered in the description of stationary systems in the one-dimensional case (see Chapter 13), as well as in the analysis of nonstationary systems (see Chapters 10, 11, 12) and variational problems (see Chapter 16).

Consider the linear non-homogeneous second order differential equation

$$u''(x) + au(x) = 1, \ 0 < x < \pi \tag{19.2}$$

with boundary conditions

$$u(0) = 0, \ u(\pi) = 0, \tag{19.3}$$

where a is a positive number, $u'' = d^2u/dx^2$.

From differential equations theory, the general solution of an inhomogeneous equation is the sum of the general solution of the homogeneous equation and a particular solution of the non-homogeneous equation. Equality (19.2) corresponds to the homogeneous equation $u'' + au = 0$, which resembles the equation of a harmonic oscillator, see Chapter 3. Its general solution is determined by the formula $u(x) = c_1 \sin(\sqrt{a}x) + c_2 \cos(\sqrt{a}x)$ with arbitrary the constants c_1 and c_2. A particular solution of equation (19.2) is easiest to look for in the form of a constant. Obviously, it is equal to $u(x) = 1/a$. Thus, the general solution of equation (19.2) is

$$u(x) = c_1 \sin(\sqrt{a}x) + c_2 \cos(\sqrt{a}x) + 1/a. \tag{19.4}$$

To find specific values of the constants, we use the boundary conditions (19.3). Let us find these values for different coefficients a.

For $a = 2$, formula (19.4) takes the form $u(x) = c_1 \sin(\sqrt{a}x) + c_2 \cos(\sqrt{a}x) + 1/2$. Setting

$x = 0$ and $x = \pi$ and using relations (19.3), we establish the system of linear algebraic equations

$$c_2 + 1/2 = 0, \quad c_1 \sin(\sqrt{2}\pi) + c_2 \cos(\sqrt{2}\pi) + 1/2 = 0.$$

Now we find the constants c_1 and c_2 and the unique solution of the boundary value problem.

For $a = 1$, formula (19.4) takes the form $u(x) = c_1 \sin x + c_2 \cos x + 1$. From the boundary conditions, it follows $u(0) = c_2 + 1 = 0$, $u(\pi) = -c_2 + 1 = 0$. These equalities cannot be fulfilled simultaneously. Therefore, the boundary value problem cannot have a solution.

Finally, for $a = 4$ equality (19.4) takes the form $u(x) = c_1 \sin 2x + c_2 \cos 2x + 1/4$. Using equalities (19.3), we get $u(0) = c_2 + 1/4 = 0$, $u(\pi) = c_2 + 1/4 = 0$. We find $c_2 = -1/4$, but the number c_1 is not definite. Thus, the function $u(x) = c_1 \sin 2x + (1 - \cos 2x)/4$ is the solution of the problem (19.2), (19.3) with $a = 4$ for any c_1.

Thus, depending on the values of the parameter a, the considered boundary value problem may turn out to be both solvable and unsolvable. The existing solution may or may not be the only one. The problem has qualitatively different properties for different values of its coefficient[6]. In practice, the coefficients of the equations are related to the process conditions. Thus, if the problem really describes some real phenomenon, then it turns out that under certain conditions, the process does not proceed at all, and under other conditions, events can develop according to different scenarios. Boundary value problems for nonlinear equations have even more difficult properties, see Appendix.

Task 19.1 *Second boundary value problem.* Analyze the second boundary problem for equation (19.2).

Solution of boundary value problems for differential equations
may not exist and may not be unique.

3　Boundary value problems for the heat equation

Consider partial differential equations. We restrict ourselves to the analysis of the ***nonlinear heat equation***, which describes a large class of various phenomena, see Chapters 10 and 11. We have the equation

$$v_t = v_{xx} + f(t, v), \quad 0 < x < L, \, t > 0 \tag{19.5}$$

with initial condition

$$v(x, 0) = u_0, \quad 0 < x < L \tag{19.6}$$

and second order boundary conditions

$$v_x(0, t) = 0, \quad v_x(L, t) = 0, \quad t > 0, \tag{19.7}$$

where u_0 is a constant, f is a given function.

Determine the function v by the formula $v(x, t) = u(t)$, where u satisfies the equalities

$$\dot{u}(t) = f(t, u(t)), \quad t > 0; \quad u(0) = u_0.$$

This is the Cauchy problem (19.1). Obviously, the derivative v_x is zero. Then equalities (19.7) are satisfied trivially, and equalities (19.5), (19.6) reduce to problem (19.1). Thus, any solution of the Cauchy problem (19.1) turns out to be a solution of the boundary value problem (19.5)–(19.7). Consequently, all previously established properties of the Cauchy problem for an ordinary differential equation can be observed for a given boundary value problem with an appropriate choice of its parameters[7].

Consider now the problem

$$v_t = v_{xx} + av - 1, \ 0 < x < \pi, \ t > 0, \tag{19.8}$$

$$v(x,0) = v_0(x), \ 0 < x < \pi; \ v(0,t) = 0, \ v(\pi,t) = 0, \ t > 0, \tag{19.9}$$

where a is a positive constant, v_0 is a given function. Suppose this problem has an equilibrium position $u = u(x)$. Then the derivative v_t tends to zero, and the limit function u satisfies the equalities

$$u''(x) + au(x) = 1, \ 0 < x < \pi; \ u(0) = 0, \ u(\pi) = 0.$$

This is problem (19.2) and (19.3).

For $a = 2$, this problem has a unique solution. Therefore, system (19.8), (19.9) has a unique equilibrium position. For $a = 1$, problems (19.2) and (19.3) do not have a solution at all, which means that the studied system does not reach an equilibrium position due to its absence. Finally, for $a = 4$ the boundary value problems (19.2) and (19.3) have an infinite set of solutions, each of which is the equilibrium position of the original system. The output of the system state to a specific equilibrium position is determined by the value of the initial state v_0. Thus, if the considered boundary value problem really describes some natural phenomenon, then, depending on the values of one of the parameters of the system (i.e., on the conditions of the process), the evolution of the system over large time intervals can occur in qualitatively different ways.

Task 19.2 *Equilibrium position for the heat equation.* Using the finite difference method (see Chapter 11), numerically solve problems (19.8) and (19.9) for different values of the parameter a and the initial state v_0 over large time intervals.

> *Solution of boundary value problems for the heat equation*
> *may not exist, be local and not unique.*
> *Heat equation can have one or many equilibrium positions or none at all.*

4   Hadamard's example and well-posedness of problems

Consider the two-dimensional **Laplace equation** describing a large class of stationary systems, see Chapter 13

$$u_{xx} + u_{yy} = 0 \tag{19.10}$$

in half-plane $y > 0$. We have also the boundary conditions

$$u(x,0) = 0, \ u_y(x,0) = z_k(x), \ 0 < x < \pi, \tag{19.11}$$

where $z_k(x) = k^{-1} \sin kx$. This is the **Hadamard's example**.

The solution of this problem is

$$u_k(x,y) = \frac{\sin kx \sinh kx}{k^2}.$$

Passing to the limit in equality (19.11) as k tends to infinity, we get

$$u(x,0) = 0, \ u_y(x,0) = z_\infty(x), \ 0 < x < \pi, \tag{19.12}$$

where $z_\infty(x) = 0$. Problems (19.10) and (19.12) have the solution $u_\infty = 0$.

As k grows, the sequence $\{u_k\}$ of solutions to problems (19.10) and (19.11) do not converge to the function u_∞. For sufficiently large values of k, the parameters z_k and z_∞ of the considered problems differ by an arbitrarily small amount. At the same time, the corresponding solutions are very far from each other. Thus, in problems (19.10) and (19.11), there is no continuous dependence of the solution on the data at the boundary $y = 0$. Note that in practice, the process parameters (in particular, boundary data) are determined experimentally, and therefore with an error. If the solution of a problem continuously depends on the parameters of the process or, as one says, is **stable in parameters**, then a small error in the data causes a small error in solution of the problem. In the absence of this property, significant distortions of the results are possible. Indeed, let us assume that the value z_∞ corresponds to the true state of the system at the boundary, and z_k characterizes the results of the experiment. For a sufficiently large k, these quantities are arbitrarily close, i.e., the accuracy of the experiment is quite high. However, when analyzing the model, instead of the true value of the state function u_∞, we obtain a very far from it value u_k. Naturally, in this case, the results of mathematical modeling have no practical value. Thus, when solving problems of mathematical physics, the stability of the solution with respect to the parameter is extremely desirable.

As a result, we have the concept of well-posedness of the problem[8]. A problem is **well-posed** if it has a unique solution that continuously depends on the parameters of the system[9]. The considered Hadamards example corresponds to the **ill-posed problem**. Ill-posed boundary problems of mathematical physics are also obtained in the case when two boundary conditions are specified for the heat equation or vibrations of a string at one end of the considered spatial interval, and not one at the other. The boundary value problems for the heat equation are also ill-posed, in which, instead of the initial condition, the condition is specified at the final time[10].

Task 19.3 *Heat equation with inverse time.* Using the finite difference method, solve the first boundary value problem for the heat equation. Carry out calculations for the same parameters in the opposite direction with the data at the end point in time obtained at the previous stage. Compare the results.

> *A boundary problem for the Laplace equation with two conditions*
> *on one of the boundaries is ill-posed, i.e.,*
> *there is no continuous dependence of the solution on the boundary data.*

5   Classical and generalized solution of problems

In mathematics, when analyzing a problem, the statement about the existence of its solution does not make sense without indicating the class to which the desired solution belongs. Even for the simplest algebraic equations, the solution may be absent on one set and exist on a larger set. Thus, the equation $x + 2 = 1$ has no solution on the set of natural numbers, but is solvable on the set of integers; the equation $x^2 = 2$ is unsolvable in the class of rational numbers, but has an irrational solution; there is no real solution for the equation $x^2 = -1$, but its solution exists on the set of complex numbers[11].

In problems of mathematical physics, the concepts of classical and generalized solutions are distinguished. Consider, for example, the **Poisson equation**, see Chapter 13

$$\Delta u(x) = f(x), \tag{19.13}$$

which holds for all points x from some n-dimensional domain Ω and vanishes on its boundary.

The **classical solution** of the considered **Dirichlet problem** is such a twice continuously differentiable function in the entire considered domain, including the boundary[12], which at each point of the domain Ω satisfies equation (19.13) and vanishes on its boundary.

For the existence of a classical solution of the problem, sufficiently stringent requirements[13] are imposed on the function f and the domain Ω, which are far from always being implemented in practice. As a consequence, the solution of the problem is often understood in a weaker sense, i.e., as an element of a larger class. The same logic was used when for the equation $x + 2 = 1$ (respectively, $x^2 = 2$ and $x^2 = -1$), instead of a sufficiently natural set of natural numbers (respectively, rational and real numbers), a much larger set of integers (respectively, real and complex numbers).

The **generalized solution** of the considered Dirichlet problem is a function $u = u(x)$ that is **square-integrable**[14] (i.e., its square is integrable) in the domain Ω with its first order derivatives, vanishes on its boundary and satisfies the equality

$$-\int_\Omega \sum_{i=1}^n \frac{\partial u}{\partial x_i} \frac{\partial \varphi}{\partial x_i} dx = \int_\Omega f\varphi dx \qquad (19.14)$$

for all functions φ, possessing the above set of functional properties[15] for the function u.

Let us establish the relationship between the classical and generalized solutions of the Dirichlet problem. Suppose that the function u is a classical solution of the considered Dirichlet problem. Then it is square integrable along with its first derivatives. Multiply equality (19.13) by an arbitrary function φ with the corresponding set of properties and integrate the result over the domain Ω. After simple transformations[16], we obtain equality (19.14). Thus, any classical solution of the Dirichlet problem is its generalized solution. It is shown in a similar way that a twice continuously differentiable generalized solution of the Dirichlet problem is its classical solution[17]. Naturally, to ensure the existence of a less regular generalized solution of the problem, weaker requirements are imposed on the function f and the domain Ω than for the existence of its classical solution[18]. Thus, the generalized solution of the problem serves as a generalization of its classical solution, since in some cases it turns out to be classical, although in the general case it is not, while the classical solution invariably turns out to be generalized, see Figure 19.3.

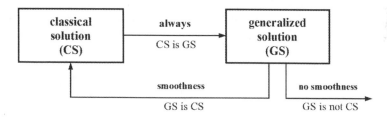

FIGURE 19.3: Relationship between classical and generalized solutions.

The generalized solution is presented as a mathematical abstraction. However, in reality, it can have a direct physical meaning, being a special form of a mathematical model, see Appendix.

Classical solution of the problem is a smooth function,
satisfying the given equation everywhere.
Generalized solution of the problem is a function
possessing weak enough properties, satisfying an integral equality.
Classical solution is always generalized.
Smooth generalized solution is classical.

Direction of further work. We have considered various forms of mathematical models that relate to different subject areas. Each mathematical model is obtained based on the laws of the corresponding subject area and describes not a single phenomenon, but a whole class of processes characterized by these laws. All these processes differ in the parameters that make up the model. The next chapter will investigate the problem of choosing these parameters in such a way that the process under consideration runs in a certain sense in the best possible way.

Appendix

Boundary value problems for linear differential equations are non-trivial. However, problems defined by nonlinear equations have even more surprising properties. Below are two such problems, one of which has an infinite number of solutions, and the number of solutions to the second problem is determined by a coefficient of the equation. Similar effects arise in the Euler and Bénard problems, which have a direct physical meaning. The lack of uniqueness of the solution of the problem has serious meaning[19] and practical importance[20]. In the final two sections, the physical meaning of a generalized solution is explained and the procedure for constructing mathematical models is justified.

1   Nonlinear boundary value problems

In Section 2 of the lecture, a boundary value problem for a second order linear differential equation was considered. Depending on the choice of a parameter, it can have one solution, infinite set, or none at all. Consider now the **nonlinear boundary value problem**

$$u''(x) + u(x)^3 = 0,\ 0 < x < 1;\ u(0) = 0,\ u(1) = 0. \tag{19.15}$$

It can be shown to have an infinite set of solutions[21]. Particularly, the existence of a nonzero solution u_1 of this problem is first proved. Then the function $z_k = z_k(t) = k u_1(kt)$ is defined, where k is an arbitrary natural number. By direct verification, one can verify that other solutions u_k of problem (19.15) are such that the value $u_k(t)$ is equal to $z_k(t)$ if $0 < t < 1/k$, $-z_k(2/k - t)$ if $1/k < t < 2/k$, $z_k(t - 2/k)$ if $2/k < t < 3/k$, $-z_k(4/k - t)$ if $3/k < t < 4/k$, etc. see Figure 19.4.

An even more surprising example is the **Chafee–Infante problem**

$$u''(x) + \mu u(x) - \nu u(x)^3 = 0,\ 0 < x < \pi;\ u(0) = 0,\ u(\pi) = 0,$$

where μ and ν are positive constants. It can be shown[22] that under the inequality $(k-1)^2 < \mu \leq k^2$, the Chafee–Infante problem has exactly $2k - 1$ solutions, where k is an arbitrary natural number. The critical values of the parameter μ, at which the number of solutions to the problem changes, are called **bifurcation points**[23]. In this case, these are the values $\mu_k = k^2$. The phenomenon of bifurcation is widely used in practice, see below.

2   Euler's elastic problem

A thin elastic inhomogeneous rod of unit length is considered. Suppose one its end is fixed and the other is acted upon by a force, compressing the bar, see Figure 19.5*a*. Let us introduce the x-axis, directed along the bar, with the origin at its fixed point. The

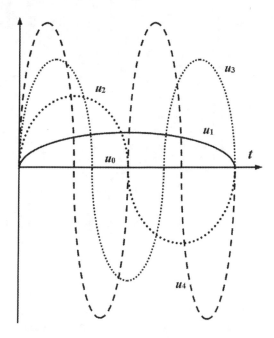

FIGURE 19.4: Solutions of the problem (19.15).

state function here is the deviation $u = u(x)$ of the rod from the equilibrium position corresponding to the coordinate axis.

This system is described by the boundary value problem[24]

$$u''(x) + \mu\rho(x)\sqrt{1 - (u'(x))^2}u(x) = 0, \ 0 < x < 1; \ u(0) = 0, \ u(1) = 0, \qquad (19.16)$$

where ρ is the density of rod, a parameter μ is proportional to the force.

Obviously, problem (19.16) has a trivial solution identically equal to zero. Thus, under the action of the force, the rod, in principle, can remain in its natural state. It can be shown that for sufficiently small values of the parameter μ there are no other solutions. However, if the number μ exceeds some critical value μ' (bifurcation point), then the problem has a positive solution u_1, see Figure 19.5b. Thus, the rod bends under a relatively large force. Besides, the function u_2 is also a solution, which differs from u_1 only in sign. Consequently, the bending of the rod can occur in both the positive and negative directions. Thus, if for $\mu \geq \mu'$ the boundary value problem has three solutions that are positive, negative and zero.

This result may cause some confusion. In this case, a really existing phenomenon is being investigated. On the one hand, we are confident that the boundary value problem (19.16) describes the state of the system with some degree of accuracy. On the other hand, it obviously has more than one solution. At the same time, one can set up an experiment and measure the position of the rod at the appropriate value of the acting force. The result of the experiment is a specific shape of the rod, but not three at once. However, setting up the experiment a second time, under the same conditions, we can observe a different position of the rod, corresponding to a different solution of the boundary value problem. Having a sufficiently large amount of experimental information, we note that in about half of the cases, the rod bends down, and in half bends up. We encountered similar properties in Chapter 18 when analyzing stochastic models. The third solution of the boundary value

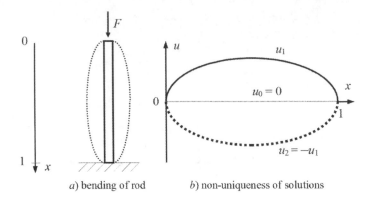

a) bending of rod   *b*) non-uniqueness of solutions

FIGURE 19.5: Eulers problem has non-unique solution.

problem (trivial) at large values of the force characterizes the unstable state of the rod, which is practically not realized.

The experimental results show that under these conditions the system can be in two states, which are solutions to the boundary value problem. However, we cannot predict which of these states will be realized as a result of a particular single experience. Both possible states are achieved with equal probability. Thus, the considered stationary system itself is not deterministic. The mathematical model that characterizes it has similar properties. Consequently, the ambiguous solvability of the considered boundary value problem is an advantage of the model, since this property corresponds to the physical meaning of the studied system.

3   Bénard problem

A layer of fluid is considered, located in a gravitational field between two infinite parallel plates. Let us assume that the temperature u_1 is maintained on the lower plate, and u_2 on the upper one, with $u_1 > u_2$. The state of the system is characterized by temperature and velocity vector and is described by the corresponding partial differential equations, see Chapters 10 and 14. If the temperature difference between the plates is small enough, and the distance between them is relatively large, then the system of considered equations in a stationary state have a unique solution. In this case, the velocity of the liquid is zero, and the temperature gradually changes from the value u_1 in the lower part of the liquid layer to the value u_2 in its upper part. With a relatively large temperature difference between the plates and a small thickness of the liquid layer, the situation changes significantly.

It is known that when heated, bodies expand, i.e., increase in volume. This means that the density of a substance decreases when heated. Taking into account the temperature difference between the plates, we conclude that the upper layers of the liquid are denser, and the lower ones are less dense. Liquid molecules are in constant chaotic motion. As a result, some part of the molecules of the upper layer may be lower, i.e., in the lighter layer. Due to the difference in density, this liquid droplet will have a greater weight compared to neighboring droplets at the same level. If the difference in weight turns out to be large enough (this is observed just at a large temperature difference between the plates and a small thickness of the liquid layer), then the drop falls down, see Figure 19.6*a*. On the other hand, in the lower part, some liquid drop may appear higher, in a denser layer. In this regard, according to Archimedes' law, a buoyant force directed upward will act on it. This force can cause hot droplets to float, see Figure 19.6b.

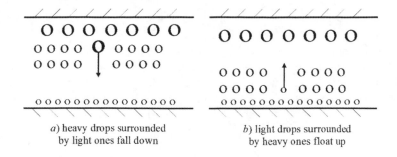

a) heavy drops surrounded
by light ones fall down

b) light drops surrounded
by heavy ones float up

FIGURE 19.6: Formation of convective fluid flows.

Thus, under certain conditions, convective flows appear in the layer, i.e., heat fluxes due to the mechanical movement of fluid. The result is fluid circulation. Along its length, isolated cells of circulating fluid are formed, called **Bénard cells**, see Figure 19.7. This effect is observed both in the study of the mathematical model and experimentally. Depending on the values of the process parameters, the appearance of Bénard cells and even their sizes can be predicted. However, neither natural nor a numerical experiment makes it possible to predict the direction of fluid circulation in a particular cell. Cells with different circulation directions correspond to different system solutions[25].

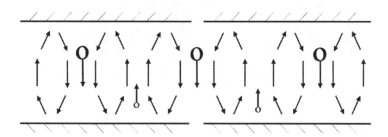

FIGURE 19.7: Bénard cells.

4 Generalized model of stationary heat transfer

Earlier, a generalized solution of the Dirichlet problem for the Poisson equation was defined. This was understood as a certain function with a corresponding set of properties that satisfies the integral relation (19.14). A natural question arises: is this concept a mathematical abstraction that is of interest exclusively to specialists in the theory of partial differential equations, or can it have a direct physical meaning?

One gets the impression that the first statement is true. Indeed, the integral equality (19.14) appears to be the result of some manipulations with the Poisson equation. Even the very term "generalized solution of the Dirichlet problem" suggests that the direct object of research is the Dirichlet problem for the Poisson equation, and equality (19.15) is only a means for its analysis.

Let us consider an example. Determine the temperature distribution in a thin long body in a stationary case under the action of a heat source, provided that zero temperature is maintained at the ends of the body. In Chapter 13, it was noted that the mathematical model of this system is the equation

$$\frac{d}{dx}\left(\lambda\frac{du}{dx}\right) = -F, \ \ 0 < x < L \tag{19.17}$$

with homogeneous boundary conditions, where u is temperature, λ is thermal conductivity coefficient, F is density of heat sources. The **classical solution** to the problem is a twice continuously differentiable function on the segment $[0, L]$ that satisfies the indicated conditions. Its **generalized solution** is a function u that is square-integrable together with its derivative on a given interval, vanishes at its ends and satisfies the equality

$$\int_0^L \lambda(x)\frac{du(x)}{dx}\frac{d\varphi(x)}{dx}dx = \int_0^L F(x)\varphi(x)dx \tag{19.18}$$

for any function φ, which has the same functional properties as u.

Equation (19.17) was obtained (see Chapter 13) as a result of passing to the limit as $\Delta x \to 0$ in the equalities

$$\frac{q(x) - q(x + \Delta x)}{\Delta x} + \frac{1}{\Delta x}\int_x^{x+\Delta x} F(\xi)d\xi = 0, \ \ q(x) = -\lambda(x)\frac{u(x) - u(x - \Delta x)}{\Delta x}, \tag{19.19}$$

directly describing the physical process, where q is the heat flux. We multiply the first equality in (19.19) by a sufficiently smooth function φ that vanishes at the boundaries of a given interval, and integrate the result over this interval. The result can be written as

$$\int_0^L q(x)\frac{\varphi(x) - \varphi(x - \Delta x)}{\Delta x}dx + \frac{1}{\Delta x}\int_0^{\Delta x} q(x)\varphi(x - \Delta x)dx-$$

$$-\frac{1}{\Delta x}\int_L^{L+\Delta x} q(x)\varphi(x - \Delta x)dx + \int_0^L \frac{1}{\Delta x}\int_x^{x+\Delta x} F(\xi)d\xi\varphi(x)dx = 0.$$

Passing here to the limit as $\Delta x \to 0$ taking into account the mean value theorem, the second condition in (19.19), and the vanishing of the function φ on the boundary, we obtain relation (19.18). Thus, the equality underlying the definition of the generalized solution of the problem can be obtained without referring to the differential equation (19.17) directly from conditions (19.19). Thus, the generalized solution is not a consequence of this equation, but has a direct physical meaning. As a result, relation (19.18) itself is a mathematical model of this system, which is naturally called its **generalized model**. Then equation (19.17) with the corresponding boundary conditions can be called the **classical model**[26] of this system.

5   Sequential model of stationary heat transfer

Discussing the mathematical problems associated with modeling various processes, we can still ask ourselves the question of how justified the procedure for constructing a mathematical model in each specific case is. In particular, for stationary heat transfer, the above were given a classical model characterized by a boundary value problem for equation (19.17) and a generalized model defined by equality (19.18). Both of these relations were obtained as the results of passing to the limit in equalities (19.19). Passage to the limit is a fundamental concept of mathematical analysis and in each specific situation needs a rigorous justification. The question arises, under what conditions can we really perform the above actions?

It is easy to verify that for this it is sufficient that the considered function u is in the first case twice continuously differentiable, and in the second one is integrable with the square together with its derivative, i.e., met exactly the requirements that are included in the definition of a classical and generalized solution. These properties can indeed be established by means of the theory of differential equations with appropriate restrictions on the parameters of the problem, i.e., functions F and λ. However, these results are established in relation to the already obtained relations (19.17) and (19.18), and not at the stage of their construction. We can justify the passage to the limit and obtain model data if the corresponding solutions have the required properties. However, these properties are justified after the models are obtained.

It should be noted that a similar situation arises when defining the very concept of a **limit**. As is known, the sequence $\{x_k\}$ converges to a value x if all elements x_k with sufficiently large numbers turn out to be arbitrarily close to x. Following this definition, one can check whether the elements of a given sequence will approach a particular x without bound. However, in practice, we most often have a sequence, but do not know its limit. Moreover, we cannot even be sure of its convergence. The result is a situation similar to that described above. In order to be convinced of the convergence of a sequence, one must know its limit in advance. But we can get this limit only as a result of sequence convergence.

In mathematical analysis, this difficulty is circumvented using the **Cauchy criterion**. The concept of a **Cauchy sequence** is introduced, the defining property of which is the unlimited convergence of its elements with each other. According to the Cauchy criterion, any Cauchy sequence of real numbers has a limit. In order to make sure whether we are dealing in a particular case with the Cauchy sequence, one should operate exclusively with the elements of the sequence itself. Knowledge of the limit and even the fact of convergence is not needed here. But if we really have a Cauchy sequence, then the existence of the limit is guarantee[27].

Let us try to use a similar method to determine a mathematical model of stationary heat transfer. Divide the segment $[0, L]$ into M equal parts with a step $h = L/M$, which is chosen as Δx. In equalities (19.19) we determine $x = x_i = ih$, $i = 1, ..., M - 1$. The result is the **discrete stationary heat transfer model** discussed in Chapter 17

$$\delta_x\big[\lambda(x_i)\delta_{\bar{x}}u(x_i)\big] = f_i, \ i = 1, ..., M - 1, \tag{19.20}$$

where

$$f_i = -\frac{1}{h}\int\limits_{x_i}^{x_{i+1}} F(\xi)d\xi, \ \delta_x q_i = \frac{q(x_{i+1}) - q(x_i)}{h}, \ \delta_{\bar{x}}u(x_i) = \frac{u(x_i) - u(x_{i-1})}{h}.$$

The system of linear algebraic equations (19.20), supplemented by the boundary $u_0 = 0$, $u_M = 0$, can be solved using the tridiagonal matrix algorithm, see Chapter 11. Denote by $u_h = u_h(x)$ the result of linear interpolation of values $u_0, u_1,...,u_M$, determined by the step h. In particular, as h one can choose elements of some sequence $\{h_k\}$ of positive numbers tending to zero. Then, for any number k, one can define the corresponding function v_k, equal to the interpolation u_h for $h = h_k$. Note that the construction of the functional sequence $\{v_k\}$ is carried out without any a priori assumptions. If now it is possible to show that this sequence is in some sense a Cauchy sequence, then we can establish its convergence[28]. In this case, this sequence can be considered a***sequential model*** of the system under consideration. It can be shown[29] that, under certain restrictions on the functions F and λ, the sequence $\{v_k\}$ converges in a sense to the solution of problem (19.18). Then the procedure for constructing a generalized model of the system under consideration turns out to be justified. If the corresponding limit turns out to be twice continuously differentiable, then from equality (19.18) one can go over to equation (19.17), justifying the process of constructing a classical model.

The above reasoning can be explained as follows. We have a sequence of rational numbers $\{x_k\}$. Perhaps its elements do not approach each other at all, for example, for $x_k = k$. Then we do not have a Cauchy sequence, and $\{x_k\}$ does not serve as a "sequential model" of any object. It is possible that the elements of the considered sequence approach without limit, for example, for $x_k = 1/k$. Moreover, this sequence converges to zero, which serves as an analogue of the solution to problem (19.20), i.e., generalized state of the system. Thus, this sequence can be considered as a "sequential model" of zero, as long as its elements with sufficiently large numbers can approximate zero with an arbitrary degree of accuracy. However, we now consider a sequence with elements $x_1 = 3$, $x_2 = 3.1$, $x_3 = 3.14$, $x_4 = 3.141$, $x_5 = 3.1415$, $x_6 = 3.14159$, etc. The elements of this sequence also converge indefinitely, but the sequence does not converge on the set of rational numbers, which serves analogous to the situation when we cannot obtain a generalized model of the system using the previously described sequence $\{v_k\}$. Nevertheless, the given sequence $\{x_k\}$ is associated with a number π that lies outside the set of rational numbers[30]. It can be considered a sequential model, since its elements (rational numbers) approximate the number π (irrational number) with any degree of accuracy. This corresponds to the case when a sequential model of the system exists, but is not reduced to its generalized model.

Notes

[1] For the properties of the Cauchy problem for differential equations, see, for example, [14], [59], [132], [168], [212], [275], [346]. Similar problems for partial differential equations are considered, for example, in [141], [192], [193], [215], [235], [274], [345], [361].

[2] The set of all possible solutions of the considered Cauchy problem is not just infinite, but even uncountable, i.e., the elements of this set cannot be enumerated.

[3] The uniqueness of the global solution to the Cauchy problem is certainly realized in the linear case, including for systems of linear differential equations and for linear equations of higher orders.

[4] Indeed, the absolute majority of real numbers are irrational, the absolute majority of functions are discontinuous, and the absolute majority of continuous functions are nonsmooth.

[5] In particular, chain chemical reactions lead to a system of differential equations having only a local solution, see [176], [213]. In this case, the mathematical model describes the process under study only up to the moment of explosion.

[6] In this case, everything is determined by whether the number belongs to the spectrum of the operator corresponding to the given boundary value problem. If it does not belong to the **spectrum**, then the boundary value problem is uniquely solvable. Otherwise, the problem either has no solution at all, or has an infinite set of solutions. For the spectrum of boundary value problems for differential equations see [132], [153], [283]. Similar problems arise for different problems. For example, the system of linear algebraic equations $x + y = 1$, $2x + y = 2$ has a unique solution $x = 1$, $y = 0$. Changing only one numerical parameter, we get the system $x + y = 1$, $2x + 2y = 2$, which has an infinite set of solutions. If one numerical parameter is changed here, then the system $x + y = 1$, $2x + 2y = 3$ is obtained, which has no solution at all.

[7] About more meaningful examples of non-unique solutions or non-global existence of solutions for nonlinear partial differential equations, see [141], [216].

[8] About ill-posed problems and methods for solving them, see [4], [19], [170], [123], [133], [207], [216], [350]. Most of optimal control problems and inverse problems of mathematical physics are among the ill-posed ones, see Lectures 20 and 21.

[9] Chapter 21 defines the concept of well-posedness of optimal control problems.

[10] Chapter 21 considers a method for solving this problem, based on its interpretation as an inverse problem. In contrast to the equation of heat conduction, the vibration string equation with final (rather than initial) states gives a well-posed problem. If the process of heat conduction is accompanied by an increase in entropy and, therefore, is not reversible in time, then the string vibration (without regard to friction, etc.) occurs at a constant entropy. In the second case, a dynamic group of transformations stands behind the mathematical model (see Chapter 5), while in the first case we are dealing only with a dynamic monoid of transformations.

[11] It is extremely important that the expansion of numerical classes has historically been carried out precisely with the aim of ensuring the existence of a solution in a situation when there is no solution understood in the "natural" sense. Particularly, if necessary, for example, dividing one apple into three parts, it was necessary to substitute integers numbers by fractional ones. If it was necessary to register debt in trade operations, then one changes positive numbers by negative ones. When determining the length of the hypotenuse of an isosceles right-angled triangle, irrational numbers had to be introduced. Thus, whenever in practice difficulties arose in determining the solution to a problem on a known class of objects, this class was extended so that the problem being solved became solvable.

[12] Classical solution is an element of the $C^2(\bar{\Omega})$ *space* of twice differentiable functions on the closure $\bar{\Omega}$ of the set Ω.

[13] About the existence of a classical solution to the Dirichlet problem for the Poisson equation see, for example, [193].

[14] More exact, one considers the *space* $L_2(\Omega)$ space of Lebesgue measurable functions u such that $\int_{\Omega} |u(x)|^2 dx < \infty$, where the integral is understood in the sense of Lebesgue. To be more precise, it should be noted that two measurable functions coinciding on a set of zero Lebesgue measure are assumed to be equivalent and are understood as the same element of the space $L_2(\Omega)$. The space $L_2(\Omega)$ can be found in any functional analysis course, see, for example, [153], [185], [283].

[15] The set of functions belonging to the space $L_2(\Omega)$ with all their first derivatives is called the *Sobolev space* $H^1(\Omega)$. In this case, both the generalized solution of the problem and an arbitrary function φ are objects of the $H_0^1(\Omega)$ *space* for which, in addition, the values of the function on the boundary surface must be equal to zero. This set is also called the Sobolev space. About Sobolev spaces see [215], [330]. Note that, since the considered functions are only integrable, their derivatives should be understood in a generalized sense defined using distribution theory, see [215], [330].

[16] Indeed, we have the equality

$$\int_{\Omega} \Delta u(x)\varphi(x)dx = \int_{\Omega} f(x)\varphi(x)dx.$$

The integral on the left is transformed using Green's formula

$$\int_{\Omega} \Delta u(x)\varphi(x)dx = -\int_{\Omega} \sum_{i=1}^{n} \frac{\partial u}{\partial x_i} \frac{\partial \varphi}{\partial x_i} dx + \int_{S} \frac{\partial u}{\partial n} dx,$$

where S is the boundary of Ω. Taking into account that the second integral on the right-hand side is equal to zero due to the properties of the function φ, we derive formula (19.14) from the previous equality.

[17] Indeed, if a twice continuously differentiable function u satisfies equality (19.14), then as a result of applying Green's formula, taking into account the equality of the function φ on the boundary, we obtain the equality

$$\int_{\Omega} (\Delta u - f)\varphi dx = 0.$$

Hence, since φ is arbitrary, we derive the Poisson equation.

[18] About the existence of the generalized solution to the Dirichlet problem for the Poisson equation see [193], [215], [235], [361].

[19] The non-uniqueness of the solution to the problem is evidence of the lack of determinism of the system. In this direction, mathematical modelling comes into contact with the probability theory. So, when you flip a coin, there are two possible realizations of the state of the system. The situation here is similar to that which arises in the Euler or Bénard problems. Given the enormous influence of random phenomena in the world around us, one can come to the conclusion that the non-uniqueness of the solution should be encountered quite often in practice. The ambiguity of solving mathematical models is a form of manifestation of the relative, but not absolute, cognizability of the surrounding world. By analyzing a complex model, we can indicate its possible solutions, i.e., predict admissible system state realizations. We know that when a coin is tossed, either an "heads" or a "tails" appears, that under a certain action on the rod it will bend, that under

certain conditions Bénard cells of certain sizes appear in the liquid layer. However, in principle, we cannot predict in advance which side the coin fell, in which direction the rod bends, in which direction the liquid circulates. The ambiguity of the admissible states of the system has another important aspect. If the solution to the problem were always unique, if only one state of the system were possible in any practical situation, then the world would be strictly deterministic. In such a world, everything would be predetermined. There would be no room left for the freedom of human will. Only the presence of non-uniqueness of the solution allows us to realize our choice of one state from a certain number of possible ones, and to be responsible for our actions. In a strictly determined world, there can be no question of any choice. Another aspect of the non-uniqueness of the solution to the problem is related to the problem of its numerical solution. Naturally, when solving the problem numerically, we can get any one solution. Finding a specific solution from several possible ones is due to the peculiarities of the applied algorithm. In particular, in the case of using iterative methods for solving the problem, it is possible that for some initial approximations we observe one solution, and for others, another. The situation here is similar to setting up a natural experiment.

[20] Numerous practical applications such as Euler and Bénard are discussed in [128], [237], [239], [256], [277]. These questions are related to non-equilibrium thermodynamics, see [122], [277].

[21] A proof of the non-unique solvability of boundary value problem (19.15) in a much more general multidimensional case is given in [216].

[22] The analysis of the Chafee–Infante problem is carried out in [53], [141].

[23] The phenomenon of bifurcations is discussed, for example, in [141], [153], [157].

[24] See, for example, [351].

[25] As in the Euler problem, the considered model also admits a trivial solution in which there is no circulation at all. However, under the conditions of the existence of a nontrivial solution, the zero state of the system corresponds to an unstable equilibrium position of the corresponding nonstationary problem and is not realized in practice.

[26] An argument against the recognition of relation (19.18) as a mathematical model of the considered phenomenon could be the presence in its composition of the function φ, chosen arbitrarily. What arbitrariness can there be when it comes to describing a specific physical phenomenon? However, the properties of a system can be judged not only by directly measuring its state, but also by its response to external influences. An arbitrary function in the considered equality can be interpreted as an impact on the system, and the equality itself for a concrete function φ can be the system's response to this impact. Another question requiring clarification is related to the practical application of equality (19.18). As is known, the corresponding boundary value problem can be solved numerically on the basis of the finite difference method, see Chapter 11. The method is based on the approximate differentiation formulas. They are derived using the expansion of a function in a Taylor series, which means that they are related to its differentiability. In the definition of the Sobolev spaces, with which the concept of a generalized solution is associated, not classical, but generalized derivatives are used. In this regard, there may be doubts about the possibility of a practical solution of problem (19.18) without addressing the corresponding boundary value problem. However, it can be shown that for generalized derivatives the formulas of numerical differentiation remain valid, and the finite difference method is applicable also for the direct solution of problem (19.19), see [314].

[27] In practice, one has to work not only with sequences of real numbers. In this connection, the question arises about the applicability of the Cauchy criterion in the general case. All mathematical spaces are divided into two classes. There are **complete spaces**, where the Cauchy criterion is applicable, and incomplete spaces, where we cannot use it. For example, spaces of rational and positive numbers, as well as spaces of continuous and Riemann integrable functions with integral metric, are not complete. Spaces of Lebesgue integrable functions and Sobolev spaces are complete. The notion of completeness underlies **Banach** and **Hilbert spaces**, which are widely used in practice. In the case of an incomplete space, one can proceed to their **completion**, establishing the convergence of any Cauchy sequence on a larger set. In particular, the completion of the set of rational numbers is the set of real numbers, and the completion of the space of continuous and Riemann integrable functions with an integral metric is the set of Lebesgue integrable functions. Completeness is one of the most important in functional analysis, see [153], [185], [283].

[28] Naturally, the corresponding limit is an object of the space on which the Cauchy sequence is considered if this space is complete. Otherwise, it turns out to be an element of the corresponding completion.

[29] The corresponding proof is given in [314]. Moreover, the procedure for this proof is similar to the justification of the convergence of the finite difference method, see [280], [301], [325], [340], [347].

[30] The number π belongs to a much larger set of real numbers, which is the completion of the original set of rational numbers.

Chapter 20

Optimal control problems

As noted many times, mathematical models include state functions, independent variables, and system parameters. In the process of analyzing mathematical models, we tried to determine the dependence of the system state functions on independent variables for given values of the system parameters. From a practical point of view, this allows solving forecasting problems, i.e., predict the behavior of the studied system under certain external conditions. However, mathematical modelling provides additional opportunities that are of extremely important universal significance.

Note that the development of mankind has been largely extensive. With the steady increase in the population and the constant growth of its needs, new lands were developed, which provided a person with everything he needed, i.e. food, clothing, building materials, etc. Over time, the zones of cattle breeding and agriculture expanded. New mineral deposits were developed. More and more powerful sources of energy were mastered. With the development of science and technology, new materials have been used. Transport and communication facilities were modernized. In a negligible period of time by historical standards, a person moved from a cart to a car and an airplane, mastered electricity and the energy of the atom, defeated many deadly diseases, mastered a TV and a computer, went into outer space and landed on the moon. All this instilled confidence in the infinity of human capabilities.

However, over time, it became more and more clear that the possibilities of extensive development of society are very limited, that it is necessary to soberly assess our forces and means and try to reasonably dispose of them. Our problem is to choose among all possible scenarios for the development of events the one that best suits our interests. Returning to mathematical modelling, this corresponds to such a selection of permissible system parameters that provided the best result. As a result, we get optimal control problems[1]. From a mathematical point of view, we would like to minimize or maximize some quantities that depend on the system state functions. The problems of finding extrema have already been considered in Chapter 16. However, there the problem of extremum itself turned out to be a mathematical model of the process under study, and the question of choosing the conditions for its course was not posed. In this case, a mathematical model of the system has already exist, and we are talking about its use for active intervention in the considered system in order to find its optimal state[2].

The lecture begins with an analysis of the well-known problem of maximizing the shell flight range, which is reduced to finding the maximum of a function of one variable. As a consequence, its solution can be found without using special methods. Its generalization is the problem of maximizing the missile flight range[3], which is based on the mathematical model described in Chapter 1. Starting from this problem, the general problem of optimal control is posed. After that, one of the most important methods for solving such problems is presented, that is the maximum principle. It is used to solve the problem of maximizing the missile range, as well as the time-optimal control problem. In the Appendix, another applied optimal control problem is formulated, and practical methods for solving such problems are described.

DOI: 10.1201/9781003035602-20

Lecture

1 Maximizing the shell flight range

We begin the consideration of optimal control problems with a fairly simple well-known example related to maximizing the range of a shell. The shell is launched at an angle to the horizon. It is required to choose this angle in such a way that it flew away as far as possible.

Consider the movement of the shell in the vertical plane, with its launch point chosen as the origin of coordinates, the horizontal coordinate x in the direction of its movement, and the vertical coordinate y in the vertical direction. It is assumed that at the initial moment of time $t = 0$ the shell is imparted with velocity v, and it is launched at an angle φ to the horizon. Thus, the initial state of the system is characterized by the equalities

$$x(0) = 0, \ y(0) = 0, \ \dot{x}(0) = v \cos \varphi, \ \dot{y}(0) = v \sin \varphi. \tag{20.1}$$

In the process of movement, only the gravitational force (weight) P acts on the shell, directed vertically downward. So, the shell movement is described by the equations

$$m\ddot{x} = 0, \ \ m\ddot{y} = -P, \tag{20.2}$$

where m is the mass of shell. Solving equations (20.2) with initial conditions (20.1), determine the position of the shell, i.e., its coordinates at any time.

The shell moves until it lands, i.e., until such moment of time T, when its vertical coordinate equals zero, i.e., we get

$$y(T) = 0. \tag{20.3}$$

Then the flight range L is determined by its horizontal coordinate at the moment of landing, see Figure 20.1. Thus, the optimal control problem is to find a number φ that maximizes the value $L = x(T)$, where x is a solution to problem (20.1), (20.2), and T is determined from equality (20.3). Note that the dependence of the value of L on the sought value φ is not direct. By changing the control φ, by means of the equations of state of the system, we change the state function x, which directly determines the optimality criterion L.

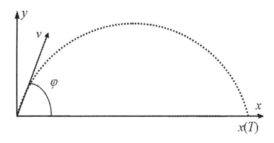

FIGURE 20.1: Flight of shell.

Because of the significant simplicity of the state equations and the optimality criterion, we can find a solution to this problem without using any special mathematical apparatus. Indeed, as a result of integrating equations (20.2) taking into account the second pair of initial conditions (20.1), we find the derivatives $\dot{x} = v \cos \varphi$, $\dot{y} = -gt + v \sin \varphi$, where g is the gravity acceleration. Repeated integration taking into account the first pair of initial conditions (20.1) leads to the equalities $x(t) = tv \cos \varphi$, $y(t) = gt^2 + tv \sin \varphi$. Using condition (20.3), we obtain a quadratic equation for the shell flight time

$$y(T) = gT^2 + Tv \sin \varphi = 0.$$

Since the zero solution has no physical meaning[4], we define the time $T = 2v \sin \varphi/g$. As a result, we determine the explicit dependence of the shell flight range on the launch angle

$$L = L(\varphi) = x(T) = Tv \cos \varphi = 2v \sin \varphi \cos \varphi/g = v \sin 2\varphi/g.$$

Thus, the considered problem has been reduced to finding the maximum of a function of one variable. To find it, we equal to zero the derivative of this function. As a result, we have the equality $2v \cos 2\varphi/g = 0$. This implies the desired value $\varphi = \pi/4$. Thus, in order for the shell to fly away as far as possible, it should be launched at an angle of 45^0.

If it is possible to define an explicit dependency
of the optimality criterion from the desired parameter,
then the optimization problem is solved without using special methods.

2   Maximizing the missile flight range

The optimization problem from the previous section was reduced to finding the maximum of a function of one variable, which is not typical. Consider its natural generalization, particularly, maximizing the missile flight range, which is a much more complicated problem. If we can influence the flight of the shell only at the initial time, then the missile can be controlled during the entire time of its active flight. There is a fixed time interval, due to the fuel reserves, when the missile flies under the action of the thrust force. At this time, one can choose the course of the missile. After the engine stops working, the missile continues on its way by inertia and lands after a while. As with the shell, our goal is to maximize the range of flight.

The mathematical model of this process was considered in Chapter 1. As in the case of the shell, we assume that the missile moves in a vertical plane, with the origin corresponding to the launch point, the vertical coordinate upward, and the horizontal coordinate in the direction of the movement, see Figure 20.2. The entire missile flight consists of two stages. At the first stage, the movement is carried out under the action of the traction force F, and at the second, there is no traction force. We neglect the air resistance, as well as the mass of the fuel. The duration of the active flight of the missile T is known.

By Newton second law, the movement of a missile over the time interval $(0, T)$, i.e., in its active flight is characterized by the equations

$$m\ddot{x} = F_x, \quad m\ddot{y} = F_y - P,$$

where m is the missile, P is its weight, F_x and F_y are horizontal and vertical components of the traction force, see Figure 20.2. We get $F_x = F \cos u$, $F_y = F \sin u$, and $P = mg$, where u is the missile heading angle, and g is the gravitational acceleration. Then we obtain the equations of movement

$$\ddot{x} = a \cos u(t), \quad \ddot{y} = a \sin u(t) - g, \tag{20.4}$$

where $a = F/m$ is the acceleration of the traction force that is the system parameter.

The missile is at the origin and is at rest at $t = 0$. So, we get the initial conditions

$$x(0) = 0, \ y(0) = 0, \ \dot{x}(0) = 0, \ \dot{y}(0) = 0, \tag{20.5}$$

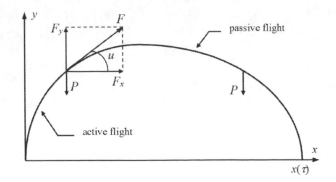

FIGURE 20.2: Flight of missile.

Choosing a function $u = u(t)$ and solving equations (20.4) with initial conditions (20.5), one can find the coordinates and the velocity vector of a moving missile. This makes it possible to determine these characteristics at the time T of switching off the motors.

Further, at $t > T$, the missile flies by inertia with a constant acting gravitational force. Its free flight is described by the equations

$$\ddot{x} = 0, \ \ddot{y} = -g.$$

Its general solution is $x(t) = \alpha_1 t + \beta_1$, $y(t) = -gt^2/2 + \alpha_2 t + \beta_2$, where unknown constants α_1, β_1, α_2, β_2 are determined by the state system at the time $t = T$. Now we find

$$\alpha_1 = \dot{x}(T), \ \beta_1 = x(T) - \dot{x}(T)T, \ \alpha_2 = \dot{y}(T)T + gT, \ \beta_2 = y(T) - \dot{y}(T)T - gT^2/2.$$

Thus, the movement of the missile in its free flight is characterized by the equalities

$$x(t) = x(T) + \dot{x}(T)(t - T), \ y(t) = y(T) + \dot{y}(T)(t - T) - g(t - T)^2/2.$$

At the moment of the missile lands T_l, its vertical coordinate is equal to zero. Setting in the last equality $t = T_l$, we obtain the quadratic equation

$$g(T_l - T)^2 - \dot{y}(T)(T_l - T) - y(T) = 0.$$

Find its solution

$$T_l - T = \frac{\dot{y}(T) \pm \sqrt{\dot{y}(T)^2 + 2gy(T)}}{g}.$$

Since the moment of landing occurs after the engines are turned off, the difference $T_l - T$ is positive. Therefore, in the last ratio, we should choose the plus sign in front of the root.

The missile flight range L is equal to its horizontal coordinate at the time of landing. As a result, we find the value

$$L = x(T_l) = \frac{\dot{y}(T)}{g}\left[\dot{y}(T) + \sqrt{\dot{y}(T)^2 + 2gy(T)}\right]. \tag{20.6}$$

Thus, we get the problem of finding such a function $u = u(t)$, which, in accordance with relations (20.4), (20.5), maximizes the value L. Note that in the absence of the problems of finding the extremum considered in Chapter 16, the dependence of the maximized value (functional) L of the unknown function u (control) is indirect. The functional L explicitly depends on the system state functions, which are related to control by means of the equations of state (20.4) and (20.5). This situation is typical for optimal control problems.

Before turning to the solution of this problem, we define a general optimal control problem and describe a method for its analysis.

The problem of maximizing the missile flight range
is in maximizing a functional that depends on the state functions
depending on the control function included in the state equations.

3   General optimal control problem

Let us give a general statement of the optimal control problem. First of all, it is necessary to have a mathematical model of the process, which is the basis for the practical application of any mathematical methods. In this case, the considered system must certainly have a degree of freedom. This means that there should be a choice of possible options for the evolution of the system, i.e., the investigated process is **controllable**[5]. As applied to a mathematical model, this means that the equations of state of the system include one or more parameters that can be changed at the will of the researcher. The variable parameter can be a number, vector, function, vector function, etc. In any case, it is called the **control**.

Thus, we have the equation of state of the controlled system, which in the general case can be written as the formula

$$A(u)x = 0,$$

where $A(u)$ is the state operator depending on the control u, x is the system state function. In particular, in the problem of maximizing the missile flight range, the state of the system is described by equations (20.4) with initial conditions (20.5), the state functions are the missile coordinates x, y, and the control is the function u that is an angle that can be changed during the entire active flight time.

As a result of changing the control in accordance with the equation of state, various realizations of the considered system are obtained. In order to choose the best of them, an **optimality criterion** I is specified, which depends on the state and control function, i.e., $I = I(x, u)$. For the problem of maximizing the missile flight range, the missile flight range L, calculated by formula (20.6), is chosen as the optimality criterion.

As a rule, the choice of control is due to technical, technological, and economic constraints. Thus, we have a **set of admissible controls**, within which the control can be changed. Constraints, in principle, can be imposed both on the control itself and on the system state. For the problem of maximizing the missile flight range, we do not impose any restrictions on the system, although we can assume that the values of the angle $u(t)$ lie in the interval $[0, \pi/2]$. So, the general **optimal control problem** is to choose an admissible control that delivers the minimum or maximum to the optimality criterion.

Consider the system described by the vector function $x = x(t) = (x_1(t), x_2(t), ..., x_n(t))$, which satisfies the system of differential equations

$$\dot{x} = f(t, x(t), u(t)) \tag{20.7}$$

on a time interval $[0, T]$ with initial condition

$$x(0) = x_0, \tag{20.8}$$

where the vector function $u = u(t) = (u_1(t), u_2(t), ..., u_r(t))$ is a control, f is a given vector function, x_0 is a given n-th order vector. We assume that constraints are given, characterized by a set $U = \{u| \ u(t) \in V(t), \ t \in (0, T)\}$, where $V(t)$ is some r-dimensional domain. By changing the control u within a given set, we obtain different solutions to the Cauchy problems (20.7) and (20.8), i.e., possible variants of the evolution of the system under study. Define the optimality criterion

$$I = \int_0^T F(t, x(t), u(t))dt + G(x(T)),$$

where F and G are given functions. The optimal control problem consists in finding a control u from the set of admissible controls U that minimizes[6] the functional I on this set, and the state of the system x included in its definition is determined from problems (20.7) and (20.8).

Let us describe a method for solving this problem. Define the function

$$H(t, x, \lambda, u) = -F(t, x, u) + \sum_{i=1}^n \lambda_i f_i t, x, u),$$

where $\lambda = \lambda(t) = (\lambda_1(t), \lambda_2(t), ..., \lambda_n(t))$ is the vector function satisfying the equations

$$\dot{\lambda}_i = -\frac{\partial}{\partial x_i} H\big(t, x(t), \lambda(t), u(t)\big), \ t \in (0, T) \tag{20.9}$$

with final conditions

$$\lambda(T) = -\frac{\partial}{\partial x_i} G(x(T)), \tag{20.10}$$

where $i = 1, ..., n$. System (20.9), (20.10) is called the **adjoint system**, and its solutions are called the **Lagrange multipliers**[7]. Under certain restrictions[8], according to the **maximum principle**, the solution of the considered problem, i.e., **optimal control** satisfies the following equality[9]

$$H\big(t, x(t), \lambda(t), u(t)\big) = \max_{v \in V(t)} H\big(t, x(t), \lambda(t), v\big). \tag{20.11}$$

Since the direct dependence of the function H on the control is known, equality (20.11) is a problem for the conditional extremum of the function[10]. As a result, we obtain a problem that includes state equations (20.7), (20.8), adjoint system (20.9), (20.10), and maximum condition (20.11), from where the solution to the problem can be found[11]. The method for solving such problems will be described in the Appendix.

> *The optimal control problem is to choose an admissible control*
> *that minimizes the optimality criterion, depending on the system state,*
> *satisfying the equation, containing the control as a parameter.*
> *Optimal control problems are solved using the maximum principle,*
> *which is a problem for the conditional extremum of a function*
> *that depends on the state of the system and the Lagrange multipliers.*

4 Solving of the maximization problem of the missile flight range

To solve the problem of maximizing the missile flight range, we transform it to the standard form described in the previous section. First of all, note that the number of controls r in this case is equal to one, i.e., the function u is scalar. Determine the state functions $x_1 = x$, $x_2 = y$, $x_3 = \dot{x}$, $x_4 = \dot{y}$. Thus, the state of the system has dimension $n = 4$, and the components of the vector function f in equation (20.7) are defined as follows $f_1 = x_3$,

$f_2 = x_4$, $f_3 = a\cos u$, $f_4 = a\sin u - g$. It follows from equalities (20.5) that the vector x_0 included in initial condition (20.8) is zero. Since there are no explicit restrictions on controls, U will mean the set of all continuous functions on the interval $[0, T]$. To determine the functional I, we note first of all that the maximum of the length L corresponds to the minimum of the value $-L$. Then the functional to be minimized is reduced to the standard form with the functions

$$F(t, x, u) = 0, \ G(x) = -x_1 - \frac{x_3}{g}\left(x_4 + \sqrt{x_4^2 + 2gx_2}\right).$$

In accordance with the method described above, define the function

$$H = \lambda_1 x_3 + \lambda_2 x_4 + \lambda_3 a\cos u + \lambda_4(a\sin u - g).$$

Determine adjoint system (20.9)

$$\dot{\lambda}_1 = 0, \ \dot{\lambda}_2 = 0, \ \dot{\lambda}_3 = -\lambda_1, \ \dot{\lambda}_4 = -\lambda_2. \tag{20.12}$$

The corresponding final conditions (20.10) are

$$\begin{cases} \lambda_1(T) = 1, \ \lambda_2(T) = \dfrac{x_3(T)}{\sqrt{x_4(T)^2 + 2gx_2(T)}}, \\[2mm] \lambda_3(T) = \dfrac{x_4(T) + \sqrt{x_4(T)^2 + 2gx_2(T)}}{g}, \ \lambda_4(T) = \dfrac{x_3(T)}{g}\left[1 + \dfrac{x_4(T)}{\sqrt{x_4(T)^2 + 2gx_2(T)}}\right]. \end{cases} \tag{20.13}$$

Maximum principle (20.11) in this case is the problem of minimizing the concrete function H with respect to control u. Equal its derivative to zero, we get the equality $a(\lambda_4\cos u - \lambda_3\sin u) = 0$. The resulting relation is an algebraic equation for the control u. Solving it, we find the value

$$\tan u(t) = \frac{\lambda_4(t)}{\lambda_3(t)}. \tag{20.14}$$

Define the functions λ_3 and λ_4 for finding the optimal control. As can be seen from the last two equations (20.12), they depend on λ_1 and λ_2, which, according to the first two equations, are constant. Determining their values from the adjoint system, we get[12]

$$\frac{\lambda_4(t)}{\lambda_3(t)} = \frac{x_3(T)}{\sqrt{x_4(T)^2 + 2gx_2(T)}}.$$

This implies a far from trivial conclusion that the optimal control u does not depend on time. For its final finding, it is required to determine the functions x_2, x_3, x_4.

Solving the Cauchy problems (20.4) and (20.5) with constant control, we obtain

$$x_3(t) = ta\cos u, \ x_4(t) = t(a\sin u - g), \ x_2(t) = t^2(a\sin u - g)/2.$$

As a result, equality (20.4) takes the form

$$\frac{\sin u}{\cos u} = \frac{a\cos u}{\sqrt{a^2\sin^2 u - ag\sin u}}.$$

Squaring the left and right sides of this equality, we have $g\sin 3u - 2a\sin^2 u + a = 0$. Denote $v = \sin u$, $b = a/g$. Then we get the cubic equation

$$\varphi(v) = v^3 - 2bv^2 + b = 0. \tag{20.15}$$

Thus, the optimal control is determined solely by the acceleration of the traction force, or rather, its relation to the gravity acceleration.

Let's analyze this equation. Note that the initial acceleration of the missile is positive. Based on the second equation (20.4), we find the value $\ddot{y} = a \sin u - g$. Obviously, for the acceleration to be positive, at least the inequality $a > g$ must hold, and hence $b > 1$. Find the derivative $\varphi'(v) = 3v^2 - 4bv$. Equaling it to zero, we establish that the function φ has two local extrema. In particular, at $v = 0$, a local maximum is achieved, and at $v = 4/3b$, a local minimum, see Figure 20.3.

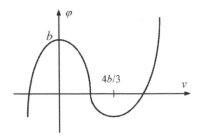

FIGURE 20.3: Graph of the function φ.

We have

$$\varphi(0) = b > 0, \quad \varphi(4/3b) = \frac{4b}{3}\left(1 - \frac{16b^2}{9}\right) < 0.$$

Thus, equation (20.15) has three roots, one of which is negative, the second belongs to the interval $(0, 4/3b)$, and the third is greater than $4/3b$, see Figure 20.3. Note that the value of v, being a sine, cannot exceed unity in absolute value. Since the missile is launched in the direction of increasing horizontal coordinate, the angle u, and hence its sine v, must be positive. Thus, we may be interested only in those values of the parameter v that belong to the unit interval. Naturally, the minimum and maximum roots of equation (20.15) do not satisfy this condition.

Taking into account the inequalities $\varphi(0) > 0$, $\varphi(1) = 1 - b < 0$, we conclude that the function φ changes sign on the interval $(0,1)$. Then equation (20.15) has a unique solution $v^* = v^*(a)$ here. We conclude that the solution to the problem of maximizing the missile flight range is determined by the formula $u(t) = \arcsin v^*$ for all t. So, the optimal control is unique, it is constant and depends exclusively on the acceleration of the thrust force a. Its specific value can be found approximately using an iterative algorithm for equation (20.15).

Task 20.1 *Maximizing the shell flight range.* The following steps are required.

1. Solving algebraic equation (20.15), find the optimal control for different values of the acceleration of the traction force.

2. Construct a graph of the dependence of optimal control on the traction force.

3. Solving problem (20.3), (20.4) for the optimal value of the angle and for the angle 45^0 find the corresponding values of the missile flight range.

For the problem of maximizing the missile flight range,
optimal control is constant and is determined by the acceleration of traction force.

5 Time-optimal control problem

A body moving under the action of a force is considered. The movement process is described by the equation

$$\ddot{x} = u.$$

It is necessary to choose a control $u = u(t)$ (acceleration) satisfying the condition $|u(t)| \leq 1$, so as to transfer the body from the initial state

$$x(0) = a, \ \dot{x}(0) = v$$

to the equilibrium position (origin)

$$x(T) = 0, \ \dot{x}(T) = 0$$

in the minimum time T.

First of all, let us transform the problem to the standard form. Introduce the notation $x_1 = x$, $x_2 = \dot{x}$. Then the state equations take the form

$$\dot{x}_1 = x_2, \ \dot{x}_2 = u. \tag{20.16}$$

We have also the initial conditions

$$x_1(0) = a, \ x_2(0) = v \tag{20.17}$$

and the final conditions

$$x_1(T) = 0, \ x_2(T) = 0. \tag{20.18}$$

The considered functional (time) can be written as

$$I = \int_0^T dt.$$

The optimal control problem is to minimize it on the set of functions taking values from the interval $[-1, 1]$.

To apply the maximum principle in the case under consideration, one question needs to be clarified. In contrast to the general problem formulated in Section 3, in this case the final state of the system is fixed, which corresponds to equalities (20.18). It can be shown that for a fixed final state for the conjugate equation (20.9) no boundary conditions are required.

Determine the function $H = -1 + \lambda_1 x_2 + \lambda_2 u$. Now we have the adjoint system

$$\dot{\lambda}_1 = -\frac{\partial H}{\partial x_1} = 0, \ \dot{\lambda}_2 = -\frac{\partial H}{\partial x_2} = -\lambda_1. \tag{20.19}$$

According to the maximum principle, the solution of this problem delivers the maximum of the function H on the interval $[-1, 1]$. This function is linear. Consequently, its maximum is reached at the boundary of the considered segment. Obviously, the result is determined by the sign of the function λ_2. Thus, from the maximum condition, one can find the control

$$u(t) = \begin{cases} 1, & \lambda_2(t) > 0, \\ -1, & \lambda_2(t) < 0. \end{cases} \tag{20.20}$$

Thus, we have system (20.16)–(20.20) with respect to five unknown functions, that are control, two state functions and two Lagrange multipliers. According to the first equation

(20.19), the function λ_1 is constant, i.e., $\lambda_1(t) = c_1$, where c_1 is a constant. Then it follows from the second equation in (20.19) that the function λ_2 is linear, in particular, $\lambda_2(t) = c_1 t + c_2$, where c_2 is a constant. Any linear function on an interval can change sign at most once at some point τ. Thus, in accordance with equality (20.20), the control can take only two values 1 and -1, and there is at most one switching point from one of these values to another.

For $u = 1$, solutions to problem (20.16), (20.17) are determined by the formulas $x_1(t) = a + vt + t^2/2$, $x_2(t) = v + t$. Eliminating the parameter t from these equalities, we establish that the values of the functions x_1 and x_2 are related by the equality

$$x_1(t) = \left(a - \frac{v^2}{2}\right) + x_2(t)^2.$$

In the plane x_1, x_2 (**phase plane**, see Chapter 5), for each pair of parameters a and v (initial position and initial velocity), the set of all possible points $(x_1(t), x_2(t))$ forms a parabola (phase curve), see Figure 20.4a, where arrows indicate the direction of change of the considered functions with increasing time t. Similarly, $u = -1$ solutions of system (20.16), (20.17) are equal to $x_1(t) = a + vt - t^2/2$, $x_2(t) = v - t$. Hence it follows that the values of these functions are related by the equality

$$x_1(t) = \left(a + \frac{v^2}{2}\right) - x_2(t)^2.$$

These values also form a parabola in the x_1, x_2, plane, see Figure 20.4b. According to the results obtained above, the fastest movement from an arbitrary point of the phase plane at the origin is possible only along these parabolas. In this case, the desired result is obtained either by moving along one parabola (there is no switching point), or by a single switch from one parabola to another (there is a switching point).

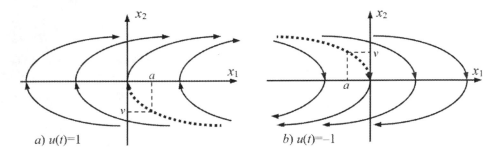

FIGURE 20.4: Phase curves for the system (20.16).

Obviously, there are only two main parabolas (they are highlighted in the figure), which enter the origin corresponding to conditions (20.18). One of them corresponds to the control $u(t) = 1$ and is realized for positive values of the parameter a and $v = -\sqrt{2a}$ the second one corresponds to the control $u(t) = -1$ and is realized for negative values of a and $v = \sqrt{-2a}$, see Figure 20.4. These options exhaust the cases of initial states for which the optimal control has no discontinuity points.

Consider now the initial position and initial velocity such for which the corresponding point A with coordinates a and v does not lie on the indicated two main parabolas, moving along which one can get to the origin see Figure 20.5. Therefore, from this point it is impossible to get to the origin of coordinates, moving along one of these parabolas. Obviously, two parabolas pass through point A, one of which corresponds to the control $u(t) = 1$, and the second to the control $u(t) = -1$. Moving along one of them, we do not get to the main parabola at all. And as a result of moving along the second of them, one can eventually get to a point B, which lies on the main parabola. Switching to it, i.e., changing the control sign, one can get to the origin of coordinates, i.e., ensure the fulfillment of conditions (20.18). Thus, for any initial position and initial velocity of the body, in accordance with the described procedure, a solution to the problem of optimal performance is found.

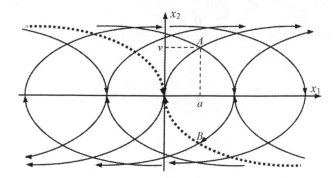

FIGURE 20.5: Solutions of time-optimal control problem.

Task 20.2 *Time-optimal control problem*. Find solutions to the optimal control problem for various initial states of the system. In this case, the specified parameters should be selected in such a way as to detect all four possible options, i.e., when the optimal control always takes the value 1, -1, it switches from 1 to -1 and vice versa. For all variants, find the time of movement, and in the last two cases, also the moment of switching from one control value to another.

The time-optimal problem is solved using the maximum principle.
Its solution is either always equal to 1, or always equal to -1,
or it switches from one of these values to another.

Direction of further work. Mathematical models include various parameters that are often unknown. To find them, you can use some additional information about the state of the system under consideration. These issues will be addressed in the final chapter, using the optimization techniques described in this chapter.

Appendix

We began this lecture by solving the problems of maximizing the range of the shell and missile. Another optimization problem of flight dynamics is associated with maximizing the probe's ascent height[13]. In addition, below will consider some practical methods for solving optimal control problems, both related to the previously considered optimality conditions, and some others[14].

1 Maximizing the probe's ascent height

Chapter 1 discussed the problem of lifting a probe. The probe moves vertically upward under the action of the traction force caused by the combustion of the fuel, in the presence of air resistance and gravity. Over time, the fuel burns out. The probe flies upward for some time by inertia, and then descends downward. The fuel consumption should be selected so that the probe rises as high as possible.

The entire movement of the probe consists of three stages. During active flight, which continues from the launch of the probe from the surface of the earth and until the moment of fuel combustion T, the probe moves upward in the presence of forces of thrust, resistance and gravity. At the second stage, from T to a certain moment τ, the probe continues its ascent even in the absence of the thrust force. The third stage, in which the probe goes down, is of no interest to us.

At the initial moment of time, the probe is on the surface of the earth, at rest and has a launch mass, which is the sum of the usable probe mass m_p and the fuel mass m_f. Thus, we have the initial conditions

$$y(0) = 0, \ \dot{y}(0), \ m(0) = m_p + m_f.$$

By Newton second law, the equation of movement of the probe in active flight has the form

$$\frac{d}{dt}\left(m\frac{dy}{dt}\right) = cu - \mu\dot{y}^2 - mg, \ 0 < t < T,$$

where y is the height of the probe above the ground, t is the time, m is the mass of the probe, u is the fuel consumption, c is the coefficient of thrust, μ is the coefficient of air resistance, g is the acceleration of gravity. In this case, the change in the mass of the probe occurs according to the law

$$\dot{m} = -u, \ 0 < t < T.$$

The fuel combustion rate selected as a control does not exceed the maximum permissible value u_{max}. Thus, there are the following restrictions $0 \le u(t) \le u_{max}, \ 0 < t < T$.

The fuel combustion time corresponds to the case when the probe mass is equal to its useful mass $m(T) = m_p$. Upon reaching this value, the mass of the probe remains unchanged, and there is no thrust. Thus, the movement of the probe at the second stage is characterized by the equation

$$m_p\ddot{y} = -\mu\dot{y}^2 - m_pg, \ T < t < \tau.$$

The end of the ascent of the probe occurs at the moment when its velocity becomes equal to zero. As a result, we have the equality $\dot{y}(\tau)$ The height of the rise of the probe h corresponds to the height of the probe at the moment of the termination of its rise, i.e., $h = y(\tau)$.

The optimal control problem consists in choosing such a fuel combustion rate $u = u(t)$ from a given interval so that the probe lift h is maximum[15].

Task 20.3 *Optimal glider flight control.* In Chapter 1, along with the movement of the missile and the probe, the movement of the glider was also considered. Using the mathematical model of this process given there, give the formulation of the problem of optimal control of the glider flight.

2 Approximate methods for solving optimality conditions

The optimal control problems described in the lecture were reduced to optimality conditions in the form of the maximum principle, amenable to direct analysis. However, in a more difficult case, the optimality conditions are problems, the solution of which can be found exclusively with the help of some iterative methods. Return to the optimal control problem discussed in Section 3.

It is required to choose such a vector function u from the set of admissible controls

$$U = \{u|\, u(t) \in V(t),\ t \in (0, T)\},$$

which minimizes on this set the functional

$$I = \int_0^T F(t, x(t), u(t))dt + G(x(T)),$$

where the vector function x is determined from the equations of state

$$\dot{x}(t) = f(t, x(t), u(t)),\ t \in (0, T);\ \ x(0) = x_0.$$

As noted earlier, the solution u to this problem satisfies the maximum principle

$$H(t, x(t), \lambda(t), u(t)) = \max_{v \in V(t)} H(t, x(t), \lambda(t), v),$$

where x is the corresponding solution to the equations of state, and the vector function λ is the solution to the adjoint system

$$\dot{\lambda} = -\frac{\partial}{\partial x} H(t, x(t), \lambda(t), u(t)),\ t \in (0, T);\ \ \lambda(T) = -\frac{\partial}{\partial x} G(x(T)).$$

Consider, for example, the following system

$$\dot{x} = u,\ t \in (0, T);\ \ x(0) = 0$$

the set of admissible controls is determined by the formula

$$U = \{u|\, |u(t)| \le 1,\ t \in (0, 1)\},$$

and the optimality criterion is

$$I = \frac{1}{2} \int_0^1 (u^2 + x^2)dt.$$

Then the function H is determined by the formula $H(u) = \lambda u - (u^2 + x^2)/2$. We get the adjoint system

$$\dot{\lambda} = x,\ t \in (0, T);\ \ \lambda(1) = 0$$

and the maximum condition

$$H(u) = \max_{|v| \le 1} H(v).$$

The maximum principle is the problem of maximizing a function of many variables (in terms of the number of control components) on a given set, which includes functions x and λ as parameters. From this problem, in principle, one can find control as a function of x and λ. In particular, for the given example, we are talking about finding the maximum of the function H by one variable u on the interval $[-1, 1]$. It is easy to verify the result[16]

$$u(t) = \begin{cases} -1, & \lambda(t) < -1, \\ \lambda(t), & -1 < \lambda(t) < 1, \\ 1, & \lambda(t) > 1. \end{cases} \tag{20.21}$$

Substituting the found dependence of the control on the functions x and λ into the state equations and the adjoint system, we obtain a system of differential equations with the corresponding boundary conditions. Unfortunately, for its practical solution it is not possible to use the previously described approximate methods for solving differential equations, since the boundary conditions for the equations of state are specified at the initial moment of time, and the adjoint state is given at the final moment. As a consequence, to solve the problem, one can use the **method of successive approximations.**[17]

Task 20.4 *Maximizing the missile flight range.* Solve the problem of maximizing the missile flight range in accordance with the method of successive approximations. Compare the results with the results of Task 20.1.

In Section 5, the optimal control problem with a fixed final state was solved. It was noted that in this case the maximum condition and the adjoint equation are also established. However, for the latter, the boundary condition is absent, while the state function is known both at the initial and at the final time. For the practical solution of the obtained system of optimality conditions, it is no longer possible to use the described method of successive approximations, since, firstly, it is not clear why in the process of solving equations of state with initial conditions at the current iteration the existing final condition will be satisfied, and, secondly, it is not clear how to solve the conjugate system in the absence of boundary conditions.

In this case, for an approximate solution of the problem, you can use the **shooting method**. It implies the artificial introduction of the initial or final condition for the adjoint system, for example, $\lambda(T) = a$, where a is an unknown quantity, a vector of the same dimension as the state function of the system. As a result, the algorithm for solving the system of optimality conditions is determined as follows. First, the initial approximations of the control and the vector a are specified. Now let their corresponding approximations u_k and a_k known at the k-th iteration. Substituting the value u_k into the equations of state with the given conditions, we find the corresponding state x_k. Substituting the known values $u = u_k$ and $x = x_k$ into the adjoint system considered with the final condition $\lambda_k(T) = a_k$, we determine its solution λk. The next approximation of the control u_{k+1} is determined from the maximum condition for the known values $x = x_k$ and $\lambda = \lambda_k$. Finally, there is also a final condition for the state of the system, which has the form $x(T) = x_T$ with a known final state x_T. This equality can be interpreted as an algebraic equation $\Phi(a) = x(T) - x_T = 0$. For its approximate solution, one can use some iterative method for solving algebraic equations[18], for example, $a_{k+1} = a_k - \theta_k \Phi(a_k)$, where θ_k is an algorithm parameter. Then the new approximation of the vector a is found by the formula $a_{k+1} = a_k - \theta_k[x_k(T) - x_T]$. After that, a new approximation of the system state function is found, etc.

Task 20.5 *Time-optimal control problem.* Solve the time-optimal control problem by the shooting method. Compare the results with the results of Task 20.2.

3   Gradient methods

In the final part of the previous section, approximate methods for solving algebraic equations were mentioned. However, the simplest extremum problems can also be reduced to algebraic equations. In particular, a necessary condition for the minimum of the function f at a point x is the equality to zero of its derivative at this point $f'(x) = 0$, which is an algebraic equation with respect to x. For its approximate solution, one can use the algorithm described above, which corresponds to the equality

$$x_{k+1} = x_k - \theta_k f'(x_k), \ k = 0, 1, \dots . \tag{20.22}$$

In contrast to the corresponding method for general algebraic equations, there is one essential circumstance here. The problem of minimizing the function f is solved. If its derivative at the point x_k is positive, then the function increases at this point, as a result of which we would like the value of x_{k+1} to be less than x_k. If the derivative at this point is negative, then the function f decreases here, and it is desirable to obtain x_{k+1} greater than x_k. In both cases (see Figure 20.6) parameter θ_k in the given algorithm should be chosen positive.

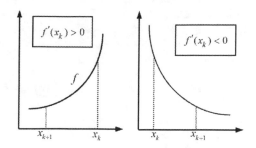

FIGURE 20.6: Direction of the algorithm is determined by the sign of the derivative.

The above algorithm can be easily extended to functions of many variables. In this case, the argument x of the function f is a vector, and equality (20.22) should be written componentwise, replacing the derivative with the corresponding partial derivatives. Taking into account that the vector whose components are all the partial derivatives of a given function f is its gradient ∇f, we obtain the vector form of the algorithm (20.22)

$$x_{k+1} = x_k - \theta_k \nabla f(x_k), \ k = 0, 1, \dots .$$

As a consequence, the considered algorithm is called the ***gradient method***[19]. There is also a natural generalization of the gradient method to the case of functionals, when the sought value x is a function or even a vector function[20].

The gradient method can also be extended to problems for conditional extremum. Let, in particular, it is required to minimize the function f on a set U. In this case, one can proceed as follows. At the k-th step of the algorithm, using the known value of x_k, the value x_{k+1} is found in accordance with formula (20.22). If it belongs to the given set, then it is chosen as the next approximation. Otherwise, the point of the set U closest to x_{k+1} is chosen as a new approximation. The corresponding algorithm is written as follows

$$x_{k+1} = P_U \left[x_k - \theta_k f'(x_k) \right], \ k = 0, 1, \dots .$$

This is called the ***gradient projection method***. Here, P_U denotes a transformation that assigns to an arbitrary point the nearest point of the set U, called a ***projector***. In particular, equality (20.21) ensures that the value $\lambda(t)$ is projected onto the interval $[-1, 1]$. There are other approximate methods for solving conditional extremum problems[21].

Notes

[1] General methods for solving optimal control problems are considered, for example, in [103], [156], [178],

[213], [290]. Optimal control problems for distributed parameter systems are considered, for example, in [106], [215], [252], [313]. Optimal control applications are considered in [151], [210], [252] for physical and technical systems, in [356] for chemical systems, in [57], [82] for biological systems, in [75], [155], [184], [311], [364] for economic systems, see also [230].

   [2] An extremely important application of optimization methods is inverse problem theory, see Chapter 21. This is not about choosing the optimal conditions for the studied process, but about finding the unknown parameters of the system and the problem of minimizing the standard deviation of the state of the system from the measurement results.

   [3] The problem of maximizing the range of a missile flight is considered in [210].

   [4] At $T = 0$, the shell lands without starting.

   [5] The concept of controllability is used in a different sense as well. A system is called ***controllable*** if there is such control that transforms it from one state to another, see, for example, [331].

   [6] If it is required to find the maximum of the functional I, then we can go to the problem of minimizing the functional $-I$.

   [7] We have already encountered Lagrange multipliers in Chapter 16 when considering problems in the calculus of variations with constraints.

   [8] On the conditions of applicability of the maximum principle [156], [290].

   [9] Like the extremum conditions discussed in Chapter 16, the maximum principle is a necessary condition for optimality, i.e., a situation is possible when its solution is not optimal. [312] gives extremely simple examples of optimal control problems in which the maximum principle is necessary but not sufficient optimality conditions. There are also considered examples of optimal control problems that have no solution or have a non-unique solution, as well as ill-posed optimization problems. Chapter 21 provides an example of an optimal control problem that is Tikhonov ill-posed.

   [10] The problem on the conditional extremum of a function (in this case, a function of many variables) refers to ***nonlinear programming***. Practical methods for solving such problems are given, for example, [52], [102], [112], [271], [328].

   [11] Chapter 21 gives an example of the degeneration of the maximum principle.

   [12] Indeed, using equalities (20.13), we find the values

$$\lambda_1(t) = 1, \ \lambda_2(t) = \frac{x_3(T)}{\sqrt{x_4(T)^2 + 2gx_2(T)}}.$$

From conditions (20.12), (20.13), determine the function

$$\lambda_3(t) = \lambda_3(T) + (T - t)\lambda_1(t) = \frac{x_4(T) + \sqrt{x_4(T)^2 + 2gx_2(T)}}{g} + (T - t).$$

Now we find the function

$$\lambda_4(t) = \lambda_4(T) + (T - t)\lambda_2(t) = \frac{x_3(T)}{g}\left[1 + \frac{x_4(T)}{\sqrt{x_4(T)^2 + 2gx_2(T)}}\right] + (T - t)\frac{x_3(T)}{\sqrt{x_4(T)^2 + 2gx_2(T)}} =$$

$$\frac{x_3(T)}{\sqrt{x_4(T)^2 + 2gx_2(T)}}\left\{\frac{\sqrt{x_4(T)^2 + 2gx_2(T)} + x_4(T)}{g} + (T - t)\right\}.$$

Finally, we get

$$\frac{\lambda_4(t)}{\lambda_3(t)} = \frac{x_3(T)}{\sqrt{x_4(T)^2 + 2gx_2(T)}}.$$

   [13] The problem of maximizing the probe height is considered in [210].

   [14] Methods for the approximate solution of extremal problems are given, for example, in [52], [97], [102], [112], [271], [328].

   [15] In contrast to the previously considered problem of maximizing the missile flight, the time of the controlled flight of the probe is not fixed in advance, but is determined by the law of control variation.

   [16] Indeed, equating zero the derivative of the function H with respect to u, we find $u(t) = \lambda(t)$. Since the corresponding second derivative is negative, we conclude that we are indeed dealing with a maximum point. However, the function H is minimized on the segment $[-1, 1]$. For $\lambda(t) < -1$, this function decreases on this segment, which means that its maximum there is attained at point -1. For $\lambda(t) > 1$, the function H increases on this segment and reaches its maximum at point 1. Thus, the above formula holds.

   [17] Indeed, the initial approximation of the control u_0 probably lies on the interval $[-1, 1]$. Then the corresponding state function satisfies the inequality

$$-t < x_0(t) \leq \int_0^t u_0(\tau)d\tau \leq t.$$

Integrating the adjoint system, we find

$$-\frac{1}{2} \leq \frac{t^2 - 1}{2} \leq p_0(t) = -\int_t^1 x_0(\tau)d\tau \leq \frac{1 - t^2}{2} \leq \frac{1}{2}.$$

Thus, the next control approximation will lie on the interval $[-1/2, 1/2]$. Repeating the above reasoning, we can establish that after iteration the interval to which the control will belong will again decrease by two. As a result, the estimate $|u_k(t)| \leq 2^{-k}$ is established. Hence it follows that the control converges to zero, which will be optimal due to the nonnegativity of the optimality criterion.

[18] For approximate methods for solving algebraic equations, see [143].

[19] There are different variants of the gradient method, differing in the way of choosing the iterative parameter θ_k, see, for example, [52], [112], [271], [328].

[20] In the next chapter, the gradient method will be used in the problems of minimizing functionals. This use is based on the definition of the derivative of the functional.

[21] For algorithms for solving constrained extremum problems, see [52], [97], [102], [112], [271], [328].

Chapter 21

Identification of mathematical models

In the previous chapters, it was shown how, on the basis of the general laws of physics, chemistry, economics, etc., one can obtain mathematical models of the considered phenomena of nature and society. The objects included in the mathematical models can be divided into three different classes. First, these are the state functions, which are the most important characteristics of the system. Actually, a mathematical model is a problem that consists in determining the state functions. Secondly, these are independent variables, which are most often time and spatial variables. They act as direct arguments to state functions. In particular, in the process of analyzing the mathematical model, the values of the state functions are found for any considered independent variables, for example, the state of the system at the corresponding points of the given region and at different times. All other characteristics included in the mathematical model we call the parameters of the system. Analysis of the mathematical model involves the determination of the dependence of state functions on their arguments for various combinations of system parameters. The presence in the mathematical model of some freedom of choice of parameters gives it considerable flexibility. As a result, it describes not only a concrete unique system, but a fairly wide class of systems of the same type, differing in the conditions of the process. As a result, the application of mathematical modeling to study a specific system is preceded by the adaptation of the model to the immediate specific situation, i.e., *identification* of the model[1]. If the derivation of the mathematical model itself is based on the corresponding theoretic laws of physics, chemistry, etc., then the parameters of the system, in principle, are determined experimentally. Unfortunately, a situation often arises when individual parameters of the system do not lend themselves to direct experimental determination. In this case, they are found indirectly, starting from the results of measuring the state functions of the system, solving the corresponding inverse problems[2]. The final chapter of the book is devoted to these ideas.

In the first section of the lecture, examples of fairly simple mathematical models considered in the previous chapters are analyzed, where the problem of determining the parameters of the system naturally arises. Then, a general formulation of the inverse problem is given and approaches to its solution related to the application of the optimization methods described in the previous chapter are described. Section 3 poses the problem of reconstructing the prehistory of the system described by the heat equation, which in the natural setting is ill-posed. However, it can be interpreted as an inverse problem, which opens up the possibility of its solution. Section 4 describes gradient methods for minimizing functionals, which are widely used in the practical solution of inverse problems. The final section shows how these ideas are applied to solve the problem posed in the previous section 3[3]. In the Appendix, a similar technique is used for the inverse boundary problem for the heat equation and for the problem of determining the air resistance coefficient. In addition, the inverse problem of gravimetry is described and the problem of well-posedness of optimization problems associated with the considered inverse problems is discussed.

DOI: 10.1201/9781003035602-21

Lecture

1 Problem of determining the system parameters

Let us start with the mathematical models that were described in the previous chapters. The first example is related to the process of **body falling** under its weight, see Chapter 1. Based on Newton second law, the law of movement is characterized by the equality $F = ma$, where a is an acceleration of a moving body, m is its mass, and F is a force acting on it. Acceleration is the second derivative of the x coordinate (height) of the body. Force F is the sum of the gravity force F_g directed in the direction of movement, i.e., decrease of the x coordinate, and the air resistance force F_r, which slows down the movement. The gravity force is proportional to the mass of the body, and the force of resistance is proportional to the velocity of body movement, i.e., the derivative of the x coordinate, i.e., the equalities $F_g = mg$, $F_r = k\dot{x}$ are true, where g is the gravity acceleration, and k is the coefficient of air resistance. As a result, we have the equation

$$m\ddot{x} = k\dot{x} - mg.$$

Among the objects included in this equality, time is an independent variable, body height is a state function. All other objects, i.e., mass, coefficient of air resistance and gravity acceleration, one refers to the parameters of the system. However, they are far from being equal. The gravity acceleration can be considered constant with a sufficiently high accuracy, i.e., a value known in advance. Body mass can be measured relatively easily, i.e., determine based on direct experiment. The situation with the air resistance coefficient is much more difficult. A direct experiment only says that with an increase in the velocity of a body's movement, air resistance probably also increases, but what exactly is the corresponding coefficient of proportionality is not so easy to determine.

Let us now consider the chemical process, in particular, to the **synthesis of hydrochloric acid**. We have the reaction $H_2 + Cl_2 \rightarrow 2HCl$, according to which, as a result of the combination of hydrogen and chlorine molecules, two molecules of hydrochloric acid are obtained. Denote by x, y and z the concentrations, respectively, of hydrogen, chlorine and hydrochloric acid. Then, using the law of mass action, we obtain the equations, see Chapter 7

$$\dot{x} = -kxy, \ \dot{y} = -kxy, \ \dot{z} = 2kxy.$$

The reaction rate constant k entering these equations can be found by the **Arrhenius law** $k = A\exp(-E/RT)$, where A is the preexponential factor, E is the activation energy, R is the universal gas constant, and T is the temperature. Among these quantities, the universal gas constant is a well-known constant. Temperature is relatively easy to measure. However, finding the preexponential factor and activation energy is a serious problem.

Now consider a biological system in which, for example, predators and prey coexist on the same territory. The process is characterized by time-varying numbers of prey x_1 and predators x_2. It is known that the velocity of change in the number of each species is proportional to the value of the number itself. Moreover, in the absence of predators, the number of preys grows, and in the absence of prey, the number of predators decreases. In the presence of predators, the increase in the number of preys decreases the more, the greater the number of predators, and in the presence of prey, the increase in the number of predators increases the more, the greater the number of preys. As a result, we have the **Volterra–Lotka equations**, see Chapter 7

$$\dot{x}_1 = (a_1 - b_1 x_2)x_1, \ \dot{x}_1 = (a_2 - b_2 x_1)x_2.$$

These equalities include the parameters a_1 and a_2, which characterize the increase in the number of preys in the absence of predators and the decrease of predators in the absence of preys, and the parameters b_1 and b_2, which characterize the decrease of the growth in the number of preys in the presence of predators and the increase of the growth in the number of predators in the presence of preys. Experimental determination of all of them is far from obvious.

We choose the final example from economics. Consider the **competition** between two firms producing the same product. The state of this system can be described as changing with the capital of both firms x_1 and x_2. When deriving a mathematical model of this process, it is assumed that under conditions of unlimited demand for manufactured products, the capital of firms would grow at a velocity. However, due to limited demand, the decrease in capital growth is observed, the more, the larger the total volume of output, proportional to the sum of the capitals of both firms. As a result, we get the equations, see Chapter 8

$$\dot{x}_1 = \big[a_1 - b_1(x_1 + x_2)\big]x_1, \ \dot{x}_2 = \big[a_2 - b_2(x_1 + x_2)\big]x_2.$$

The parameters a_1 and a_2 included in these equalities characterize the capital gain of firms under conditions of unlimited demand for the product, and b_1 and b_2 describe a decrease in this growth due to limited demand. All these coefficients are not amenable to direct experimental determination.

A serious advantage of mathematical models is the fact that each of them describes not a specific single process, but a whole class of processes that differ in the conditions of their occurrence. The possibility of using the same model to analyze various processes is precisely due to the presence in their composition of some parameters, which in each specific case take on certain values. Thus, the practical use of the existing mathematical model for a specific situation is preceded by its adaptation to specific conditions, consisting in the proper selection of the model parameters, which corresponds to its identification. However, these values cannot always be found on the basis of direct experiment, like the mass of a moving body or the temperature in a chemical reactor. In these cases, these parameters are determined indirectly based on the measurement of the state of the system.

Indeed, it is not entirely clear how the coefficient of air resistance can be measured, but it is possible to establish the position of a moving body at certain points in time. If experimental measurement of the preexponential factor and activation energy is difficult, then the concentrations of reactants can, in principle, be measured. Determining the number of biological species is much easier than finding the coefficients included in the Volterra–Lotka equations. It is definitely easier to obtain information about the capital of firms than about the parameters of the competition equation. As a result, there is a real possibility of adapting the model to specific conditions, i.e., solving the identification problem.

Presence of parameters in the mathematical model
makes it possible to use it to analyze a class of processes of the same type.
Application of the model in a specific situation is preceded by its identification.
Parameters of the model are not always found from direct experiment.
Additional information about the state of system can be used to identify the model.

2 Inverse problems and their solving

Give a general statement of the identification problem. There is a system equation

$$A(u)x = 0, \tag{21.1}$$

where $A(u)$ is a **state operator** depending on a parameter u, x is the system state function. This parameter can be a number, a vector, a function, a vector function, etc. In this case, we are only interested in those parameters that are not possible to determine directly. In particular, the process of falling of a body is described by the corresponding equation with initial conditions, the state function is the height of the body, and u is the coefficient of air resistance. For the chemical reaction considered earlier, the process is described by the corresponding kinetic equations, the state functions are the concentrations of all substances, and the parameter u is a second-order vector, the components of which are the preexponential factor and the activation energy. For the predator–prey model, the state equation is characterized by the Volterra–Lotka equations with initial conditions, and the parameter u is a fourth-order vector, the components of which are all the coefficients of the equations. A similar situation is observed for the competition equations.

The **direct problem** is understood as finding the state function for known values of the system parameter. The inverse problem, on the contrary, consists in finding the parameter of the system based on the available information about its state. This statement needs clarification. Thus, there is some information about the state of the system. This fact can be written as the equality

$$Bx = z, \tag{21.2}$$

where B is the measurement operator, and z is the measurement result, i.e., experimental data. In particular, for the process of falling of a body, we can measure its position at some moments of time t_1, t_2,...,t_n. Then condition (21.2) corresponds to the equalities $x(t_i) = x_i$, $i = 1,...,n$. The measurement operator here transforms the function x into the n-dimensional vector with components $x(t_i)$, and z is the vector $(x_1, x_2, ..., x_n)$. If we measure the position of the body on a certain time interval $[a, b]$, then equality (21.2) takes the form $x(t) = z(t)$, $t \in [a, b]$, the operator B associates the function x with itself[4], but only on the interval $[a, b]$, and z is the corresponding function defined on this interval. In the general case, the inverse problem is to find a parameter u such that the state of the system x corresponding to it to equality (21.2) satisfies condition (21.2). The analysis of both inverse problems of restoration of the air resistance coefficient is given in the Appendix.

The question arises, how can inverse problems be solved? Consider, for example, the problem of reconstructing the air resistance coefficient k from the results of measuring body height at certain points in time. Determine the standard deviation

$$I(k) = \sum_{i=1}^{n} \left[x(t_i) - x_i\right]^2$$

of the solution of the corresponding movement equation at the considered points for a given value of the coefficient k from the measurement results. Obviously, for any k, this value is not negative. However, if k is a solution to the corresponding inverse problem, then $I(k) = 0$. If at least one of the conditions $x(t_i) = x_i$ is violated, then the value of $I(k)$ turns out to be positive. In this regard, it is natural to try to determine the parameter k from the condition of the minimum of the corresponding standard deviation. As a result, we get an optimal control problem, the methods of solving which were considered in the previous chapter[5].

Suppose now that we want to determine the coefficient of air resistance based on the measurement of body position at a certain interval $[a, b]$. In this case, it is also possible to determine the standard deviation of the solution of the equation of state from the measurement results. It can be specified using the integral

$$I(k) = \int_{a}^{b} \left[x(t) - z(t)\right]^2 dt.$$

Again, for any k, this value is not negative, and the equality to zero is realized for those k for which the solution of the equation of state will be fulfilled exactly coincides with the measurement results on the considered segment[6]. Thus, the solution to the corresponding inverse problem can be sought by minimizing the above integral, i.e., solving the optimal control problem.

A similar way can be done in the general case, when the system describes by equation (21.1), and measurement is determined by equality (21.2). In this case, the **standard deviation** is written in the following form

$$I(u) = \left\| Bx(u) - z \right\|^2,$$

where $x(u)$ is the solution to equation (21.1) corresponding to the parameter u, and the value $\|f\|$ is called the **norm** of f. In particular, if f is a number, then its norm corresponds to the absolute value of number; if f is an n-th order vector, then $\|f\|$ is the modulus of the corresponding vector or, which is the same, its Euclidean norm, equal to the square root of the sum of the squares of its components; if f is a function defined on the interval $[a, b]$, then its norm is the square root of the integral of the square of this function[7]. Thus, the previously given formulas for the standard deviation in the case of a body fall are special cases of the last formula.

The norm takes on exclusively non-negative values, and the norm $\|f\|$ is equal to zero only when f is zero[8]. Consequently, the solution to the inverse problem is sure to be the solution to the corresponding optimization problem. Now let there be some solution to the optimal control problem. If, in this case, the corresponding value of the functional I is equal to zero, then we have a solution to the inverse problem as well. If the minimum value of the functional is positive, then this inverse problem has no solution at all[9]. Thus, if the inverse problem has a solution, then it can be found by solving the corresponding inverse problem.

Task 21.1 *Predator–prey model*. For the predator–prey model, all the coefficients of the equation are unknown. The number of predators is measured at certain points in time, and the number of preys at a certain time interval. Give the formulation of the inverse problem and its corresponding optimal control problem.

> *Direct problem is to determine the state functions with known system parameters.*
> *Inverse problem is to determine the model parameters*
> *with known information about the state of the system.*
> *Inverse problem is reduced to minimizing the standard deviation*
> *of the corresponding values of the system state from the measurement results.*

3   Heat equation with data at the final time

In practice, the problems of restoring the system history often arise. In this case, the state of the system at a given time is known, and it is necessary to find out how the system came to this state. It seems most natural to solve such a problem in its natural setting, i.e., to set the condition for the corresponding equation not at the initial, but at the final time. Solving the problem in the opposite direction of time, we thereby restore the prehistory of the system. However, serious difficulties can arise along the way.

Consider, for example, the heat equation

$$v_t = v_{xx}, \ 0 < x < \pi, \ t > 0 \tag{21.3}$$

with boundary conditions

$$v(0, t) = 0, \ v(\pi, t) = 0, \ t > 0 \tag{21.4}$$

and final condition

$$v(x,T) = z_n(x),\ 0 < x < \pi, \tag{21.5}$$

where $z_n(x) = n^{-1}\sin k\pi x$, n is a number. This ***time inverse problem*** has the solution

$$v_n(x,t) = n^{-1}\sin n\pi x \exp[n^2(T-t)].$$

Passing to the limit in equality (21.5), we obtain

$$v(x,T) = z_\infty(x),\ 0 < x < \pi, \tag{21.6}$$

where $z_\infty(x) = 0$. Obviously, problems (21.3), (21.4) and (21.6) have a unique solution v_∞, identically equal to zero. Note that for sufficiently large values of the parameter n, the final data z_n and z_∞ are arbitrarily close, while the corresponding solutions v_n and v_∞ can differ by an arbitrarily large amount. Thus, the solution to the heat equation is not continuous with respect to the data at the final moment of time, i.e., problem (21.3)–(21.5) is ill-posed, see Chapter 19. As a result, the direct solution of the heat equation in the opposite direction of time turns out to be ineffective, since it can lead to significant distortions of the results.

This circumstance raises the question of finding other approaches to solving the problem of restoring the system's prehistory. The most natural solution method is based on the use of the apparatus of the theory of inverse problems. We considered earlier, for example, the problem of restoring the coefficient of air resistance for the process of falling body from the results of measuring the position of the body. At the same time, the parameter (resistance coefficient) included in the mathematical model was unknown, but there was additional information about the state of the system (position of the body at certain points in time or at a certain time interval). In this case, we can assume that we have a mathematical model of the heat propagation process, which is the first boundary value problem for the equation of heat conduction in the natural direction of time with an unknown parameter that is the initial state of the system. To find it, there is additional information about the system state is the temperature at the final time[10].

Thus, the mathematical model of the considered system is the heat equation (21.3) with boundary conditions (21.4) and the initial condition

$$v(x,0) = u(x),\ 0 < x < \pi, \tag{21.7}$$

where the initial temperature $u = u(x)$ is unknown. The inverse problem is to find this function in such a way that the corresponding solution of problems (21.3), (21.4) and (21.7) satisfy condition (21.5). In accordance with the previously described technique, the standard deviation is

$$I(u) = \int_0^\pi \left[v(x,T) - z_n(x)\right]^2 dx. \tag{21.8}$$

where v is a solution to the well-posed boundary value problems (21.3), (21.4) and (21.7) corresponding to the initial state u. The solution to this problem can be found using well-known optimization methods, see Chapter 20. However, for their successful application, it is necessary to adapt these methods to the considered situation.

Boundary value problem for the heat equation with data at a final time is ill-posed. This problem is reduced to an inverse problem in which the unknown initial state, and an additional condition is the state function at the final time.

$p_0(t) = (1 - e^{-kt})^m$. It is initially equal to zero, and increases with time, tends to unity. Thus, it can be argued with all certainty that over time the entire population dies out. Thus, the stochastic model, in general, leads to the same result as the Malthus model. The probability distribution for the death process is shown in Figure 18.3.

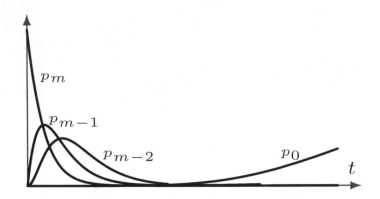

FIGURE 18.3: Distribution of probabilities in the death model.

Find the expected value

$$\mathbf{E}[x(t)] = \sum_{n=0}^{m} n p_n(t) = m e^{-kt}.$$

Thus, on average, the extinction of the population is observed, which is consistent with the properties of the deterministic model. The corresponding variance is determined by the formula

$$\mathbf{Var}[x(t)] = m e^{-kt}(1 - e^{-kt}).$$

Note that at the initial moment of time the variance is equal to zero, since the initial state of the system is deterministic. With an unlimited increase in time, the variance also tends to zero, since over time the population will certainly die out. However, at any time $t > 0$, the variance is positive. Then at some point in time it reaches its maximum value. To find it, we vanish the derivative

$$\frac{d}{dt}\mathbf{Var}[x(t)] = -kme^{-kt} + 2kme^{-2kt} = 0.$$

This equation has a unique solution $t = k^{-1}\ln 2$. It corresponds to the variance value $V = m/4$. Thus, the maximum expected deviation of the population size from its expected value is equal to a quarter of the initial population size and occurs at a time point characterized exclusively by the coefficient characterizing the probability of death of one individual. A graph of the time variation of the mean and variance of the population size is shown in Figure 18.4.

Among the physical analogues of the random process of population death, the process of **radioactive decay** can be noted[29]. Here the decay of one unstable atomic nucleus at a certain moment of time can serve as a random event. Then the number $x(t)$ of unstable atoms at time t turns out to be a random variable. The probability of decay of one nucleus within a sufficiently short time interval is the higher, the more unstable nuclei are present and the considered time interval. As a result, the probability distribution of the number

4 Differentiation of functionals and gradient methods

In the previous chapter, the gradient method was described for solving various extreme problems. However, it was used to minimize functions of one or many variables. In particular, to find the minimum of the function $f = f(x)$, one can use the following algorithm

$$x_{k+1} = x_k - \theta_k f'(x_k), \ k = 0, 1, ..., \tag{21.9}$$

where θ_k is a positive constant. If a function of many variables is minimized, then the argument x is a vector, and the latter should be understood in the vector sense. In this case, instead of the derivative of the considered function, its gradient is used. In the optimization problem obtained in the previous section, the unknown quantity is no longer the vector, but the function u, and it is not the function that is minimized, but the functional I, which depends on u by means of the equations of state.

We could extend the gradient method to such problems if we gave a meaningful definition of the derivative of a functional. Note that the formal definition of the derivative here as the ratio of the increment of values to the increment of the argument, when the latter tends to zero, is not possible already for functions of many variables. Indeed, it is completely incomprehensible how to divide the increment of this function by the increment of the argument, which is a vector?

To find the derivative of the functional I at the point u, which has the meaning of a function, let us try to consider the difference $I(u + \sigma h) - I(u)$, where σ is a number, and h is an object from the domain of this functional. In particular, for the optimization problem from the previous section, h is a function of the spatial variable x. This difference is divisible by σ since the division by the number of objects of a sufficiently general nature (numbers, vectors, functions, vector functions, etc.), as a rule, does not cause much difficulty. After that, the passage to the limit is carried out as σ tends to zero. If the corresponding limit exists, then it is called the ***derivative*** of the functional I at the point u ***in the direction*** h. Obviously, this limit depends on both the point and the direction. The most interesting case is when the specified limit exists for any h and is linear with respect to this object, i.e., has the form Lh, where L is a linear transformation. In this case, L is called the ***Gâteaux derivative*** of the considered functional at a given point[11]. The standard notation for it is $I'(u)$. Note that for a function of one variable, the Gâteaux derivative coincides with the ordinary derivative, and for functions of many variables, with a gradient. It is characteristic that a necessary condition for the extremum of a functional at some point is the equality to zero of its Gâteaux derivative at this point[12].

Having determined the derivative of the functional, we, in principle, can write down an analogue of the gradient method (21.9). However, we restrict ourselves to the case when, in the domain of the functional, one can introduce a dot product so that the square of the norm of an object is its dot product by itself[13]. This covers optimization problems that go back to inverse problems, where the functional is the standard deviation, i.e., just the square of the norm. In particular, for numbers (minimization of a function of one variable), the dot product is the usual product, for vectors (minimization of a function of many variables), the dot product is the sum of the squares of the products of their components, and for functions (minimization of integral functionals), the dot product is the integral of the product of these functions[14]. In this case, the Gâteaux derivative of the functional I at the point u can be defined as an object $I'(u)$ that satisfies the equality in the domain of this functional

$$\lim_{\sigma \to 0} \frac{I(u + \sigma h) - I(u)}{\sigma} = \big(I'(u), h\big)$$

for any h. Thus, to find the derivative of the functional, it is required to go to the limit in the indicated relation and obtain the limit in the form of the dot product of two objects, one of which is h. Then the second factor is the desired Gâteaux derivative. Now we can write the **gradient method**

$$u_{k+1} = u_k - \theta k I'(u_k), \ k = 0, 1, \dots . \tag{21.10}$$

Task 21.2 *Function of many variables.* In accordance with the described technique, find the Gâteaux derivative for a function of many variables.

> *Gradient method is extended to the problem of minimizing functionals using the Gâteaux derivative.*

5 Solving of the heat equation with reversed time

Let us establish an algorithm for solving the problem of reconstructing the initial temperature of a body from its known final temperature, which reduces to minimizing functional (21.8). This requires finding its derivative and substituting it into formula (21.10).

Functional (21.8) includes the solution v of the boundary value problem (21.3), (21.4), (21.7), which includes the desired parameter (control) u. To find its derivative, one should determine the difference $\Delta I = I(u + \sigma h) - I(u)$, divide the result by σ and go to the limit. If we then write the resulting limit in the form of the dot product of h by some other object, then the latter is the desired derivative. Let v be the indicated boundary value problem corresponding to the initial value u, and let w denote the solution of the same problem corresponding to $u + \sigma h$. Find the increment

$$\Delta I = \int_0^\pi 2\big[v(x, T) - z(x)\big]\varphi(x, T)dx + \int_0^\pi \big[\varphi(x, T)\big]^2 dx, \tag{21.11}$$

where $\varphi = w - v$, and we denote by z the known final state of the system, omitting the subscript n for brevity.

The last formula does not include the number σ by which it should be divided explicitly, and the dependence of the difference ΔI on it is carried out by means of the function φ. The latter is a solution to the following boundary value problem[15]

$$\varphi_t = \varphi_{xx}, \ 0 < x < \pi, \ t > 0, \tag{21.12}$$

$$\varphi(0, t) = 0, \ \varphi(\pi, t) = 0, \ t > 0, \tag{21.13}$$

$$\varphi(x, 0) = \sigma h(x), \ 0 < x < \pi. \tag{21.14}$$

Multiply equality (21.12) by an arbitrary function $\lambda = \lambda(x, t)$ and integrate the resulting value. We have

$$\int_0^T \int_0^\pi \varphi_t \lambda \, dx \, dt = \int_0^T \int_0^\pi \varphi_{xx} \lambda \, dx \, dt. \tag{21.15}$$

Integrating by parts, we get

$$\int_0^T \varphi_t \lambda \, dt = \varphi \lambda \Big|_0^T - \int_0^T \varphi \lambda_t \, dt,$$

where the value $\varphi(x, 0)$ is determined by formula 21.14. Similarly, after two-fold integration by parts, we have

$$\int\limits_0^\pi \varphi_{xx}\lambda dt = \varphi_x\lambda\big|_0^\pi - \varphi\lambda_x\big|_0^\pi + \int\limits_0^\pi \varphi\lambda_{xx}dx,$$

where, by the conditions (21.13), the function φ on the boundary is equal to zero. As a result, equality (21.15) is reduced to the form

$$-\int\limits_0^T\int\limits_0^\pi \varphi(\lambda_t + \lambda_{xx})dxdt + \int\limits_0^\pi \varphi(x, T)\lambda(x, T)dx+$$

$$\int\limits_0^T \big[\varphi_x(\pi, t)\lambda(\pi, t) - \varphi_x(0, t)\lambda(0, t)\big]dt = \sigma\int\limits_0^\pi h(x)\lambda(x, 0)dx.$$

The obtained equality is valid for any functions λ. We choose as λ the solution to the following problem, called the **adjoint system**[16]

$$\lambda_t + \lambda_{xx} = 0, \ \ 0 < x < \pi, \ t > 0, \tag{21.16}$$

$$\lambda(0, t) = 0, \ \lambda(\pi, t) = 0, \ \ t > 0, \tag{21.17}$$

$$\lambda(x, T) = 2\big[v(x, T) - z(x)\big], \ \ 0 < x < \pi. \tag{21.18}$$

Then, in the preceding equality, the first and third integrals on the left-hand side are equal to zero, and the second integral is equal to the first term on the right-hand side of equality (21.11). Then this equality is reduced to the form

$$\Delta I = \sigma\int\limits_0^\pi h(x)\lambda(x, 0)dx + \int\limits_0^\pi \big[\varphi(x, T)\big]^2 dx.$$

It follows from relations (21.12)–(21.14) that the function $\psi = \varphi/\sigma$ does not depend on σ. Then we get

$$\frac{I(u + \sigma h) - I(u)}{\sigma} = \int\limits_0^\pi h(x)\lambda(x, 0)dx + \sigma\int\limits_0^\pi \big[\psi(x, T)\big]^2 dx.$$

As a result of passing to the limit as σ tends to zero, we have

$$\lim_{\sigma \to 0} \frac{I(u + \sigma h) - I(u)}{\sigma} = \int\limits_0^\pi h(x)\lambda(x, 0)dx.$$

The right-hand side of this equality contains the integral of the product of two functions, one of which is h, i.e., their dot product. Consequently, the second multiplier is the required derivative of the functional, i.e., $I'(u) = \lambda(x, 0)$.

Now one can use the gradient method (21.10). Suppose that at the k-th iteration the corresponding approximation of the initial state of the system u_k is known. Then, from relations (21.3), (21.4), (21.7) for $u = u_k$, we can find the next approximation $v = v_k$ of the state function. Solving problem (21.16)–(21.18) for $v = v_k$ the k-th approximation of the function λ, denoted by λ_k. As a result, we find the derivative $I'(u_k) = \lambda_k(x, 0)$. Thus, a new approximation of the initial state of the system in accordance with equality (21.10) is found by the formula

$$u_{k+1} = u_k - \theta_k \lambda_k(x,0), \ k = 0, 1, \dots .$$

Thus, the solution to the problem of restoring the initial state for the heat transfer process from the known final state of the system can be found iteratively using the gradient method[17].

Task 21.3 *Time reversed heat equation.* Set the function $u = u(x)$ as the initial state of the system. Solving numerically the boundary value problems (21.3), (21.4) and (21.7)) up to a time instant $t = T$, determine the function $z(x) = v(x,T)$. In accordance with the described algorithm, find a solution to the problem of finding a solution to the heat equation with homogeneous boundary conditions of the first kind, which at $t = T$ takes the value z. Compare the results with the selected function u.

Solution of the problem of determining the initial body temperature at a known temperature at the final time can be found using the gradient method.

Appendix

In the lecture, the gradient method was used to solve the problem of restoring the initial state for the heat equation from a known final state. A similar method is applied below to solve the boundary inverse problem when two boundary conditions are given at one end of the body and there are no boundary conditions at its other end. Further, the problem posed in the lecture is considered to restore the coefficient of air resistance for the body fall problem. The main thing here is the fact that measurements of the state of the system are carried out on a part of the considered time interval or at separate points in time. In addition, the inverse problem of gravimetry is described. The final section discusses some of the problems that arise in the practical application of optimization methods for solving inverse problems.

1   Boundary inverse problem for the heat equation

When studying heat transfer processes, a situation may also arise when we do not have any information on one part of the boundary. At the same time, on the other part of the border, it is possible to measure both temperature and heat flux. Consider the one-dimensional case for simplicity. The system is described by the heat equation

$$v_t = v_{xx}, \ 0 < x < \pi, \ t > 0, \tag{21.19}$$

The initial temperature distribution is known, which corresponds to the condition

$$v(x,0) = a(x), \ 0 < x < L. \tag{21.20}$$

At the left end, the temperature and heat flux are set, which corresponds to the conditions

$$v(0,t) = b(t), \ v_x(0,t) = c(t), \ t > 0 \tag{21.21}$$

Here the functions a, b, and c are known.

It can be shown that the boundary value problem (21.19)–(21.21) is ill-posed. As a consequence, condition (21.21) is replaced by more boundary conditions

$$v(0, t) = b(t), \ v(L, t) = u(t), \ t > 0 \tag{21.22}$$

with unknown function u. It is chosen in such a way that the second condition in (21.21) is satisfied. The resulting boundary inverse problem is reduced to the problem of minimizing the functional

$$I(u) = \int_0^T \left[v_x(0, t) - c(t) \right]^2 dt,$$

where v is the solution of the problems (21.19), (21.20) and (21.22).

Task 21.4 *Boundary inverse problem for the heat equation*. Find the derivative of the functional by analogy with the time reversal problem. The numerical solution of the inverse problem is carried out in the same way as in Task 21.3.

2 Inverse problem for the falling of body

We return to the process of the body falling. It is described by the equation

$$m\ddot{x} = k\dot{x} - mg$$

on a time interval $[0, T]$. The initial position x_0 and the velocity v_0 are assumed to be known, i.e., there are initial conditions. At the same time, the coefficient of air resistance k is unknown here. One can use body position measurements to find it. Measurements are carried out either at some time interval $[a, b]$, where $0 < a < b < T$, or at points in time t_1, $t_2,...,t_n$ from the considered time interval. In the first case, the additional condition takes the form $x(t) = z(t)$, $t \in [a, b]$. We have the equalities $x(t_i) = z_i$, $i = 1, ..., n$ for the second case. As already noted, in the first case, the inverse problem reduces to minimizing the integral

$$I_1(k) = \int_a^b \left[x(t) - z(t) \right]^2 dt.$$

For the second case, we get

$$I_2(k) = \sum_{i=1}^n \left[x(t_i) - z_i \right]^2.$$

Thus, we have two optimization problems, for the solution of which we use the described gradient method[18]. The key step here is again finding the derivative of the functional at an arbitrary point k, which in this case is a number. We denote by x the solution of the equation of movement with the given conditions, corresponding to the considered value of k, and by y the solution of the same problem, corresponding to $k + \sigma h$. Then the increments of the considered functionals are determined by the formulas[19]

$$I_1(k + \sigma h) - I_1(k) = 2 \int_a^b \left[x(t) - z(t) \right] \varphi(t) dt + \int_a^b \left[\varphi(t) \right]^2 dt, \tag{21.23}$$

$$I_2(k + \sigma h) - I_2(k) = 2 \sum_{i=1}^n \left[x(t_i) - x_i \right] \varphi(t_i) + \sum_{i=1}^n \left[\varphi(t_i) \right]^2, \tag{21.24}$$

where $\varphi = y - x$ satisfies the equation

$$m\ddot{\varphi} = k\dot{\varphi} = \sigma h\dot{x} + \sigma h\dot{\varphi}$$

with zero initial conditions.

Multiply this equality by an arbitrary function λ and integrate. After integration by parts, taking into account the initial conditions, we obtain

$$\int\limits_0^T (m\ddot{\lambda} + k\dot{\lambda})\varphi dt + \dot{\varphi}(T)\lambda(T) - [\dot{\lambda}(T) + \lambda(T)]\varphi(T) = \sigma \int\limits_0^T \lambda\dot{x}h dt + \sigma \int\limits_0^T \lambda\dot{\varphi}h dt.$$

The question arises, how to relate this equality with the above increments of functionals? For this, we would like to equate the first integral in its left-hand side with the first term on the right-hand sides of equalities (21.23) and (21.24).

In the first case, it is sufficient to introduce the function $\theta = \theta(t)$, which is equal to one on the segment $[a, b]$ and zero outside it. Now equality (21.23) takes the form

$$I_1(k + \sigma h) - I_1(k) = 2 \int\limits_0^T [x(t) - z(t)]\theta(t)\varphi(t)dt + \int\limits_a^b [\varphi(t)]^2 dt.$$

We define as λ the solution of the equation

$$m\ddot{\lambda} + k\dot{\lambda} = 2(x - z)\theta,$$

supposing that this function and its first derivative are equal to zero for $t = T$. Then from the last two equalities after dividing by σ we obtain

$$\frac{I_1(k + \sigma h) - I_1(k)}{\sigma} = h \int\limits_0^T \lambda\dot{x} dt + \sigma^{-1} \int\limits_a^b [\varphi(t)]^2 dt + h \int\limits_0^T \lambda\dot{\varphi} dt.$$

It can be shown that as σ tends to zero, the second and third terms on the right-hand side tend to zero. Thus, the derivative of I_1 is a value that is multiplied by h in the first term[20]

$$I_1'(k) = \int\limits_0^T \lambda\dot{x} dt.$$

To find the derivative of the second integral, it is required to represent the value of the function at f at a specific point τ from the interval $(0, T)$ in terms of the corresponding integral. This is achieved through the use of the δ-***function*** concept[21]. This means an object that is characterized by the equality

$$f(\tau) = \int\limits_0^T f(t)\delta(t - \tau)dt.$$

As a result, equality (21.24) can be written in the form

$$I_2(k + \sigma h) - I_2(k) = 2 \sum_{i=1}^n \int\limits_0^T [x(t) - z(t)]\delta(t - t_i)\varphi(t)dt + \sum_{i=1}^n [\varphi(t_i)]^2,$$

where z is an arbitrary function[22], satisfying the equalities $z(t_i) = z_i$, $i = 1, ..., n$. Further, repeating the previous reasoning, we can establish that the derivative of I_2 has the same form as the derivative of I_1, but with a function λ satisfying the equation

$$m\ddot{\lambda} + k\dot{\lambda} = 2\sum_{i=1}^{n} [x(t) - z(t)]\delta(t - t_i)$$

with same condition for $t = T$.

Having determined the derivatives of the functionals, in both cases, we can apply the gradient method.

Task 21.5 *Restoration of the air resistance coefficient.* Using the gradient method, find a solution to both inverse problems described above.

3 Inverse gravimetry problem

Chapter 13 discussed the mathematical model of the gravitational field, which is based on the Poisson equation

$$\Delta u = 4\pi G\rho,$$

where u is the field potential, ρ is the density, G is the gravitational constant. On the basis of this equation with the corresponding boundary conditions, it is possible to find the distribution of the potential of the gravitational field in a certain region from the known distribution of the density of matter in this region. However, from a practical point of view, of greater interest is the ***inverse gravimetry problem***, in which, based on the results of measuring the gravitational field, it is required to establish what exactly lies in the earth[23].

In the process of gravimetric measurements, the acceleration of gravity is usually measured, which corresponds to the vertical derivative of the potential of the gravitational field and is carried out most often on the earth's surface. The structure of matter underground can be judged by the density distribution. Thus, the inverse problem of gravimetry is to restore the density distribution, i.e., the free term of the Poisson equation from the results of measuring the vertical derivative of the field potential.

In general, such a problem usually has a significantly different solution, since different density distributions in depth can cause the same responses on the earth's surface (the meaning of this property is clarified in the next section). As a result, various particular inverse problems of gravimetry can be considered. In the simplest cases, the presence of one or several uniform gravity anomalies of the correct form is assumed underground. In this case, the problem is to determine their density, location (in particular, depth) or size. In this case, one can often restrict oneself to an analysis of the two-dimensional case, i.e., consideration of the problem in the vertical plane. Considering that the potential of the gravitational field of any object decreases with distance from it, the area in which the gravitational anomaly is located is artificially expanded so that at the boundary of the expanded area the field potential can be set equal to zero. The corresponding inverse problems can be solved using the gradient method or other numerical optimization methods[24].

4 Well-posedness of optimal control problems

As already noted, the practical solution of inverse problems is reduced to the minimization of functionals characterizing the standard deviation of the system state function from the measurement results. However, such optimization problems do not have very good properties. Let us clarify them with the simplest example

Consider the system described by the system

$$\dot{x}(t) = u(t), \ t \in (0, \pi); \quad x(0) = 0$$

with unknown function u. Imagine an ideal case when the system state function is given: $x(t) = z(t)$, $t \in (0, \pi)$. It is clear that the function u can be found as a derivative of z. If, for example, $z = 0$, then the solution to the inverse problem will also be zero. However, we will investigate this problem in a standard way, i.e., by minimizing the functional

$$I(u) = \int_0^\pi x(t)^2 dt.$$

Naturally, the minimum of the functional is equal to zero, which is achieved exclusively for $u = 0$.

Consider now the sequence $\{u_k\}$ characterized by the equality $u_k(t) = \cos kt$. It corresponds to the state function $x_k(t) = k^{-1} \sin kt$. Let us calculate the integral

$$I(u_k) = \frac{1}{k^2} \int_0^\pi \sin^2 kt \, dt = \frac{\pi}{2k^2}.$$

Obviously, with unlimited growth of the number k, the value of $I(u_k)$ tends to zero, i.e., to a minimum of functionality. At the same time, the function u_k itself is periodic, and its frequency increases indefinitely with increasing k. Thus, for sufficiently large k, the function uk is extremely different from zero, which is the only solution to the problem under consideration, while the corresponding value of the functional $I(u_k)$ turns out to be arbitrarily close to its minimum[25].

In optimal control theory, the concept of the **Tikhonov well-posed problem** is introduced[26]. It is such if it has a unique solution, and any minimizing sequence (a sequence of admissible controls, the value of the minimized functional on which converges to its minimum) converges to this solution. In this case, the considered optimization problem corresponding to the simplest inverse problem turns out to be Tikhonov ill-posed. In this regard, in the process of numerically solving inverse and related optimization problems, certain difficulties may arise[27].

Let us note one more nuisance that arises in the study of such optimization problems. For this example, suppose we want to use the maximum principle described in the previous chapter. For this, the function $H = \lambda u - x^2$ is introduced, where λ is the solution of the adjoint system

$$\dot{\lambda}(t) = 2x(t), \ t \in (0, \pi); \quad \lambda(\pi) = 0.$$

Then the optimal control satisfies the maximum condition

$$\lambda u - x^2 = \max_v (\lambda v - x^2).$$

However, it is difficult to find the function u from this. In this case, we are dealing with the degeneration of the maximum principle, and the corresponding solution is called the **singular control**[28]. The presence of a singular control is another manifestation of the ill-posedness of the problem.

Notes

[1] General principles of identification are discussed, for example, in [93], [117].

[2] For various applied inverse problems, their properties and methods for solving them, see [4], [19], [133], [170], [207].

[3] In [207], to solve this problem, the quasi-inversion method is used, which is associated with the regularization of the considered equation.

[4] In this case, B is said to be the **restriction** of the function to the subset $[a, b]$ of its domain.

[5] Naturally, in this case, we do not solve the control problem, i.e., we are not trying to choose the best possible evolution of the system. The point is only that the resulting problem of minimizing the corresponding standard deviation belongs to the class of optimal control problems, which means that it can be solved, in principle, by the methods described in Chapter 20.

[6] Here one should pay attention to one circumstance related to the properties of the integral. It is known that the **Riemann integral** of some function does not depend on its value in a finite set of points. Thus, the equality to zero of the standard deviation is also realized in the case when the measurement condition is violated for a finite set of points. However, modern analysis and its applications usually use **Lebesgue integral**, with **measurable functions** being integrated. In this case, it is assumed that as a result of changing the function on a set of zero **Lebesgue measure** (for example, on any finite or countable set of points) the same function is obtained, i.e., changing the function on a set of zero measure, in essence, does not affect anything. Thus, if in the considered standard deviation the integral is understood in the Lebesgue sense and is equal to zero, then the measurement condition, now understood in the sense of measure theory, will be satisfied for sure. In other words, as in the discrete case, the standard deviation will be zero if and only if the measurement condition is satisfied.

[7] Naturally, in this case it is assumed that the corresponding integral exists. The possibility of determining a norm with an appropriate set of properties for any element f of a set X indicates that this set has the structure of a **normed vector space**. In particular, if f is a number, then the space X is the set of real numbers $\mathbb{R}$; if f is an n-dimensional vector, then X is an n-dimensional **Euclidean space** $\mathbb{R}^n$ endowed with the Euclidean norm; if f is a function defined on the interval $[a, b]$, then in our case X is the set of measurable functions $L_2(a, b)$, square-integrable in the sense of Lebesgue on a given interval.

[8] This property is included in the definition of the concept of the norm.

[9] In this case, the solution to the optimization problem is given such a value of the parameter that best ensures the fulfillment of the measurement condition.

[10] A similar idea was used in the shooting method described in the previous chapter.

[11] Naturally, the above reasoning requires some restrictions on the domain of the functional definition. In particular, the considered objects can be added and multiplied by numbers, after which the sum $x + \sigma h$ becomes meaningful. This leads to the concept of **vector space**. In addition, on this set, one can go to the limit, and this action is somehow consistent with the existing algebraic operations. As a consequence, the domain of definition of the functional turns out to be **topological vector spaces**. Sometimes stronger requirements are imposed on the derivative, which leads to the concept of the **Fréchet derivative**. The properties of the Gâteaux and Fréchet derivatives are described in any course on functional analysis, see [153], [185], [283]. In order to expand the class of optimization problems to be solved, weaker definitions of derivatives are also considered, for example, the **subdifferential** [88], the **Clarke derivative** [58], the **extended derivative** [313], etc. There are also algorithms for solving optimization problems that do not require the calculation of derivatives at all, for example, the **Nelder–Mead method** [253] and **genetic algorithms** [236].

[12] Indeed, if u is a minimum point of the functional I, then the inequality $I(u + \sigma h) - I(u) \geq 0$ holds. Choosing here the number σ positive, after dividing by it and passing to the limit as σ tends to zero, we obtain $I'(u)h \geq 0$. If we choose σ negative, then after dividing by it in the previous inequality and passing to the limit, we will establish that $I'(u)h \leq 0$. The last two inequalities imply the equality $I'(u)h = 0$, which is valid for any h. The Gâteaux derivative is, by definition, a linear transformation, which in this case maps an arbitrary h to the number 0. This property is the definition of a zero transformation, i.e., we are really dealing with the equality $I'(u) = 0$. However, it should be borne in mind that on its right side there is a null object, which is the number 0, a null vector, a null function, etc. depending on the domain of the given functional.

[13] This case is typical for **Hilbert spaces** that play an extremely important role in functional analysis and its many applications, see [153], [185], [283]. In the general case, by definition, the Gâteaux derivative is a linear transformation of an arbitrary direction h into a number, i.e., is a linear functional. Linear functionals on some space form the corresponding **adjoint space**. However, for Hilbert spaces, the elements

of the adjoint space can in a sense be identified with the objects of the original space, as a result of which the Gâteaux derivative can be understood as an object of the original space. This allows the gradient method to be written in a natural way.

[14] Naturally, here it is assumed that the integral of the square of the function exists, which corresponds to the previously mentioned function space L_2.

[15] To obtain this problem, we write relations (21.3), (21.4), (21.7) for the initial states $u + \sigma h$ and u with term-by-term subtraction of the corresponding values.

[16] This boundary value problem is an analogue of the adjoint system considered in the previous chapter, which is part of the optimality conditions in the form of the maximum principle.

[17] Let us pay attention to the analogy of the above algorithm with the method of successive approximations described in the previous chapter for the practical implementation of optimality conditions in the form of the maximum principle.

[18] Naturally, other numerical optimization methods can also be used here.

[19] In fact, the parameter k is a number, as a result of which I_1 and I_2 turn out to be ordinary functions of k. However, in this case, we are interested in the possibilities of working with inverse problems, with data determined on a part of the investigated time interval or at its individual points.

[20] For this, it is necessary to investigate the existing Cauchy problem for the function φ.

[21] Contrary to its name, the δ-function is not an ordinary function, but belongs to the class of distributions, see [283], [330].

[22] One might get the impression that there is some arbitrariness in the choice of the function z. However, when working with the δ-function, only its value at the considered points is actually taken into account.

[23] On inverse problems of gravimetry see [26], [162], [315].

[24] It can be shown that even for the problem of determining the location of a single homogeneous anomaly of a given structure, shape, and size, the corresponding functional turns out to be non-differentiable. In the problem of determining the density of the anomaly for known shapes, sizes and locations, the functional is differentiable. However, if the area was expanded in order to set zero boundary conditions, then the surface where the gravimetric measurements were carried out would be inside the expanded area. Since it is not the state function itself (potential) that is measured, but its derivative, when determining the derivative of the functional, the conjugate system will include not just the δ-function, as in the previous section, but its derivative. In this regard, to solve many inverse problems of gravimetry, it is necessary to use optimization methods that do not require the calculation of derivatives, see above.

[25] One can establish the convergence of the sequence $\{u_k\}$ in a weak topology to a function that is identically zero.

[26] The problem of well-posedness of optimal control problems, examples of ill-posed problems and methods of their investigation are considered in [312], [350], [360].

[27] For the practical solution of such problems, regularization methods are often used, see [350], [360].

[28] The degeneration of the maximum principle consists in the fact that its solution corresponds to the case when the coefficient before the control in the definition of the function H (in this case, this is the function λ) is equal to zero. Singular control is not necessarily optimal. In particular, in the problem of maximizing the same functional, the zero control is again singular, but naturally it is not optimal. For singular controls see [107].

Epilogue

Our book has ended. Beyond its borders, there are a great many phenomena, for which the apparatus of mathematical modelling can and should be applied. Perhaps a reader will need to determine his own model of some process. I will be very glad if acquaintance with this book will help him a little.

More than two and a half thousand years have passed since Pythagoras announced the existence of a real world, perceived by the senses, and an ideal world of human ideas. They are connected by number, which is a striking ideal concept through which one can comprehend the harmony of the reality around us. This is how mathematical modelling was born at the very origins of philosophy and science.

The essence of modelling is clarified in two old parables. One of them, owned by Plato, tells of a deep cave, in front of the entrance to which a fire is burning. If something happens between the fire and the cave, then vague shadows appear on the walls of the cave. For people inside the cave, they serve as the only information about events taking place outside of it. Another parable tells of the blinds who met an elephant. Comprehending the world with the help of touch, one of them stumbled upon the tail of an elephant, another on a tusk, and the third on a leg. As a result, each of them developed their own idea of the elephant. These hazy representations, like the shadows on the wall of Plato's cave, are models of the studied object.

Mathematical modelling should be treated from the same standpoint. To somebody, this situation may seem disappointing. However, the experience of the last millennia testifies to the opposite. As the famous English historian Arnold Toynbee said, our understanding of the world will always be just a glimpse, but this does not discourage us, and we continue this endless search.

Bibliography

[1] S.J. Aarseth. *Gravitational n-Body Simulations*. Cambridge University Press, New York, 2003.

[2] D. Acemoglu. *Introduction to Modern Economic Growth*. Princeton University Press, 2009.

[3] P. Alexandroff. *An Introduction to the Theory of Groups*. Dover Publications, 2012.

[4] O.M. Alifanov, O.M. Artyukhin, and S.V. Rumyantsev. *Extreme Methods for Solving Ill-posed Problems*. Science, Moscow, 1988.

[5] H.R. Alker, K. Dyutsch, and A.H. Stoetzel. *Mathematical Approaches to Politics*. San Francisco: Jossey, Bass, 1973.

[6] K.T. Alligood, T. Sauer, and J.A. Yorke. *Chaos: An Introduction to Dynamical Systems*. Springer-Verlag, New York, 1997.

[7] E.S. Allman and J.A. Rhodes. *Mathematical Models in Biology. An Introduction*. Cambridge University Press, 2004.

[8] L.A. Anchordoqui and T.C. Paul. *Mathematical Models of Physics Problems*. Nova Science Publishers, University of Wisconsin-Milwaukee, 2013.

[9] J.D. Anderson. *Computational Fluid Dynamics. The Basics with Applications*. McGraw-Hill, 1995.

[10] D. Andreucci. *Lecture Notes on the Stefan Problem*. Università di Roma La Sapienza, 2004.

[11] J.G. Andrews and R.R. McLone. *Mathematical Modelling*. Butterworths, London, 1976.

[12] A.A. Andronov, A.A. Witt, and S.E. Khaikin. *Theory of Oscillations*. Gos. Izd. Phys. Math. Lit., Moscow, 1959.

[13] T. Apostol. *Mathematical Analysis*. Addison-Wesley, New York, 1974.

[14] V.I. Arnold. *Ordinary Differential Equations*. Springer-Verlag, Berlin, Heidelberg, 1992.

[15] V.I. Arnold. *What is Mathematics?* MCNMO, Moscow, 2004.

[16] M. Ascher and L. Petzold. *Computer Methods for Ordinary Differential Equations and Differential-Algebraic Equations*. SIAM, 1998.

[17] U.M. Ascher, R.M.M. Mattheij, and R.D. Russell. *Numerical Solution of Boundary Value Problems for Ordinary Differential Equations*. SIAM, Philadelphia, PA, 1995.

[18] A.S. Ashmanov. *Mathematical Models and Methods in Economics.* Moscow State University, Moscow, 1980.

[19] R. Aster, B. Borchers, and C. Thurber. *Parameter Estimation and Inverse Problems.* Elsevier, 2018.

[20] K. Atkinson. *The Numerical Solution of Integral Equations of the Second Kind.* Cambridge Monographs on Applied and Computational Mathematics, 1997.

[21] J.P. Aubin. *Mathematical Methods of Game and Economic Theory.* Amsterdam, 1979.

[22] R. Baierlein. *Thermal Physics.* Cambridge University Press, 1999.

[23] N.T.J. Bailey. *The Mathematical Theory of Infectious Diseases.* Griffen, London, 1975.

[24] M. Bailyn. *A Survey of Thermodynamics.* American Institute of Physics Press, New York, 1994.

[25] S. Banerjee. *Mathematical Modeling: Models, Analysis and Applications.* Chapman and Hall/CRC, New York, 2014.

[26] R. Barzaghi and F. Sansó. The Integrated Inverse Gravimetric-Tomographic Problem: a Continuous Approach. *Inverse Problems,* 14(3):499–520, 1998.

[27] L.M. Batuner and M.E. Pozin. *Mathematical Methods in Chemical Kinetics.* GIT and ChL, Leningrad, 1963.

[28] R. Beckley, C. Weatherspoon, M. Alexander, M. Chandler, A. Johnson, and G. Batt. *Modeling Epidemics with Differential Equations.* Tennessee State University Internal Report, 2013.

[29] R. Bellman and K.L. Cooke. *Differential-Difference Equations.* Academic Press, New York, London, 2012.

[30] E.J. Beltrami. *Mathematics for Dynamic Modeling.* Academic Press, 1998.

[31] E.A. Bender. *An Introduction to Mathematical Modeling.* Dover, New York, 2000.

[32] J. Berger, W. Buhler, R. Repges, and P. Tautu (eds.). *Mathematical Models in Medicine.* Springer-Verlag, New York, 1976.

[33] L. Bers. *Mathematical Aspects of Subsonic and Transonic Gas Dynamics.* Dover, New York, 2016.

[34] C. Bertulani. *Nuclear Physics in a Nutshell.* Princeton University Press, 2007.

[35] A.T. Bharucha-Reid. *Elements of the Theory of Markov Processes and Their Applications.* Dover Publications, 2010.

[36] N. Biggs. *Discrete Mathematics.* Oxford Science Publications, 2002.

[37] P. Billingsley. *Probability and Measure.* New York, Toronto, London: John Wiley and Sons, 1979.

[38] F. Black and M. Scholes. The Pricing of Options and Corporate Liabilities. *Journal of Political Economy,* 81(3):637–654, 1973.

[39] G. Bliss. *Calculus of Variations*. Open Court Pub. Co., Illinois, 1944.

[40] P. Bonacich and P. Lu. *Introduction to Mathematical Sociology*. Princeton University Press, 2012.

[41] J.A. Bondy and U.S.R. Murty. *Graph Theory*. Springer, New York, Amsterdam, Oxford, 2008.

[42] A. Borovik. *Mathematics under the Microscope?* American Mathematical Society, 2009.

[43] N. Bourbaki. *Elements of the History of Mathematics*. Springer, Berlin–Heidelberg, 1998.

[44] R. Brady and H. Enderling. Mathematical Models of Cancer: When to Predict Novel Therapies, and When Not to. *Bull Math Biol*, 81:3722–3731, 2019.

[45] A.S. Bratus, A.S. Novozhilov, and A.P. Platonov. *Dynamic Systems and Models of Biology*. Physmathlit, Moscow, 2010.

[46] F. Brauer (ed.) *Mathematical Epidemiology*. Springer, 2008.

[47] N.N. Buchholz. *The Basic Course of Theoretical Mechanics*. Science, Moscow, 1972.

[48] I.R. Buchler and H.G. Nutini. *Game Theory in the Behavioral Sciences*. University of Pittsburgh Press, 1968.

[49] M.G. Bulmer. *Principles of Statistics*. New York, Dover Publ., Inc., 1979.

[50] D.N. Burghes and A.D. Wood. *Mathematical Models in the Social, Management and Life Sciences*. Wiley, New York, 1980.

[51] E. Burmeister and A.R. Dobell. *Mathematical Theories of Economic Growth*. Macmillan, New York, 1970.

[52] J. Céa. *Optimization Théorie at Algorithmes*. Dunod, Paris, 1971.

[53] N. Chafee and E.F. Infante. Bifurcation and Stability for a Nonlinear Parabolic Partial Differential Equation. *Bull. Amer. Math. Soc.*, 80(1):49–52, 1974.

[54] J. Chaskalovic. *Finite Elements Methods for Engineering Sciences*. Springer Verlag, Berlin, Heidelberg, 2008.

[55] M. Chester. *Primer of Quantum Mechanics*. John Wiley and Sons, New York, 1987.

[56] I. Chorkendorff and J.W. Niemantsverdriet. *Concepts of Modern Catalysis and Kinetics*. Wiley, 2017.

[57] C. Clark. *Mathematical Bioeconomics: the Optimal Management of Renewable Resources*. Wiley, New York, 1990.

[58] F. Clarke. *Optimization and Nonsmooth Analysis*. Wiley, 1983.

[59] E. Coddington and N. Levinson. *Theory of Ordinary Differential Equations*. McGraw-Hill, New York, 1955.

[60] D. Coldwell. *Mathematical Models in Microeconomics*. Allyn and Bacon, Boston, 1970.

[61] M. Comenetz. *Calculus: The Elements*. World Scientific, New Jersey, London, Singapore, Hong Kong, 2002.

[62] J.B. Conway. *Functions of One Complex Variable*. Springer-Verlag New York, 1978.

[63] W. Cook. *In Pursuit of the Traveling Salesman: Mathematics at the Limits of Computation*. Princeton University Press, 2012.

[64] R. Courant and H. Robbins. *What is Mathematics?* Oxford University Press, London–New York–Toronto, 1941.

[65] I.A. Crass. *Mathematical Models of Economic Dynamics* . Sov. Radio, Moscow, 1976.

[66] F. Crawford. *Waves*. McGraw-Hill, 1968.

[67] E.L. Crow and K. Shimizu (eds.). *Lognormal Distributions, Theory and Applications*. New York, Marcel Dekker, Inc., 1988.

[68] E.V. Danilina. *Models and Methods for Assessing Anthropogenic Systems*. Science, Novosibirsk, 1986.

[69] P.A. Davidson. *An Introduction to Magnetohydrodynamics*. Cambridge University Press, Cambridge, 2001.

[70] S. de Marchi. *Computational and Mathematical Modeling in the Social Sciences*. Cambridge University Press, 2005.

[71] N. Deo. *Graph Theory with Applications to Engineering and Computer Science*. Englewood, New Jersey: Prentice-Hall, 1974.

[72] A. Deutsch et al. (ed). *Mathematical Modeling of Biological Systems. Vol. I*. Birkhauser, Boston, 2007.

[73] A. Deutsch et al. (ed). *Mathematical Modeling of Biological Systems. Vol. II*. Birkhauser, Boston, 2008.

[74] O. Diekmann, H. Heesterbeek, and T. Britton. *Mathematical Tools for Understanding Infectious Disease Dynamics*. Princeton University Press, Princeton, 2013.

[75] A.K. Dixit. *Optimization in Economic Theory*. Oxford University Press, 1990.

[76] J.L. Doob. *Stochastic Processes*. Wiley, New York, 1990.

[77] A.E. Douglas. *Symbiotic Interaction*. Oxford University Press, 1994.

[78] S. Dunbar. *Mathematical Modeling in Economics and Finance: Probability, Stochastic Processes, and Differential Equations*. MAA Press, 2019.

[79] W. Dunham. *The Calculus Gallery*. Princeton University Press, 2005.

[80] J. Dunning-Davies. *Concise Thermodynamics: Principles and Applications*. Horwood Publishing, 1997.

[81] G. Duvaut and J.L. Lions. *Inequalities in Mechanics and Physics*. Springer Verlag, Berlin, Heidelberg, New York, 1976.

[82] C.L. Dym and S.I. Elizabeth. *Principles of Mathematical Modeling*. Academic Press, New York, 1980.

[83] F.J. Dyson. *Birds and Frogs: Selected Papers, 1990–2014*. World Scientific Publishing Company, 2015.

[84] K. Ebert, H. Ederer, and T.L. Isenhour. *Computer Applications in Chemistry: An Introduction for PC Users*. VCH Verlagsgesellschaft, Weinheim, 1989.

[85] L. Edelstein-Keshet. *Mathematical Models in Biology*. Society for Industrial and Applied Mathematics, 2005.

[86] G. Eilenberger. *Solitons. Mathematical Methods for Physicists*. Springer-Verlag, Berlin, Heidelberg, 1981.

[87] I. Ekeland. *Éléments d'Économie Mathématique*. Hermann Paris, 1979.

[88] I. Ekeland and R. Temam. *Convex Analysis and Variational Problems*. Amsterdam, Oxford. North-Holland Publ. Company, 1976.

[89] L.E. Elsgolc. *Calculus of Variations*. Pergamon Press Ltd, London, 1962.

[90] N.M. Emanuel and D.G. Knorre. *Course in Chemical Kinetics*. High School, Moscow, 1984.

[91] P. Érdi and J. Tóth. *Mathematical Models of Chemical Reactions: Theory and Applications of Deterministic and Stochastic Models (Nonlinear Science)*. Princeton University Press, 1989.

[92] A. Etheridge. *A Course of Financial Calculus*. Cambridge University Press, 2002.

[93] P. Eykhoff. *System Identification: Parameter and System Estimation*. John Wiley and Sons, New York, 1974.

[94] A. Faghri, Y. Zhang, and J. Howell. *Advanced Heat and Mass Transfer*. Columbia, MO: Global Digital Press, 2010.

[95] G. Falkovich. *Fluid Mechanics, a Short Course for Physicists*. Cambridge University Press, 2011.

[96] S. Farlow. *Partial Differential Equations for Scientists and Engineers*. John Wiley and Sons, New York, 1982.

[97] R.P. Fedorenko. *Approximate Solution of Optimal Control Problems*. Science, Moscow, 1978.

[98] W. Feller. *An Introduction to Probability Theory and Its Applications, Vol. 1*. Wiley, 1968.

[99] LF. Fernandez and H.S. Bierman. *Game Theory with Economic Applications*. Addison-Wesley, 1998.

[100] R.J. Field and M. Burger (Eds). *Oscillations and Traveling Waves in Chemical Systems*. John Wiley and Sons, Inc., 1985.

[101] G.S. Fishman. *Monte Carlo: Concepts, Algorithms, and Applications*. Springer-Verlag, New York, 1996.

[102] R. Fletcher. *Practical Optimization Methods*. John Wiley and Son, Chichester, 1987.

[103] C. Floudas and P. Pardalos (Eds.). *Encyclopedia of Optimization*. 2008, Springer.

[104] J. France and J.H.M. Thornley. *Mathematical models in agriculture.* Betterworths, London, 1984.

[105] E. Frenkel. *Love and Mathematics. The Heart of Hidden Reality.* Peter, Moscow, 2016.

[106] A.V. Fursikov. *Optimal Control of Distributed Systems. Theory and Applications.* Providence, Amer. Math. Soc., 1999.

[107] R.F. Gabasov and F.M. Kirillova. *Singular Optimal Control.* Nauka, Moscow, 1973.

[108] G. Gandolfo. *Mathematical Methods and Models in Economic Dynamics.* North-Holland, Amsterdam, 1971.

[109] I.M. Gelfand and S.V. Fomin. *Calculus of Variations.* Courier Corporation, New York, 2000.

[110] N. Gershenfeld. *The Nature of Mathematical Modeling.* Cambridge University Press, 1999.

[111] S.R. Gidrovich and I.M. Syroezhkin. *Game Modeling of Economic Processes .* Economics, Moscow, 1976.

[112] Ph.E. Gill, W. Murray, and Wright M.H. *Practical Optimization.* Academic Press, London, 1981.

[113] R. Glowinski, J.-L. Lions, and R. Tremolieres. *Numerical Analysis of Variational Inequalities.* North-Holland, Amsterdam, 1981.

[114] B.V. Gnedenko. *On Mathematics.* Editorial URSS, Moscow, 2000.

[115] A. Gorban, N. Kazantzis, I. Kevrekidis, H.C. Öttinger, and C. Theodoropoulos. *Model Reduction and Coarse-Graining Approaches for Multiscale Phenomena.* Springer-Verlag, Berlin, Heidelberg, 2007.

[116] H. Gould, J. Tobochnik, and W. Christian. *An introduction to Computer Simulation Methods: Applications to Physical Systems.* Pearson/Addison Wesley, 2006.

[117] D. Graupe. *Identification of Systems.* Van Nostrand Reinhold, New York, 1972.

[118] A.A. Gukhman. *Introduction to the Theory of Similarity.* Higher School, Moscow, 1973.

[119] S.C. Gupta. *The Classical Stefan Problem.* Elsevier, 2017.

[120] A. Gut. *Probability: A Graduate Course.* Springer-Verlag, 2005.

[121] H. Taha. *Operations Research: An Introduction.* Pearson, 2016.

[122] R. Haase. *Thermodynamik der irreversiblen Prozesse.* Steinkopff, Darmstadt, 1963.

[123] J. Hadamard. *Le problème de Cauchy et les Equations aux Derivées Partielles Lineaires Hyperbolic.* Hermann, Paris, 1932.

[124] J. Hadamard. *Essai sur la Psychologie de l'Invention dans le Domaine Mathematique.* A. Blanchard impr. Jouve, Paris, 1959.

[125] E. Hairer and G. Wanner. *Solving Ordinary Differential Equations I: Nonstiff Problems.* Springer-Verlag, Berlin, 1993.

[126] E. Hairer and G. Wanner. *Solving Ordinary Differential Equations II: Stiff and Differential-algebraic Problems.* Springer-Verlag, Berlin, New York, 1996.

[127] O. Hájek. *Dynamical Systems in the Plane.* Academic Press, 1968.

[128] H. Haken. *Synergetics, an Introduction: Nonequilibrium Phase Transitions and Self-Organization in Physics, Chemistry and Biology.* Springer–Verlag, New York, 1983.

[129] P.R. Halmos. *Measure Theory.* Springer, New York, 2014.

[130] R. Hanneman. *Computer-assisted Theory Building. Modeling Dynamic Social Systems.* SAGE, New York, 1988.

[131] W.D. Hart (ed.). *The Philosophy of Mathematics.* Oxford University Press, Oxford, 1996.

[132] P. Hartman. *Ordinary Differential Equations.* SIAM, Providence, 2002.

[133] A. Hasanoglu and V. Romanov. *Introduction to Inverse Problems for Differential Equations.* Springer, 2017.

[134] B. Haubold and T. Wiehe. *Introduction to Computational Biology. An Evolutionary Approach.* Birkhäuser, 2006.

[135] C. Hayashi. *Nonlinear Oscillations in Physical Systems.* Princeton University Press, 1964.

[136] P. Hayes. *Mathematical Methods in the Social and Managerial Sciences.* Wiley, New York, 1975.

[137] W.H. Hayt. *Engineering Electromagnetics.* McGraw-Hill Education, New York, 1989.

[138] C.V. Heer. *Statistical Mechanics, Kinetic Theory, and Stochastic Processes.* Academic Press, 1972.

[139] W. Heisenberg. Meaning and Signification of Beauty in the Exact Sciences. *Philosophy Issues*, (12):49–60, 1979.

[140] W. Heisenberg. *Physics and Philosophy: The Revolution in Modern Science.* Harper-Collins, 2007.

[141] D. Henry. *Geometric Theory of Semilinear Parabolic Equations.* Springer-Verlag, Berlin, 1981.

[142] R.C. Hilborn. *Chaos and Nonlinear Dynamics: An Introduction for Scientists and Engineers.* Oxford University Press, 2000.

[143] F.B. Hildebrand. *Introduction to Numerical Analysis.* Dover Publ., 2013.

[144] F.S. Hillier and G.J. Lieberman. *Introduction to Operations Research.* McGraw-Hill: Boston MA, 2014.

[145] J.O. Hirschfelder, C.F. Curtiss, and R.B. Bird. *Molecular Theory of Gases and Liquids.* Wiley, New York, 1954.

[146] R.V. Hogg, A. Craig, and McKean. J.W. *Introduction to Mathematical Statistics.* Pearson, 2012.

[147] L. Hörmander. *The analysis of Linear Partial Differential Operators. I.* Springer-Verlag, Berlin, Heidelberg, 2003.

[148] J.M. Howie. *Real Analysis.* Springer, London, 2005.

[149] R. Huckfeldt, C. Kohfeld, and T. Likens. *Dynamic Modeling: An Introduction.* SAGE Publications, 1982.

[150] J.L. Hudson and J.C. Mankin. Chaos in the Belousov–Zhabotinskii Reaction. *J. Chem. Phys.*, 74(11):6171–6177, 1981.

[151] D.G. Hull. *Optimal Control Theory for Applications.* Springer-Verlag New York, 2003.

[152] A.S. Huseynova, Yu.N. Pavlovsky, and V.A. Ustinov. *Experience of Simulation of the Historical Process.* Science, Moscow, 1984.

[153] V. Hutson, J.S. Pym, and Cloud M.J. *Applications of Functional Analysis and Operator Theory.* Elsevier Science, 2005.

[154] F.P. Incopera et al. *Fundamentals of Heat and Mass Transfer.* Wiley, London, 2012.

[155] M.D. Intriligator. *Mathematical Optimization and Economic Theory.* Phi Learning, 2013.

[156] A.D. Ioffe and V.M. Tihomirov. *Theory of Extremal Problems.* Nauka, Moscow, 1974.

[157] G. Iooss and D.D. Joseph. *Elementary Stability and Bifurcation Theory.* Springer-Verlag Berlin and Heidelberg, 1980.

[158] R. Isaacs. *Differential Games: A Mathematical Theory With Applications to Warfare and Pursuit, Control and Optimization.* Dover Publications, New York, 1999.

[159] A. Iserles. *A First Course in the Numerical Analysis of Differential Equations.* Cambridge University Press, 1996.

[160] Yu. Ivanilov and A. Lotov. *Mathematical Methods in Economics.* Moscow State University, Moscow, 1980.

[161] M.H. Jacobs. *Diffusion processes.* Springer-Verlag, New York Inc., 1967.

[162] W. Jacoby and Smilde P. *Gravity Interpretation: Fundamentals and Application of Gravity Inversion and Geological Interpretation.* Springer-Verlag Berlin Heidelberg, 2009.

[163] J.E. Greivenkamp. *Field Guide to Geometrical Optics.* SPIE Press, 2004.

[164] A. Jenkins. Self-oscillation. *Physics Reports*, 525(2):167–222, 2013.

[165] J.L. Kelley. *General Topology.* Springer Verlag, 1975.

[166] P.E. Johnson. *Formal Theories of Politics: Mathematical Modelling in Political Science.* Elsevier, 2014.

[167] R. Johnsonbaugh. *Discrete Mathematics.* Prentice Hall, 2008.

[168] D.W. Jordan and P. Smith. *Nonlinear Ordinary Differential Equations – An Introduction for Scientists and Engineers.* Oxford University Press, 2007.

[169] J. Jost. *Partial Differential Equations.* Springer-Verlag, New York, 2002.

[170] S.I. Kabanikhin. *Inverse and Ill-posed Problems: Theory and Applications.* De Gruyter, 2012.

[171] V.V. Kafarov and M.B. Glebov. *Mathematical modeling of the main processes of chemical production.* Higher school, Moscow, 1991.

[172] O. Kallenberg. *Foundations of Modern Probability.* Springer, 2002.

[173] J.G. Kemeny and J.L. Snell. *Mathematical Models in the Social Sciences.* MIT Press, 1978.

[174] J. Kepner, J. Gilbert. *Graph Algorithms in The Language of Linear Algebra.* SIAM Philadelphia, Pennsylvania, 2011.

[175] A.A. Kharkevich. *Self-oscillations.* GITTL, Moscow, 1954.

[176] L.N. Khitrin. *Combustion and Explosion Physics.* Moscow University, 1957.

[177] D. Kinderlehrer and G. Stampacchia. *An Introduction to Variational Inequalities and Their Applications.* Academic Press, Boston, 1980.

[178] D.E. Kirk. *Optimal Control Theory: An Introduction.* Englewood Cliffs, New Jersey, 2004.

[179] C. Kittel, W. Knight, and M. Ruderman. *Berkeley Physics Course. Mechanics. Vol. 1.* McGraw-Hill Book Company, 1973.

[180] M. Kline. *Mathematics: The Loss of Certainty.* Oxford University Press, 1980.

[181] M. Kline. *Mathematics and the Search for Knowledge.* Oxford University Press, 1985.

[182] S.J. Kline. *Similitude and Approximation Theory.* McGraw-Hill, New York, 1965.

[183] J. Knudsen and P. Hjorth. *Elements of Newtonian Mechanics.* Springer Science and Business Media, 2012.

[184] V.A. Kolemaev. *Economic and Mathematical Modelling.* UNITI-DANA, Moscow, 2005.

[185] A.N. Kolmogorov and S.V. Fomin. *Elements of the Theory of Functions and Functional Analysis.* Dover Publications, 1999.

[186] A.V. Kolobov, A.A. Anashkina, V.V. Gubernov, and A.A. Polezhaev. Mathematical Model of Tumor Growth Taking into Account the Dichotomy of Migration and Proliferation. *Computer Research and Modeling*, 1(4):415–422, 2009.

[187] S. Körner. *The Philosophy of Mathematics, An Introduction.* Harper Books, 1960.

[188] I. Kovacic and eds. Brennan, M.J. *The Duffing Equation: Nonlinear Oscillators and their Behaviour.* Wiley, 2011.

[189] M. Krasnov, A. Kiselev, and G. Makarenko. *Problems and Exercises in Integral Equations.* Mir Publishers, Moscow, 1971.

[190] A.P. Kuznetsov, S.P. Kuznetsov, and N.M. Ryskin. *Nonlinear Oscillations.* Fizmatlit, Moscow, 2005.

[191] A.P. Kuznetsov, I.R. Sataev, N.V. Stankevich, and L.V. Tyuryukina. *Physics of Quasiperiodic Oscillations.* Science, Saratov, 2013.

[192] O.A. Ladyzhenskaya. *The Boundary Value Problems of Mathematical Physics.* Springer Verlag, Berlin, Heidelberg, New York, 1985.

[193] O.A. Ladyzhenskaya and N.N. Ural'tseva. *Linear and Quasilinear Elliptic Equations.* New York and London, Academic Press, 1968.

[194] K.J. Laidler. *Chemical Kinetics.* Harper and Row, 1987.

[195] J.D. Lambert. *Numerical Methods for Ordinary Differential Systems.* John Wiley and Sons, Chichester, 1991.

[196] K. Lancaster. *Mathematical Economics.* Dover Publications, 2011.

[197] K. Lánczos. *Variational Principles of Mechanics.* Dover Publications, 2012.

[198] L.D. Landau and E.M. Lifshits. *Mechanics.* Butterworth-Heinemann, 1976.

[199] L.D. Landau and E.M. Lifshitz. *Quantum Mechanics: Non-Relativistic Theory.* Pergamon Press, New York, 1977.

[200] L.D. Landau and E.M. Lifshitz. *The Classical Theory of Fields.* Butterworth-Heinemann, 1987.

[201] L.D. Landau and E.M. Lifshitz. *Fluid Mechanics.* Pergamon, London, 1987.

[202] L.D. Landau and E.M. Lifshitz. *Statistical Physics.* Butterworth-Heinemann, 2013.

[203] L.D. Landau, E.M. Lifshitz, A.M. Kosevich, and Pitaevskii L.P. *Theory of Elasticity.* Elsevier, 1986.

[204] L.D. Landau, L.P. Pitaevskii, and E.M. Lifshitz. *Electrodynamics of Continuous Media.* Butterworth-Heinemann, 1984.

[205] R.J. Larsen and M.L. Marx. *An Introduction to Mathematical Statistics and Its Applications.* Prentice Hall, 2012.

[206] R. Larson and B. Edwards. *Calculus of a Single Variable.* Cengage Learning, Boston, 2010.

[207] R. Lattès and J.L. Lions. *Méthodes de Quasi-Réversibilité et Applications.* Dunod, Paris, 1967.

[208] C.A. Lave and J.G. Marsh. *An Introduction to Models in the Social Sciences.* Harper and Row, New York, 1975.

[209] B.H. Lavenda. *Nonequilibrium Statistical Thermodynamics.* John Wiley and Sons, 1985.

[210] D.F. Lawden. *Optimal Trajectories for Space Navigation.* London, Butterworth, 1963.

[211] A. Leach. *Molecular Modelling: Principles and Applications.* Prentice Hall, 2001.

[212] S. Lefschetz. *Differential Equations: Geometric Theory.* Interscience Publishers, New York, 1963.

[213] F.L. Lewis. *Optimal Control.* John Wiley and Son, New York, 1986.

[214] R.L. Liboff. *Introductory Quantum Mechanics.* Addison Wesley, 2002.

[215] J.L. Lions. *Quelques méthodes de résolution des problèmes aux limites non linéaires.* Paris, Dunod, 1969.

[216] J.L. Lions. *Control of distributed singular systems.* Gauthier-Villars, 1985.

[217] Z. Liu and C. Yang. A Mathematical Model of Cancer Treatment by Radiotherapy Followed by Chemotherapy. *Mathematics and Computers in Simulation,* 124(June):1–15, 2016.

[218] L.G. Loitsyanskii. *Mechanics of Liquids and Gases.* Pergamon Press, 1966.

[219] J.C.C. López, L.A.J. Sánchez, and R.J.V. Micó (Eds). *Mathematical Modeling in Social Sciences and Engineering.* NOVA, 2014.

[220] D.G. Luenberger. *Introduction to Dynamic Systems: Theory, Models, and Applications.* New York: John Wiley & Sons, 1979.

[221] G. Mackey. *The Mathematical Foundations of Quantum Mechanics.* Dover Publications, New York, 2004.

[222] A. Maitland and M.H. Dunn. *Laser physics.* North-Holland Pub. Co, 1969.

[223] L.I. Mandelstam. *Lectures on the Theory of Oscillations.* 1972, Science, Moscow.

[224] M.N. Marchenko (ed.). *Politology.* Zerzhalo, Moscow, 1997.

[225] G.I. Marchuk. *Mathematical Modeling in Environmental Problems.* Science, Moscow, 1982.

[226] G.I. Marchuk. *Mathematical Models in Immunology.* Optimization Software, Incorporated, Publications Division, 1983.

[227] J.E. Marsden. *Vector Calculus.* W.H. Freeman and Company, 1976.

[228] M.J. Matteson and C. Orr. *Filtration: Principles and Practices.* Marcel Decker, New York, Basel, 1986.

[229] D. McMahon. *Quantum Field Theory.* McGraw-Hill, 2008.

[230] M. Meerschaert. *Mathematical Modeling.* Academic Press, 2007.

[231] A.M. Meirmanov. *The Stefan Problem.* De Gruyter, Berlin – New York, 1992.

[232] A. Michel, W. Kaining, and Bo H. *Qualitative Theory of Dynamical Systems.* Taylor and Francis group, 2001.

[233] A. Miele. *Flight Mechanics: Theory of Flight Paths.* Dover Publications, 2016.

[234] N.M. Mikhin. *External Friction of Solids.* Science, Moscow, 1977.

[235] S.G. Mikhlin. *Linear Equations of Mathematical Physics.* Holt, Rinehart and Winston, New York, 1967.

[236] M. Mitchell. *An Introduction to Genetic Algorithms.* Cambridge, MA: MIT Press, 1996.

[237] S. Moghadas and M. Jaberi-Douraki. *Mathematical Modelling: A Graduate Textbook.* Wiley, 2018.

[238] N.N. Moiseev. *Mathematics Makes an Experiment*. Science, Moscow, 1979.

[239] F.C. Moon. *Chaotic Vibrations: An Introduction for Applied Scientists and Engineers*. John Wiley and Sons, Inc., Hoboken, New Jersey, 2004.

[240] D.D. Mooney and R. Swift. *A Course in Mathematical Modeling*. Mathematical Association of America Textbooks, 1999.

[241] A. Morely and E. Hughes. *Principles of Electricity*. Longman, 1994.

[242] P.M. Morse and K.U. Ingard. *Theoretical Acoustics*. Princeton University Press, 1986.

[243] K.W. Morton and D.F. Mayers. *Numerical Solution of Partial Differential Equations, An Introduction*. Cambridge University Press, 2005.

[244] J. Müller and C. Kuttler. *Methods and Models in Mathematical Biology: Deterministic and Stochastic Approaches*. Springer-Verlag Berlin Heidelberg, 2015.

[245] J.D. Murray. *Lectures on Nonlinear Differential Equations in Biology*. Oxford University Press, 1978.

[246] A.D. Myshkis. *Elements of the Theory of Mathematical Models*. KomKniga, Moscow, 2007.

[247] M. Nahvi and J. Edminister. *Electric Circuits*. McGraw-Hill, 1965.

[248] Yu.I. Naimark. Simple Mathematical Models and their Role in Comprehending the World. *Soros Educational Journal*, (3):139–143, 1997.

[249] F. Nani and H.I. Freedman. A Mathematical Model of Cancer Treatment by Immunotherapy. *Mathematical Biosciences*, 163(2):159–199, 2000.

[250] A.H. Nayfeh and D.T. Mook. *Nonlinear Oscillations*. Wiley, New York, 1995.

[251] S. Neidle. *Cancer Drug Design and Discovery*. Elsevier Inc. All, 2008.

[252] P. Neittaanmaki and D. Tiba. *Optimal Control of Nonlinear Parabolic Systems. Theory, Algorithms, and Applications*. Marcel Dekker, New York, 1994.

[253] J.A. Nelder and Mead R. A Simplex Method for Function Minimization. *Computer Journal*, 7(4):308–313, 1965.

[254] D.E. Neuenschwander. *Emmy Noether's Wonderful Theorem*. Johns Hopkins University Press, 2010.

[255] A.C. Newell. *Solitons in Mathematics and Physics*. SIAM, 1985.

[256] G. Nicolis and I. Prigogine. *Exploring Complexity: An Introduction*. New York, W. H. Freeman, 1989.

[257] J. Nilsson and S. Riedel. *Electric Circuits*. Prentice Hall, 2007.

[258] J.T. Oden. *An Introduction to Mathematical Modeling: A Course in Mechanics*. Wiley, 2018.

[259] D. Olander. *General Thermodynamics*. CRC Press, New York, 2007.

[260] M.J. Osborne. *An Introduction to Game Theory*. Oxford University Press, 2004.

[261] L. Padulo and M.A. Arbib. *System Theory: a Unified State-Space Approach to Continuous and Discrete Systems.* Saunders, 1974.

[262] H.J. Pain. *The Physics of Vibrations and Waves.* Wiley, 2005.

[263] C.H. Papadimitriou and K. Steiglitz. *Combinatorial Optimization: Algorithms and Complexity.* Dover Publ., 1998.

[264] C. Parsons. *Philosophy of Mathematics in the Twentieth Century: Selected Essays.* Cambridge, MA: Harvard University Press, 2014.

[265] R. Peierls. Model-Making in Physics. *Contemporary Physics*, 21(1):3–17, 1980.

[266] V.V. Penenko and A.E. Aloyan. *Models and Methods for Environmental Tasks.* Science, Novosibirsk, 1985.

[267] L.A. Petrosyan and V.V. Zakharov. *Introduction to Mathematical Ecology.* Leningrad State University, 1986.

[268] J.R. Pierce. *Almost All About Waves.* Dover Publ., 2006.

[269] Y. Pinchover and J. Rubinstein. *An Introduction to Partial Differential Equations.* Cambridge University Press, Cambridge, 2005.

[270] H. Poincaré. *The Foundations of Science.* Science Press, New York, 1921.

[271] E. Polak. *Computational Methods in Optimization. A Unified Approach.* Academic Press, New York, London, 1971.

[272] R.A. Poluektov, Yu.A. Peh, and I.A. Shvytov. *Dynamic Models of Ecological Systems.* Gidrometeoizdat, Leningrad, 1980.

[273] A. Polyanin and A. Manzhirov. *Handbook of Integral Equations.* CRC Press, Boca Raton, 1998.

[274] A.D. Polyanin. *Handbook of Linear Partial Differential Equations for Engineers and Scientists.* Chapman and Hall/CRC Press, 2002.

[275] A.D. Polyanin and V.F. Zaitsev. *Handbook of Exact Solutions for Ordinary Differential Equations.* Chapman and Hall/CRC Press, Boca Raton, 2003.

[276] K.N. Ponomarev. *Formation of Differential Equations.* High Scool, Minsk, 1973.

[277] I. Prigogine. *Advances in Chemical Physics.* Wiley InterScience, New York, 2002.

[278] E.M. Purcell and D.J. Morin. *Electricity and Magnetism.* Cambridge University Press, 2013.

[279] T.P. Pushkareva and A.V. Peregudov. *Mathematical Modeling of Chemical Processes.* Krasnoyarsk Ped. Univ., 2013.

[280] J. Randall. *Finite Difference Methods for Ordinary and Partial Differential Equations.* SIAM, 2007.

[281] N. Rashevsky. *Some Medical Aspects of Mathematical Biology.* Springfield, IL: Charles C. Thomas, 1964.

[282] A. Rasmuson, B. Andersson, Olsson L., and R. Andersson. *Mathematical Modeling in Chemical Engineering.* Cambridge University Press, 2014.

[283] M. Reed and B. Simon. *Functional Analysis*. Academic Press, 1980.

[284] F. Reif. *Fundamentals of Statistical and Thermal Physics*. Waveland Press Inc., 2008.

[285] K.F. Riley, M.P. Hobson, and S.J. Bence. *Mathematical Methods for Physics and Engineering*. Cambridge University Press, 2006.

[286] J. Riordan. *An Introduction to Combinatorial Analysis*. Dover, 2002.

[287] G.Yu. Riznichenko. *Lectures on Mathematical Models in Biology*. Moscow, Izhevsk, RHD, 2011.

[288] B. Roehner and G. Valent. Solving the Birth and Death Processes with Quadratic Asymptotically Symmetric Transition Rates. *SIAM Journal on Applied Mathematics*, 42(5):1020–1046, 1982.

[289] Yu.M. Romanovsky, N.V. Stepanova, and D.S. Chernavsky. *Mathematical Modeling in Biophysics*. Science, Moscow, 1975.

[290] M. Ross. *A Primer on Pontryagin's Principle in Optimal Control*. Collegiate Publishers, 2015.

[291] J. Rotman. *An Introduction to the Theory of Groups*. Springer-Verlag, New York, 1994.

[292] B.L. Rozhdestvensky and N.N. Yanenko. *Systems of Quasilinear Equations and their Applications to Gas Dynamics*. Science, Moscow, 1978.

[293] L.I. Rubinshteĭn. *The Stefan Problem*. American Mathematical Soc., 1971.

[294] Yu.B. Rumer and M.Sh. Ryvkin. *Thermodynamics, Statistical Physics, and Kinetics*. Novosibirsk University, 2000.

[295] B. Russell. *On the Philosophy of Science*. The Bobbs–Merrill Company, 1965.

[296] M.H. Sadd. *Elasticity: Theory, Applications, and Numerics*. Elsevier, Oxford, 2005.

[297] M. Sadiku. *Elements of Electromagnetics*. Saunders College Publishing, Orlando, Florida, 1989.

[298] H. Sagan. *Introduction to the Calculus of Variations*. Dover, New York, 1992.

[299] J.J. Sakurai. *Modern Quantum Mechanics*. Addison Wesley, 1994.

[300] A.A. Samarski and A.P. Mihailov. *Mathematical Modelling. Ideas. Methods. Examples.* Physmathlit, Moscow, 2001.

[301] A.A. Samarsky. *Theory of Difference Schemes*. Marcel Dekker, Inc., New York, 2001.

[302] P. Samuelson and W. Nordhouse. *Economics*. McGraw-Hill/Irwin, 2004.

[303] H. Schlichting. *Boundary-Layer Theory*. McGraw-Hill, New York, 1979.

[304] T. Schlick. *Molecular Modeling and Simulation*. Springer-Verlag New York, 2002.

[305] A. Schrijver. *Combinatorial Optimization*. Berlin, New York, Springer, 2003.

[306] L. Schwartz. Sur l'Iimpossibilité de la Multiplication des Distributions. *C. r. Acad. Sci. Paris*, 239(15):847–848, 1954.

[307] W.R. Scott. *Group Theory*. Dover, New York, 1987.

[308] L.I. Sedov. *A Course in Continuum Mechanics*. Wolters-Noordhoff Publishing, Netherlands, 1971.

[309] K. Seeger. *Semiconductor Physics: An Introduction*. Springer-Verlag Berlin Heidelberg, 2004.

[310] L. Segel and L. Edelstein-Keshet. *A Primer on Mathematical Models in Biology*. Society for Industrial and Applied Mathematics, 2013.

[311] A. Seierstad. *Optimal Control Theory with Economic Applications*. Paperback North Holland, 2012.

[312] S. Serovajsky. *Counterexamples in the Optimal Control Theory*. Brill Academic Press. Netherlands, Utrecht, Boston, 2004.

[313] S. Serovajsky. *Optimization and Differentiation*. CRS Press. Taylor and Francis group, 2017.

[314] S. Serovajsky. *Sequential Models of Mathematical Physics*. CRC Press, Taylor and Francis Group, Boca Raton, London, New York, 2019.

[315] S. Serovajsky, A. Azimov, M. Kenzhebayeva, D. Nurseitov, A. Nurseitova, and M. Sigalovskiy. Mathematical Problems of Gravimetry and its Applications. *International Journal of Mathematics and Physics*, 10(1):29–35, 2019.

[316] S. Serovajsky, A. Azimov, and D. Nurseitov. Mathematical Model of Hormonal Treatment of Cancer Under Hormonal Resistance. In *11th ISAAC Congress. Book of Abstracts, 14–18 August 2017*, page 12. Växjö, Linnaeus University, Sweden, 2017.

[317] S. Serovajsky, D. Nurseitov, S. Kabanikhin, A. Azimov, A. Ilin, and R. Islamov. Identification of Mathematical Model of Bacteria Population under the Antibiotic Influence. *Journal of Inverse and Ill-posed Problems*, 26(5):565–576, 2018.

[318] S. Serovajsky, D. Zhakebaev, and O. Turar. Mathematical Model of the Epidemic Propagation with Limited Time Spent in Exposed and Infected Compartments. *Mathematics, Mechanics and Computer Science*, 2021.

[319] K. Shakenov. Solution of Equation for Ruin Probability of Company for Some Risk Model by Monte Carlo Methods. In *Advances in Intelligent Systems and Computing. Contributions from AMAT 2015*, pages 169–182. Springer, Heidelberg, 2015.

[320] K.K. Shakenov, I.K. Shakenov, and A.K. Khikmetov. *Probabilistic Theory and Mathematical Models of Epidemics. Tutorial*. Kazakh University, 2021.

[321] D.I. Shvitra. *Physiology Dynamical Systems*. Vilnius, Moxlas, 1989.

[322] R. Sikorski. *Boolean Algebras*. Springer-Verlag, Berlin, 1969.

[323] W.T. Silfvast. *Laser Fundamentals*. Cambridge University Press, 2008.

[324] D.V. Sivuhin. *Electricity*. Physmathlit, Moscow, 2004.

[325] G.D. Smith. *Numerical Solution of Partial Differential Equations: Finite Difference Methods*. Oxford University Press, 1985.

[326] J.M. Smith. *Mathematical Ideas in Biology*. Cambridge University Press, 1968.

[327] J.M. Smith. *Models in Ecology*. Cambridge University Press, 1978.

[328] J.A. Snyman and D.N. Wilke. *Practical Mathematical Optimization: Basic Optimization Theory and Gradient-Based Algorithms*. Springer, Berlin, 2018.

[329] I.M. Sobol'. *Monte Carlo Method*. Moscow, Nauka, 1968.

[330] S.L. Sobolev. *Some Applications of Functional Analysis in Mathematical Physics*. American Mathematical Society, 1991.

[331] E.D. Sontag. *Mathematical Control Theory: Deterministic Finite Dimensional Systems*. Springer Science and Business Media, 2013.

[332] B.Ya. Sovetov and S.A. Yakovlev. *Modeling of Systems*. Higher School, Moscow, 2001.

[333] C. Sparrow. *The Lorenz Equations: Bifurcations, Chaos, and Strange Attractors*. Springer-Verlag New York, 1982.

[334] W.M. Stacey. *Nuclear Reactor Physics*. Waley-VCN, 2018.

[335] J. Stachurski. *Economic Dynamics: Theory and Computation*. MIT Press, 2009.

[336] H. Steinhaus. *Mathematical Snapshots*. Oxford, 1951.

[337] I. Stewart. *Concepts of Modern Mathematics*. Dover Publications, 1995.

[338] I. Stewart and D. Tall. *The Foundations of Mathematics*. OUP Oxford, 2015.

[339] U. Strawinska-Zanko and L.S. Liebovitch (Eds.). *Mathematical Modeling of Social Relationships*. Springer, 2018.

[340] J. Strikwerda. *Finite Difference Schemes and Partial Differential Equations*. SIAM, Philadelphia, 2004.

[341] S.H. Strogatz. *Nonlinear Dynamics and Chaos: With Applications to Physics, Biology, Chemistry, and Engineering*. Addison Wesley, 1994.

[342] Yu.M. Svirezhev and D.O. Logofet. *Stability of Biological Communities*. Moscow, Nauka, 1978.

[343] I.E. Tamm. *Fundamentals of the Theory of Electricity*. Mir Publishers Moscow, 1979.

[344] S.A. Telyakovsky. *Course of Lectures on Mathematical Analysis*. MIAN, Moscow, 2009.

[345] R. Temam. *Navier–Stokes Equations Theory and Numerical Analysis*. North Holland, 1979.

[346] G. Teschl. *Ordinary Differential Equations and Dynamical Systems*. American Mathematical Society, Providence, 2012.

[347] J.W. Thomas. *Numerical Partial Differential Equations: Finite Difference Methods*. Springer-Verlag, Berlin, New York, 1995.

[348] J.H.M. Thornley. *Mathematical Models in Plant Physiology*. Academic Press, London, New York, 1976.

[349] A.N. Tihonov and Samarsky A.A. *Equations of Mathematical Physics*. Nauka, Moscow, 1977.

[350] A.N. Tihonov and V.Ya. Arsenin. *Methods of Solving of Ill-posed Problems*. Nauka, Moscow, 1979.

[351] S.P. Timoshenko and J.M. Gere. *Theory of Elastic Stability*. McGraw-Hill, 1961.

[352] P.A. Tipler and G. Mosca. *Physics for Scientists and Engineers: Electricity, Magnetism, Light, and Elementary Modern Physics: 2*. W.H. Freeman, 2003.

[353] W.F. Trench. *Introduction to Real Analysis*. Pearson Education, San Antonio, 2013.

[354] P.V. Trusov (Ed.). *Introduction in Mathematical Modelling*. Logos, Moscow, 2004.

[355] P. Turchin. *Complex Population Dynamics: a Theoretical/Empirical Synthesis*. Princeton University Press, 2003.

[356] S.R. Upreti. *Optimal Control for Chemical Engineers*. 2017, CRC Press.

[357] V. Uspensky. *Preface to Mathematics*. Amphora, St. Petersburg, 2015.

[358] V.A. Ustinov and A.F. Felinger. *Historical and Social Researches. Computers and Mathematics* . Think, Moscow, 1973.

[359] J. van Lint and R. Wilson. *A Course in Combinatorics*. Cambridge University Press, 2001.

[360] F.P. Vasiliev. *Optimization Methods. Vol. II*. MCNMO, Moscow, 2011.

[361] V.S. Vladimirov. *Equations of Mathematical Physics*. M. Dekker, New York, 1971.

[362] V. Volterra. *Leçons sur la Théorie Mathématique de la Lutte pour la Vie*. Gauthier-Villars, Paris, 1990.

[363] J. von Neumann and O. Morgenstern. *Theory of Games and Economic Behavior*. Princeton University Press, 2007.

[364] T.A. Weber. *Optimal Control Theory with Applications in Economics*. MIT Press, 2011.

[365] G.H. Weiss. *Aspects and Applications of the Random Walk*. North-Holland, New York, 1994.

[366] H. Weyl. *Handbuch der Philosophie*. Druck; Verlag von R. Oldenbourg, München–Berlin, 1926.

[367] J. A. Wheeler, C. Misner, and K.S. Thorne. *Gravitation*. W.H. Freeman and Co., 1973.

[368] F.M. White. *Viscous Fluid Flow* . McGraw-Hill, New York, 1974.

[369] J.E. Whitesitt. *Boolean algebra and its applications*. Courier Dover Publications, 1995.

[370] E.H. Wichmann. *Quantum Physics*. McGraw-Hill Book Company, New York, 1967.

[371] N. Wiener. *Cybernetics: or the Control and Communication in the Animal and the Machine*. MIT Press, Cambridge, 1961.

[372] M. Williamson. *Biological Population Analysis*. Edward Arnold, 1972.

[373] A.T. Winfree. The Prehistory of the Belousov–Zhabotinsky Oscillator. *Journal of Chemical Education*, 61(8):661–663, 1981.

[374] T. Witelski and M. Bowen. *Methods of Mathematical Modelling: Continuous Systems and Differential Equations*. Springer, 2015.

[375] S.V. Yablonsky. *Introduction to Discrete Mathematics*. Science, Moscow, 1979.

[376] I.M. Yaglom. *Mathematical Structures and Mathematical Modelling*. Gordon and Breach Science Publishers Ltd, Routledge, 1984.

[377] N.N. Yanenko. *The Method of Fractional Steps: The solution of Problems of Mathematical Physics in Several Variables*. Springer-Verlag, Springer-Verlag, Berlin, 1971.

[378] L. Young. *Lectures on the Calculus of Variations and Optimal Control Theory*. W.B. Saunders Co., Philadelphia–London–Toronto, 1969.

[379] Y. Yu. *The Index Theorem and the Heat Equation Method*. World Scientific Inc., 2001.

[380] L.A. Zadeh. *The Concept of a Linguistic Variable and its Application to Approximate Reasoning*. EECS Department, University of California, Berkeley, 1973.

[381] V.E. Zakharov, S.V. Manakov, S.P. Novikov, and L.P. Pitaevskii. *Theory of Solitons. The Inverse Scattering Method*. Nauka, Moscow, 1980.

[382] O.O. Zamkov, Yu.A. Cheremnykh, and A.V. Tolstopyatenko. *Mathematical Methods in Economics*. Business and Service, Moscow, 1999.

[383] D. Zwillinger. *Handbook of Differential Equations*. Academic Press, Boston, 1997.

Index

$C^2(\bar{\Omega})$ space, 379
$H^1(\Omega)$ space, 379
$H_0^1(\Omega)$ space, 379
L_2 space, 292, 379
δ-function, 410

absolute minimum, 315
acoustic equations, 275
acoustics, 232
action of the system, 303
adjoint
    space, 413
    system, 386, 407
algebra, 85
allied relations model, 157
analytic
    function, 258
    solution, 41
angular
    displacement, 39
    velocity, 40
antibiotic resistance model, 122
Arrhenius equation, 354
asymmetric traveling salesman problem, 325
asymptotically stable equilibrium position, 77
attractor, 68
autocatalytic reaction, 97
autonomous system, 74

Bénard cell, 375
Banach space, 380
barometric formulas, 267
barotropic fluid, 263
bending an elastic plate, 254
Bernoulli equation
    (differential equation), 102
    (fluid dynamics), 266
bifurcation point, 372
bimolecular reaction, 91
binomial
    coefficients, 359
    distribution, 360

Biot number, 193
bipartisan system model, 151
body falling model, 7, 124
Bolza problem, 311
Boolean algebra, 14
boundary
    condition, 173
    inverse problem, 408
    layer model, 273
    value problem, 36, 174, 367
Boussinesq approximation, 275
brachistochrone problem, 297
brussellator, 83
Burgers equation, 271

calculus of variations, 15, 297, 315
cancer, 207
Cauchy
    criterion, 377
    problem, 27, 188, 221, 365
    sequence, 377
center (equilibrium), 80, 86
central limit theorem, 344
certain event, 358
Chafee–Infante problem, 372
chaotic system, 321
characteristic, 271
    equation, 42
    of equation, 221
chemical
    chain reaction, 343
    competition model, 101
    kinetics, 89
    niche model, 101
    reaction, 89
Clapeyron equation, 267
Clarke derivative, 413
classical
    mechanics, 20
    model, 376
    solution, 371, 376
Cobb–Douglas function, 146
combinatorics, 14

complete
    graph, 325
    space, 380
completion, 380
composite function, 237
conditional probability, 358
continuity equation, 253, 263
continuous
    random variable, 344
    system, 14
control, 385
controllable
    process, 385
    system, 396
convex functional, 317
coordinate system, 12
Couette flow, 277
Couette–Taylor flow, 277
Courant condition, 197
Crank–Nicholson scheme, 211
cross product, 245
cylindrical coordinates, 243

D'Alembert
    formula, 223
    method, 223
de Broglie wave, 281
derivative, 28
    of the functional, 315
deterministic
    model, 14
    process, 71
    system, 71
differentiable process, 72
differential, 187
    equation, 8
    geometry, 211
differential-difference system, 341
diffusion
    equation, 186
    of chemical reactants, 197
dimensionless variables, 125
direct problem, 15, 402
directed graph, 325
Dirichlet problem, 241, 371
discrete
    heat transfer model, 323, 377
    Malthus model, 320
    mathematics, 14, 319
    model of epidemic propagation, 329
    system, 14

Verhulst equation, 321
dissipation of energy, 69
dissipative system, 69
distributed-parameter system, 14, 72
distribution, 291
divergence, 247
    theorem, 262
dominance of strategies, 335
dot product, 245
Duffing
    equation, 52
    spring, 52
dynamic
    system, 15, 71
    transformation
        group, 73
        monoid, 85, 189

ecological niche model, 141
economic
    competition model, 131
    cooperation model, 143
    niche model, 134
edge of graph, 324
eigenfunction, 175
eigenvalue, 175
Einstein formula, 282
elasticity theory, 233
electrical
    circuit, 57
    conductivity, 203
electromagnetic waves, 234
electrostatic field, 247
elliptic equation, 188, 237, 258
epidemic propagation, 203, 343
equation
    of harmonic oscillator, 41
    of pendulum oscillations, 41, 313
        nonlinear, 41
    of spring oscillations, 51
    of state, 263
equipotential surface, 248
establishment method, 257
Euclidean space, 413
Euler
    elastic problem, 372
    equation
        (calculus of variations, 301
        (fluid mechanics), 265
    formula, 281
    method, 29

evolutionary process, 71
expected value, 292, 341
explicit
    difference scheme, 196, 238
    method, 36
extended derivative, 413
external boundary value problem, 241
extremum theory, 299

Faraday law, 60, 234
Fermat's principle, 307
Fick's law, 186
filtration, 203
finite
    difference method, 195
    element method, 211
finite-dimensional
    process, 71
    system, 14
first
    boundary value problem, 173, 182, 217
    integral, 307
    type boundary condition, 173
fluid and gas mechanics, 261
flux of the vector field, 246
focus (equilibrium), 86
force field, 244
forced pendulum oscillations, 49
Fourier
    coefficient, 189
    law, 172
    method, 178
    number, 193
    series, 176, 189
Fréchet derivative, 413
fractional step method, 211
free market model, 138
Froude pendulum, 52
function of a complex variable, 258
functional, 299
    analysis, 86
fundamental
    lemma of calculus of variations, 300
    solution to the Laplace equation
        in space, 259
        on the plan, 259

Gâteaux derivative, 314, 405
game
    in normal form, 335
    theory, 15, 327, 335

Gauss law, 234
general solution, 20
generalized
    model, 376
    solution, 371, 376
genetic algorithm, 413
geometric optics, 307, 316
glider flight model, 19, 31
global solution, 366
goods transfer equation, 195
gradient, 235, 244
    method, 314, 395, 406
    projection method, 395
graph, 324
    theory, 14
gravity field, 250
Green function, 259
grid
    function, 196
    node, 196
group, 73, 85

Hadamard's example, 369
Hamilton's principle, 303
Hamiltonian cycle, 325
harmonic function, 258
heat
    equation, 172
    transfer, 102
Heisenberg uncertainty principle, 280
Helmholtz equation, 254
Hilbert space, 316, 380, 413
homogeneous
    boundary value problem, 174
    equation, 53, 102
Hooke's law, 50
hormone resistance model, 207
hydrodynamic equations for ideal fluid, 265
hydrodynamics, 261
hydrostatic problem of incompressible fluid, 267
hyperbolic equation, 187, 237

ideal fluid, 264
identification, 399
identity operator, 85
ill-posed problem, 370
implicit
    difference scheme, 205, 238
    method, 37
impossible event, 358

improper integral, 292
incompressible fluid, 263
inconsistent events, 358
independent
    random variables, 359
    variable, 12
infinite-dimensional
    process, 72
    system, 14
inflation model, 143
initial condition, 8
input information, 13
integral, 187
    curve, 73
    equation, 14
internal boundary value problem, 241
inverse
    gravimetry problem, 411
inverse problem, 15
iterative process, 260

Kirchhoff law, 58
Korteweg–de Vries equation, 273

L'Hôpital's rule, 360
Lagrange
    multiplier, 386
    problem, 300
Lagrangian, 303
laminar flow, 277
Laplace
    equation, 241, 369
    operator, 182
laser, 101
law
    of conservation
        of energy, 44, 60, 231, 307
        of momentum, 216, 263
    of mass action, 90
    of universal gravitation, 37
Lebesgue
    integral, 413
    measure, 413
limit, 377
    cycle, 68, 81
linear
    equation, 53
    programming, 335
    space, 53
    system, 15
local

minimum, 315
    solution, 366
logistic equation, 107, 321
lognormal distribution, 360
Lorenz attractor, 87
lumped-parameter system, 14, 72
Lyapunov stable equilibrium position, 77

macroeconomics, 147
magnetohydrodynamics, 276
Malthus model, 106, 130
Markov chain, 360
mass center, 34
material field equations, 234
mathematical
    analysis, 14
    model, 3
    pendulum, 39
    statistics, 358
matrix, 319
maximizing
    the missile flight range, 383
    the shell flight range, 382
maximum principle, 386
Maxwell equations, 234
mean
    free path, 202
    value theorem, 172
measurable function, 413
measure of set, 210
method
    of characteristics, 223, 271
    of separation of variables, 174, 190, 255
    of successive approximations, 394
metric, 86
metropolis–colony model, 161
missile flight model, 18, 31
model, 1
modulus of the vector, 245
moments of random variable, 341
monomolecular reaction, 90
monopolized market model, 140
Monte Carlo method, 344
movement
    equation of ideal fluid, 265
    of dynamic system, 73

Nash equilibrium, 327, 335
Navier–Stokes equations, 268
necessary minimum condition, 315
neighborhood, 86

Nelder–Mead method, 413
Neumann problem, 241
neutron
　diffusion, 203
　passage through a plate, 356
Newton
　second law, 16, 304
　third law, 34
node (equilibrium), 86
Noether's theorem, 316
non-homogeneous
　equation, 49
　heat equation, 178
non-stationary field, 259
nonlinear
　boundary value problem, 372
　equation, 41, 53
　heat equation, 368
　programming, 396
　system, 15
norm, 86
normal
　distribution, 344
　to the boundary, 183
normalized wave function, 289
normed vector space, 413
nuclear chain reaction, 343

Ohm law, 61
one-phase Stefan problem, 200
one-step method, 36
operation, 85
operations research, 319, 336
optimal
　control, 386
　　problem, 385
　　theory, 15
optimality criterion, 385
order
　of accuracy, 36
　of approximation, 209
　of chemical reaction, 90
　of differential equation, 21, 53
　of system of differential equations, 37
oregonator, 100
output parameters, 13

parabolic equation, 188, 237
Pareto optimality, 327, 336
partial differential equation, 14, 172, 187
particular solution, 20

path of graph, 324
pendulum oscillation with friction, 46
periodic function, 54
phase
　curve, 73
　flow, 73
　plane, 72
　space, 72
　velocity, 73
physical
　chemistry, 89
Poiseuille law, 269
Poisson equation, 241, 370
political
　competition model, 149
　niche model, 151
population
　dynamics, 105
　migration, 203
potential
　barrier, 284
　field, 245
　method, 333
　of electrostatic field, 247
　of field, 245
　of gravitational field, 252
　of velocity field, 253
Prandtl equations, 273
predator–prey model, 35, 113
principle of least action, 303
prisoner's dilemma, 327
probability, 291, 358
　density, 291, 344
　distribution, 358
　function, 358
　theory, 14, 358
　wave, 281
probe movement model, 16, 30
production
　function, 146
　planning problem, 335
projector, 395

quantum
　field theory, 291
　mechanics, 279
quasiperiodic oscillations, 66

radioactive decay, 348
random
　event, 358

process, 358
processes, 211
value, 358
recurrence formula, 36
reduced mass, 35
refraction of light, 309
relaxation time, 45, 61
resonance, 49
Riccati equation, 102
Riemann integral, 413
rotor, 234, 245
Runge–Kutta method, 33
running wave, 223

saddle (equilibrium), 79, 86
scalar
    field, 259
Schrödinger equation, 283, 284
second
    boundary value problem, 173, 183, 217
    law of thermodynamics, 189
    type boundary condition, 173
SEIR epidemic model, 126
self-oscillations, 68, 82
sequence, 319
    convergence, 320
sequential model, 377
set of admissible controls, 385
shock wave, 271
shooting method, 394
similarity theory, 192
simply connected set, 259
singular
    control, 412
    point, 73
SIR epidemic model, 121
situation of game, 327
Snell's law, 309
Sobolev space, 379
solenoidal field, 247
soliton, 273
Solow model, 146
spectrum, 378
spherical coordinates, 242
square-integrable function, 371
stability of difference scheme, 197
stable
    focus, 79
    in parameter solution, 370
    node, 79
standard deviation, 403

state
    function, 12
    operator, 402
stationary
    condition, 299, 314
    fluid flow, 253
    system, 15
steady-state oscillations, 254
Stefan
    condition, 200
    problem, 200
stiff differential equation, 37
stochastic
    Malthus model, 349, 352
    missile flight model, 354
    model, 14
        of population death, 346
        of pure birth, 339
        of selling goods, 355
    SIR model of epidemic, 355
    synthesis reaction model, 354
    system, 339
stoichiometric
    coefficient, 90
    equation, 90
strange attractor, 86
strategy of game, 327
streamline, 266
Sturm–Liouville problem, 175
subdifferential, 413
sufficient minimum condition, 315
symbiosis model, 116
symmetric traveling salesman problem, 325
symmetry, 85
synthesis reaction, 91
system
    identification, 15
    parameter, 12

Taylor series, 209
telegraph equations, 227
thermal insulation condition, 173
third type boundary condition, 173
Thomas algorithm, 211
three-sigma rule, 344
Tikhonov well-posed problem, 412
time inverse problem, 404
time-optimal control problem, 389
topological vector space, 413
topology, 86, 259
Torricelli formula, 267

trade union activity model, 156
translational symmetry, 37
transonic gas dynamics, 188
transportation problem, 324
transversality condition, 311
traveling salesman problem, 325
tribe
    demarcation model, 159
    struggle model, 158
Tricomi equation, 188
tridiagonal matrix algorithm, 205
tunneling effect, 287
turbulent flow, 277
two-body problem, 33
two-phase Stefan problem, 210

undirected graph, 325
unstable
    equilibrium position, 77
    focus, 79
    node, 79
utility function, 327

Van der Pol
    circuit, 67
    equation, 68
variance, 342
variation
    of the function, 315
    of the functional, 315
variational inequality, 15, 317
vector, 319
    analysis, 244
    field, 73, 244
    space, 53, 413
velocity field, 246
Verhulst
    equation, 107, 130
    model, 107
vertex of graph, 324
vibrating
    beam equation, 233
    membrane equation, 232
    string equation, 216, 306
Volterra–Lotka equations, 97, 113

wave
    equation, 232, 275
    function, 280
wave-particle duality, 280
well-posed problem, 370